Compressive Sensing for the Photonic Mixer Device

Miguel Heredia Conde

Compressive Sensing for the Photonic Mixer Device

Fundamentals, Methods and Results

Springer Vieweg

Miguel Heredia Conde
Siegen, Germany

Dissertation at University of Siegen, Department of Electrical Engineering and Computer Science, 2016.

First Supervisor: Prof. Dr.-Ing. Otmar Loffeld
Second Supervisor: Prof. Dr.-Ing. Andreas Kolb
Date of the oral examination: 14.11.2016

ISBN 978-3-658-18056-0 ISBN 978-3-658-18057-7 (eBook)
DOI 10.1007/978-3-658-18057-7

Library of Congress Control Number: 2017938168

Springer Vieweg

Printed on acid-free paper

This Springer Vieweg imprint is published by Springer Nature
The registered company is Springer Fachmedien Wiesbaden GmbH
The registered company address is: Abraham-Lincoln-Str. 46, 65189 Wiesbaden, Germany

To my parents, my brother and the memory of my grandparents.

"In omnibus requiem quaesivi, et nusquam inveni nisi in angulo cum libro."
Thomas à Kempis, 1380–1471

Acknowledgement

I would like to thank Prof. Otmar Loffeld for the vision on applying Compressive Sensing (CS) to the Photonic Mixer Device (PMD) and for trusting me for such an exciting mission. Without him and his support on CS theory and our many discussions, the results in this thesis would not have been achieved. I would like to extend my gratitude to Prof. Andreas Kolb, who showed an uninterrupted interest in the results of my work from the very beginning, contributing to it with his valuable supervision. Thanks to Dr. Klaus Hartmann for directing the project and, specifically, for finding always some time for discussion and supervision tasks, of capital importance during my first months at ZESS. I am very thankful to Dr. Benjamin Langmann for the time we shared our office as colleagues, during which he restlessly answered my many questions regarding the PMD operation principle and our PMD-based ToF hardware. Even after he left ZESS already as a doctor, he continued supporting me whenever I needed it. Special thanks to Dr. Wolfgang Weihs, not just for the uncountable discussions and advices on PMD technology, but also for the time invested in turning a PMD-19k frontend into the CS-PMD module presented in Section 5.2.2. Without his work, the CS-PMD prototype in Chapter 5 could not have been completed. Also impossible to oversee is the help of Dr. Rainer Bornemann, specially during the preparation of the experiments destined to the characterization of the spatial response of PMD pixels (Section 4.3.1). The excellent spatial resolution achieved in the experiments is, up to a great extent, thanks to his assistance. Thanks to Dr. Alexander Seel for the many discussions on CS and for the thorough linguistic revision of the German version of the abstract. Thanks also to Dr. Stefan Lammers, for the discussions and support, mostly related to C++ developments.

I want to thank all the members of the GRK 1564 *Imaging New Modalities* and the ZESS that helped me along these years in one way or another. Thanks to Dr. Joanna Czajkowska for the support during the first days, the discussions on optimal representations and some novel ideas, e. g., the use of complex Zernike moments as 2D complete orthogonal basis. Thanks to Stefan Hoberg, not just for the multiple discussions on PMD technology during his time at the GRK, but also for his assistance when constructing the frontend for our CS-PMD module. Without his support, building the frontend by ourselves would have been a much more challenging task, if possible at all. Thanks to Damien Lefloch and Hamed Sarbolandi for some informal discussions on the Kinect technology. Special thanks to Davoud Shahlaei for the time he spent in our joint work on inverse lighting of human faces using CS. At this point I must also thank the GRK itself and its coordinator, Dr. Christian Köhler, for providing a mean of knowledge exchange via periodic internal presentations. Without such a system, cooperations like [H10], with Davoud Shahlaei and Prof. Volker Blanz, could not have been conceived.

Thanks to the technical staff of the ZESS for their support, namely: Thanks to Arne Stadermann for the thorough support with computers and servers. Thanks to Wolf Twelsiek for the uncountable occasions when I made use of his scarce time and wide experience. Thanks to Peter Hof for the multiple pieces and mechanical setups that he prepared for the experiments in this work, including those in Section 4.3.1 and our CS-PMD prototype (Section 5.2). Thanks also to Rolf Wurmbach and Bodo Ohrndorf for the everyday help they provided.

I would also like to have some words of gratefulness towards the undergraduate students who helped me and other researchers at ZESS. Thanks to Omar Ismail for his help acquiring the data in the experiments presented in Appendix A.6. Thanks to Amitkumar Kumbhar for his efforts in providing reliable empirical data to characterize the frequency response of the NIR LEDs used in our illumination systems. Also thanks to Christoph Klingspor for the DLL implementation of my calibration algorithm for Dominik Scarpin.

I do not forget the enormous amount of bureaucratic work that my research has indirectly occasioned to the GRK and ZESS administrative teams. Thanks to them everything ran smoothly at the bureaucratic level and I could concentrate in my tasks. Specifically, thanks to Willi Gräfrath, as cordial as helpful with me from the first minute to my very last days at the GRK. I cannot think of any issue, from the many for which I required his help, that he did not solve readily. The same applies to our secretaries at ZESS, Silvia Niet-Wunram, *die Stimmung des ZESS*, Katharina Haut, Susanne Hoffmann, who restlessly cared about the improvement of my German skills, and Caterina Sinner. Thanks also to the film-scanner team of the Universitätsbibliothek Siegen for the assistance when digitalizing the microfilms of [421].

Out of the University of Siegen, I want to express my gratitude to all other researchers that showed interest in my work, specially to those who tried to establish a cooperation with us. In this regard, I would like to mention the cooperation with Prof. Keiichiro Kagawa on enhancement of *multitap* and *multiaperture* images in low-light conditions [H8]. Thanks to Prof. Kagawa and Bo Zhang for their thorough work and for granting the use of their own datasets, both synthetic and acquired with their own multiaperture camera. Thanks also to Lars Meinel and his colleagues at the Technische Universität Chemnitz, for considering the ZESS MultiCam technology as starting point for a future camera development that fits their own requirements better than commercial solutions and for initiating preliminary discussions towards an eventual future cooperation.

Last, but absolutely not least, I want to express my greatest gratitude towards my family, specially to my parents José Antonio and Leonor and my brother José Manuel, for their unconditional support, understanding and even interest in my research.

Abstract

In the last years, Time-of-Flight (ToF) imaging has experienced a fast growth. Its areas of application have changed from especial applications in industry or research to commercial products, oriented to a general public, in most cases as a mean for natural interaction (NI) between the user and a computer or gaming station. A common method for ToF imaging is to avoid the need for the extreme hardware capabilities required for gathering real per-pixel time measurements by means of a continuous wave (CW) phase-shift-based ToF sensor. In this domain, the Photonic Mixer Device (PMD), born at ZESS, was the first implementation of phase-shift-based ToF imaging to be made publicly available and still is a reference technology. Many later ToF sensors are directly or indirectly based on the *smart* pixels introduced by PMD Technologies®. During the development of this work new commercial ToF sensors have arrived to the market that widely outperform the capabilities of previous ToF sensors. Known examples are not only the new PMD chips, manufactured by Infineon, but also off-the-shelf systems, such as the Intel Creative camera, the Hamamatsu distance sensor or, doubtlessly, the new version of the Microsoft® Kinect® sensor, commercialized as the Xbox One sensor. Despite the notable increase of lateral resolution and depth accuracy, ToF sensors are still far away from the typical resolution of a conventional color camera, while the depth accuracy is rarely below the centimeter range.

The recent and vertiginously-growing theory of Compressive Sensing (CS) brings a new light to problems where compressible signals are to be estimated from incomplete, bandlimited or noise-corrupted measurements. In very short terms, CS theory assures that a signal that admits a sparse or compressible representation in an appropriate domain can be recovered from few measurements if certain conditions regarding the way the measurements are performed, e. g., incoherence between the sensing matrix and the sparsity basis, are satisfied. In this context, *few measurements* means fewer than those suggested by the Shannon criterion. CS provides, in principle, a powerful and widely applicable signal recovery framework, since most signals, both natural and man-made, admit a sparse or compressible representation in an appropriate basis or, in the worst case, in a tight frame.

The aim of this work is to find the ways to fit the valuable mathematical results of CS theory to the specific case of the PMD to improve the final depth image. The potential improvement is to be pursued in several directions, as several are the domains where the signal gathered by the PMD sensor is sparse. On the one hand, since the readouts of the sensor are 2D images, compressibility in, e. g., wavelet domain can be exploited. Depending on the direction we apply CS, we might be able to recover the same information with fewer measurements, or to recover more information with a number of measurements increased in a lower amount than that of the information recovered. An example of the first kind in this domain would be to recover the raw images from a lower number of measurements than the number of pixels, eventually reducing noise through a synergistic joint recovery of the several raw images needed per depth image. An example of the second kind would be to superimpose custom patterns on the image section projected onto a single pixel, in a similar way to the well-known *single pixel camera*, and recover a super-resolved image, where the number of measurements needed for recovery is lower than the size of the super-resolved image. In the specific case of superresolution, modeling the pass from a high-resolution scene—infinite, in the continuous case— to a low-resolution image is a crucial step towards the construction of a realistic sensing matrix. The complex structure of the PMD pixels makes conventional downsampling approaches (e. g., Gaussian blur plus regular decimation) completely invalid for modeling the PMD sensing process, since they rely on implicit hypotheses such as a high fill factor and uniformity of the pixel response. We show that most of the imaging area of a PMD chip is non-sensitive and that the pixel response within the sensitive area is highly non-uniform in spatial domain. Consequently, a thorough characterization of the PMD pixel response in spatial domain, with resolution up to $1\,\mu$m, is one of the central pillars of this work.

CS requires a linear sensing model, which cannot be achieved if the sensor undergoes saturation. Pixel saturation is common in PMD imagers, due to the relatively low dynamic range of the pixels and the eventual presence of highly-reflective objects or light sources in the scene. The Suppression of Background Illumination (SBI) system is an *in pixel* hardware solution that removes a constant current from both PMD pixel channels when any of them is reaching saturation levels. This way, the effective dynamic range of the pixel can be artificially extended. The main drawback is the appearance of high-frequency noise in the raw images, which severely harms the image sparsity and precludes the application of CS. A method for extending the dynamic range through several acquisitions and avoiding the SBI-related noise is presented, which adapts a set of low exposure times to accurately sense eventual *saturating* objects that might come into the scene. Further sparsity is achieved by linear calibration of the raw images, which are originally affected by the different offsets and gains of the ADCs used for digitalization.

Another domain where sparsity is to be considered is the time-frequency domain. A pulsed ToF system exhibits sparsity in time domain, so the exact locations of the pulses in time can be estimated from several frequency measurements. Complementary, a CW ToF system, such as PMD sensors, continuously emits a periodic signal, which is either sparse or highly compressible in frequency domain and can, therefore, be recovered from few measurements in time domain. Adopting the latter focus, we propose a framework for CS-based multifrequency depth estimation that recovers the reflected signal in frequency domain, which allows using high frequency harmonics to improve the depth resolution and opens the door for multipath estimation.

We also exploit the possibility of combining the sparse recovery both in spatial and frequency domain, provided that the measurements in time domain produce raw images sharing the same sparse support in the compressed spatial domain (e. g., wavelet basis), since they are images of the same scene. A hardware implementation of such a joint sensing, together with the corresponding recovery software, have been developed as a demonstrator and to evaluate the feasibility of a CS-based PMD camera. The measurements in temporal domain are gathered as the scalar product of a period of the incoming signal and pseudorandom binary codes. This requires a modification of the signal generation module of the PMD frontend. In spatial domain, an external spatial modulator whose dimensions (in pixels) are one order of magnitude larger than those of the PMD chip is used in a reflective setup to superimpose high-resolution binary patterns to the image before reprojecting it onto the PMD chip. This paves the way for superresolution, at the cost of a higher number of acquisitions.

Kurzfassung

In den letzten Jahren hat die Time-of-Flight-Methode (Fluglaufzeit, kurz ToF) ein rasantes Wachstum erlebt. Ihr Anwendungsgebiet hat sich dabei von besonderen Anwendungen in der Industrie und Forschung hin zu kommerziellen Produkten für die breite Öffentlichkeit verlagert—in den meisten Fällen als Hilfsmittel zur natürliche Interaktion (NI) zwischen Anwendern und Computern bzw. Spielekonsolen. Eine verbreitete Methode für die ToF-Bildgebung ist der Einsatz eines Sensors, der auf dem Prinzip von Phasenverschiebungen einer kontinuierlich ausgesandten Welle im so genannten Dauerstrich-Modus (CW) beruht. Der am ZESS entwickelte Photonen-Misch-Dedektor (PMD) war für dieses Prinzip die erste frei verfügbare Umsetzung einer ToF-Bildgebung, die noch immer als Schlüsseltechnologie gilt. Dabei werden extreme Hardware-Anforderungen vermieden, die sonst für die reale pro-Pixel-Zeitmessung notwendig wären. Viele später entwickelte ToF-Sensoren beruhen direkt oder indirekt auf diesen *intelligenten*, von PMD Technologies®entwickelten Pixeln. Während des Entstehens dieser Arbeit wurden am Markt neue kommerzielle ToF-Sensoren eingeführt, die die Fähigkeiten früherer Sensoren bei weitem überflügeln: Dazu zählen nicht nur die von Infineon gefertigten PMD-Chips, sondern auch seriengefertigte Systeme wie etwa die Intel Creativ Camera, der Hamamatsu Abstands-Sensor oder die neue Version des Microsoft®Kinect®-Sensors, die als Xbox-One-Sensor vertrieben wird.

Das relativ junge und sich schnell entwickelnde Gebiet des Compressive Sensings (komprimierte Abtastung, kurz CS) eröffnet eine neue Sichtweise auf Problemstellungen, bei denen komprimierbare Signale aus unvollständigen, bandbegrenzten oder verrauschten Messung zu extrahieren sind. Zusammengefasst sagt die CS-Theorie aus, das ein Signal in einer nur dünn besetzten oder komprimierbaren Darstellung aus wenigen (linearen) Messungen rekonstruiert werden kann, wenn gewisse Voraussetzungen den Messvorgang betreffend erfüllt sind. Hierzu zählt z.B. die Inkohärenz zwischen dieser Basis-Darstellung und den Messmatrizen zur Beschreibung des Messvorgangs. In diesem Zusammenhang bedeutet *wenige Messungen* eine geringere Anzahl als z.B. nach dem Shannon-Nyquist-Kriterium zu erwarten wäre. Damit stellt CS prinzipiell einen mächtigen und weithin anwendbaren Rahmen zur Signalrekonstruktion zur Verfügung, da die meisten natürlichen und künstlich erzeugten Signale entweder eine nur dünn besetzte oder komprimierbare Darstellung in einer Basis oder, im schlimmsten Falle, nur in einem so genannten straffen 'Frame' besitzen.

Das Ziel dieser Arbeit ist die Anwendung der mathematischen CS-Ergebnisse auf den PMD-Fall um das resultierende Tiefenbild zu verbessern. Die potenziellen Verbesserungen müssen dabei in verschiedene Richtungen verfolgt werden, da es verschiedene Bereiche gibt, in denen das aufgenommene Signal als dünn besetzt angesehen werden kann. Da die ausgelesenen Daten einerseits 2D-Bilder sind, kann z.B. die Komprimierbarkeit im Wavelet-Bereich ausgenutzt werden. Je nachdem wie CS angewandt wird, vermag man andererseits entweder den gleichen Informationsgehalt mit weniger Messungen zu rekonstruieren oder es lässt sich der Informationsgehalt in einem größeren Maße steigern als die Zahl der dazu nötigen Messungen. Ein Beispiel für den ersten Fall wäre die Rekonstruktion von Rohdaten aus weniger Messungen als Pixel vorhanden sind und letztendlich eine Rauschunterdrückung durch die gemeinsame Rekonstruktion einzelner Rohdaten je Tiefenbild. Ein Beispiel für den zweiten Fall wäre die Überlagerung eines speziell zugeschnittenen Musters mit dem Ausgangsbild, die—wie in der Ein-Pixel-Kamera—gemeinsam auf einen einzigen Pixel projiziert werden. Das Ergebnis ist ein super-aufgelöstes Bild mit einer geringeren Anzahl von Messungen im Vergleich zur Größe des super-aufgelösten Bildes. Im Sonderfall der Super-Auflösung ist die Modellierung des Übergangs von der—im Prinzip unendlich—hoch-aufgelösten realen Szene zum erhaltenen Bild endlicher niedriger Auflösung entscheidend für die Konstruktion einer realistischen Abtastmatrix. Durch die komplexe Struktur der PMD-Pixel können konventionelle down-sampling-Methoden (z.B. Gaußsche Unschärfe zusammen mit äqudistanter Dezimierung) den PMD-Messprozess nicht modellieren, da sie auf implizite Hypothesen wie etwa einem hohen Füllfaktor oder Gleichförmigkeit der Pixelantwort angewiesen sind. Wir zeigen, dass meisten Bildbereiche des PMD-Chips unempfindlich sind, und die Pixelantwort in den empfindlichen Bereichen räumliche sehr ungleichförmige ist. Infolgedessen ist eine vollständige räumliche Charakterisierung der PMD-Pixelantwort mit $1\,\mu$m Maximalauflösung ein zentraler Punkt dieser Arbeit.

CS beruht auf einem linearen Modell für die Messungen, das nicht mehr gültig ist, wenn der Sensor in den Bereich der Sättigung kommt. Die Sättigung bei Pixeln ist aufgrund des niedrigen Dynamik-Bereichs der Pixel und der Anwesenheit von hoch-reflektierenden Objekten oder Lichtquellen in der Szene nicht ungewöhnlich. Das System zur Unterdrückung der Hintergrundbeleuchtung (SBI) ist im Pixel hardwaremäßig gelöst und entfernt von beiden PMD-Pixel-Kanälen einen konstanten Strom, wenn einer von beiden den Bereich der Sättigung erreicht. Dadurch kann der effektive Dynamik-Bereich des Pixels künstlich erweitert werden. Die größte Beeinträchtigung entsteht aber durch das Auftreten von hochfrequentem Rauschen in den Rohdaten, das die Eigenschaft der Signale in einer Darstellung dünn besetzt zu sein stark beeinträchtigt und die Anwendung von CS einschränkt. Wir stellen eine Methode vor, die den Dynamikbereich durch Nutzung verschiedener Aufnahmen erweitert und das SBI-bezogene Rauschen durch eine Abfolge kurzer Belichtungszeiten vermeidet. Dadurch lassen sich auch Objekte sorgfältig ausmessen, die beim Auftauchen in einer Szene zu Sättigungseffekten führen würden. Durch eine lineare Kalibrierung der Rohdaten, die ursprünglich durch die unterschiedlichen ADC-Offsets und Verstärkungsfaktoren für die Digitalisierung beeinflusst werden, kann eine noch dünnere Besetzung des Signals erreicht werden.

Ein anderer Bereich, in dem die Eigenschaft eines Signal in einer Darstellung dünn besetzt zu sein in Erwägung gezogen werden muss, ist der Zeit-Frequenz-Bereich. Ein gepulstes ToF-System ist in der Zeitdarstellung dünn besetzt. Daher kann die Position des Pulses aus verschiedenen Frequenzessungen rekonstruiert werden. Im Gegensatz dazu ist ein kontinuierlich arbeitendes ToF-System wie der mit einem periodischen Signal sendende PMD-Sensor im Frequenzbereich dünn besetzt (oder zumindest komprimierbar) und kann daher durch verschiedene Zeitmessungen rekonstruiert werden. Speziell für den zweiten Fall stellen wir eine CS-basierte Tiefen-Rekonstruktion vor, die auf der Grundlage verschiedener Frequenzen das reflektierte Signal im Frequenzbereich rekonstruiert und durch die Nutzung höherer Harmonischer die Tiefenauflösung verbessert und den Weg f"ur die Einbeziehung verschiedener Trajektorien bereitet.

Wir nutzen auch die Möglichkeit, die Rekonstruktion dünn besetzter Signal sowohl im Zeit- als auch Frequenzbereich zu kombinieren—vorausgesetzt alle Messungen im Zeitbereich erzeugen Rohdaten mit dem gleichen dünn besetzten Träger im komprimierten räumlichen Bildbereich (z.B. Wavelet-Basis), da sie Abbildungen der gleichen Szene sind. Eine hardwaremäßige Umsetzung dieser gemeinsamen Messdatenaufnahme wurde zusammen mit der Rekonstruktionssoftware als Vorfürgerät entwickelt, um die Machbarkeit einer CS-basierten PMD-Kamera auszuloten. Die Messungen im Zeitbereich werden als sich periodisch wiederholende Skalarprodukte des einfallenden Signals mit einem pseudo-zufälligen Binärsignal gebildet. Dazu ist eine Modifizierung des Signalgenerators im PMD Frontend erforderlich. Im räumlichen Bereich wird ein Modulator eingesetzt, dessen Ausdehnung (in

Pixeln) eine Größenordnung über der des PMD-Chips liegt. Er wird benutzt, um dem eigentlichen Bild in Reflektion hochaufgelöste binäre Muster auf dem PMD-Chip zu überlagern. Damit wird der Weg für die Superauflösung geebnet, allerdings auf Kosten einer größeren Anzahl an Messungen.

Contents

Nomenclature

Abbreviations

AC Alternating Current. In general, said of a zero-mean periodic signal

ADC Analog-to-Digital-Converter

AHDR Adaptive High Dynamic Range, see [H4]

AMCW Amplitude Modulated Continuous Wave

APEG Adaptive Progressive Edge-Growth, see [H9]

BASC Best Antipodal Spherical Codes

BCASC Best Complex Antipodal Spherical Codes

BGL Background Light

BPBD Binary Permuted Block Diagonal

BP Basis Pursuit

BSC Best Spherical Codes

CBS Correlated Balanced Sampling

CCD Charge-Coupled-Device

CDMA Code Division Multiple Access

CMOS Complementary Metal-Oxide-Semiconductor

CoSaMP Compressive Sampling Matching Pursuit

CPLD Complex Programmable Logic Device

CRB Cramér-Rao Bound

CS Compressive Sensing

CS&H Current-Sample-and-Hold

DB4 Daubechies-4. Refers to the Daubechies wavelets with four vanishing moments

DC Direct Current. In general, said of a constant signal

DCT Discrete Cosine Transform

DFT Discrete Fourier Transform

DLP Digital Light Processing

DMD Digital Micromirror Device

DOF Degree of Freedom, in general, or Depth of Field for lenses and optical systems

EOM Electro-optic Modulator

ERDW Equilibrium of Rescaled Differences of Code Words

ERW Equilibrium of Rescaled Code Words

EZW Embedded Zerotree Wavelet Algorithm

FDCT Fast Discrete Curvelet Transform

FFT Fast Fourier Transform

FLC Ferroelectric Liquid Crystal

FLIM Fluorescence Lifetime Microscopy

FOCUSS FOcal Underdetermined System Solver

FOV Field Of View

FPGA Field Programmable Gate Array

FPN Fixed-Pattern Noise

FTIR Frustrated Total Internal Reflection

FTR Frustrated Total Reflection

GPCA Generalized Principal Component Analysis

GT	Ground Truth
HDRC	High Dynamic Range CMOS
HDR	High Dynamic Range
HTP	Hard Thresholding Pursuit
ICS	Illumination Control Signal
IHT	Iterative Hard Thresholding
IRLS	Iteratively Reweighted Least Squares
ISD	Image Signature Dictionary
JPEG	Joint Photographic Experts Group, denotes the compression standard created by them
KLT	Karhunen-Loève Transform
LCoS	Liquid Crystal on Silicon
LDPC	Low-Density Parity-Check
LDPD	Lateral Drift-Field Photodiode
LED	Light-Emitting Diode
LEFM	Lateral Electric Field Charge Modulator
MCP	Microchannel Plate
MCT	Minimum Charge Transfer
ME	Metal-Electrode
MEMS	Microelectromechanical Systems
MMV	Multiple Measurement Vector
MPI	Multipath interference
MP	Matching Pursuit
MSM	Metal-Semiconductor-Metal
NIR	Near Infrared

NN	Neural Network
NSP	Null Space Property
OFDM	Orthogonal Frequency-Division Multiplexing
OMP	Orthogonal Matching Pursuit
ORFI	Optical Radio Frequency Interferometry
ORMP	Order Recursive Matching Pursuit
PAL	Programmable Array Logic
PCA	Principal Component Analysis
PEG	Progressive Edge-Growth
PG	Photo-Gate
PLA	Programmable Logic Array
PMD	Photonic Mixer Device
PMT	Photo-Multiplier Tube
PN	Pseudo-Noise
PR	Pseudo-Random
PSF	Point Spread Function
PVDF	Polyvinylidene Fluoride
PWI	Pixelwise Integration
PWM	Pulse-Width Modulation
PTZ	Piezoelectric Transducer
PZT	Lead Zirconate Titanate
QCIF	Quarter Common Intermediate Format
QE	Quantum Efficiency
RGB-D	Red Green Blue Depth

RGB Red Green Blue

RIP Restricted Isometry Property

RMSE Root Mean Square Error

ROI Region Of Interest

SBHE Scrambled Block Hadamard Ensemble

SLE System of Linear Equations

SLM Spatial Light Modulator

SMV Single Measurement Vector

SNR Signal-to-Noise Ratio

SPAD Single-Photon Avalanche Diode

SP Subspace Pursuit

SR SuperResolution

SVD Singular Value Decomposition

TCSPC Time-Correlated Single Photon Counting

THD Total Harmonic Distortion

TIR Total Internal Reflection

ToF Time-of-Flight

TV Total Variation

USFFT Unequispaced Fast Fourier Transform

VGA Video Graphics Array

WNN Wavelet Neural Network

ZESS Zentrum für Sensorsysteme. Center for sensor systems (University of Siegen)

Greek Letters

Γ Pixel channel. For PMD pixels, $\Gamma \in \{A, B\}$

μ Coherence

$\boldsymbol{\Phi}$ Sensing matrix

$\boldsymbol{\Psi}$ Sparsity dictionary or basis, by columns

ρ Reflectivity

$\rho_{\boldsymbol{\Psi}}$ Redundancy of the dictionary Ψ

$\sigma_k(\cdot)_p$ k-term approximation error in l_p

Roman Letters

A Signal amplitude or Area

A A-channel of a PMD pixel

$\boldsymbol{A}$ Measurement matrix

B B-channel of a PMD pixel

c Speed of light in vacuum: 299792458 m/s

d Depth or distance

$\delta(t)$, $\delta(\phi)$ Dirac delta function

$e(t)$, $e(\phi)$ Scene response in time and phase domain, respectively

f Frequency

f_{CLK} Clock frequency

f_{lens} Focal length of a lens or lens system

f_{mod} Modulation frequency

$f\#$ F-number of a lens or lens system

$g_{\text{illu}}(t)$, $g_{\text{illu}}(\phi)$ Convolution function modeling the low-pass filtering effect of the light emitters (LEDs) on the ICS, in time and phase domain, respectively

$g_{\text{pix}}(t)$, $g_{\text{pix}}(\phi)$ Convolution function modeling the low-pass filtering effect of PMD circuitry and pixel on the reference signal, in time and phase domain, respectively

$i(t)$, $i(\phi)$ Emitted illumination signal in time and phase domain, respectively

m Number of measurements

n Signal dimensionality

n_{atoms} Number of *atoms* in a dictionary, when different from n

n_{rows}, n_{cols} Number of rows and columns of the sensor array, respectively. Their product yields the number of pixels of the sensor

NA Numerical Aperture

$q(t)$, $q(\phi)$ PMD reference signal at pixel level, in time and phase domain, respectively

$q_{\text{th}}(t)$, $q_{\text{th}}(\phi)$ Theoretical (square) PMD reference signal at pixel level, in time and phase domain, respectively

r Radius or range

$r(t)$, $r(\phi)$ Reflected light signal, as received at the pixel surface, in time and phase domain, respectively

T Period

t Time

t_{exp}, t^{exp} Exposure time

$\boldsymbol{X}$ Unknown matrix of sparse vectors, by columns

$\vec{x}$ Unknown sparse vector or pixel coordinates

$\boldsymbol{Y}$ Matrix of measurement vectors, by columns

$\vec{y}$ Vector of measurements

List of Figures

List of Tables

1. Introduction

Nowadays, computers, machines and, in general, any kind of automatic system is required to interact with the environment in a natural way, including human-machine interaction. Natural Interaction (NI) is intended to be bidirectional, meaning that the machine is not only required to provide the right information or actions to an external user or environment, but also to extract the needed information from them in a natural way, that is, without forcing the environment or the humans within it to be or behave in a way that meets some non-natural machine expectations.

A clear witness of this trend is the field of mobile robotics, which vast area of research includes robots destined to operate in poorly structured and unstructured environments and to interact with humans in a natural way. This requires a higher degree of sophistication both in control and in the on-board sensors. The control has to move from reactive to deliberative, where extraction of high-level information and an appropriate task planning play a key role. The requirement for high-level information, which allows the robot to recognize locations, objects, people, behaviors, etc. comes together with the need for novel sensors, providing data from which such information can be extracted. Traditional sensor setups, oriented to solve the localization and navigation problems and based on rotary-mirror laser scanners and low-resolution ultrasonic sensors, have become obsolete to deal with the current challenges of mobile robotics. Color cameras and 3D sensors are necessary sensors of any actual mobile robot that aims to extract the most information from its environment while minimizing disturbance. It is clear that the extraction of human-like information allows using human-like concepts, which lead to human-like decisions, which eventually enable human-like behavior and interaction. The success of a real-world robotics application is, therefore, strongly coupled with the observability of high-level information. This suggests the key role that 2D/3D sensors are playing in nowadays robotics.

Another area that might be closer to the general public than mobile robotics is gaming. Computer games require tight interaction between user and machine. The conventional solution was to use a physical Human-Machine Interface (HMI), which could be a keyboard, a mouse or specific gaming devices featuring several buttons, scrolls, etc. As games evolve in realism and complexity, users are able to perform a wide range of actions in some virtual-space domain. Consequently, means for natural interaction allowing a faster and more accurate control are highly desirable. Conventional cameras can be used as a natural HMI, but at least two cameras are required if 3D locations of relevant features in the images are to be estimated. In this scenario, RGB-D sensors arise as a natural solution, providing both RGB and depth images at video frame rates and allowing for a compact construction, difficult to achieve using medium-range stereo systems or laser scanners. It is, therefore, understandable that the development of RGB-D sensors has been driven by the gaming industry. The Kinect sensor is probably the most well-known RGB-D sensor to date. The first model was developed by Microsoft® Research and intended to be a NI device for the Xbox®[0] 360 video game console and Microsoft Windows®[0]-based computers. The Kinect®[0] sensor was able to capture both gestures and spoken commands of the player. The hardware was based on the structured light range camera technology provided by PrimeSense®[0] . Despite its originally restricted application area, the sensor was soon adopted by research groups on mobile robotics for being able to gather 3D information of the environment at high frame rate and at a very low cost, much lower than that of laser scanners. The interest on RGB-D sensors grew rapidly once the Kinect capabilities were publicly demonstrated by scientists in robotics and computer graphics and individual developers.

The success of the Kinect sensor awoke the interest of a wide research community in RGB-D sensors and depth sensing, but also pointed the main limitations of structured light depth sensing, namely low effective lateral resolution, quadratic decay of depth accuracy with the distance and inability to operate in presence of strong ambient light. The Kinect manufacturer, as well as other world-class companies—e. g., the Intel-Creative consortium—, set their eyes on Time-of-Flight (ToF) depth imaging technology, as an alternative to structured light approaches. The principle of operation of ToF imaging is to emit pulsed or continuously-modulated light to the scene and focus the reflected light on a special imaging array. In the conceptually simplest case, each pixel measures the time between light emission and reflection reception. Alternatively, *intelligent* pixels can be used to compute the phase shift between emitted and received light. In both cases the depth can be directly computed, provided that the speed of light is a constant.

[0]Microsoft®, Xbox®, Windows®, Kinect® and PrimeSense® are registered trademarks and will be denoted in the following without the ® symbol for notation simplicity.

Despite ToF imaging being a known technique, at the time the first Kinect model reached the market, ToF devices were expensive and mostly confined to special applications, where dense depth data were desirable, but no strict requirements in terms of lateral resolution and depth accuracy were to be met. The result of the Microsoft research in ToF depth imaging was the advent of a ToF-based Kinect sensor, which arrived to the market as *Xbox One* Kinect, mostly targeted for use with the Xbox gaming console, but also presented to developers as a general-purpose NI device, regardless of the application domain. The ToF-based Kinect implemented an improved depth imaging technology based on that previously developed by Canesta and at the time it reached the stores in July 2014, it could be considered state of the art for ToF imaging, in terms of native lateral resolution, depth accuracy, stability and robustness to motion and sunlight. The new Kinect sensor exhibits depth errors in the millimeter range after linear calibration, one order of magnitude lower than that of direct competitors and low enough to become a realistic alternative to laser scanners in applications requiring high-accuracy depth sensing.

The advent of Microsoft low-cost RGB-D sensors have unveiled a large number of potential application areas for ToF imaging technology. Thousands of scientific publications have appeared, providing comparative evaluations and novel applications of the Kinect sensors in the most diverse fields of science. In order to give an overview of the wide potential of ToF imaging, we provide the following list of research areas where the use of the Kinect sensor has been demonstrated:

1. Pattern recognition and classification: Object recognition, object and scene classification, posture recognition, gesture recognition, action recognition, human recognition, face recognition, sign language translation, vegetation characterization.

2. Computer graphics and HMI: augmented reality applications, 3D scene reconstruction and modeling, motion capture, 3D animation, skeleton tracking, articulations tracking, hand tracking, fingerprints and palm tracking, head pose estimation, stereo generation.

3. Mobile robotics: localization and mapping (both decoupled and simultaneous), teleoperation.

4. Medicine and health care: medical imaging, rehabilitation systems, elderly people assistance, fall detection, gait analysis, visually-impaired assistance, respiratory motion monitoring, surgical instrument tracking.

5. Civil security: human detection, people-flow estimation.

Most of the thousands of scientific papers exploiting the capabilities of the Kinect sensor are based on the structured light version. Their promising results would experience a significant improvement upgrading to the ToF version. The broad range of applications introduced above will widen even more as the researchers adopt the ToF Kinect. With this promising future to come, it is clear that fundamental research on the core ToF technology is a keystone to boost the transition from *proof of concept* experimental prototypes to commercial solutions implementing low-cost ToF imaging technology.

The core depth sensing technology of the ToF Kinect was developed by Canesta and implements the concept of multichannel *intelligent* pixels, originally introduced by PMD Technologies® with the Photonic Mixer Device (PMD). Despite the PMD concept was originally conceived in 1996, phase-shift-based ToF was not able to get into the application areas later unveiled by the Kinect sensor. One can argue that the price of PMD cameras and, in general, of any commercial ToF camera used to be at least one order of magnitude higher than that of the Kinect sensor, placing them out of the sphere of low-cost applications. Nevertheless, one of the main reasons PMD cameras did not get the degree of acceptance one could expect was the relatively low quality of the depth images, mostly in terms of lateral resolution and depth accuracy. Despite the fact that the ToF version of the Kinect sensor offers a significant improvement with respect to PMD devices, the basics of the technology remain the same and, consequently, both share the same fundamental issues. Section 1.1 gives an overview on how we try to overcome the main limits of PMD technology, which are shared, eventually in a lower degree, by the recent ToF-based Kinect sensor.

1.1. Motivations and Contributions

In this work we focus on the PMD ToF imaging technology, for being a reference technology for phase-shift-based ToF imaging. However, the methods we present apply, if not directly, with minor modifications, to any phase-shift-based ToF imaging system, such as, e. g., the new Microsoft Kinect sensor.

The current limits of the PMD technology will be analyzed in Section 2.4. From the image processing point of view, one of the clearest issues is the

[0]PMD Technologies® is a registered trademark and will be denoted in the following without the ® symbol for notation simplicity.

relatively low lateral resolution of PMD depth images. PMD chips with 120×160 and 200×200 are available, while the last Kinect sensor already features 512×424 pixels. In general, these resolutions are far lower than those of conventional color cameras. This poses a fundamental problem when both modalities are to be combined, which is the goal of RGB-D imaging. Previous works exist that combine a high-resolution color image with the low-resolution depth image in order to achieve a high-resolution depth image. Such approaches rely on certain hypothesis such as the fact that texture and depth are correlated up to certain extent, which might be completely false in practice, and do not provide real depth superresolution.

Another main motivation of this work is the relatively low effective depth accuracy of ToF cameras. While the ToF-based Kinect sensor is able to achieve millimeter accuracy, most commercial systems, specially those based on PMD technology, show average accuracy in the centimeter range. Excessive confident measurement models that do not account for any of the signal distortions and superposition effects present in real operation are in most cases responsible of the limited capabilities of ToF imaging systems when operating in non-ideal conditions.

In general, the problems we face can be reduced to specific instances of the general problem of signal recovery from incomplete and eventually noisy measurements. In the case of the poor lateral resolution, motivated by a large pixel size together with large blind areas between pixels, it is clear that the pixels provide partial measurements in spatial domain. Even ignoring blind areas, pixels integrate the light according to a certain measurement function in spatial domain, which could be non-constant. Consequently, we always get *measurements* of a variable, not punctual samples of the variable itself. The actual values can only be recovered if the measurement process is properly characterized and an appropriate recovery method for solving the—typically ill-posed—inverse measurement problem is found. In the case of poor depth accuracy in phase-shift-based ToF systems, it must be pointed out that the phase shift is obtained from cross-correlation measurements, i. e., the pixel acts as a correlator between the received light signal, reflected by the scene, and an internal reference signal. For simplicity, both signals are typically supposed sinusoidal, which allows retrieving the phase shift from very few measurements. In reality, none of the signals is sinusoidal and, consequently, only partial information is conveyed by the measurements. Sticking to the ideal case neglects the contribution of signal harmonics, which produce, in turn, noisy measurements and, consequently, inaccurate depth measurements. Once again, we face again the problem of signal recovery—or, at least, phase recovery— from partial measurements, which requires an accurate measurement model to solve the inverse problem.

From the simplified high-level problem formulation given above, the recent theory of Compressive Sensing (CS) naturally arises as an adequate framework to deal with the main limitations of ToF imaging. CS theory offers the possibility of exact or accurate estimation of a compressible signal from incomplete, bandlimited or noise-corrupted measurements. Most natural signals are, if not exactly sparse, at least compressible in an appropriate domain: natural images in wavelet domain, periodic signals in frequency domain, etc. In the worst case, if an appropriate basis cannot be found where the signal of interest is sparse, a dictionary can be trained from a large amount of signals of interest, in order to find the basis or tight frame that best represents a specific kind of signal. Consequently, CS can be applied, in a higher or lower degree, to almost every sensing system. The aim of this thesis is to adapt the general theory of CS to the specific case of PMD-based ToF imaging in order to achieve depth maps of higher quality.

CS requires a linear sensing model and the signal recovery is typically very sensitive to noise. Consequently, one of our basic contributions is to provide means to obtain PMD raw data which is free of non-linearities, such as local pixel saturation, and free of systematic noise, such as the fixed-pattern noise produced by the ADCs.

Regardless of the domain where CS is applied, a realistic sensing model is a necessary condition for accurate signal recovery. Ignoring the complexity of the sensing process might lead to a simplification of the phase shift estimation procedure, but at the cost of implicitly casting the ignored phenomena as measurement noise, leading to wasting the capabilities of the sensor, in terms of maximum recoverable resolution or accuracy. Therefore, one of our main contributions is an accurate discrete characterization of the response of the PMD pixels in spatial domain and in spatial-phase domain, with subpixel resolution [H1]. This contribution is of capital importance, since no superresolution method can claim that the obtained high resolution image is physically meaningful, unless the function modeling the pass from the high resolution to the low resolution domain is known. One can argue that this necessity can be avoided if the pixel shows a constant response all over its area, which might be the case in conventional imagers. To the best of our knowledge, we are the first proving such hypothesis wrong for PMD pixels and giving a high-resolution characterization of the PMD pixel response, which allows modeling downsampling—or, conversely, superresolution—factors up to 45.

We show that accurate recovery of PMD raw images from few measurements is possible in wavelet domain, where images admit a compressible representation. Additionally, we illustrate with a real application that our characterization of the PMD subpixel spatial response can be used to achieve real superresolution. Our main contribution in this field is the development of a greedy algorithm for joint recovery of all PMD raw images from few measurements that takes into account the degree of redundancy among them and exploits the tree-structure of wavelet representations [H7].

Increasing the lateral resolution might lead to an eventual improvement in depth accuracy due to the reduction of uncertainty in spatial domain, but it does not help in reducing the error due to non-sinusoidal illumination and reference signals if conventional methods for phase estimation, such as the so called *four phases algorithm* are applied. In this regard, our main contribution is to provide a general framework for signal reconstruction and phase retrieval that applies CS to recover the light signal as received by the pixel from few measurements, exploiting the fact that the illumination waveforms are periodic and, therefore, compressible in frequency domain [H3].

Finally, we combine the application of CS in spatial domain and in time-frequency domain in a synergistic manner, providing an experimental CS-based PMD camera, which allows for superresolution in spatial domain and does not impose any restriction on the modulation waveform, other than being periodic. The prototype includes both software and specific hardware developments intended to overcome the current limits of ToF image resolution, given by the last Kinect sensor, offering a maximum resolution of 1200×1600, i. e., a superresolution factor of 10 with respect to the native resolution of the PMD sensor we use.

Extensions of our CS methods to closely-related research areas have also been considered, eventually leading to solutions that outperform the state of the art in those areas. We refer here to two examples of such success cases, which are not explicitly included in this thesis due to space constraints. The first problem we consider is image enhancement for fast ToF imaging and fast multiaperture systems. Fast imaging becomes necessary when several acquisitions of the scene are required in a very short period of time, e. g., when imaging very fast phenomena or, in the case of applying CS to ToF in time-frequency domain, when several (still few) measurements are needed to recover the signal. Fast imaging or, equivalently, low-light imaging, produces raw data that is largely corrupted by Poisson noise. Our CS-based methods have shown to yield outstanding noise reduction in multiaperture data. We refer to [H8] for details. The second contribution deals with the problem of inverse lighting of a single face image, that is, estimating the location and RGB intensities of some light sources that would produce the illumination that is present in a certain image of a human face. We use CS to provide an efficient and robust estimation framework that is respectful with physical constraints and favors the solution with lower number of active light sources. The algorithm, together with a thorough experimental evaluation, are given in [H10].

1.2. Outline

The outline of the thesis is as follows: Chapter 2 is intended to provide an introduction to ToF imaging systems, with main focus on phase-shift-based ToF, an area where PMD is a reference technology. In this chapter we also analyze the specific case of the PMD ToF chips and explain the principle of operation of its *intelligent* pixels. Chapter 3 introduces the CS theory. Different sensing matrices are presented and their respective strengths and weaknesses are pointed out. We focus on some deterministic alternatives to random constructions, which have been shown to reach and, eventually, outperform the latter in terms of sparse recovery performance. Different sparsity bases and, more generally, dictionaries, are also considered, in order to achieve the highest sparsity. A review of most common methods for sparse recovery is also included in this chapter.

Chapter 4 is the core of this thesis and aims to bring CS to the field of PMD-based ToF imaging, enunciating the domains where sparsity or compressibility can be exploited. In this chapter we present the main issues that prevent the application of compressive sensing and our methods to overcome them. An accurate sensing model is a necessary condition for successful sparse recovery in a CS framework. We provide a high resolution characterization of the response of PMD pixels in spatial domain, which allows generating physically meaningful CS sensing matrices with subpixel resolution, that is, modeling the transition from a higher resolution domain to the low resolution PMD pixel domain. Such a model is required if an image with higher resolution than that of the PMD sensor is to be recovered from the compressed measurements. Extending the concept of spatial response of a pixel to an area wider than the pixel itself, i. e., covering neighboring pixels, an accurate crosstalk characterization can be obtained. Apart from a complete study of the crosstalk phenomena in a PMD chip, we also provide a method for crosstalk compensation that exploits this knowledge. The last sections of the chapter are dedicated to the main contribution of this work: the application of CS and sparse recovery in spatial domain and in time-frequency domain.

Since there are two domains where signal sparsity can be exploited, an idea that arises naturally is to combine the sensing and the sparse recovery framework in both domains in a synergistic manner. This idea is further elaborated in Chapter 5. A prototype is presented that operates gathering compressed measurements using a modified PMD camera and an additional spatial modulator. We provide a detailed description of the modification carried out on the PMD frontend and the additional hardware. Specific software is developed to control the spatial and temporal compressive sensing, gather the compressed measurements from the camera and generate the final depth images out of them.

Chapter 6 concludes the thesis, summarizing the achievements. It also offers an overview of future research topics, whose interest and feasibility are backed by the results of this work.

2. Phase-Shift-Based Time-of-Flight Imaging Systems

Time-of-Flight (ToF) is a widely spread technique for depth imaging. The term *depth imaging* is often used as a synonym of *3-dimensional (3D) imaging*, but *sensu stricto* the former is a specific case of the latter. In general terms, 3D imaging aims to capture 3D information of a scene in a 2D array of data (an image). The addition of a third dimension does not imply that the data are dense in all three dimensions of the scene space. We speak of dense 3D imaging if we are actually able to measure the value of a certain property—e. g., reflectivity—all along the three dimensions of the scene space. Examples of this kind are active systems using radiation that is able to go through objects up to a certain extent, allowing dense full 3D reconstructions. A well-known example is the Magnetic Resonance Imaging (MRI) [228] technique, widely used as non-invasive method to detect malfunctions within alive biological systems.

In most cases, the radiation used for imaging does not penetrate the object and we get sparse data in the third dimension. In this category fall conventional depth imaging systems, which provide a collection of distances from the camera to the scene points, in the shape of a depth image. Such cameras also provide an intensity value per pixel. If a 3D intensity map is reconstructed, we obtain a surface in 3D-space, not a dense volume. For this reason, depth cameras are often called 2.5D sensors, instead of 3D sensors, since they can only capture surfaces in 3D.

More recently, the development of novel transient imaging techniques [445] has allowed recovering an approximate time profile of the returning light signal in ToF imaging, opening the possibility of sensing several depths per pixel when dealing with translucent objects in the scene, recovering its volume. This concept, in combination with the PMD chip, allows for low-cost systems able to face the ToF multipath problem [248] and sense occluded objects [220].

Apart from these exceptions, most 3D imaging devices sense the scene as a discrete 2D-array, offering a single depth value per pixel. The interested reader can find a survey of the multiple possibilities that consumer depth cameras offer in the field of computer vision in [181]. In this chapter we provide an introduction to conventional depth imaging methods, pointing out the main strengths and weaknesses of each technology. We describe the principle of operation of ToF imaging and focus on the Photonic Mixer Device (PMD), as reference technology for phase-shift-based ToF imaging. The understanding of the PMD principle of operation naturally unveils the main limitations of the technology, which are enunciated.

2.1. Introduction to Depth Imaging

This section provides a general overview on 3D imaging [404]. Four fundamental systems widely used for 3D imaging [49] are introduced: laser scanners, stereo systems, light coding and ToF cameras. Some basic background on the principles of operation and comparative remarks are provided. Some techniques that, even being previous to ToF imaging, can be classified as a special type of phase-shift-based ToF technique—e. g., interferometric techniques—are considered in Section 2.2.

2.1.1. Laser Scanners

Laser scanners are the preferred option when high accuracy in depth measurements is required. Laser scanners are either based on the Time-of-Flight principle, like ToF cameras, or in the triangulation principle, which is closely related to the concept of stereo vision. Generally, laser scanners gather data sequentially, therefore no 2D-array of distances is acquired at once and their classification as 3D-*imaging* device is questionable.

ToF laser scanners measure the time the emitted light takes to reach the scene point and return to the sensor. Consequently, the accuracy of ToF-based scanners is directly dependent on the precision of their time measurements and is typically constant along the measurement range [61]. The time measurement is often substituted by a measurement of difference of phase between emitted and received light, which is also proportional to the distance traveled.

Triangulation scanners use one or two—typically CCD—cameras to receive the returning signal, i. e., the receiver is no longer a single element but a line or an array of pixels. In the simpler setup, using a single camera, the laser beam is projected onto the scene with a known direction and is received by the camera, which is placed at a known distance from the emitter. The distance to the object can be easily determined by solving the triangle between emitter, object and reception position of the point or line formed on the camera image. Note that, in this setup, the operation range of the scanner is bounded by the length of the baseline, or distance between emitter and camera. The accuracy of triangulation scanners decreases with the square of the distance between scanner and object [61]. When measuring optically rough surfaces, speckle will be observed in the image. The specle noise limits the uncertainty of the spot localization and, consequently, the uncertainty of the distance measurement [155]. These characteristics make ToF-based scanners suitable for long distances, while triangulation-based scanners are preferred for short range measurements. For a quantitative evaluation on the accuracy and resolution of commercial laser scanners the interested reader is referred to [60].

Laser scanners make use of rotating mirrors to change the direction of the beam. The lateral resolution of laser scanners—number of pixels of the depth *image*—is typically high, much higher than most other 3D-imaging devices, and is governed by the minimum step size of the mirror rotation system. Due to the sequential acquisition, a large number of points slows down the acquisition process. Typical acquisition rates are in the order of thousands to tens of thousands of points per second. Consequently, for very high resolutions, the acquisition time might grow up to several seconds or even minutes. This gives rise to one of the main limitations of laser scanners: the need for a strictly static scene and no external motion of the scanning device. This requirement is often impossible to fulfill in real scenes and, therefore, severe motion artifacts come up when gathering dense data from non-static environments. A way to overcome this limit is using many laser sources and receivers simultaneously in the same scanner (e. g., 64 in the Velodyne lidar [224]), leading to bulky and expensive systems that can reach the frame rate of ToF cameras.

Typically, laser scanners gather the measurements following sequential raster lines, i. e., a regular acquisition pattern. In applications such as 3D modeling this may lead to an insufficient number of points in critical areas, where high detail is required, or too many points in unimportant areas. Using custom acquisition patterns, which can be adapted to the object to sense, are an appropriate solution in such cases. The laser scanner presented in [49] uses Lissajous curves as acquisition patterns. They also develop a recursive optimization method that allows obtaining high-resolution 3D reconstructions from the sparse range data.

2.1.2. Stereo Systems

One of the most widely-used methods for depth imaging is stereo vision [46]. The hardware normally consists on two cameras sensing the scene from two different and typically known positions. Using two images simultaneously taken from both cameras, it is possible to extract 3D information of the scene. Note that stereo vision is, in its basic concept, a passive method, since no energy is emitted and no camera movement is needed. Conventional stereo systems often use two cameras of the same model, pointing in the same direction but displaced horizontally a certain distance (baseline) from one another, similarly to the human binocular vision system. In such case, and in absence of any distortion, corresponding pixels between the two images lie on the same horizontal row and a disparity map can be directly computed from the positions of corresponding pixels in the images. The disparity is inversely proportional to the distance from the system to the object, therefore, a depth value can be computed from the disparity if the camera parameters are known. In most practical cases, there exist lens distortions, that might be different for each camera, and the cameras might not be perfectly aligned. Consequently, images must first be cleaned from distortions and later both images of the stereo pair must be projected to a common plane (rectification process), in order to ensure that corresponding pixels are actually in the same row in both images.

In general terms, the depth estimation is more accurate for objects close to the stereo system (higher disparity) than for far objects (disparities tending to zero). Nevertheless, for very close objects we might encounter overlapping problems if the object is being observed mostly by only one camera of the stereo pair. From the description of the principle of operation it is easy to observe one major problem of stereo depth estimation, the so-called parallax problem. For large parallaxes (cameras placed at largely different viewpoints) an accurate depth estimation can easily be computed, even for objects that are relatively far from the system. A large baseline implies a reduction of the SNR in depth estimation. On the other hand, the matching problem becomes more difficult due to the occlusion of features. Another critical point of this technique is the search for correspondences. Depending on its location in 3D-space, an object point might be only present in one camera view, in which case no match is possible. Additionally, in many real applications, the system faces areas without texture—e. g., long corridors without distinctive elements on the walls—, where useful correspondences cannot be established. The projection of structured light [331, 397], known as active stereo, is a common solution in such situations. The optimal design of the patterns to project has been studied in [227]. Other active stereo solutions include camera motion [126, 442], allowing fixation in different scene objects and accurate surface reconstruction. Passive stereo systems usually cannot provide a dense depth image due to lack of correspondences. Coarse-to-fine approaches and certain hypothesis on depth smoothness can partially solve the problem but cannot totally overcome the effect of occlusions.

2.1.3. Light Coding Technology: The Kinect Sensor

The so-called *light coding* technology evolves from the structured light range imaging systems [177]. In general terms, structured light systems project a light pattern on the scene and observe it through a camera. The different depths of the scene points induce different distortions in the observed projection and the depths can be estimated if univocal correspondences between the emitted and observed patterns are established. If several cameras are used, the pattern can be used to ease the search for correspondences between the different camera images and the depth can be estimated from disparity as in a stereo system. In early approaches [376] several masks featuring different spatial codes are used to generate different projection patterns. One of the most common options is to use a set of Gray codes [453], which are robust and can be efficiently generated [48]. Other binary codes can also be used as projection patterns [319]. Later approaches tend to rely on a single acquisition [379] and the code design becomes a critical issue. Using projection of color codes [97, 482] in combination with a color camera allows reducing the acquisition time or, conversely, acquiring more information, resulting in a better depth estimation.

The light coding method we refer from now on is that used in the first version of the widely-known Kinect sensor, developed by Microsoft. The hardware of the Kinect sensor is based on the range camera technology provided by PrimeSense. Focusing in the depth estimation hardware, the sensor features an infrared (IR) laser projector and an IR camera, pointing in the same frontal direction and separated a few centimeters from each other in horizontal direction. The existence of an emitter and a receiver could induce to think about a ToF-based system but the emitted light is DC and no time or phase measurements are performed. Instead of modulating in time domain, the light is modulated in spatial domain, according to a fixed pattern. The pattern, which looks like small dots randomly distributed, is projected onto the scene. The pattern is composed by a sub-pattern, which is repeated 3×3 times. Fig. 2.1b shows a schema of the sub-pattern, where each bright dot in the pattern is represented by a black square. The reflection of the pattern is captured by the IR camera. The relative position of the dots in the infrared image, with respect to a reference image of the pattern at a known distance, depends on the distance between the sensor and the object that reflected the dot. The full description of the system is given in [188], which seems to be an implementation of the more general system presented in [418], where it is explicitly stated that groups of spots in the IR image are compared to those in the reference image, using image correlation or some other image matching technique. Fig. 2.1a is a diagram of the Kinect sensor observing a wall. It has been shown that the raw measurements delivered by the sensor depend linearly on the angle θ [447] and the actual depth, conceived as the distance from the imaginary image plane to the object, can be calculated through a tangent transformation.

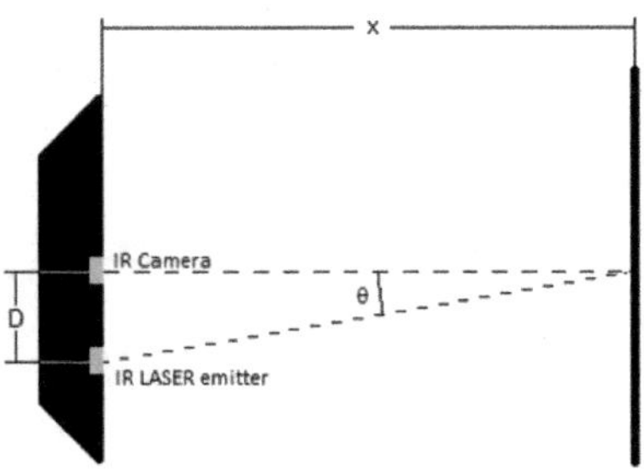

(a) Kinect Operation

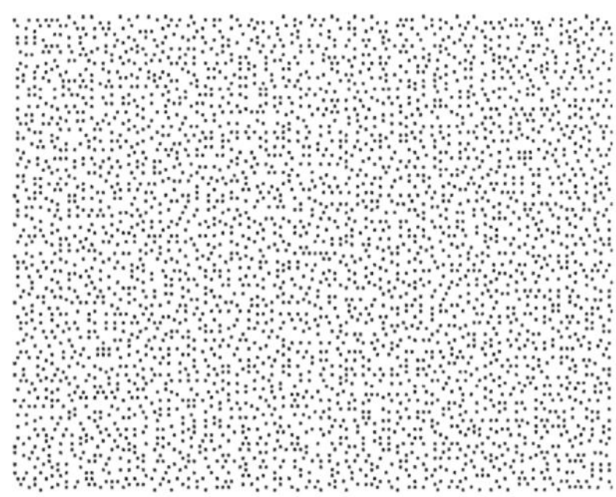

(b) Kinect Pattern

Figure 2.1.: Schema of the Kinect sensor (a). A pattern of dots is projected onto the scene and the reflection is captured by an infrared camera. The disparity of the acquired IR image with respect to a reference image of the pattern depends on the angle θ and this, in turn, on the distance to measure, x. (b): the basic sub-pattern used to generate the pattern, from [386]. Black squares represent the bright spots. The pattern is obtained by repeating this sub-pattern 3×3 times.

The depth images are delivered as monochrome, with 11-bit VGA resolution (640×480 pixels), streamed at 30 Hz frame rate. The value of the sensor response is calculated internally from the disparity between the reflected pattern recorded by the IR camera and a reference pattern, which is stored within the sensor [260]. Objects whose distance to the sensor is higher or lower than the default distance used to obtain the reference pattern will reflect the pattern points displaced in the direction of the baseline between the laser emitter and the IR camera, with respect to their original position in the reference pattern. Using simple image correlation between the captured pattern and the reference pattern the disparity image can be calculated. This means establishing correspondences between two images, for what multiple pixels are used. As a consequence, the real lateral resolution of the depth images is lower than that given by the number of pixels [282] (118 pixels in width, instead of 640). A theoretical limit of the lateral resolution is given by the size of the pattern, which is composed by 220×170 bright dots. Refined formulas to compute the depth value from raw data are given in [282].

According to the data provided by PrimeSense, the field of view (FOV) of the cameras is 58° horizontal, 45° vertical and 70° diagonal. The spatial x/y resolution is 3 mm and the depth resolution is 1 cm, both at 2 m distance from the sensor [447]. PrimeSense gives an operation range for the Kinect sensor between 0.8 m and 3.5 m, but the real range depends on the required precision. Ranges up to 5 m [447] are found in some works, but it has to be taken into account that the depth resolution is proportional to the square of the depth. The Kinect also implements an internal lower limit in depth of 0.4 m, probably to avoid a severely poor overlap with the RGB image. Known limitations of the first version of the Kinect are the inability of operating under sunlight illumination and the absence of depth estimation for dull objects, such as computer screens.

The Kinect sensor has often been used as an inexpensive substitute of laser scanners in mobile robotic applications. A comparison between the first version of the Kinect and laser scanners is given in [488]. One of the many examples of this trend is found in [233], where a Kinect sensor is mounted on a quadcopter and used to perform visual odometry and mapping. A considerable amount of research has been oriented to adapt the Simultaneous Localization And Mapping (SLAM) algorithms used in mobile robotics to the data provided by this new sensor, giving birth to the so-called RGB-D SLAM [171, 168, 143], which was first introduced in [223]. Not only the area of robot navigation has profited from this sensor. Often the interest is moved towards scene reconstruction [349, 239, 257, 167, 290], in which case the crucial point is how to fuse redundant measurements to reduce noise and increase reconstruction accuracy.

2.1.4. Time-of-Flight Cameras

Time-of-Flight (ToF) cameras [179, 282] can be seen as an extension of ToF laser scanners, where the beam has been substituted by a dense light projection over the scene and the receiver is a planar array of pixels. As in the case of laser scanners, the principle of operation can be the measure of the time a light signal needs to travel from the camera to the object and back to the camera, or the measure of the phase shift between emitted and received signal, using some high-frequency periodic waveform. An image of the first phase-shift-based ToF camera [279] is provided in Fig. 2.2a.The main advantage of ToF cameras over laser scanners is the absence of scanning motion: all the points are acquired simultaneously. This softens the static scene restriction of laser scanners, since only movements that are fast enough to produce changes in the image during the exposure time might lead to motion artifacts in the depth map. On the other hand, the lateral resolution of commercial ToF cameras is generally low (ranging from 64×48 to 640×480 pixels), significantly lower than that achievable with laser scanners. Depth resolution strongly depends on the estimation method and the hardware capabilities and can reach subcentimetric theoretical values. Nevertheless, the depth accuracy, strongly influenced by environmental conditions, is typically of several centimeters (one to five for most commercial cameras).

When compared to stereo systems [425], ToF cameras offer several advantages. One of them is that they perform direct depth—or phase—data acquisition, using methods such as the pixel-level correlation in PMD chips (Section 2.3), instead of applying heavy algorithms on pairs of conventional intensity images. Additionally, ToF cameras can operate in textureless scenes, where stereo approaches typically fail. Another strong point of ToF imaging is that the depth image is generally dense, while stereo approaches can only offer a depth value where a pair of correspondent pixels are found. As a drawback, ToF cameras are active systems, that is, an illumination source is needed, typically IR light modulated at several megahertz frequency. Consequently, interference phenomena might come up when several ToF cameras operate simultaneously. The signal-to-background ratio, i. e., effective dynamic range, may be dramatically decreased when a source of IR light is present in the scene or when operating under sunlight illumination. For phase-shift-based ToF cameras, the modulation frequency of the illumination signal establishes a tradeoff between depth resolution and unambiguous range: high frequencies are needed to obtain good depth resolution, but the higher the frequency, the shorter the unambiguous range of the camera.

A widely-spread technology for phase-based ToF imaging is the so-called Photonic Mixer Device (PMD), developed by PMD Technologies. At the time the author started to conduct the research presented in this thesis commercial PMD sensors offered 19k (120×160) and 41k (200×200) resolutions. The manufacturer also offered complete ToF cameras based on these chips, e. g., the PMD CamCube (41k). Since 2015 the production of PMD sensors has been externalized to Infineon, whose sensor family REAL3™ [385] offers ToF imagers based on PMD pixels. The sensitivity has been improved by means of microlenses. Differently from the previous generation of PMD chips, the control signals are internally generated in the chip, as well as the ADCs. This allows for higher modulation frequencies (up to 100 MHz) and lower latency, at the cost of reducing the flexibility of the system. Additionally, two higher resolutions are added to the 19k standard: 38k (224×172) and 100k (352×288). In fact, the recent PMD CamBoard pico flexx reference design already supports the 38k chip. Due to the importance of PMD technology both for ToF imaging in general and for this work in particular, it is analyzed independently in Section 2.3. An image of a state-of-the-art PMD camera is provided in Fig. 2.2b. Another ToF sensor that also requires special mention is the new Kinect sensor (the so-called second-generation Kinect, or Kinect v2 or Xbox One sensor [471]), which is depicted in Fig. 2.2c. This camera is—as well as the first Kinect—an RGB-D camera, delivering both RGB and depth images. It offers a depth lateral resolution of 512×424, very high for a ToF device. Problems derived from sunlight illumination and illumination interference with other ToF cameras have been solved using a complex modulation signal, whose frequencies range from 10 to 130 MHz. The demodulation contrast (68% at 50 MHz) is also high when compared to state-of-the-art technology (Section 2.3). An introduction to the new Kinect technology is presented in [362] and further detailed in [22], where numerical values of key parameters are given (Table 1).

In order to provide a visual overview of the performance of some well-known commercial solutions for depth imaging, we extend the depth camera comparison in [282] by including the recent Intel-Creative®[1] ToF-based RGB-D camera and the new Kinect sensor. The other sensors in the comparison are the first generation Kinect sensor, the PMD CamCube, featuring a 41k PMD chip, a PMDTec camera featuring an earlier 3k chip, two ZESS MultiCams featuring 3k and 41k PMD chips, and the Softkinetic sensor. In the case of the Softkinetic sensor, different light powers (denoted in the legend as percentage of the maximum illumination power) and a special *close* mode (probably using a higher modulation frequency) are tested. Fig. 2.3a shows the depth measurements provided by each one of the cameras we consider after applying a linear correction. Fig. 2.3b depicts the error with respect to the ground truth for all the linearly corrected measurements

[1] Intel® , Creative® and SwissRanger® are registered trademarks and will be denoted in the following without the ® symbol for notation simplicity.

(a) Lange's Camera **(b)** CamBoard Pico S **(c)** Kinect v2

Figure 2.2.: Evolution of phase-shift-based time-of-flight imaging. From left to right: The first ToF camera (a), presented in [279] and based on *intelligent* pixels, similar to those of the current PMD cameras (image taken from [278]). The PMD CamBoard pico S (b), the latest PMD 3D camera reference design. The second-generation Kinect sensor (c) (image from [471]), commercialized together with the Xbox One gaming console, provides state-of-the-art depth sensing at a very low cost. The basic principle of operation of these three cameras is the *in pixel* cross-correlation.

shown in Fig. 2.3a. Both graphs are in meters and 100 measurements were averaged per position. The two variants of the Kinect v1 (simple and tangent) make use of the two different formulas for depth calculus given in [282], respectively. Note that the Intel-Creative camera operates accurately at its theoretical maximum range (99 cm) but cannot sense over 2 m. The new Kinect sensor obtains surprisingly good results, offering a very low depth error along the whole range of operation. According to our results, the sensor delivers high accuracy depth measurements between 0.56 and 4.44 m. The average depth error within this range is 4.8 mm, one order of magnitude lower than most competitors, including the first Kinect version. Additionally, it was observed that the new Kinect provides no measurements out of its optimal range (0.50 to 4.5 m). As it is easy to deduce from Fig. 2.3b, where the Kinect v2 errors are the lower bound of all the other sensors considered, this new sensor is a serious competitor for any 3D-imaging device. Standard deviation measurements reveal that the Kinect v2 is the most stable sensor among all considered here, with an average standard deviation of 2.2 mm along its operative range. A comparison between a ToF camera (the SwissRanger®[1] SR4000, from MESA Imaging®) and the structured light Kinect sensor is provided in [213]. For a thorough performance comparison between the first and second versions of the Kinect sensor, the interested reader is referred to [396].

After this expanded comparison was completed, new ToF sensors from different manufacturers arrived to the market, which are, consequently, not analyzed here. The author is aware of the existence of new models of ToF depth sensors from Panasonic® and Texas Instruments®[2] .

Despite the D-IMager[3] ToF cameras from Panasonic [1] had a resolution of 120×160 pixels, i. e., equivalent to a PMD or Swiss Ranger SR-2 camera, for instance, the newest Panasonic ToF chips reach VGA resolution, namely, 640×480. The depth accuracy of the D-IMager cameras was 3 cm in absence of ambient light and as poor as 14 cm with ambient light, both for the models EKL3104 (test with 2×10^4 lx ambient illumination) and EKL3106 (test with 10^5 lx ambient illumination). Interestingly, for the model EKL3105 Panasonic reported improved accuracies of 2 cm and 5 cm, without and with ambient light (2×10^4 lx), respectively. An example of ToF depth camera featuring the recent VGA Panasonic chip MN34902BL is the Basler ToF camera (first model, named 6m) [32], presented in mid-2015. The peculiarity of the new generation of Panasonic ToF chips (series MN349XXXX) is that they are not just a NIR ToF sensor, but also a color sensor, that is, they simultaneously provide a depth image and a color image of the scene, pixelwise registered. The manufacturer of the camera claims an accuracy of ± 1 cm in a range of 0.5 to 5 m in ideal conditions, but, obviously, the real value will depend, at least, on the depth being measured. In any case, this accuracy is superior to that of the earlier D-IMager cameras. Also, according to the number of pixels in the sensor, the Basler ToF camera might be state of the art in terms of lateral resolution. Additionally, the manufacturer plans three other models implementing the Panasonic ToF chips MN34903TL, MN34922BL and MN34923TL, which are to appear under the model names 6c, 13m and 13c, respectively. We just want to point the reader's attention towards the future model 13m, implementing the Panasonic MN34922BL chip, which would deliver color and depth images, both of SXGA resolution (1280×1024), with a single image sensor (see [31]).

Texas Instruments offers the OPT8320 and the OPT8241 [358] ToF sensors. The first is a complete system-on-chip (SoC) that delivers depth images directly, without the need of any external circuitry. This SoC even integrates an illumination driver. Unfortunately, the resolution is as low as 80×60 pixels, i. e., QQQVGA. The second sensor is a ToF sensor that only integrates

[2] Panasonic® and Texas Instruments® are registered trademarks and will be denoted in the following without the ® symbol for notation simplicity.

[3] The production of the Panasonic D-IMager cameras was discontinued in April 1st 2015

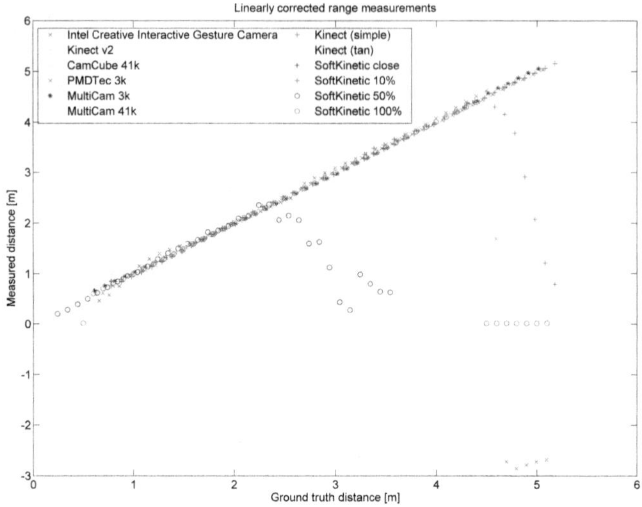

(a) Corrected Depth

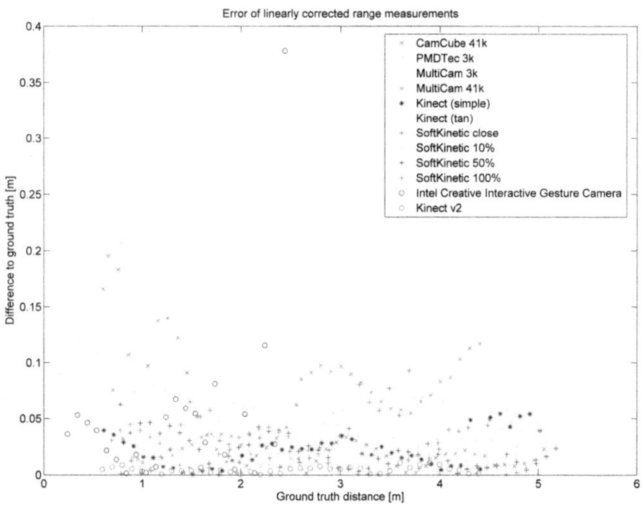

(b) Depth Error

Figure 2.3.: Linearly corrected depth measurements averaged over 10 frames per position (a) and absolute error with respect to the ground truth (b). ©2015 IEEE.

the timing generator and the ADCs on chip, but offers a superior resolution of 320 × 240 pixels, i. e., QVGA.

2.2. Phase-Shift-Based Time-of-Flight Imaging Systems

As indicated in the previous section, a ToF imaging system, *sensu stricto* measures the time the light needs to travel from the illumination system to the scene and from the scene to the camera. Unfortunately, a depth sensor based on this principle requires picosecond resolution in the time measurements to achieve millimeter resolution in depth. Building an array of such sensors operating simultaneously as an imaging system is challenging even for low resolutions. The pulse counting ToF system of [19] is an example of such a pure ToF imaging sensor, but a crucial limitation is imposed by the maximum frequency of the clocking signal, which is suggested to be of 2 GHz frequency, leading to only 500 ps ToF resolution. This is the reason why most ToF cameras measure phase shift instead of time.

In order to measure phase shift, a periodic signal is required. This brings the concept of unambiguous range, which is the maximum distance that can be measured without ambiguity and is determined by the frequency of the signal, f. Provided that the light has to cover the depth twice, the depth d can be computed from the phase shift θ_{depth} as

$$d = \frac{c}{4\pi f}\theta_{\mathrm{depth}} \tag{2.1}$$

and, in the simplest case, if only one frequency is considered, the unambiguous range is

$$d_u = \frac{c}{2f} \tag{2.2}$$

At this point we make a distinction between pure *optical interferometry* and what was originally called *optical radio frequency interferometry* (ORFI) [474]. These two techniques are often presented separately, being the first one named simply *interferometry* and the second one *phase-shift-based* ToF or simply *ToF*. The truth is that both techniques share the same principle of operation and should be presented together, since the ORFI mimics the optical interferometry, aiming to obtain a much larger unambiguous range by means of an intensity modulation of the light at radio frequencies. Supposing a periodic modulation scheme, in both cases the depth can only be estimated from the phase shift and, therefore, Eq. 2.1 and Eq. 2.2 hold, being $f = f_{\text{mod}}$ the modulation frequency in the ORFI case and the actual electromagnetic frequency of the light in the optical interferometry case.

2.2.1. Interferometry

The human eye is typically sensitive to light of wavelengths between 390 and 700 nm [424], which correspond to a frequency band between 430 and 790 THz. Even light of relatively large wavelength, such as near infrared light (NIR), leads to frequencies in the range of hundreds of terahertz. For instance, a wavelength of 850 nm is equivalent to a frequency of 350 THz, which leads to an unambiguous range of 425 nm, i. e., half the wavelength.

Interferometry [214, 215] is based on the superposition of light waves (interference), which can be constructive or destructive. For instance, if at a certain point in space, which we denote by the vector $\vec{r} = [x, y, z]$, two electromagnetic waves of equal amplitude and frequency interfere, the amplitude of the resulting wave may be higher or lower than that of the interfering waves, depending on their relative phase shift. In general, consider two plane light waves, E_1, E_2, traveling along the same direction, say z, and polarized with their field vectors in the same plane:

$$\begin{aligned} E_1(\vec{r}, t) &= E_1(z, t) = e_1 \cos(2\pi f t - \frac{2\pi}{\lambda} z) \\ &= \Re[e_1 \exp(-i\theta) \exp(i\omega t)] = \Re[e_1' \exp(i\omega t)] \\ E_2(\vec{r}, t) &= E_2(z, t) = e_2 \cos(2\pi f t - \frac{2\pi}{\lambda} z + \Delta\theta) \\ &= \Re[e_2 \exp(-i(\theta - \Delta\theta)) \exp(i\omega t)] = \Re[e_2' \exp(i\omega t)] \end{aligned} \tag{2.3}$$

where e_k is the amplitude of the wave $E_k, k \in \{1, 2\}$, f and λ are the frequency and the wavelength of the radiation, respectively, and $\Delta\theta$ is the relative phase shift between waves. In the complex formulation, the following changes of variables was applied: $\theta = 2\pi z/\lambda$, $\omega = 2\pi f$. In the last expression, e_k' is the complex amplitude, dependent on the phase shift. The intensity of a wave is defined as $I_k = |e_k'|^2$ and, consequently, the resultant intensity is

$$\begin{aligned} I = |e'|^2 &= (e_1' + e_2')(e_1'^* + e_2'^*) \\ &= |e_1'|^2 + |e_2'|^2 + e_1' e_2'^* + e_1'^* e_2' \\ &= I_1 + I_2 + 2\sqrt{I_1 I_2} \cos\Delta\theta \end{aligned} \tag{2.4}$$

Note that Eq. 2.4 is equivalent to the cross-correlation function between two sinusoidal signals, as far as they both result in another sinusoidal function, which depends on a relative phase shift. If the two waves share the same amplitude, the resultant intensity would vary from zero to twice that of the interfering waves, depending exclusively on the phase shift.

An interferometer is a system that captures the interference pattern, to retrieve the phase shift between the interfering waves. If both waves share the same frequency and their intensities are known, the phase shift can be calculated directly from Eq. 2.4. If there are additional unknowns, e. g., the light intensities I_1, I_2 or an additional DC offset due to ambient light I_0, more constraints are required to estimate the phase shift. If it is possible to delay one of the waves with respect to the other, custom delays can be superimposed and different resultant intensities will be obtained. Since Eq. 2.4 is a sinusoidal function, at maximum three parameters are to be estimated: the DC intensity offset $(I_0 + I_1 + I_2)$, the amplitude of the sinusoid $(I_1 I_2)$ and the phase shift $\Delta\theta$. Consequently, three acquisitions using three different custom shifts suffice to estimate the phase shift. Obviously, one can profit from gathering more than three acquisitions if noise in the measurements is to be expected. In such case, solving the overdetermined system derived from the multiple acquisitions may help to minimize the error produced by the measurement noise. The optimal solution, in the least squares sense is given by a formula of the form of Eq. 2.5 [314], which is based on the general least squares separation procedure of [206].

$$\Delta\theta = \arctan\left(-\frac{\sum_{k=1}^{N} I_k \left(A_{1,1} + A_{1,2}\cos\theta_k + A_{1,3}\sin\theta_k\right)}{\sum_{k=1}^{N} I_k \left(A_{2,1} + A_{2,2}\cos\theta_k + A_{2,3}\sin\theta_k\right)} \right) \tag{2.5}$$

where $I_k, k \in [1, N]$ denotes the kth measurement, gathered at the phase shift θ_k. The coefficients $A_{i,j}$ are constants which depend only on the sampling scheme, i. e., which θ_k are considered. For the definition of these constants, the reader is referred to Eq. 5.6-5.9 of [314]. Clearly, in this and later formulas implying the arctangent function, the specific sign of both numerator and denominator is to be taken into account in order to resolve the correct quadrant. If the measurements are taken at equally spaced intervals, Eq. 2.5 further simplifies to Eq. 2.6 [332], which is the so-called *synchronous detection algorithm* or *diagonal least-squares algorithm* [314].

$$\Delta\theta = \arctan\left(-\frac{\sum_{k=1}^{N} I_k \sin\theta_k}{\sum_{k=1}^{N} I_k \cos\theta_k} \right) \tag{2.6}$$

Before diving into implementation details, which are obviously wavelength-dependent, it is worth having a last look at Eq. 2.6, this time considering the minimalistic case in which two orthogonal phases are acquired, say $\theta_1 = 0°$ and $\theta_2 = 90°$. Then, provided that $\sin\theta_1 = \cos\theta_2 = 0$ and $\cos\theta_1 = \sin\theta_2 = 1$, the argument of the arctan boils down to the quotient $-I_2/I_1$, which is totally expectable, since this is nothing else than calculating the phase of a complex number, given its real and complex components. This special case of two-phases interferometry is the classical coherent sensing implemented in, e. g., radar systems, yielding the so-called *in-phase* and *quadrature* components, which fully define both the amplitude and the phase of the received echo. Differently from optical interferometry, radar systems operate in radio frequencies, at which silicon is insensitive and, therefore the arrays of detectors are not miniaturized dense arrays of pixels, but bulky arrays of very few antennas. This induces to think that radar technology cannot produce high-resolution depth images and is relegated to target detection and coarse ranging, thus unrelated to the core technology in this work. Nevertheless, if the antenna(s) are mounted on a moving vehicle, e. g., an aircraft, the motion can be used to synthetically increase the aperture, without actually increasing the antenna size. This is the basic idea behind Synthetic Aperture Radar (SAR), a technique that is able to produce high resolution images from comparatively small antennas. Furthermore, SAR interferometry (often shortened IFSAR) [23] makes use of two coordinated vehicles (aircrafts or spacecrafts), or, alternatively, two passes of a single vehicle over the scene to image, in order to achieve outstandingly large baselines, e. g., between 100 and 300 m. Relatively short wavelengths can be used, e. g., $\lambda = 3$ cm for the X band and very fine range resolution can be attained, e. g., around 1 m. Despite the phase images are extremely noisy, the difference of phase between the interferometric pair accurately retains the information on the scene structure. IFSAR yields high resolution interferometric images with height accuracy in the order of 1 to 10 m.

The different (optical) interferometer setups can be classified in two main groups, regarding whether they use two or a single optical path. Classic interferometers, such as the Michelson and the Mach-Zehnder interferometer belong to the double path group. One example of common path interferometer is the Sagnac interferometer. Conventional fiber optic gyroscopes are not more than a Sagnac interferometer where the optical circuit is confined in an optical fiber. For completeness, we provide simplified schematics of the Michelson and Mach-Zehnder interferometers in Fig. 2.4. For a complete description of these and other interferometers, the reader is referred to [215].

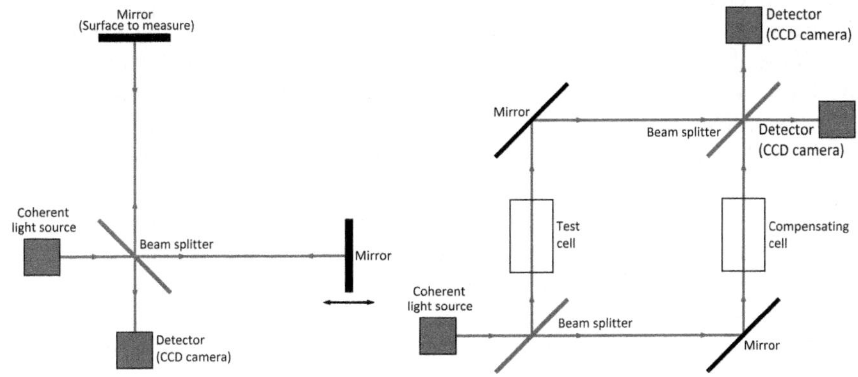

(a) Michelson Interferometer **(b)** Mach-Zehnder Interferometer

Figure 2.4.: Michelson (a) and Mach-Zehnder (b) interferometers. In the Michelson interferometer the light traverses the light paths twice, before and after reflecting on each mirror. The Mach-Zehnder interferometer offers a more flexible setup, with separated light paths for the reference and the probe beams. Consequently, in a Mach-Zehnder interferometer each optical path is traversed only once.

One could say that the simple setup of the Michelson interferometer is adequate for measurements in reflective mode, where the sample to measure acts as one of the mirrors. In such setup, if the addition of custom phase shifts was required to retrieve the actual phase shift induced by the sample, they could be generated, e. g., by precise displacement of the mirror of the reference beam in the direction of the beam. A piezoelectric transducer (PTZ) might be used to achieve the required precision. The Mach-Zehnder interferometer is the setup of choice for working in transmissive mode. There are two independent paths, one for the reference beam and the other for the probe beam. The reference beam traverses the so-called *compensation cell*, while the other traverses the *test cell*. The function of the compensation cell is just to mimic the properties of the test cell and assure that the final measurements are representative of the sample, exclusively. In general, regardless of the interferometer setup, if a 2D sample or scene is to be sensed, lenses are to be placed after the laser source, in order to open the beam, and before the sensing array, in order to form the image on it.

Until now we implicitly sticked to the homodyne case, i. e., the interference occurs between light of the same wavelength, resulting in light of the same wavelength whose intensity depends on the relative phase shift of the interfering radiations. In the heterodyne case, the interfering signals are of different frequencies and the result of the interference is formed by two periodic signals of frequencies equal to the difference and the sum of those of the interfering signals. In most cases, only one of those frequencies is measured, typically the lowest (beat frequency), and the other is filtered out.

Despite the good precision that can be achieved measuring depth with interferometric techniques (two to three orders of magnitude lower than the wavelength, i. e., in the nanometer range), the short unambiguous range of half a wavelength limits their fields of application. The naïve solution to this problem is to use several frequencies instead of only one [303]. This method is often called multi-wavelength interferometry (MFI) and the unambiguous range is one half of the least common multiple of the used wavelengths, i. e., half the lowest value that is simultaneously a multiple of all the wavelengths considered. The *half* is due to operation in reflective mode, as in ToF imaging, but is to be omitted in transmissive mode. Clearly, wavelengths that are multiples from each other are to be avoided, since they bring no advantage over using simply the highest. A common approach is the use of closely spaced wavelengths. A study on frequency selection and a method for selecting the optimal set of frequencies in MFI is given in [457]. Using closely spaced wavelengths leads to an unambiguous range equivalent to that that would be obtained using radiation of a frequency equal to the beat frequency, i. e., the low frequency given by the difference of frequencies of the emitted radiations. This technique often called Synthetic Wavelength Interferometry (SWI). A recent work [458] has shown that micrometric resolution with an unambiguous range of more than one meter can be achieved in an SWI setup using four closely spaced wavelengths.

Other methods exist that solve the ambiguity problem. The so-called electronic speckle pattern interferometry (ESPI) [415, 240], which was first introduced in [289, 65], is based on the fact that the interference pattern generated between a reference and a speckle wave contains the phase information of the speckle wave. It differs from holography in that there is no need for reconstructing the 3D complex amplitude of the wave. Note that the autocorrelation function of the complex amplitude and intensity of a speckle wave is close to a Dirac delta function, due to the fact that displacements of the diffusive surface lead to rapid random variations of the speckle pattern and any random pattern is, by definition, poorly correlated with another random pattern. One of the reasons of including this short note on ESPI is to give the reader the opportunity of confronting it to recent methods based on the projection of pseudorandom patterns, such as the depth-varying pattern used in [418] for depth sensing.

A natural extension of multiple wavelength interferometry is to use a band of frequencies, instead of a finite number. If the visible spectrum is chosen, the technique is called *white light* interferometry. A review of white light interferometry is given in [469], where the possible setups are classified into three categories: diffraction grating interferometers, vertical scanning or coherence probe interferometers and white light scatterplate interferometers. From these, the second one is of special interest for being the preferred option in most industrial applications requiring surface metrology. In its simpler setup, a vertical scanning white light interferometer may adopt the form of a Michelson interferometer (Fig. 2.4a), where the mirror is located at the same distance than the probe from the beam splitter and can be precisely displaced in the beam direction, typically obtaining equally-spaced samples of the white light correlogram. A maximum in the intensity of the interference signal will be achieved when the optical paths of the reference and probe beams have equal length (constructive interference). As in the original Michelson setup, the maximum depth accuracy is bounded by the positioning accuracy of the positioning stage. If further magnification is required, a Mirau interferometer is used instead ($10\times$ to $50\times$). For even higher magnification (more than $50\times$), a Linnik interferometer, which includes additional optics also in the reference path, is the appropriate setup.

2.2.2. Optical Radio Frequency Interferometry or Modulation Interferometry

White light interferometry solves the problem of the very short unambiguous range of single wavelength optical interferometry, often in the range of hundreds of nanometers. Nevertheless, there remain some issues that restrict optical interferometry to metrology in high-end industrial applications. First, one or several sources of coherent light, providing appropriate coherence length, are required. Laser sources and, specially, the control electronics, may be too expensive for some applications. Furthermore, an interferometer setup is required, which demands a very stable structure, isolated from any vibration or external disturbance and machined with fine tolerances. The same standards of quality and care apply to all elements of the optical path—lenses, mirrors, beam splitters, etc.—since even submicrometer irregularities affect the resulting interference pattern. In general, interferometers cannot be presented as a competitive solution to the rest of depth imaging techniques, designed to sense scenes with depths in the meter range or more. This is not only because of physical limitations of the measurement principle, but also because of the cost and the delicateness of the equipment they require.

Using larger wavelengths is an immediate way of increasing the unambiguous range. Radio wavelengths in the range of tens to hundreds of megahertz lead to unambiguous ranges of several meters. The Ultra High Frequency band (UHF) is also known as *decimeter* band, since the wavelength range goes from 1 dm to 1 m. Radiation in such band would be optimal for close range depth imaging in terms of unambiguous range. Another well-known frequency band for television broadcasting is the Very High Frequency (VHF), which includes wavelengths from 1 to 10 m, well-suited for medium range unambiguous depth imaging. Unfortunately, radiation of such *lower* frequencies does not produce photoelectric effect in silicon and, therefore, cannot be detected with a conventional CCD or CMOS camera. The longest wavelength for which photoelectric effect occurs is around 1.1 μm, i. e., still in the near infrared (NIR) range. Using radio frequencies would require building an array of antennas as detectors and would no longer be any kind of ToF *imaging* technique, but a radar system. In this context, Optical Radio Frequency Interferometry (ORFI) arises as a technique that aims to bring together the benefits of using visible or NIR light, i. e., photoelectric effect in silicon, and the large unambiguous range associated with radio frequencies. ORFI also avoids the need for coherent light sources and an actual interferometer.

The basic idea behind ORFI is to modulate the intensity of the light which is projected onto the scene to sense. The light, normally incoherent, is intensity modulated in continuous wave mode (IM-CW), in the simplest case according to a periodic signal of the desired radio frequency. If such a signal is split in two beams and one of them is used as reference beam and the other as probe beam in an interferometer setup, no interference will take place and the false "interference" image formed in the camera would be only determined by the different reflectivity of the scene points, but independent from their depths. A *demodulation* step, coherent with the modulation scheme, is required to mimic optical interference. This can be done using an electro-optic modulator (EOM), which is an optical device able to modulate an input beam according to an electrical signal. The same signal used to control the illumination may be used to control the EOM for demodulation of the light reflected by the scene, before focusing it on the camera chip. If the light source is DC, two EOMs can be used, one for modulation before projection to the scene and another for demodulation. For clarity, the block diagram of a general ORFI setup is given in Fig. 2.5a. In the diagram the generation of the modulated light is obtained either using an EOM over DC light or directly by driving a fast light emitter, e. g., fast LEDs or laser diodes. The latter option is preferred, since it is more efficient, compact and cheaper than the former. There exists a variety of devices that can be used as EOMs, which will be briefly discussed in the following. As well as in the light modulation block it is preferred to perform the modulation at the light generation and not after, in the demodulation part it is of interest to perform the demodulation within the detector, i. e., when the carriers are being generated by photoelectric effect, and not before. This would reduce energy losses, allowing for a reduction of the optical power of the illumination system. The bulkiness and cost of the system would also be reduced, by eliminating the EOM. For a complete comparative analysis between this setup and that using an external EOM the reader is referred to [474]. Such a system requires the use of an array of *smart* pixels, able to perform the mixing of the incoming light signal with some reference signal during the integration process. A block diagram adopting this internal demodulation method is provided in Fig. 2.5b, where the light is modulated by driving a fast emitter, as in commercial ToF cameras, and the demodulation is performed *in pixel*, by means of *smart* pixels, such as those introduced by PMD Technologies (Section 2.3). In the following, the mixing of incident light signal and reference is assumed to be perfect. In case of non-ideal mixing due to, e. g., pixel imperfections, its effect is implicitly modeled in the shape of the *effective* reference signal acting at pixel level, which will differ from the ideal one.

In both cases presented in Fig. 2.5, the reference signal used for demodulation, i. e., the signal the reflected illumination signal is mixed with, is a shifted version of the so-called illumination control signal (ICS). In other words, for each pixel a sample of the cross-correlation between the two signals is obtained at the end of the integration process, at a phase shift that depends exclusively on the phase shift induced by the depth θ_{depth} and the custom phase shift θ_k, which is different for each one of the $k \in [1, N]$ acquisitions. If the waveform of the light signal, as emitted, coincides with that of the reference signal used for demodulation, the problem of phase estimation is reduced to finding the unknown relative phase shift from samples

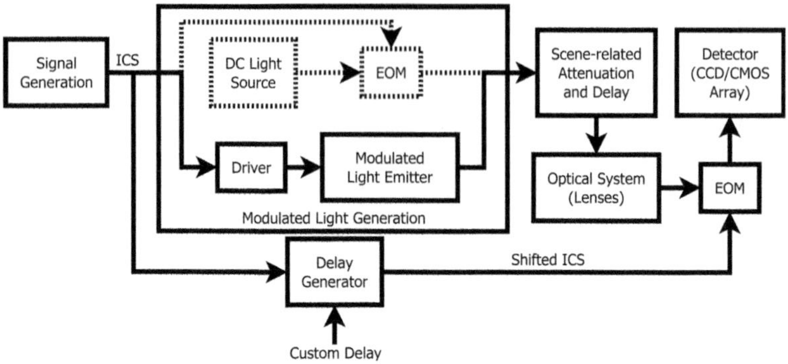

(a) External Demodulation

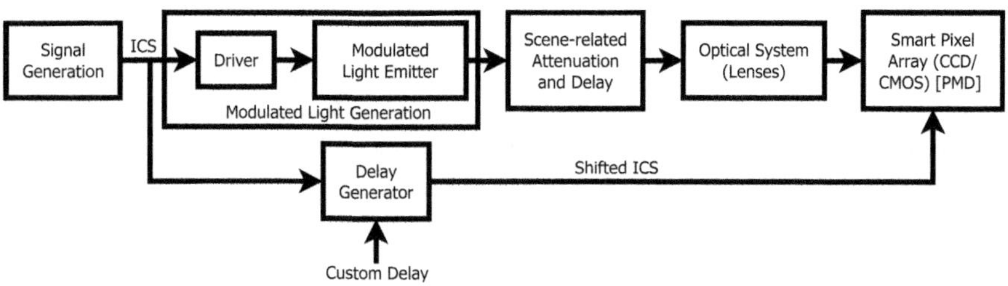

(b) PMD Demodulation

Figure 2.5.: Two possible setups for Optical Radio Frequency Interferometry (ORFI). A classic setup (a), where the demodulation is carried out by an external electro-optic modulator (EOM). In (b) the demodulation is performed internally in the pixel array, by means of *smart* pixels, able to control the integration process according to an electric signal. An example of such pixels, able to sample the cross-correlation between the modulated light arriving to the pixel and some reference signal, are the PMD pixels.

of the autocorrelation function at custom phase shifts. For simplicity and analogy with pure optical interferometry, let the emitted light signal and the reference signal be a sinusoidal signal

$$i(t) = A_0 + A\cos(\omega t - \theta_0) \tag{2.7}$$

where A_0 is some DC offset, A is the amplitude of the signal and θ_0 is the initial offset, which can be assumed zero without loss of generality. Now suppose that each point of the scene being illuminated by such signal reflects it with a certain attenuation a and produces a return that arrives to the pixel array with a certain phase shift θ_{depth}, which is exclusively given by the distance the light had to cover from the illumination system to the scene point and from it to the sensor (recall Eq. 2.1). Then we can formulate the environment response at that point as

$$e(t) = a\delta(t - t_{\mathrm{depth}}) \tag{2.8}$$

where t_{depth} is the equivalent in time-of-flight to the depth to measure and $\delta(t - t_{\mathrm{depth}})$ is a Dirac delta function centered at $t = t_{\mathrm{depth}}$. Consequently, the reflected light signal that reaches the sensor is given by

$$r(t) = (i * e)(t) = A_0^r + A^r \cos(\omega t - \theta_r) \tag{2.9}$$

where $A_0^r = aA_0 + A_0^e$ accounts for the attenuated DC component of the signal, plus some unknown level of environmental light, also called background illumination, A_0^e. The phase offset is mostly due to the environmental response, $\theta_r = \theta_0 + \theta_{\mathrm{depth}}$, but may contain additional delays produced by the electronics, wires and light emitters, which should be accounted for in θ_0. The reference signal is, at no additional phase shift, just a linear transformation of that in Eq. 2.7, that is

$$q(t) = A_0^q + A^q \cos(\omega t - \theta_q) \tag{2.10}$$

where the terms A_0^q and A^q simply model the EOM or the *in pixel* mixing, since the original signal might experience some rescaling or offset addition or subtraction. In the simplest case, $\theta_q = \theta_0$. Note that we implicitly suppose that the *residual* delays are the same in the *probe* path, i. e., the one carrying the signal for modulation, and the *reference* path, i. e., the one carrying the reference signal, similarly to an interferometer, being in both cases θ_0. In real cases this cannot be always ensured and an unknown phase offset might appear, which has to be fixed by calibration. The result of the cross-correlation between the reflected signal and the reference signal is defined as

$$
\begin{aligned}
c_{q,r}(\tau) &= \lim_{T\to\infty} \frac{1}{2T} \int_{-T}^{T} q^*(t) r(t+\tau) dt \\
&= \lim_{T\to\infty} \frac{1}{2T} \int_{-T}^{T} [A_0^q + A^q \cos(\omega t - \theta_q)][A_0^r + A^r \cos(\omega(t+\tau) - \theta_r)] dt
\end{aligned}
\tag{2.11}
$$

Eq. 2.11 can be simplified using known trigonometric equalities. For readability, we omit the mathematic demonstration here, but the interested reader is referred to Appendix A.1, where we demonstrate that Eq. 2.11 yields Eq. 2.12, which is a generalization to arbitrary sinusoidal signals of the formula already presented in [368] for the case of signals of equal offset and amplitude.

$$
c_{q,r}(\tau) = A_0^q A_0^r + \frac{A^q A^r}{2} \cos\left(\omega\tau - \theta_{\text{depth}}\right) \tag{2.12}
$$

where $\theta_{\text{depth}} = \theta_r - \theta_q$, by definition. Note that sampling the cross-correlation function at custom values of τ, $\tau_k = \theta_k/\omega$ is equivalent to performing optical interferometry (cf. Eq. 2.4). The cross-correlation samples are obtained by shifting the reference signal in θ_k. From the measurement point of view, we are in an equivalent case to interferometry and, consequently, the same methods for phase calculus from the measurements apply. In fact, the general multiphase depth estimation method described in [476], which seems to be that implemented in the ToF version of the Kinect sensor, is not more than that given in Eq. 2.6 and used in interferometry [332]. Obviously, the intensity values of Eq. 2.6, I_k, correspond to the cross-correlation samples $c_{q,r}(\tau_k)$ given by the pixels in an ORFI setup. If the illumination signal is not exactly sinusoidal, a careful choice of the phase shifts θ_k may be used to cancel the main harmonics at sensing. In [363] phase steps of 45° are used to cancel the third and fifth harmonics, which are the most dominant if the signals are square (recall that even harmonics are null in a square signal). Despite the method was implemented with PMD hardware, it is based on Eq. 2.6 and, therefore, also valid in ORFI setups with external EOM or pure interferometric setups.

In real cases, it is hard to generate a pure sinusoidal illumination signal using low-cost light emitters, such as LEDs. Using a coherent light source, together with an interferometric EOM, e. g., a Mach-Zehnder modulator, yields an accurate sinusoidal modulation, at the cost of adding bulky and expensive components. Consequently, most commercial solutions rely on direct modulation of the light source. LEDs have to be driven in current and, therefore, require a driver, which is typically controlled by a binary ICS. The ICS is a square signal, alternating between a low level (typically zero) and a high level, and, in consequence, also a square signal controls the demodulation. Then, if the mixing occurs between two square signals, the main question would actually be why ToF systems based on ORFI work at all without tweaking the sampling as in [363], since the model (Eq. 2.9-2.10) is not satisfied and, consequently, Eq. 2.6 does not hold. Indeed, a square wave has a Total Harmonic Distortion (THD) of 48.3%, i. e., around half of the signal power is contained in the harmonic components. In general, for periodic, yet not sinusoidal signals, the reflected and reference signals may be written according to their Fourier representation:

$$\begin{aligned} r(t) &= A_0^r + \sum_{n=1}^{\infty} \left(A_{1,n}^r \sin(n\omega t - \theta_r) + A_{2,n}^r \cos(n\omega t - \theta_r)\right) \\ q(t) &= A_0^q + \sum_{n=1}^{\infty} \left(A_{1,n}^q \sin(n\omega t - \theta_q) + A_{2,n}^q \cos(n\omega t - \theta_q)\right) \end{aligned} \tag{2.13}$$

Substituting the expressions in Eq. 2.13 in Eq. 2.11, we obtain:

$$\begin{aligned} c_{q,r}(\tau) = \lim_{T\to\infty} \frac{1}{2T} \int_{-T}^{T} &\left\{ A_0^q + \sum_{n=1}^{\infty} \left[A_{n,1}^q \sin\left(n\omega t - \theta_q\right) \right.\right. \\ &\qquad \left.\left. + A_{n,2}^q \cos\left(n\omega t - \theta_q\right) \right] \right\} \\ &\left\{ A_0^r + \sum_{n=1}^{\infty} \left[A_{n,1}^r \sin\left(n\omega t + \omega\tau - \theta_r\right) \right.\right. \\ &\qquad \left.\left. + A_{n,2}^r \cos\left(n\omega t + \omega\tau - \theta_r\right) \right] \right\} dt \end{aligned} \tag{2.14}$$

which is the most general expression for mixing of two periodic signals, be by means of an external EOM or by means of a *smart* pixel. Simplifying Eq. 2.14 is not immediate and, for the sake of brevity and readability, we provide the full process in Appendix A.2. We mostly exploit the orthogonality of Fourier basis, i. e., the product of two sinusoidal signals of different frequencies, which are multiples of some fundamental frequency, leads to a null contribution when integrated along a period of the lowest frequency, and well-known trigonometric identities. We obtain that Eq. 2.14 can be reduced to Eq. 2.15.

$$c_{q,r}(\tau) = A_0^q A_0^r + \frac{1}{2}\left\{ \left[\sum_{n=1}^{\infty} A_{n,1}^q A_{n,1}^r + A_{n,2}^q A_{n,2}^r\right] \cos\left(\omega\tau - \theta_{\text{depth}}\right) + \left[\sum_{n=1}^{\infty} A_{n,2}^q A_{n,1}^r - A_{n,1}^q A_{n,2}^r\right] \sin\left(\omega\tau - \theta_{\text{depth}}\right) \right\} \tag{2.15}$$

As already pointed out in [304], it would be sufficient to ensure that one of the signals involved in the correlation (Eq. 2.11) is a pure sinusoidal signal to completely remove the disturbing effect of the harmonics present in the other signal. In other words, if a perfect sinusoidal illumination signal is achieved, the shape of the reference signal, as used for the mixing at pixel level, is irrelevant, as far as it is a periodic signal of the same fundamental frequency.

Modulation and Demodulation at Different Frequencies Despite we have focused on mixing periodic signals of identical fundamental frequency, as well as in the case of optical interferometry, mixing two signals of different frequencies is also possible. The signal frequencies are to be kept high, in order to maintain good phase resolution, while the frequency difference has to be low enough to allow performing all the necessary acquisitions per period of the beat frequency. Obviously, the number of acquisitions per period has to be an integer and a perfect synchronization between the acquisition rate and the beat frequency is crucial. If a conventional camera is used to sample the beat signal, the frequency difference may fall on the hertz range, while typical modulation frequencies for imaging a real-world scene are in the megahertz range. The mixing can be carried out externally, e. g., using a Pockels cell or an image intensifier as external mixer, as done in [154]; or internally, profiting from the existing arrays of *smart* pixels, such as the PMD, which will be analyzed in depth in Section 2.3. A heterodyne camera using a PMD sensor for mixing is presented in [117].

Pseudonoise Modulation The pseudonoise (PN) modulation [67] is a *sui generis* modulation-demodulation scheme, since it cannot be classified according to the homodyne/heterodyne distinction. A PN sequence is used both for modulation of the illumination and for demodulation of the reflected signal. Interestingly, a PN sequence exhibits a wide frequency spectrum, including very high frequencies, which are always a requirement for good depth precision. Additionally, this modulation scheme allows simultaneous operation of ToF cameras, since two different PN sequences have a null cross-correlation function. Nevertheless, the most significant advantage of a PN modulation is a very sharp autocorrelation function. In fact, the autocorrelation approaches a Dirac delta function as the length of the PN sequence tends to infinity. Summarizing, the sample cross-correlation between PN sequences is

$$c_{n_i,n_j}(\tau) = \lim_{T\to\infty} \frac{1}{2T} \int_{-T}^{T} n_i^*(t) n_j(t+\tau) dt = \delta_{i,j}(\tau) \qquad (2.16)$$

where n_i, n_j are normalized PN sequences of infinite length and $\delta_{i,j}(\tau) = \delta_{i,j}\delta(\tau)$ is an hybrid between the Dirac and Kronecker delta functions and, consequently, takes the value 1 only if $i = j$ and $\tau = 0$. Despite its attractiveness, the use of PN modulation for range measurements typically requires dense sampling of the cross-correlation function in order to find the unique maximum. Multiscale and adaptive sensing schemes may leverage this requirement by progressively restricting the feasible region where such maximum can be located.

Multiple Frequencies to Increase the Unambiguous Range The use of high modulation frequencies is a requirement to achieve low depth uncertainty. On the other hand, relatively low frequencies are required to achieve a sufficiently large unambiguous range (Eq. 2.2, where $f = f_{\text{mod}}$). As well as in the case of optical interferometry, a common solution is to use several frequencies to remove the ambiguity. Similarly, the achieved unambiguous range is the least common multiple of the unambiguous ranges corresponding to each modulation frequency. An intelligent option is to use two close frequencies, as done in [280] for long range ToF imaging. This way, high depth precision is preserved, while the unambiguous range is greatly increased. If only two close frequencies are used, the unambiguous range is given by half of the wavelength corresponding to the beat frequency of the corresponding heterodyne setup. Note that the reasoning of the beat frequency is equivalent to that of the so-called *synthetic wavelength* [474], often found in interferometry, being the synthetic wavelength the one corresponding to the beat frequency. For instance, using two high frequencies with 1 MHz difference yields an 150 m unambiguous range.

Another option would be to combine a low frequency with a high frequency. The low frequency is used to determine how many full ranges of the high frequency have passed, i. e., a coarse estimation, while the high frequency is used to get a fine—yet ambiguous—estimation of the depth. Such a setup may not bring a large increase of the unambiguous range and it also does not enhance the depth precision with respect to that obtained using the high frequency alone.

In general, in a multifrequency setup measurements are gathered sequentially [280, 152] and the number of acquisitions scales with the number of frequencies. A careful selection of the frequencies and of the sampling points allows operating with two frequencies simultaneously—or sequentially but still within the integration period, which is equivalent—, as in [364], where the phase differences between sampling points for the lower frequency is twice that for the higher frequency. This idea allows reducing the number of acquisitions when using two frequencies but, as the authors recognize, the use of a low frequency increases the observed depth standard deviation.

In the limit, using a very large number of frequencies, one can do without phase shifting, that is, one can operate in frequency domain, acquiring intensity measurements for different frequencies, at an arbitrary phase, e. g., zero. This is the framework proposed in [249], inspired by optical coherence tomography. Similarly to phase (or time) domain, where the depth is encoded in phase shift, in frequency domain the depth is encoded in frequency. While this method overcomes depth ambiguities and permits resolving several paths per pixel, it demands a large modulation frequency bandwidth.

As presented in Fig. 2.5, despite they share the same principle of operation and, in consequence, the same formulas for phase retrieval, the setups for ORFI can be classified in two different categories, regarding whether the demodulation is carried out using an external EOM or internally by *smart* pixels. Before the development of the latter, initial 3D-vision systems [405] made use of external EOMs such as Pockels cells. The different possibilities that exist for performing electrically-driven optical modulation are shortly discussed in the following, while the *in pixel* modulation is left to Section 2.3, where the PMD technology is presented.

Methods for External Optical Modulation In this section the main techniques for performing optical intensity modulation of light are introduced. We adopt the underlying physical principle as classification criterion, as in [222]. According to this principle, we distinguish three categories: electro-optic, acousto-optic and magneto-optic, which we analyze in the following.

For further information on the topic the reader is referred to [467, Chapter 3] and [222].

Mecano-optic Mechanic methods for shuttering could be used as a mean for binary modulation. Unfortunately, conventional mechanical shutters, widely used in photography, rarely can provide exposures below one millisecond. This would lead to maximum modulation frequencies in the kilohertz range, far from the megahertz frequencies typically required in ToF imaging. Nowadays, there exist cameras providing up to 125 μs exposure. Supposing that we desire a square binary modulation (50% duty cycle), this poses a frequency limit of 4 KHz. One could think of mechanical setups allowing elements moving periodically at high speeds, which could act as shutters. One example may be a rotating cogwheel, used as in the famous Fizeau experiment, moved by a high-speed and low-torque motor, after appropriate speed multiplication, e. g., by means of a gearbox. Very high speed motors, such as those used for generating vibration, can turn at speeds up to 10^5 rpm, while having millimeter-range sizes. A prototype of the *MegaNdrive* low power motor of the ETH Zurich achieved over 10^6 rpm [490] before disintegration of the ball bearings. The Fizeau's cogwheel may be substituted by a printed disc of transparent plastic or glass, similar to those used in optical rotary encoders. While such an approach may achieve the required modulation frequencies, the presence of a rotary element turning at hundreds of thousands of rpm is reason enough to avoid this method due to maintenance considerations. Still, there are mechanical methods that suit the application of fast light modulation. We contemplate here using the so-called Frustrated Total Internal Reflection (FTIR) and microelectromechanical systems (MEMS).

FTIR, recently used for optical touchpads, is not a new technique. Under the denomination of *optical tunneling*, it has been extensively discussed in [34], where typical applications are given, such as optical beam splitters, modulators and filters. The basic principle and formulas are given in [265, Section 5.5], under the name of Frustrated Total Reflection (FTR). Consider the interface between two media of different refractive indices, such that the light is transmitted from the medium of higher refractive index to that of lower. Applying Snell's law, we trivially get that, for incidence angles higher than a critical angle, which depends exclusively on the ratio of refractive indices, all light gets internally reflected in the first medium and no light gets transmitted to the second. This phenomenon is often called Total Internal Reflection (TIR). If a third medium is added after the second, with a refractive index that is higher than that of the second, so that the width of the second medium is around the value of the wavelength of the radiation, then the light is no longer totally reflected and gets transmitted to the third medium, even when the incidence angle in the first medium is greater than the critical angle. This effect does not contradict Snell's law, since both interfaces are still totally reflecting, being the partial transmission due to an interference effect. Consequently, for widths of the second medium that are of higher order of magnitude than that of the wavelength of the radiation, the addition of the third medium after the second does not have any effect. Consequently, the modulation procedure is based on varying the width of the second medium. If the second medium is air, one way of varying the width is, e. g., displacing the interface of the third medium with respect to the first.

If NIR radiation is being used, micrometric or submicrometric displacements are enough to change the transmittance of the system from 0 to close to 100%. This makes FTIR highly recommendable for light modulation, since displacements in the range on tens of nanometers to micrometers can be achieved, even at very high repetition rates, without the need of bulky mechanical actuators. The most common way of performing this task is by means of a piezoelectric actuator. Piezoelectric actuators are based on the inverse piezoelectric effect, i. e., the ability of some materials (e. g., some crystals and ceramics) to generate a mechanical strain under the application of an electrical field. The mechanical strain may eventually deform the material and be used to generate fast short displacements. If the piezoelectric crystal is excited at frequencies close to the resonance frequency, high vibration amplitudes can be achieved with relatively low voltage levels. Since piezoelectric crystals are often presented as thin disks, two resonance frequencies are observed, corresponding to the radial and longitudinal modes, respectively, being the latter the highest, often in the megahertz range. A review of the properties of the main piezoelectric ceramics, such as the well-known lead zirconate titanate (PZT) is provided in [241]. Other suitable piezoelectric materials might include polymers, such as the polyvinylidene fluoride (PVDF). Binocular setups for a 3D camera using two coupled FTIR modulators, for modulation and demodulation, are proposed in [474, 222].

One of the principal factors to consider on any mechanical system with parts moving at high speed is the total aerodynamic drag, which depends linearly on the density of the fluid surrounding the system and the surface in contact with it and quadratically on the relative speed between the fluid and the moving part. That is, a linear increase of the speed of a moving part translates, *ceteris paribus*, into a quadratical increase in the required energy. Nevertheless, if the surface in frontal contact with the fluid, typically air, is reduced to a negligible value, then the corresponding aerodynamic drag will be also negligible. Similarly, if the reduction of surface is accompanied by a reduction of volume of the system parts, their mass will decrease and so will do related forces and torques. That is, under miniaturization conditions, physical phenomena that are dominant at macroscopic scale, such as the effect of gravity, become irrelevant in comparison to the effect of other, e. g., electrostatic fields, which are negligible at macroscopic scale. Microelectromechanical systems (MEMS) are systems made up of components of sizes ranging from one to a hundred micrometers, which may combine mechanical microstructures, microsensors, microactuators and microelectronics. Typically the volume reduction with respect to macroscopic systems is higher than the surface reduction, leading to greater ratios surface/volume. This concede even more importance to electrostatic effects. Since the inertia of the components is very low, well-designed MEMS can produce microscopic movements (e. g., rotation, tilting) that are fully controlled by a rapidly-varying electrical field, with very low response times. For instance, MEMS engines able to rotate at speeds up to 10^6 rpm have been characterized in [380], when rotating up to 5×10^5 rpm.

Even more interesting than using high speed MEMS motors for shuttering are the arrays of micromirrors, which are able to tilt independently, driven by an electrical field. The worldwide leading technology in this area is the Texas Instruments Digital Light Processing (DLP®). The core of the DLP is the so-called Digital Micromirror Device (DMD®[4]3), which was invented by Larry Hornbeck in 1987 [229]. Each DMD pixel can be in two stable micromirror states, namely, tilted $+12°$ or $-12°$ with respect to the surface normal [145]. Below each micromirror, a dual CMOS memory stores the desired state of the micromirror. Once the memory is set to the desired mirror state, a clock signal is responsible to translate the memory state into an actual mirror state. The micromirror is able to tilt thanks to a central via, which attaches it to a torsional hinge. It is clear that such a system can provide binary modulation with very large contrast. Furthermore, using two detectors would lead to a lossless modulation, according to two simultaneous binary signals displaced 180° between each other. The remaining issue is the driving speed. In normal operation, the main time constraint is imposed by the data transfer to the memories of the DMD pixels. Suppose that the DMD pixels can be all simultaneously set (not possible) and reset (possible) to one or another position. In that case, if all the DMDs can be driven by a

[3]The Texas Instruments DLP® and DMD® are registered trademarks and will be denoted in the following without the ® symbol for notation simplicity.

single signal without sequentially writing the memories, the minimum period is given by the mirror transition and settle time [158]. It takes about 18 μs to achieve a stable position, but the transition itself is completed in less than 5 μs. Summarizing, a DLP could provide a square binary modulation, with 100% modulation depth at a physically-limited maximum frequency of 100 KHz.

Electro-optic Three classical electro-optic effects that can be used for light modulation are presented here: the linear electro-optic effect, also known as Pockels effect, the quadratic electro-optic effect, or Kerr effect, and the electro-optic effect in liquid crystals. Additionally, the use of multiple quantum well (MQW) structures as light modulators exploiting the quantum confined Stark effect (QCSE) is briefly introduced.

The Pockels effect [62] refers to the linear variation of the refractive index (birefringence) that some materials experience under the application of an electric field. The electric field can be applied longitudinally or transversely to the light beam. Longitudinal Pockels cells require transparent electrodes. In general, Pockels cells may be used to rotate the polarization of a polarized light beam. Consequently, in combination with polarizers, they provide a technique for intensity modulation of light.

Within a Pockels cell, the refractive index depends on the polarization of the light beam, i. e., light polarized in a certain direction propagates faster through the crystal than light polarized in the perpendicular direction. In longitudinal Pockels cells, the phase difference between these two principal axes is directly dependent on the applied electric field, and not on the length of the crystal. The main drawback of Pockels cells is the relatively large voltage required to generate the electrical field. The KD*P (Potassium Dideuterium Phosphat) modulator studied in [222] required 5233 V to achieve the so-called half-wavelength point (180° phase shift) and 1850 V for the quarter-wavelength point (90° phase shift), for a wavelength of 630 nm. Another issue is that the transmittance of the Pockels cells depends largely on temperature [474], so that variations of tens of degrees Celsius translate into discrepancies of thousands of volts in the previous values.

In the transversal case, the phase difference is directly dependent on the modulator length. Transversal cells with small apertures can have lower voltages, but in the case of depth imaging, flat 2D modulators with relatively large apertures are required. For instance, a Pockels cell from KD*P crystal of 25 $\times$ 25 mm still requires driving voltages in the range of 1500 V.

Pockels cells behave as capacitors from the electrical point of view and, clearly, driving them with voltages of hundreds to thousands of volts and switching speeds in the nanosecond range require very fast electronics, able to deliver very high currents. Apart from space or efficiency considerations, systems driving tens of amperes at megahertz frequencies may introduce distortions in the power network that could disturb the correct operation of other devices. Electromagnetic interference (EMI) might also become an issue, both for the camera and for other electronic systems close to it.

The Kerr effect is similar to the Pockels effect, since it is also a change in the refractive index induced by an electric field. It differs from the Pockels effect in that the variation is directly proportional to the square of the applied electric field, for which reason it is also called quadratic electro-optic effect. There exist two different types of optical Kerr effects: the electro-optic and the pure optical effect.

The first one is also called DC Kerr effect and is exploited in a setup that is close to that of a Pockels cell, in that the external electric field is applied by means of electrodes at the sides of the material. The electric field is generated perpendicularly to the light beam and two crossed polarizers are placed before and after the material. Some organic polar liquids exhibit a strong DC Kerr effect and are confined in a glass cell, forming a so-called *Kerr cell*. Kerr cells can be used to achieve very fast modulation, in the gigahertz range, at the cost of driving voltage even higher than those required for Pockels cells, often in the kilovolt range.

The optical Kerr effect is also called AC Kerr effect and studies the case when the electric field is produced by the light itself. The variation of the refractive index depends linearly on the intensity of the radiation, i. e., quadratically on the amplitude of the wave, as in the DC Kerr effect. A modulator based on the DC Kerr effect would avoid the need for electrodes and high voltages, providing a mean for pure optical mixing, since the intensity of a *reference* beam can be used to change the polarization (and, in combination with the crossed polarizers, modulate the intensity) of a *probe* beam. Unfortunately, very large intensities (irradiances in the order of GW/cm^2) are required to produce the AC Kerr effect in most materials, only achievable by a well-focused laser beam.

Regarding the electro-optical effect in fluids, liquid crystals [52, 258] deserve special mention. As said above, some organic polar liquids show strong DC Kerr effect, but the principle of operation of optical modulators based on liquid crystals is slightly different. Liquid crystals are liquids in which the molecules are arranged according to a definite structure, i. e., a crystalline state, which leads to anisotropic mechanic, electric and magnetic properties, differently from the isotropic non-crystalline liquids. As pointed out in [258], the birefringence produced in the liquid crystalline state can be up to four orders of magnitude greater, with similar permanent dipole moments. Since the molecules of the crystal are dipoles, when an external electric field is applied, they orient themselves along the direction of the field. Applying a magnetic field produces a similar effect and the molecules tend to align with or against the field. Controlling the orientation of the molecules translates into controlling the rotation of polarized light. Together with crossed polarizers, liquid crystals can be used as optical modulators. The principal drawback is the low speed of the reorientation process, i. e., once the electric field is extinguished, the molecules need times in the millisecond range to return to the original orientation. Imposing special structures to the liquid crystal may lead to a reduction of the decay times. The nanocomposite presented in [238], which combines an anisotropic polymer with a liquid crystal, achieves a decay time of 15 μs. A different composite is presented in [237], achieved by photopolymerization of a mixture of a monomer and a liquid crystal. In that work nanometric pores are formed in the polymeric matrix and a decay time of 10 μs is reported. Still, supposing symmetric rising and falling times, 10 μs decay time means a frequency limit of 50 KHz, still far from the megahertz level.

One of the major issues of external demodulation, which is solved with *in pixel* demodulation, is to avoid an eventual image distortion produced by anisotropy of the materials and complex optical paths. It is, thus, desirable to perform the external demodulation as close to the pixel surface as possible. For this reason, spatial light modulators made of nematic ferroelectric liquid crystal (FLC) are an appealing solution, since they are fast and it is possible construct them over VLSI silicon (the so-called FLCoS or FLC-LCoS) [116], which can, in turn, be used for the image sensing. A schematic of a FLC-LCoS for *on pixel* ORFI demodulation in 3D imaging systems is found in [474].

Multiple quantum well (MQW) structures are formed by stacking quantum wells. They were intended to be a part of optically-controlled optical switches, basic components of a fully optical computer, which would provide faster processing and communications. A representative example of such devices is the so-called self-electro-optic device (SEED) [327], which is briefly a MQW-based switch with optical inputs and outputs. Briefly, a quantum well is a potential well where only discrete energy levels are allowed. One method to create quantum wells is to confine particles to a very thin layer, i. e., a 2D space, instead of 3D. When the well thickness becomes comparable to the *de Broglie* wavelength of the carriers, the so-called quantum confinement is achieved and the energy levels are quantized in energy subbands, which range from 10 to 100 meV. As a result, a staggered optical absorption spectrum is observed, with strong absorption peaks at the edges of the steps—the so-called exciton absorption peaks—.

Quantum wells have a sandwich structure, where the external semiconductors exhibit a wider bandgap than that in the middle. Common materials for constructing quantum wells are gallium arsenide between aluminium gallium arsenide or indium gallium nitride between gallium nitride [186]. A MQW structure is not more than a sequence of alternating thin layers of each material of the quantum well pair (e. g., 10 nm width per pair). In general, quantum well modulators are based on the quantum confined Stark effect (QCSE), which describes the effect of applying an electric field to a quantum well. If an electric field is applied perpendicularly to the MQW layers, electrons and holes are pulled toward opposite barriers of the layer, but still confined in the smaller bandgap layer (middle layer of the *sandwich*), reducing the energy of the electron-hole pair. Consequently, the optical absorption associated with the creation of the pair decreases and the absorption edge shifts towards higher wavelengths (i. e., less photon energy). This effect is sensibly stronger in quantum wells than in a bulk semiconductor, since it gets enhanced by bound excitons. The effect can be used to construct an electrically-driven absorption-based fast light modulator, either in reflective or transmissive mode. As a side effect, the refractive index also changes when applying an electrical field and, therefore, refractive modulators are also possible.

MQW modulators have several advantages when compared to other optical modulation techniques. One of them is the very low energy consumption, comparable to that of a full electronic system. Electrically, the quantum well behaves as a capacitor of very low capacitance, requiring low electrical fields (ranging between 10^4 and 10^5 V/cm) for modulation. Additionally, it is of capital importance that the modulator can be located *on pixel*, over a silicon-based detector array, e. g. a CCD array. This fact implies that better fill factors than those achievable with the *smart* pixels introduced in Section 2.3 can be achieved, thanks to the decoupling between modulation and sensing processes. A schematic of a CCD array with built-in MQW modulation layer in transmissive mode for application in ToF imaging is found in [474]. Note that the modulation operates—differently from PMD pixels—in transversal direction and therefore does not impose any restriction on the size or structure of the pixels.

Acousto-optic Acousto-optic modulators (AOMs) are based on the acousto-optic effect, which is a specific case of photoelasticity, that is, a change of the permitivity of the material under a mechanical strain. In short terms, an acoustic wave produces a periodic strain in the material, which leads to a periodic variation of the refractive index with the same frequency as the acoustic wave. Since the acoustic wave also travels through the material, the refractive index variation is a sinusoid that depends both on the time and the position. Consequently, a diffraction grating is produced, that travels through the material in the propagation direction of the sound wave, at the same speed this propagates. For simplicity, consider that the acoustic wave is of single frequency, then two different types of diffraction are produced: Raman-Nath diffraction and Bragg diffraction. The later occurs at high acoustic frequencies, often over 100 MHz, while the former occurs for lower frequencies, typically lower than 10 MHz.

AOMs based on Bragg diffraction are also called *Bragg cells* and are composed by a piece of the appropriate material, typically quartz or glass, with a piezoelectric transducer at one side and an acoustic absorber at the other. The acoustic wave travels through the material, from the piezoelectric actuator to the absorber. The light beam traverses the material and diffracted beams appear at specific angles that depend on the order of diffraction and the ratio between the wavelength of the light and that of the sound. The intensity of the diffracted beam is proportional to the amplitude of the acoustic wave [462]. Consequently, the Bragg cell can be used directly as a linear light modulator. Since the acoustic wave travels through the material, the frequency of the diffracted beam of order m will be that of the input beam plus a Doppler shift of m times the frequency of the acoustic wave; that is, a Bragg cell could also be used for optical frequency shifting [2], e. g., in a heterodyne interferometric system. Similarly, the phase of the diffracted beam is also shifted by that of the acoustic wave. Also remarkable is the fact that Bragg cells produce a rotation of the polarization angle in a linearly-polarized beam [161], an effect which is exploited by other electro-optical and magneto-optical modulation techniques.

Magneto-optic We consider four magneto-optic effects that can be used to build a magneto-optic modulator: the magneto-optic Kerr effect, the Faraday effect, the Cotton-Mouton effect and the Voigt effect. In general, magneto-optic effects are the magnetic counterpart of electro-optic effects, since they are mostly based on optical rotation, produced by an anisotropic medium. This translates into non-diagonal permitivity or permeability tensors [378].

The magneto-optic Kerr effect (MOKE) or surface magneto-optic Kerr effect (SMOKE) [459] describes the change of polarization and intensity that the light experiences when it is reflected by a magnetized surface. The effect is similar to the Faraday effect, differing in that the Faraday effect describes the changes experienced by light when it traverses a magnetic material, while the MOKE accounts only for reflection on the surface. In short terms, the MOKE is based, as well as the Faraday effect, on the anisotropic permitivity of the material. Provided that the speed of light through the material is inversely proportional to the square root of the product of the permitivity and permeability, light propagates at different speed depending on the orientation. As a consequence, variations in the phase of polarized light are observed. Depending on the direction of the magnetization vector with respect to the surface and incidence plane, the MOKE can be classified in three categories: polar, longitudinal and transversal. In the polar setup the magnetization vector is perpendicular to the reflective surface and parallel in both the longitudinal and transversal cases. The magnetization vector is also parallel to the plane of incidence in the longitudinal setup, while perpendicular in the transversal. In the longitudinal MOKE light changes its polarization from linear to elliptical, being the change proportional to the magnetization vector. In the transversal MOKE, the surface changes its reflectivity, being the change proportional to the magnetization vector.

The Faraday effect, also called Faraday rotation, is a simple magneto-optical phenomenon that describes the rotation of the plane of polarization in presence of a magnetic field, being the rotation proportional to the component of the magnetic field that is parallel to the direction of the beam. In the Faraday effect the rotation is due to a non-diagonal permeability tensor, while the permitivity tensor is supposed diagonal. The angle of rotation is given by the product of the Verdet constant of the material, the magnetic flux density in the direction of the beam and the length of the sample. Materials with high Verdet constants can be used to construct Faraday rotators. An appropriate material to construct a Faraday modulator is the terbium gallium garnet (TGG) [27], which is interesting due to its high transparency and very high Verdet constant. It is to be taken into account that the Verdet constant of TGG strongly depends on the wavelength of the radiation, decreasing dramatically with it, i. e., the material is more suitable for operation in the visible range than in the NIR. A variety of organic and inorganic liquids exist, which exhibit high Verdet constants [456], still typically one order of magnitude lower than that of TGG. As in previous cases, placing a sample of the material between crossed polarizers, together with the appropriate driving system, yields a simple optical modulator. The main drawback of the Faraday effect, as well as of the other magneto-optic effect, is its extremely low efficiency. Large magnetic fields are required to generate moderate optical rotation. For instance, if a sample of TGG of 10 mm length is used to construct a Faraday rotator, a magnetic flux density of approximately 0.5 T is needed to rotate the polarization 45°, when operating with 600 nm light. Furthermore, so intense magnetic fields cannot be generated at high frequencies, due to the highly-inductive coils used to generate them.

Apart from rotation of the plane of polarization (Faraday effect), the magnetic field also produces birefringence, which is known as Cotton-Mouton effect [121]. The optically isotropic probe becomes anisotropic in presence of a magnetic field applied perpendicular to the direction of light propagation through the probe. The refractive index, as observed by the light traversing the probe is different for light polarized parallel and perpendicular to the direction of the magnetic field. This induces a *retardation* or optical path difference between parallel and perpendicularly polarized light at the end of the probe which depends linearly on the difference of refractive indices and the length of the probe. The variation of refractive indices depends linearly on the material-dependent Cotton-Mouton constant, the wavelength of the light and the square of the magnetic field.

The Voigt effect [452] is often presented as an independent effect, but it is a magnetically-induced birefringence, observed in materials that are optically isotropic in absence of magnetic field. For this reason, it is common to find in the literature explicit references to the equivalence between the Cotton-Mouton and Voigt effects [489, 451]. The effect depends quadratically on the magnetic field and produces a change of ellipticity of circularly polarized light.

Apart from all methods described above, image intensifiers can also be used to construct an EOM. The use of an image intensifier as EOM in a ToF imaging system is reported in [154]. Previously, submillimetric depth resolution had been reported in [226], where an image intensifier was used as external EOM in combination with a CCD camera. Key points of the approach in [226] are the coaxial illumination and detection, the high modulation frequency (up to 225 MHz) and large number of phase steps (up to twelve). Image intensifiers are optoelectronic devices commonly used in medical imaging and low-light imaging applications. Several generations of image intensifiers have been developed in the last decades, driven by the military demand. A common factor among all of them is that they are based on the idea of the image intensifier tube, in which the photons are converted first into electrons, then the electrons are amplified and converted back into photons by means of a phosphorescent material. The image or beam first reaches the photocathode, where electrons are generated by photoelectric effect. A microchannel plate (MCP) at high voltage is used to accelerate the electrons, which, in turn, release multiple electrons when they hit the plate.

At the end of the tube, the electrons hit a phosphor screen, where a photon is released for every electron, generating an intensified image or beam. Clearly, decreasing the voltage of the MCP would slow down the electrons and fewer would arrive to the phosphor screen, decreasing the image intensity. That is, an image intensifier can be directly used as intensity modulator with amplification.

2.3. The Photonic Mixer Device (PMD)

This section describes the technique used to perform the ORFI demodulation or mixing of a modulated light signal with a reference electric signal—that is, electro-optical (de)modulation—at the detector. In the case of phase-shift-based ToF this implies that each pixel of the detector array has to be endowed with an apparatus to regulate the integration process according to the external reference signal. One can predict that, regardless of the specific implementation, this idea requires both *in pixel* control electronics and modifications in the structure of the active area of the pixel. Such pixels are often called *smart* or *intelligent* pixels and are introduced in Section 2.3.1. The initial concept of *intelligent* pixels aimed to integrate control functions and basic computational functions within the pixel [140], while keeping an acceptable fill factor. The first array of *smart* pixels that was made commercially available as a ToF imaging system implemented the so-called Photonic Mixer Device or PMD technology. The original idea of the PMD was conceived by Prof. Rudolf Schwarte in 1996 [402], at the *Institut für Nachrichtenverarbeitung* of the University of Siegen, today the Center for Sensor Systems (ZESS). The production of PMD chips was externalized to S-Tec Sensor GmbH, founded in 1996 and today integrated in the well-known PMD Technologies. Nowadays PMD technology remains being a reference technology worldwide and, for this reason, has been adopted as the reference phase-shift-based ToF technology in this work. A description of the PMD pixel structure and its principle of operation is given in Section 2.3.2.

2.3.1. *Smart* Pixels

A definition of *smart* pixel that fits the wide range of usages of the term found in the literature could be: a pixel whose measurement after integration is not directly the result of a photon count during the exposure time, but the result of a different process, still mainly influenced by the arrival of photons to it. According to its functionality, *smart* pixels used in ToF imaging can be classified in those actually measuring the *time of flight* of the light and those intended to provide samples of the cross-correlation between the modulated light signal that arrives to the pixel and some reference signal.

The pulse counting ToF system of [19] is an example of a time-counting system, but recent research in this area points to the single-photon avalanche diodes (SPADs) [481] as the reference technology for pure ToF imaging. In very short terms, the diode is reverse biased above breakdown voltage for a short time, so that, when a photon generates a charge carrier, an avalanche may be triggered. Obviously, unless the avalanche is quenched, the diode will be destroyed. For that reason, once the avalanche is produced, the bias voltage is immediately lowered below breakdown. This operation mode is called Geiger mode and, in theory, implies a virtually infinite gain in the sensing process. SPADs can be constructed using conventional CMOS technology [102] and, consequently, dense arrays of SPAD pixels can be generated and used as a ToF image sensor. A 32×32 CMOS array of SPADs for ToF imaging was presented in [351], where the circular shape of the pixels prioritizes correct operation of the SPAD over fill factor. Millimetric depth standard deviations are reported in [351] and submillimetric in [352]. The shape evolved from circular to octagonal and the fabrication process [103] from $0.8\,\mu$m to 65 nm. SPAD array sizes have grown up to that of conventional correlating pixels (PMD arrays), e. g., the 160×120 single-photon image sensor in [443]. Conversely, scanning mirrors may be used to obtain an acceptable depth image size, at the cost of having to scan the scene, increasing motion sensitivity and decreasing the frame rate [10]. A disadvantage of SPAD arrays is the extremely low fill factor, e. g., 1.1% in [351] or 1% in [443], which has to be effectively increased by means of microlenses. Other technical issues are the complex readout schemes and low frame rates.

Smart pixels internally performing a cross-correlation are typically multitap pixels. A *multitap* pixel features two or more integration wells, also called integration channels or *taps*. The modulated light signal that arrives to the pixel surface is transferred from optical to electrical domain due to photoelectric effect. The generated charge carriers are then stored in one of the integration wells at a time, being the active tap selected by a control signal. Clearly, the level of each pixel channel at the end of the integration is equivalent to scalar product of the modulation signal and the reference binary signal used to control the integration in that channel. If the pixel features only two channels and the integration time is divided equally for integration in each of them, the result of each channel is a sample of the cross-correlation between the modulation signal and a square signal between zero and some maximum level. The difference of both channels, provided that the reference signals are complementary, corresponds to the cross-correlation between the modulation signal and a square signal of double amplitude and no DC offset (i. e., pure AC). In case we have a large number of pixel channels, the process is equivalent to sampling the modulated light signal itself. If the number of channels tend to infinity, then each channel provides a sample of the cross-correlation between the modulation signal and a Dirac delta function, which is the definition of a conventional sampling process.

The first multitap pixels were presented in [423, 421]. The idea of a lock-in structure was immediately adopted to generate the first 2D arrays of demodulating pixels for ToF imaging [422, 278]. Note that this idea is closely related to the array of QE-modulated photodiodes presented in [20]. General introductions to lock-in pixels, together with the main challenges they face can be found in [71, 69]. The *smart* pixels can achieve very large dynamic ranges, over 100 dB [422], by means of an offset subtraction scheme, similar to that implemented in the active pixel sensor (APS) of [450, 449], able to achieve 150 dB while keeping a linear response. Other methods for increasing the dynamic range are the minimum charge transfer (MCT) and the pixelwise integration (PWI) [291]. The basis of MCT is the fact that only the differences between the charge level of the different integration gates is relevant, while the common mode part is due to unmodulated light (DC offset, background illumination) and can be discarded. Consequently only the relevant signal part is transmitted from the integration well to the sense node. To this end, the integration gate potential is lowered at a certain slew rate and the transfer is stopped when all taps become lower than a threshold. This way, the common mode stays underneath the integration gates and do not get transmitted to the sense node. This method may be seen as a precursor of the PMD suppression of background illumination (SBI), which is briefly introduced in Section 2.3.2. PWI adapts the integration time independently for each pixel. This is done by means of sequential integration steps and seems a prior hardware implementation of the *staggered integration* idea presented in [304]. Another hardware solution for increasing the dynamic range of these pixels [21] performs a reset of the common mode component before the charge reaches the saturation level, while preserving the difference between channels. This can be done as many times as the charge approaches that level during integration. Especial pixels have been designed to achieve outstanding inmunity to background light (BGL), such as that presented in [136], which uses a current-sample-and-hold (CS&H) circuit to sense the current generated by DC light. According to the authors, the pixel achieved the highest BGL-immunity reported at that time.

Multitap pixels with number of taps ranging from one to 16 are found in the literature. One-tap pixels are *sensu stricto* not multitap pixels, but a conventional pixel where the integration can be inhibited by an external signal. A one-tap pixel has been presented in [278], as an attempt of maximizing the fill factor. Note that in the one-tap configuration, charge carriers that are generated out of the integration intervals are not stored in any channel and do not contribute to any measurable signal, differently from, e. g., a two-tap configuration. If the signal controlling the integration is square of 50% duty cycle, then, supposing a random depth distribution, on average, 50% of the optical energy captured by the pixel is being wasted. Another drawback of one-tap pixels is the rise of the total acquisition time, since only one sample of the cross-correlation function is gathered per acquisition. A first design of a four-tap PMD pixel is presented in [406]. The main advantage of a four-tap structure is that only one acquisition is required, providing better robustness to motion. A clear drawback is the degradation of the fill-factor and the eventual need for longer exposure times per acquisition. Two pixel structures for a four-tap pixel are evaluated in [278]. The first one is a wheel-shaped CCD structure, where the integration channels or storage gates are located around the photogate, which is in the middle. The main disadvantage of the wheel-shaped four-tap pixel is the very low fill factor, direct consequence of the shape. Another design, with 1 : 2 aspect ratio and two output diffusions instead one, is proposed. The fill factor is three times higher than that of the wheel-shaped pixel and the symmetry of this design seems to reduce inhomogeneities between taps.

Interestingly, the first solid-state ToF camera [279] implements one-tap pixels, in order to get rid of the inhomogeneities between pixels channels that affected the first arrays of *smart* pixels and maximize the fill factor, also very low in the first pixels. Note that increasing the fill factor implies decreasing the exposure times and, in consequence, partially compensating the need for more acquisitions derived from using only one channel per pixel. Nowadays, pixels with two-taps and four-taps are common and can be reliably produced. A dedicated diffusion node and amplifier per tap allows increasing the read-out speed by parallel operation. The last generation ToF cameras, e. g., PMD cameras, the SwissRanger ToF cameras [355, 354] or the new Kinect sensor implement two-tap pixels. The pixels in the SwissRanger SR3000, for instance, are similar to the two-tap PMD pixels, but featuring three photogates instead two for a smoother potential gradient [72]. The sensor provides QCIF resolution and the pixels can operate with frequencies up to 41 MHz. Fast carrier separation and storage is achieved using CCD structures, while CMOS technology is used to allow flexible and fast readout. The first commercial ToF imaging sensor implementing these *smart* pixels was named Photonic Mixer Device (PMD), which became the reference technology for phase-shift ToF imaging and will be described in Section 2.3.2.

Especial pixels, designed for extreme operating conditions can be found in [66]. Two pixel architectures are presented, which enable modulation-demodulation frequencies of hundreds of megahertz. The first architecture is similar to that of photo-gate PMD pixels, but substituting the several conductive photogates by a single high-resistive photogate [70]. This allows a smoother, linearly-decreasing, potential in the semiconductor, instead of a staircase-shaped potential distribution. This modulated-drift-field pixel was said to allow for frequencies up to 1 GHz. Unfortunately, the high per-pixel power consumption makes it unsuitable for high-resolution sensors. The second design overcomes some issues of the first, such as high power consumption and low-pass filtering effect of the gate signal, by means of photogates with huge resistance and capacitance. The core idea of the pixel is to separate the detection from the demodulation, still happening both *in pixel*. A static drift field is applied to the detection area, allowing fast charge separation and transport to the demodulation area [68]. Instead a single photogate, many thin photogates connected in series, with intermediate resistors, are used, in order to achieve a high resistive and capacitive load. The demodulation process, which is just a switching between two storage capacitances, is carried out in a small region and can also be performed at high speed.

Recently, new pixel designs featuring three taps have been proposed, in an attempt to reduce the number of acquisitions per depth image. Recall that in a phase-shift-based ToF system with sinusoidal modulation there are three free parameters and, thus, three measurements suffice to estimate them. The LEFM proposed in [212] features four taps for three effective outputs and a drain. The three-tap structure allows automatic subtraction of the BGL, since one of the taps is integrating exclusively BGL an not the reflected light signal, while the other two integrate the sum of BGL and signal. The use of lateral drift-field for charge transport reminds the LDPD design in [420], which allows for on-chip SBI also by means of three pixel channels.

Some works exist presenting pixels that try to separate the correlation process from the photogeneration process, tightly coupled in the PMD pixels. In such pixels, a fast photodiode is used to sense the incoming light signal and the correlation process is carried out in a subsequent step, by means of additional *in pixel* circuitry. A 30×50 array of such pixels is presented in

[427]. The main drawbacks of the pixels in [427] are a large pixel pitch of $81.9 \times 81.7\,\mu$m and a low fill factor of 20% (one third less than, e. g., 19k or 41k PMD pixels). An improved *in pixel* circuitry is presented in [480], which offers outstanding immunity to ambient illumination (± 5 cm depth variation up to 100 klx) and low power consumption. They achieve a fill factor of 66%, but with a large pixel size of $125 \times 125\,\mu$m. A competitive 160×120 array of pixels of this kind is presented in [366]. Their 'Correlated Double Sampling' (CDS) emulates the natural CBS of two-tap PMD pixels. The prototype poses a serious competitor to the original PMD concept, not just because of the size of the array, equal to that of the PMD 19k, but also because of the combination of a similar fill factor (34%) and an outstandingly low pixel pitch (29.1 μm). Another pixel design that is worth mentioning is that in [236], which features *in pixel* SBI and adaptive accumulation, while offering an outstanding fill factor of 77% (cf. 60% for the Xbox One sensor, with microlenses).

A review on different ToF 3D imaging pixel structures in CMOS is provided in [159]. In [277], a simulation framework is proposed to evaluate different ToF chip layouts, namely, single and dual readout configurations. Differently from existing simulations, where the focus is on the final depth image, a physically meaningful modeling allows obtaining raw images that are not a simple intermediate result, but a realistic simulation of the real ones.

2.3.2. Principle of Operation of the Photonic Mixer Device

The Photonic Mixer Device (PMD) [77] was born with the intention of bringing the demodulation process required in ORFI ToF imaging closer to the detection process. Differently from FLC-LCoS or MQW, which provide *on pixel* demodulation, the PMD principle of operation constituted a breakthrough in that it performs *in pixel* demodulation, that is, the processes of signal mixing and detection happen simultaneously, within the same physical support and are no longer independent. The first demodulation pixels, precursors of the actual PMD pixels, should fulfill the following tasks, according to [278]: photoelectric carrier generation, fast separation of the carriers, repeated addition of the separated carriers in consecutive cycles and *in pixel* storage.

PMD pixels, originally produced in CCD technology, are currently produced in standard CMOS technology. Different PMD pixel structures have been presented and different physical implementations have been tested. Examples of optical implementations of PMD technology are [428, 403] the photo-gate PMD (PG-PMD), which was the first PMD implementation in CMOS technology and the one used in commercial PMD chips, the metal-electrode PMD (ME-PMD), the metal-semiconductor-metal PMD (MSM-PMD) and the micro-channel-plate PMD. In the following we briefly present these PMD implementations, focusing on the PG-PMD. Additionally, a non-optical implementation of the PMD pixels working with microwave radiation has also been proposed, named Microwave Mixer Device (MMD), which contributes to bridge the gap between radar and ToF imaging.

Regardless of the physical implementation, there are some common characteristics of optical PMD pixels. In general, they present a silicon substrate where charge carriers are generated from the incoming photons by photoelectric effect. One or several pairs of electrodes, typically isolated from the silicon substrate by an oxide layer, are used to generate an electrical field that is utilized to move the carriers to one of two storage areas. Normally, PMD pixels have a *finger* structure, i. e., they feature several pairs of electrodes and storage areas, one after another, being each element long and narrow, like a finger. This structure is designed to minimize the distance the carriers have to travel from the generation area to the storage area. This decreases the transport time and increases the modulation bandwidth of the pixel.

Photo-Gate PMD The photo-gate PMD (PG-PMD) was the first implementation of the PMD principle, dating back to 1997/98. The structure of the pixel can be seen as two diodes whose anodes are grounded and whose cathodes are the readout points of the two pixel channels. Therefore, the diodes are reverse-biased. The photocarriers are generated in a photosensitive area between the diodes. Over the photosensitive area a layer of silicon oxide (or, alternatively, Si_3N_4) separates it from the two conductive and transparent MOS photogates. Fig. 2.6 provides an schematic representation of a PG-PMD pixel (a) and its principle of operation with DC light (b) and modulated light (c).

The *push-pull* voltages applied to the photogates ($U_0 + u_m$, $U_0 - u_m$ in Fig. 2.6-a) modify the potential distribution in the surface and lead to what was called a *dynamic seesaw* [269] for the generated carriers, pushing them to one side of the pixel or another. In order to clarify the mixing process, suppose that the reference signal applied to the photogates is a square signal and the light arriving to the pixel surface is not modulated. Then, at the end of the integration time, the output voltages of the two readout nodes (U_a, U_b in Fig. 2.6-a) is the same, since, for each period, the photocarriers were integrated in the A channel half of the time and in the B channel the other half (Fig. 2.6-b). If the light is modulated, say according to the same reference signal applied to the photogates, then the difference of

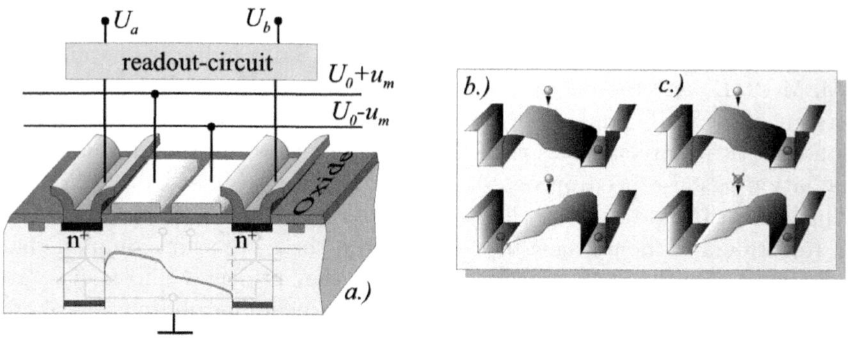

Figure 2.6.: Schematic representation of a photo-gate PMD pixel (a) and conceptual representation of the mixing process with DC (b) and modulated light (c). PMD pixels can be thought as two anode-grounded diodes, separated by a photosensitive region. A difference of potential between the transparent photogates (in yellow) induces an electric field that drives the photocarriers to one or another diode, where they are stored. The readout of each pixel channel is then performed at the cathodes of each diode. Images from [269].

potential between the two output nodes is directly dependent on the phase shift between the light signal and the reference signal applied to the pixel. It is clear that this difference provides a sample of the autocorrelation function of the reference signal (thus a triangular wave), at a phase given by the shift between light and reference signals. In the case of 0° or 180° phase shift, all the photocarriers should contribute to one channel and no one to the other (Fig. 2.6-c), i. e., a maximum or minimum of the autocorrelation function.

Depending on the voltage applied to the photogates, the diodes may operate in accumulation, depletion or inversion mode. In normal operation, they are in deep depletion mode, i. e., there is a deep space charge region or depletion region under the photogates [309] due to U_0. The modulation frequencies, typically ranging from tens to thousands of megahertz, are too fast to modify the deep depletion state, even for large values of the modulation voltage, $\pm u_m$. This region serves for fast charge transport.

Metal-Electrode PMD The metal-electrode PMD (ME-PMD) is similar to the PG-PMD, being the main difference the absence of the photogates. The transparent electrodes are substituted by the metal shields of the readout electrodes, which are widened to that end. A diagram of an ME-PMD pixel is given in [403]. The main advantage of this implementation is a more homogeneous drift field, resulting in an increased modulation bandwidth. Additionally, the absence of photogates allows for narrowing the finger structure, yielding higher bandwidth. The reference signal is not only applied through the metallic shields of the readout electrodes, but also through the cathodes themselves, by means of a coupling capacitor that acts as a barrier for the DC component of the reference signal (U_0 in Fig. 2.6-a).

Metal-Semiconductor-Metal PMD The metal-semiconductor-metal PMD (MSM-PMD) exhibits a simpler structure than the ME-PMD. There are no dedicated modulation electrodes and both the modulation and the readout are carried out through contacts on metallic stripes with the characteristic finger structure. That is, the pixel is composed by metal stripes on negatively doped silicon, i. e., metal-semiconductor junctions or Schottky diodes, all over a grounded silicon substrate. Note the similarity with the ME-PMD pixel operation, in which the modulation was applied through the metallic shielding and the cathodes simultaneously. The simplified structure of the MSM-PMD requires a low pass filter per channel before the readout circuitry, in order to block the high-frequency modulation signal ($\pm u_m$). The modulation signal is applied through coupling capacitors like in the ME-PMD, which also serve to eliminate the photocurrent due to unmodulated BGL. Small-area Schottky diodes, such as those used in RF detectors, often operate at frequencies of tens of gigahertz. One of the main advantages of MSM-PMD pixels is, therefore, an outstanding bandwidth [269], yielding better depth resolution.

Micro-Channel-Plate PMD The micro-channel-plate PMD (MCP-PMD) [485] reminds the second generation (and posterior) image intensifiers, whose central component is an MCP at high voltage. The operation of EOM based on image intensifiers was already described at the end of of Section 2.2.2. The novelty of the MCP-PMD pixel is bringing the concept of the MCP-based image intensifier to a pixel architecture, instead of simply stacking the intensifier over a pixel array. As in conventional image intensifiers, electrons are generated when the light reaches the photocathode and travel to the MCP, due to the difference of portential between the photocathode and the metalized inner layer of the MCP, which acts as a first anode. The modulation signal is applied to this layer, causing the electrons to hit the MCP in a region destined to the readout A or another destined for the B. After the electrons reach the MCP, they are conducted by finger-like microchannels where the electron multiplication takes place. Each microchannel is around 10 μm width and acts as a continuous-dynode electron multiplier. At the end of the microchannels the electrons from A-area microchannels and B-area microchannels are read in their corresponding readouts. In MCP-PMD pixels, as well as in conventional image intensifiers, voltages of more than a thousand volts between anode and cathode are required.

A related concept is that of the Photo-Multiplier-Tube PMD (PMT-PMD) [485], which is a special PMD pixel architecture including an individual PMT per pixel. The modulation signal is used to deviate the electron beam at the end of the tube to one of two anodes, which are the two channels of the pixel. The electrons are generated at the beginning of the tube, in a photocathode, as in conventional PMTs. Apart from clear concerns on the pixel size and the scalability of such a system, a problem that persists is the requirement of high differences of potential between the anodes and the photocathode.

An advantage of MCP-PMD and PMT-PMD is the large amplification of the light signal, which permits operation with very low optical power, low exposure or very large distances. The MCP-PMD could provide an amplification factor of 10^7 with 300 GHz [485] and maximum ranges of 1000 m [403].

Microwave Mixer Device (MMD) Silicon detectors are sensitive to radiation in the visible and NIR spectrum bands. Nevertheless, the PMD concept is general and can be applied to higher wavelength radiation, such as X-rays or microwaves, or even sound waves, as far as the conversion into electrical signals is possible. Once the carriers are generated, a mixing process can be carried out, as in PMD pixels. The so-called Microwave Mixer Device (MMD) is equivalent to the PMD, but using microwave radiation instead of light. There are clear advantages of using microwave radiation instead of optical radiation. One of them is the frequency of the radiation, from 300 MHz to 300 GHz, lower than that of optical radiation (wider unambiguous range in a pure interferometric setup), but high enough to achieve good depth resolution. Another advantage is the lower attenuation by meteorologic phenomena and the independence of daylight and background illumination. Unfortunately, a greater wavelength brings new issues, such as the lower lateral resolution and the requirement of large antenna apertures.

The high-frequency microwave carrier is modulated in amplitude by a lower-frequency signal, similarly to the optical case. An antenna is used to convert the received microwave signal into an electrical signal, substituting the photoelectric conversion in the PMD. The charge separation, i. e., the core of the MMD pixel, is performed by means of a pair of Schottky diodes in antiparallel configuration, connected right after the antenna output. The modulation signal (the same used for amplitude modulation of the emitted microwave radiation, $\pm u_m$) is applied to the terminals of the diode pair. Clearly, this signal will block one of the diodes while driving the other to conduction. A block diagram of the MMD and a more detailed operation description is provided in [485], together with active and pasive pixel implementations. A sketch of a 3D-MMD camera using a Fresnel lens to map the scene onto a MMD pixel array is given in [403].

Dynamic Range and Background Illumination In general, PMD pixels are able to suppress the contribution of BGL by means of the so-called "correlated balanced sampling" (CBS), which is simply exploiting the fact that, for periodic reference signals of 50% duty cycle, the DC component of the light signal contributes equally to both pixel channels and, consequently, has no influence in the correlation value given by the difference $k(\theta) = D_\theta = I_\theta^{\mathrm{A}} - I_\theta^{\mathrm{B}}$. Nevertheless, this does not avoid saturation of the pixel channels if the light intensity is excessively high, e. g., under sunlight illumination. Consider the following classical definition of dynamic range of a pixel:

$$DR = 20 \log \left(\frac{I_{\text{ceil}}}{I_{\text{floor}}} \right) \tag{2.17}$$

where I_{floor} and I_{ceil} are the minimum and maximum values that the pixel (or pixel channel) can deliver. I_{floor} is given by the noise floor and I_{ceil} by the saturation level. The dynamic range of PMD pixels has been calculated in prior investigations and found to be of 73 dB for the one-tap PMD pixel in [278], 71 dB for the two-taps PMD pixel in [309] or 70 dB in [222].

An additional mechanism to suppress the disturbance of unmodulated light is the use of narrow bandpass optical filters, centered at the wavelength of the illumination source. Unfortunately, sunlight spectrum exhibits a considerable amplitude in the NIR region, often the band of choice for ToF systems. An intelligent selection of the illumination wavelength, e. g., making it coincide with the center of some atmospheric absorption band, together with a narrow bandpass filter at that wavelength can be of great help. In general, we seek a large ratio between signal amplitude and background level. The easiest solution is to increase the optical power of the illumination system, e. g., adding more emitters or increasing their power. This has a cost and increases the size of the ToF system. Alternatively, the emitters may operate in burst mode. In burst operation, the light sources emit the same average power as in normal operation, but during a shorter time, i. e., with higher peak power. As a by-product of burst operation, fast-moving objects can be captured, due to the reduction of the exposure time.

Still, these methods do not increase the dynamic range of the PMD pixel itself. The amount of charge to be stored by some pixel channels might still be reaching the saturation threshold, while others are within the noise level. A typical case is saturation of few pixels due to background illumination. If the exposure time is reduced, saturation may be avoided, but at the cost of decreasing the SNR of the rest of the pixels. A mechanism is therefore necessary to artificially increase the dynamic range of the PMD pixels, in a local way and leaving the signal (difference between pixel channels) unaffected. The solution provided by PMD Technologies is the patented Suppression of Background Illumination system (SBI) [475, 74]. The system is a hardware solution, implemented as *in pixel* circuitry, which works independently for each pixel. The core idea of the SBI is the use of a compensation current to remove the charge due to uncorrelated light from the pixel. The system assures that exactly the same current is being removed from both pixel channels, keeping constant the difference. In principle, this means that the entire dynamic range of the pixel can be used to measure the correlated signal, leaving away the often-dominant background component. It has been shown that the SBI makes possible to operate in sunlight conditions up to 1.5×10^5 lux [387]. A model of the PMD pixels including SBI can be found in [398].

SBI operation presents several disadvantages. The intensity image cannot be recovered from the measurements, since the DC component is artificially decreased for those pixels where the SBI is active. If the current extracted from both pixel channels is not exactly the same, the SBI affects the measurements, producing a per-pixel bias, which translates into a noise-like pattern in the depth image.

The integration characteristic curves of PMD pixels are linear along the entire range of operation. The SBI introduces a severe non-linearity in the curve, changing from a positive slope to a zero slope in the pixel channel reaching saturation and a slightly-negative slope in the other. Other non-linear characteristic curves can be used to increase the dynamic range. In [11], linear, piecewise-linear, logarithmic and linear-logarithmic characteristic curves are compared in terms of dynamic range. A piecewise linear curve presents a point, for a certain charge level, from which the slope is lower than before. Logarithmic curves provide an increase of the dynamic range by a smoothly-decreasing slope. Linear-logarithmic curves combine the low slope of a logarithmic ending with a constant high slope for signals with low optical power.

2.3.3. Depth Calculus and Performance Considerations

In this section we consider that the illumination signal exhibits a close-to-sinusoidal shape and, in consequence, the cross-correlation carried out in the PMD pixels is between a sinusoid and a periodic signal of equal fundamental frequency, which is equivalent to being between two sinusoidal signals of that frequency [304]. This is the case which is implicitly assumed in PMD cameras and it is supported by the fact that the LEDs, often used as illumination sources, show a low pass filter effect in the conversion of the electrical driving signal into optical at frequencies of tens of megahertz. This is due to the characteristic rising and falling times in the range of nanoseconds. If we neglect the asymmetry between rising and falling times, the effect the LEDs have over the illumination control signal (ICS), which is used to drive them, can be modeled as a convolution with a Gaussian kernel in time domain, $g(t)$, i. e., a smooth low pass filter. Note that modeling such asymmetry would require an asymmetric convolution kernel, thus, not Gaussian. This means that the final illumination signal can be defined as $i(t) = (ICS * g)(t)$ and follows Eq. 2.7. In consequence, the return optical signal is also sinusoidal (Eq. 2.9) and the measurements provided by the PMD pixels follow Eq. 2.12 in an ideal mixing process.

If we consider as measurements the difference of the two PMD pixel channels, we are correlating against a square signal of zero mean (no offset) and, in case of ideal mixing, there is no DC component in Eq. 2.12, since $A_0^q = 0$. In other words, measurements are insensitive to the DC component of the light received by the pixel, A_0^r. Thus, only two of these differential measurements suffice to estimate the remaining parameters, namely, the amplitude and the phase shift due to depth, θ_{depth}, in case of ideal modulation and demodulation. Let $D_{\theta_k} = I_{\theta_k}^{\text{A}} - I_{\theta_k}^{\text{B}}$ denote the differential measurement obtained as difference of the PMD pixel channels (A and B) for a sampling point $\tau_k = \theta_k/\omega$ of the cross-correlation given in Eq. 2.12. Despite only two sample points suffice, in PMD systems typically four measurements at four different phases are gathered. Gathering more measurements reduces the effect of harmonics and, indirectly, also the effect of noise of random nature due to the summations in both numerator and denominator in Eq. 2.5. In order to simplify the depth calculus, the four phases are selected to be uniformly distributed in phase domain, i. e., the shifts are $\theta \in \{0°, 90°, 180°, 270°\}$ of, equivalently, the samplings points are given by $\theta_k = \frac{k-1}{N}360,\ k \in [1, N]$, for

$N = 4$ samples. Note that this sampling scheme is, in the case of an ideal mixing process, actually equivalent to acquiring only two samples, since $D_\theta = -D_{\theta+180^\circ}$, $\forall\theta$, by definition (Eq. 2.12). The four phases schema allows further simplification of Eq. 2.6 [332] due to the trivial values of sines and cosines at these phases, yielding

$$\theta_{\text{depth}} = \arctan\left(\frac{D_{270^\circ} - D_{90^\circ}}{D_{180^\circ} - D_{0^\circ}}\right) \tag{2.18}$$

which, in combination with Eq. 2.1, yields a simple formula, known as the *four phases algorithm* [330, 388], which directly computes the depth from the raw measurements:

$$d = \frac{c}{4\pi f_{\text{mod}}} \arctan\left(\frac{D_{270^\circ} - D_{90^\circ}}{D_{180^\circ} - D_{0^\circ}}\right) \tag{2.19}$$

The amplitude and the DC offset can also be estimated from the measurements. The amplitude, which refers to that of the cosine in Eq. 2.12, can be estimated from measurements of one of the channels or from differential measurements. In the former case the reference signal has an offset of half of the amplitude, while in the latter case there is no offset. Note that in the latter case the amplitude is scaled by a factor of two, due to the corresponding double amplitude of the reference signal (A^q). The offset cannot be estimated from differential measurements, since it does not contribute to them, and has to be calculated from the measurements of one of the pixel channels. The formulas are derived from Eq. 2.12 and given by Eq. 2.20 and Eq. 2.21. The proofs of Eq. 2.18, Eq. 2.20 and Eq. 2.21 are given in Appendix A.3.

$$\begin{aligned} A &= \frac{\sqrt{(I_{270^\circ} - I_{90^\circ})^2 + (I_{180^\circ} - I_{0^\circ})^2}}{2} \\ &= \frac{1}{2}\frac{\sqrt{(D_{270^\circ} - D_{90^\circ})^2 + (D_{180^\circ} - D_{0^\circ})^2}}{2} \end{aligned} \tag{2.20}$$

$$A_0 = \frac{I_{0^\circ} + I_{90^\circ} + I_{180^\circ} + I_{270^\circ}}{4} \tag{2.21}$$

Demodulation Contrast A measurement of the performance of any kind of *smart* pixel performing an *in pixel* demodulation is the so-called demodulation contrast. This complements and is based on the concept of modulation depth (or contrast) of the light source, which measures the quality of the modulation process. For completeness, consider the following classical definition of Michelson contrast [326]:

$$k_{\mathrm{mod}} = \frac{I_{\mathrm{max}} - I_{\mathrm{min}}}{I_{\mathrm{max}} + I_{\mathrm{min}}} \tag{2.22}$$

where I_{min} and I_{max} are the minimum and maximum light intensities provided by the light source. Obviously, an ideal modulation provides a contrast of 1. Note that the numerator is the amplitude of the signal, while the denominator represents twice the average, i. e., twice the offset of a periodic illumination signal. Similarly, the demodulation contrast is defined as the ratio between measured amplitude and measured offset of the light signal. For any multitap pixel that works in a four-phases basis, the amplitude and offset can be obtained from Eq. 2.20 and Eq. 2.21, respectively. For the specific case of a PMD pixel, with two integration channels denoted by A and B, we can write [330]

$$k_{\mathrm{demod}} = \frac{\text{measured amplitude}}{\text{measured offset}} = \frac{|I_{\mathrm{A}} - I_{\mathrm{B}}|}{I_{\mathrm{A}} + I_{\mathrm{B}}} \tag{2.23}$$

where I_{A} and I_{B} are the readouts of the corresponding pixel channels when the overall phase shift between the electrical signal controlling the integration and the optical signal as received at the pixel surface is $0°$ or $180°$. The demodulation contrast quantifies the ability of the pixel to perform charge separation according to the reference signal. Typical values of this parameter are 50% for two-tap PMD pixels (at 20 MHz modulation frequency) or 60% for three-tap PMD pixels. Despite it is often omitted in the literature, this measurement is a meaningful descriptor of the performance of the pixel only when gathered using a light signal with modulation contrast of 1. Otherwise a composed measurement is obtained, which in most practical cases is dominated by the signal-to-background-light ratio. The total contrast of a ToF system is given by $k_{\mathrm{total}} = k_{\mathrm{mod}} k_{\mathrm{demod}}$.

Other Operation Modes: Heterodyne Mode and Pseudonoise Modulation As already introduced in Section 2.2.2, homodyne operation is not the only possible modulation-demodulation framework in phase-shift-based ToF systems. PMD devices can be tuned to operate in heterodyne mode [117] or using pseudonoise (PN) modulation [75].

Operating in heterodyne mode brings the need for accurate frequency shifting and a perfect synchronization of the acquisitions with the low-frequency beat signal. This also means ensuring that an integer number of acquisitions are gathered per period of the beat signal. The phase shift calculus in heterodyne mode can be performed in an equivalent way to homodyne mode, using a formula which is equivalent to Eq. 2.6, but actually sampling the beat signal in time, instead of sampling the cross-correlation function at different phase shifts of the reference signal, as in homodyne PMD. The heterodyne operation makes easier to acquire a custom number of equally-spaced measurements and reduce linearity errors, caused by the presence of unexpected high order harmonics in the modulated light.

The use of PN modulation in PMD sensors has been extensively studied in [75], motivated by the spread-spectrum techniques, which had been already applied in communications time before. Originally used for military communications in the 1950s, these techniques are nowadays used everywhere in civil wireless networks. In short terms, spread-spectrum methods deliberately spread a signal, which might be bandlimited in principle, along a wide frequency band. An immediate advantage is the improvement of the resistance to interference and intentional jamming. Additionally, these techniques allow using the same channel to transmit different information to different receivers simultaneously, that is, allow multiple access without the need of time multiplexing. This is known as the code-division multiple access (CDMA) or code-division multiplexing (CDM). It is clear that, if orthogonal codes (e. g., obtained from the columns of a Hadamard matrix) are used as carriers to send the simultaneous messages, no interference will occur, if each receiver decodes the message using the right code. Similarly, in ToF imaging, using orthogonal PN sequences for illumination modulation ensures interference-free simultaneous operation, which can not be achieved in single-frequency homodyne operation. If the codes are long enough, any two different random codes can be considered orthogonal. Additionally, the sharp autocorrelation function of PN codes ensures that accurate phase measurements are possible.

Physical Considerations Commercial PMD sensors feature a daylight blocking filter, whose mission is to block all light except from that of the wavelength used for the modulated light. The filter is a low-pass filter, with transmittance close to 100% for wavelengths higher than 830 nm. That is, PMD sensors are expected to operate with NIR light. This may profit from the spectral response of silicon, which is higher in the NIR than in the visible range. Additionally, working with relatively large wavelengths reduces the thermal diffusion in the silicon and, consequently, increases the accuracy of the measurements. Large wavelengths imply lower excess of energy in the photoelectric effect. Such excess of energy generates heat in the silicon, increasing its temperature. For completeness, consider the following equations. For a certain wavelength of the radiation, λ, the excess of energy is given by Eq. 2.24,

$$\Delta E_{\text{th}} = \frac{hc}{\lambda} - E_g \tag{2.24}$$

where ΔE_{th} is the energy excess, destined to generate heat, and E_g is the band gap energy of the photosensitive material, silicon in this case. This excess should be as close to zero as possible. The wavelength for which the right side of Eq. 2.24 is zero is the lowest that will still produce photoelectric effect in silicon and it is around 1.1 μm. The equipartition theorem of the energy states that, once thermal equilibrium is achieved, energy is shared equally among all the particles in the system. This means that the total energy can be formulated as the sum of the kinetic energies of all particles moving at a certain average speed. Per particle, we obtain

$$\frac{1}{2} m_{\text{eff}} v_{\text{th}}^2 = \frac{3}{2} k_B T \tag{2.25}$$

where m_{eff} is the effective mass of the charge, v_{th} the thermal speed, k_B the Boltzmann constant and T the absolute temperature. An increase of the total energy in ΔE_{th} will add to the right side of Eq. 2.25, rising the temperature, and leading to an increase of v_{th}, $\Delta v_{\text{th}} = \sqrt{\frac{2\Delta E_{\text{th}}}{m_{\text{eff}}}}$. We are interested in the diffusion length, which is the average distance that the carriers travel until they recombine, i. e., the distance corresponding to its lifetime, τ_n, which is given by

$$L_{\text{diff}} = \sqrt{D\tau_n} \tag{2.26}$$

where D is the diffusivity of the material. Provided that D depends linearly on v_{th}^2, the energy excess, ΔE_{th}, produces an increase of the diffusion length which depends linearly on $\sqrt{\Delta E_{\text{th}}}$. Additionally, D and E_g strongly depend on the temperature, so that, the higher the temperature, the higher the resulting increase of the diffusion length. Now consider the following general definition of modulation transfer function (MTF):

$$MTF(f) = \left| \frac{T(f)}{T(0)} \right| \tag{2.27}$$

where $T(f)$ is the frequency transfer function of the system. In control theory transfer function refers to temporal frequency, but in this case it refers to spatial frequency, in the pixel domain. Eq. 2.27 describes the ability to capture spatial information. Supposing that the pixel size is small enough to consider negligible the corresponding spatial quantization error, Eq. 2.27 yields an upper bound for the lateral resolution of the ToF imager.

A strong diffusion favors crosstalk and degrades the MTF [421]. The quantum efficiency (QE) of a photosensitive element is defined as the ratio between the number of photogenerated electrons effectively collected by the element and the number of impinging photons, under uniform illumination. This ratio strongly depends on the wavelength of the light, decreasing for high wavelengths, and on the diffusion length, approaching to one for large diffusion lengths. The existence of upper layers over the depleted silicon area produces interference between shifted wavefronts, since the order of magnitude of the layers and the radiation wavelength are approximately the same. This translates into a fringe pattern in the QE in wavelength domain. Depending on the relationship between the thickness of the upper layer and the wavelength, a local minimum or maximum of QE can be achieved [421]. Thermal diffusion might also increase, as a side-effect, the effective active area of the pixels. Electrons generated out of the space charge region are, in principle, not affected by the electrical field responsible for charge transport. Nevertheless, if they diffuse into the space charge region, they can be separated by the electrical field and pushed to the corresponding storage area.

The other key parameter in Eq. 2.24 is τ_n, which depends, in turn, on the feature size of the opto-electronic processes.

Depth Measurement Uncertainty Several sources of noise exist in PMD sensors and, in general, in any image sensor. Some of the most important noise sources [434] are the optical shot noise, the thermal noise, flicker noise and the quantization noise. A rule of thumb provided in [411] says that 1 pJ of accumulated NIR photon energy in a pixel allows the measurement of the distance to a precision of around 1 cm at 20 MHz modulation frequency.

Shot noise, or quantum noise, is due to the fact that the process of photon arrival and electron generation follows a Poisson distribution, which is a limit case of the binomial distribution, when the number of tries n tends to infinity and the probability of success tends accordingly to zero, yielding a finite number of successes. A discrete random variable following a Poisson distribution, $X \sim P(\lambda)$, is characterized by a single parameter, λ, which gives both the expected value and the variance, i. e.: $\mathrm{E}(X) = \mathrm{Var}(X) = \lambda$.

The thermal noise, also known as Johnson-Nyquist noise, is the electronic noise due to thermal agitation of the charge carriers in the silicon. Thermal noise approximately follows a zero-mean white Gaussian noise distribution, with a power spectrum that is flat up to 1 THz [29] in resistive regions, such as MOS channels in strong inversion.

Flicker noise is also known as $1/f$ noise or pink noise due to its *pink* power density spectrum, i. e., decaying with frequency. This noise is found in many fields and is common in natural phenomena, where the systems tend to be more prone to low-frequency variations than violent changes. In electronics, it is related to a resistance fluctuation, which produces voltage or current variations. Due to the pink power density spectrum, at high frequencies it may be eclipsed by thermal noise.

Quantization noise is ubiquitous in electronics, since it appears there where an analog-to-digital conversion is performed. Quantization consists in mapping a large set of values into a countable smaller set. Such process also takes place in data compression, being the quantization noise the signal distortion due to the eventual information loss during compression. From a general point of view, if the discrete samples of the signal are considered measurements,—e. g., using an ideal Dirac delta function as sensing kernel,—the recent theory of Compressive Sensing (shortened CS, see Chapter 3) assures that exact recovery of the signal in a compressed domain is achievable from the measurements if the signal admits a sparse or compressible representation in that domain and its basis is incoherent with the sensing kernels. This holds even if the measurements are incomplete and affected by quantization noise, as far as the number of measurements is high enough for the given sparsity level.

Noise models for amplitude, phase and offset have already been provided in [335]. We are interested in determining how noise in the pixel readouts affects the final depth estimation. The relationship between the depth value and the measurements in a conventional PMD ToF camera is given by Eq. 2.19. Consequently, it suffices to apply error propagation to Eq. 2.19 to obtain the influence of measurement noise in the depth estimation. The result is provided in Eq. 2.28, where A is the amplitude, as defined in Eq. 2.20. The proof is given in Appendix A.4.

$$\Delta d = \frac{c}{4\pi f_{\mathrm{mod}}} \frac{1}{(2A)^2} \sqrt{(I_{270^\circ} - I_{90^\circ})^2(\Delta^2 I_{0^\circ} + \Delta^2 I_{180^\circ}) + (I_{180^\circ} - I_{0^\circ})^2(\Delta^2 I_{90^\circ} + \Delta^2 I_{270^\circ})} \tag{2.28}$$

The depth uncertainty given by Eq. 2.28 is general and valid regardless of the value of the measurements, I_θ, and the nature of the measurement noise, ΔI_θ. A realistic hypothesis is that the measurement noise is dominated by the shot noise, i. e., that follows a Poisson distribution. In that case we have that the variance is equal to the expected value and the standard deviation is $\Delta I_\theta = \sqrt{I_\theta}$. This, together with the fact that the sampling points are multiples of 90°, can be used to simplify Eq. 2.28, yielding Eq. 2.29, the proof of which is also given in Appendix A.4.

$$\Delta d = \frac{c}{4\pi f_{\text{mod}}} \frac{\sqrt{2A_0}}{2A} \tag{2.29}$$

Eq. 2.29 is equivalent to Eq. 4.11 of [278] and coincides with Eq. 10 of [411]. Other noise sources different from Poisson noise are signal-independent and can, therefore, be added to the DC component of the measurements, A_0.

Eq. 2.29 clearly shows that in cases when shot noise is dominant, e. g., due to poor illumination, the depth uncertainty is given by the ratio between the square root of the DC component and the amplitude of the cross-correlation function, which is, in turn, proportional to the amplitude of the illumination signal. If the background illumination is low, the DC offset is mainly due to pixel noise (e. g., thermal, flicker or reset noise). Therefore, it is clear that the pixel sensitivity plays a key role in the achievable depth accuracy. At the chip level, the sensitivity of a PMD sensor can be increased with larger pixel sizes or higher fill factor. If no further improvement can be achieved in this regard, the use of microlenses on the pixels to collect more light may be an alternative. The amount of photo-generated electrons also depends on the optical system. Lenses with large apertures allow collecting more light and getting a higher response in the pixels. It has been observed that, in most cases, the accuracy of the depth measurements improves linearly with the aperture, but in low-light cases, i. e., when the noise floor due to dark current and quantization is dominant, the accuracy increases quadratically with the aperture [330]. In [278, 309] an equation is provided that calculates the optical power received by a pixel in the case of Lambertian reflection, which we rewrite here as Eq. 2.30:

$$P_{\text{pixel}} = \frac{A_{\text{pixel}}}{A_{\text{image}}} P_{\text{source}} \rho \left(\frac{D_{\text{lens}}}{2r} \right)^2 k_{\text{lens}} \tag{2.30}$$

where A_{pixel} is the active area of the pixel, A_{image} is the area of the image in the sensor plane, P_{source} the optical power of the light source, ρ the reflectivity of the target, r the distance between the lens and the target, D_{lens} the aperture of the lens and k_{lens} a factor accounting for the optical losses of the lens system. Note that Eq. 2.30 relies on several assumptions that might not match the reality. First, it implicitly supposes that all the optical power of the light source is somehow transferred to the target surface, being the power reflected by the target $P_{\text{object}} = \rho P_{\text{source}}$. Additionally, it assumes that the whole scene observed by the lens is a target whose points are all at a distance r from the lens and share the same reflectivity. This might be a valid assumption for small FOVs and large observation distances, so that the observed object area is small and the reflectivity can be considered uniform, while the relative depth variations are much lower than the average distance r. Nevertheless, in general, the power received by a pixel depends on the reflectivity and distance of the object point being observed by that pixel, which might be very different from those of points observed by other pixels of the array. For this reason, we derive here a more complete formula that accounts for the light propagation from the source to the scene and calculates the power received by the pixel directly in a per-pixel basis, avoiding the need for the ratio $\frac{A_{\text{pixel}}}{A_{\text{image}}}$ to obtain the per-pixel power from the total average power received at the image plane, as done in Eq. 2.30.

$$P_{\text{pixel}} = \frac{A_{\text{pixel}} \rho k_{\text{lens}}}{8\pi (r f\#)^2 \left(1 - \cos\left(\frac{FOV_{\text{source}}}{2}\right)\right)} P_{\text{source}} \tag{2.31}$$

where r and ρ denote now the average distance and reflectivity of the object area captured by the pixel, respectively. FOV refers here to the FOV of the illumination system, which might differ from the FOV of the camera. $f\#$ denotes the f-number of the lens. The derivation of Eq. 2.30 can be found in [278], while the proof of Eq. 2.31 is given in Appendix A.5.

The number of photogenerated electrons can then be calculated from the per pixel power obtained from Eq. 2.31 as:

$$N_{e^-} = \frac{E_{\text{pixel}}}{E_{\text{photon}}} QE(\lambda) = \frac{(P_{\text{pixel}} t_{\text{exp}}) QE(\lambda)}{\frac{hc}{\lambda}} \tag{2.32}$$

where t_{exp} denotes exposure time and $QE(\lambda)$ is the quantum efficiency for the light wavelength λ.

2.3.4. Typical and New Applications

Since the advent of the PMD technology, the targeted application was phase-shift-based ToF depth imaging. Depth images of simple objects obtained with early low-resolution PMD arrays are given in [310]. A milestone in this regard was the development of the first ToF camera based on *smart* pixels [279], which used one-tap pixels, similar to the posterior two-tap PMD pixels. A real time 3D video camera featuring a PMD sensor is presented in [408]. The number of pixels of this first prototype was 16 × 16, but new designs up to the current 120 × 160 resolution were already contemplated in [408]. In that work it was suggested to combine the information from a 3D-PMD videocamera with that from a 2D-RGB videocamera with the same field-of-view (FOV). Similarly to [408], the PMD arrays have been presented as a multichannel lidar in [407], where the idea of a 3D camera with synchronized 2D-RGB image fusion is also present. The system presented in [216], and whose possibilities are thoroughly investigated in [304], fuses a PMD-based 3D camera and an RGB 2D camera into a single system. The camera, developed at ZESS, is called MultiCam [377] and allows both sensors to share the same optical path by means of a Bauernfeind prism with an integrated beam splitter, which lets the NIR light reach the PMD chip, while redirecting the visible light to the RGB sensor. The monocular setup allows achieving a perfect registration between the color and depth modalities, which is not possible in most commercial RGB-D sensors, such as the Kinect sensor. The latest model of the MultiCam implements a PMD 19k-S3 chip [374] (120 × 160 pixels) for ToF imaging and an Aptina three-megapixel color CMOS chip for RGB imaging. For further details the reader is referred to [280], where a MultiCam is used in a wide-area 2D/3D imaging system.

A linear 3D-ranging system based on the MSM-PMD pixels is presented in [76]. The large bandwidth of MSM-PMD pixels (up to 1 GHz) is exploited to generate PN modulation patterns. The same PN sequence is used for modulation and demodulation, leading to an autocorrelation function with a single sharp maximum. This allows accurate and unambiguous depth estimation, at the cost of increasing the number of acquisitions. If the method for finding the autocorrelation maximum is by regular sampling, the number of samples grows linearly with the depth resolution requirements. The system achieved depth standard deviation below 4 mm.

An appealing application, also considered from the origins of the PMD technology, is its use for optical communications [77, 75]. Arrays of PMD pixels can be used to implement CDMA in optical communications, allowing multiple users to use simultaneously the same physical medium without mutual interference by means of orthogonal PN codes. The idea of using the NIR for wireless communications has become of interest in recent times due to the saturation of the traditional radio bands and the comparatively low interference in the optical range. The development of solid state lighting technology, able to switch at high frequencies, is another factor that favors optical wireless for indoor communications [165]. In this regard, a system allowing the use of LED indoor lighting systems both for visible illumination and wireless communications has been made commercially available under the name of *pureLiFi*. The core idea is to use conventional white LED lamps as a multiple access communication system [164] and is backed by the staggered prohibition of incandescence bulbs for household use, which is expected to lead to the omnipresence of LED illumination in the domestic domain. Either using orthogonal frequency division multiplexing (OFDM) modulation or CDMA, PMD arrays are an excellent candidate for multichannel reception and decoding of the light signal in a MIMO configuration.

In the last years, the development of transient imaging techniques [264] has opened a new field of application for PMD pixels. In principle, transient imaging aims to offer a unique combination of outstanding spatial, angular and temporal resolutions. In principle, conventional cameras offer a very strict tradeoff between temporal and spatial or angular resolution. In most cases, high spatial and angular resolutions are only achieved at rates below 100 Hz. Transient imaging aims keeping the spatial resolution of conventional cameras, angular resolution as high as that of phased arrays, eventually allowing to *look around corners*, while pushing the temporal resolution down to the femtosecond range. Consequently, these techniques require a laser source operating in pulsed mode in the femtosecond range and fast photodetectors. Alternatively to the fast photodetector, a CCD camera is used in [445], in combination with an external setup that reminds an MCP-based image intensifier, with additional electrodes at the sides of the tube to change the angle of the electrons generated in the photocathode. With this setup, which is close to a phase-shift-based ToF camera with external demodulation, the time profile of the returning light signal has been recovered, which allows *seeing* how the light propagates through the scene. In [248], a PMD chip is used as low-cost substitute of the CCD camera with external electron-sweeping system in a ToF system in order to deal with multipath interference. The use of a PMD sensor for transient imaging is proposed in [220], which allows sensing the surfaces of different translucent objects occluding each other. Further work in this direction has been recently presented in [221], where the sparsity of the vector of scatters in depth (i. e., propagation time) domain is exploited to achieve depth imaging in scattering media, such as non-transparent water.

2.4. Current Limits of the PMD-Based Time-of-Flight Imaging

2.4.1. Depth Accuracy

The depth accuracy of a PMD-based ToF imaging system is not directly given by Eq. 2.29. The uncertainty given by Eq. 2.29 derives from propagating the noise expected in the raw measurements to the depth measurement, according to a certain noise model. This provides a bound on the precision of the depth measurement, which mainly depends on the modulation frequency, the amplitude of the modulated signal and the DC offset. An experimental evaluation of the dependency of the depth standard deviation and accuracy on, e. g., modulation waveform, modulation frequency, exposure time, illumination geometry, etc. is given in [284], where an optimal modulation frequency of 45 MHz was found for LED-based illumination. The accuracy involves both the aleatory and the systematic error. If the real signals or the system do not follow their respective models, a systematic error will appear, which is not of random nature, but system- and often also signal-dependent. Indeed, in real PMD ToF cameras, depth errors in the centimeter range are often registered, even under good illumination conditions, low BGL and modulation frequencies of tens of megahertz. Nevertheless, in these cases the main component of the error is of systematic nature and, therefore, cannot be predicted with Eq. 2.29 or eliminated via temporal averaging. The standard deviation of the measurements is, in favorable cases, very low, in the millimeter range, close to the uncertainty predicted by Eq. 2.29. That is, PMD cameras exhibit good precision but often poor accuracy.

Wiggling Effect A method to improve the depth accuracy is to perform a calibration of the system. The easiest way is to perform extensive experiments acquiring the depth of reference objects at known distances and to generate a look-up table (LUT) establishing correspondences between measured distance and real distance [251]. This way, the true depth can be calculated from the measured depth by interpolation. This method requires an amount of memory that grows with the desired calibration quality and, conceptually, solves the problem by *brute force*. A more elegant solution is that provided by [300], which takes profit of the oscillating shape of the depth error curves to fit a cubic B-spline to it. Retrieving the real depth only requires evaluating a cubic polynomial.

If the systematic error is due to irregularities in the modulation/demodulation process, i. e., a non-sinusoidal modulation of the light plus a non-sinusoidal demodulation at the PMD pixel, then the well-known *wiggling* effect will appear in the depth measurements. This effect is due to the non-null contribution of the harmonic content of the light signal to the measurements and consists in a slightly non-linear relationship between real depth and measured depth, which shows oscillations, as observable in Fig. 2.3b for the ToF sensors analyzed (all except from the first Kinect sensor). The frequency of the wiggling phenomenon is given by the frequency of the harmonic of greatest amplitude, typically the second or the third, depending on the signal shape. This kind of depth error can be completely eliminated by calibration. Nevertheless, a calibration does not eliminate the causes of the problem, which are well-known, but it only palliates its effects on the depth estimation. The phase retrieval method used by PMD cameras (Eq. 2.19) is derived from the hypothesis of sinusoidal modulation of the illumination or, conversely, perfect sinusoidal reference signal at the PMD pixel. If this is not the case, the phase estimation method is to be adapted to the real situation, even when it comes at the cost of increasing complexity. Otherwise, part of the power of the illumination and the reference signals will be wasted in distorting the cross-correlation values, which will generate large systematic errors that have to be removed afterwards by calibration. This concerns the efficiency of the system. Part of the limited dynamic range of the PMD pixels is wasted in storing the undesired contribution of high-order harmonics.

Illumination Irregularities In addition to the wiggling effect, there is a number of other issues that translate into depth errors. If many light sources are used to illuminate the scene, at each point of the scene the light signal is given by the sum of the light signals coming from several, if not all, light sources. That is, the signal shape, apart from being non-sinusoidal, is 3D-space dependent. Consequently, if it is desired to eliminate the subsequent depth error by calibration, such calibration has to be performed independently per pixel for a given FOV. If the optics of the camera are changed, the per-pixel depth calibration is to be repeated. Alternatively, if the FOV is reduced, the new map of calibration parameters can be generated from the existent by interpolation, but if the FOV is increased, obtaining the new calibration map by extrapolating the original one might not lead to a correct calibration.

In medium and large range ToF systems, many high-power LEDs are often used to build the illumination system. In such systems, special care is put in synchronizing the LED modules. The tracks that conduct the signals in the circuits for signal amplification and LED driving are of equal length, ensuring that all signals are equally delayed. The same applies to the wires that transport the signals between circuits. An extensive analysis of the medium-range illumination system presented in [283] has shown that, even equalizing the signal paths and using LEDs and auxiliary components of the same model, perfect synchronization is not achieved and relative phase shifts appear between the light signals emitted by different LEDs. At a modulation frequency of 20 MHz, the time delay induced by the illumination system is determined for each LED. The average of delay is around 6 ns, which means a depth offset of 90 cm. This kind of signal-propagation depth offset is always present in any ToF system and does not lead to depth errors because it is removed by a simple linear depth calibration. What generates depth inaccuracies is the slightly different delays registered for different LEDs, which is, on average, 0.66 ns. This asynchronism, which might be thought negligible at a first glance, means 9.9 cm in depth. The details of the study are omitted here for brevity, but a brief summary with the main results is provided in Appendix A.6. The asynchronism, together with slight variations in the waveform, lead to a unique depth error distribution in 3D-space, where the error does not only depend on the depth itself, but also on the 3D location of the scene point for which that depth was measured. Two points of the scene being at the same distance from the sensor can give rise to depth measurements that are, not only different from the real depth, but also different from each other, with differences lying in the centimeter range. The required calibration would have to be three-dimensional, adding two more dimensions to the depth itself, e. g., elevation and azimuth, in a polar coordinate system. Supposing that the complex illumination system is fixed and the camera can be displaced, if only the central pixel is considered, the system required to carry out the data collection for such calibration cannot be a 1D translation unit, such as that used in [251] and [282], but a 3D one.

The depth errors related to non-sinusoidality of the waveforms and inhomogeneous illumination, not just in terms of intensity, but also in terms of signal shape and delay, can be calibrated up to a certain extent. Other system-related error sources cannot be calibrated, unless additional sensors are included in the camera. For instance, the dark current, which can be included in Eq. 2.29 in the offset term, A_0, increases dramatically with the temperature, doubling its value each 8 °C, according to [330]. LEDs are also temperature-dependent and, in consequence, the shape and delay of the illumination signal with respect to the ICS change with the LED temperature, which depends on the ambient temperature and the time the LED has been in operation.

Interference Between Optical Signals

Simultaneous operation Other external source of depth error is the interference between the reflected optical signal and other unexpected modulated light signals. An example of such situation is the simultaneous operation of two or more PMD cameras with overlapping FOVs. In this case, methods such as temporal division, wavelength division, frequency division or code division have been analyzed in [304] to avoid degradation of the depth accuracy.

Temporal division simply consists in acquiring at different times and, therefore, totally avoids interference. The main advantage is also a major inconvenient, since it makes simultaneous acquisition impossible. Additionally, a synchronization mechanism is required. This method also makes the frame rate inversely dependent on the number of cameras sharing FOV.

Wavelength division does not require any synchronization, since it consists in making each camera operate with light of different wavelengths. This requires a coherent modification of both illumination systems and camera filters. In order to avoid any overlapping, relatively well-spaced wavelengths in combination with narrow filters are to be used. If the number of cameras is large, the NIR region of the spectrum might not suffice and visible wavelengths may have to be used, which might be incompatible with the application. As already commented in Section 2.3.3, the spectral response of the silicon is not flat and using short visible wavelengths may compromise the sensitivity.

Frequency division consists in using orthogonal frequencies for the different cameras. The cross-correlation between periodic signals of orthogonal frequencies is zero, so the interference is eliminated. Unfortunately, in real systems the reference signal at pixel level is not sinusoidal, but a close-to-square (trapezoidal) signal, which contains more than a single frequency. For this reason, frequency multiplexing might still produce distortion in depth measurements unless the frequencies are largely different. Apart from that, if each ToF camera is operating in single-frequency mode, the use of different frequencies will lead to different unambiguous ranges and accuracies for the different cameras. Alternatively, one could use very closely-spaced frequencies, using integration times that are higher than the period of the corresponding beat signal, in order to get an amplitude that is independent from the starting time. In [304, 302] frequency differences of several kilohertz are used to achieve simultaneous operation of several MultiCams.

Of special interest is the case of PN modulation [75], using different orthogonal codes for different cameras at a time [67]. This method does not impose any restriction on the acquisition, apart from ensuring that different codes are used for modulation in different cameras at the same time. As well as in the case of frequency division, the orthogonal illumination signals lead to a null contribution in the cross-correlation measurement, obtained as difference between pixel channels, but increase the BGL level measured by both pixel channels. This means wasting part of the dynamic range of the pixels.

Multipath Another example of unexpected interference between two or more optical signals is the multipath effect. Multipath interference (MPI) can easily cause depth errors ranging from tens of centimeters to meters in conventional ToF cameras. In short terms, the multipath effect occurs when a single pixel receives a combination of two or more reflected optical signals, which are not in phase. This can occur due to strong secondary reflections, caused by a shiny floor or walls, especially in corners [208]. A schematic representation of the effect is depicted in Fig. 2.7a. As a result, the depth measurement will be different from the real depth, the error being given by the relationship between the amplitude of the principal reflection and the amplitude(s) of the secondary reflection(s). This effect is scene-dependent and, therefore, could be seen as a *chicken and egg* problem, where the multipath cannot be corrected unless the scene structure is known and the scene structure cannot be retrieved correctly due to MPI. For this reason, most multipath-removal methods are optimization problems that jointly estimate the scene structure and the multiple paths followed by the illumination beam.

Another kind of multipath is light scattering, due to low quality optics in combination with objects that are too reflective or to close to the camera. As a consequence, the rest of the scene appears closer to the camera. In [192] a method was proposed to estimate the MPI in ToF cameras from the MPI-affected depth measurements and correct them. The method deals with diffuse multipath and was extended in an iterative fashion to deal with scenes with non-constant reflectance in [193]. Another iterative method for diffuse multipath is presented in [243], consisting in sequential scene reconstruction and ray tracing steps. In each iteration, the model of the scene is updated, until the error between the original depth map and that obtained from the scene model by multipath-affected ray tracing is lower than a threshold. Despite the impressive error reduction (e. g., around 70% in [243]), these methods [192, 193, 243] have runtimes of one to ten minutes per frame and deal exclusively with diffuse multipath, arising from perfect Lambertian reflectors. Multipath due to scattering depends also on the pixel design. Pixels with large metal-shielded areas (exhibiting high reflectivity), in combination with a filter or lenses without anti-reflective coating, lead to internal reflections that are equivalent to secondary paths. This case, together with the general diffuse multipath due to Lambertian reflectors, is illustrated in Fig. 2.7b.

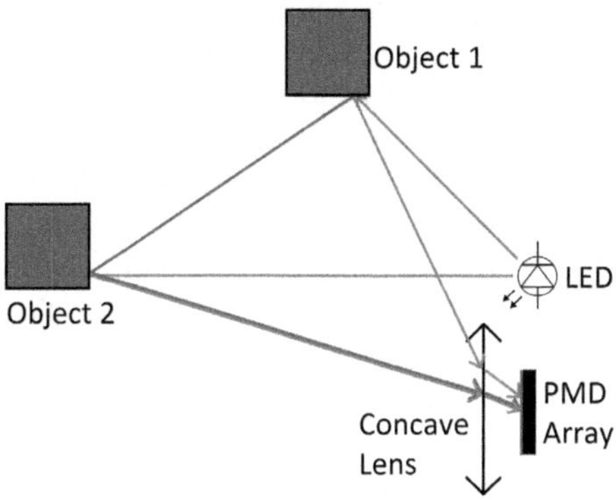

(a) Reflective Multipath

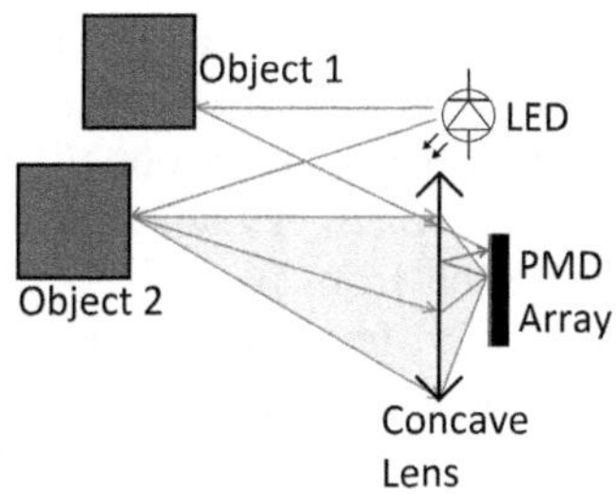

(b) Diffuse Multipath

Figure 2.7.: Schematic representation of multipath interference in a ToF camera. (a): Two-path interference due to secondary reflection. The light rays coming from the light source (LED in the figure) are depicted in blue. The expected reflections, collected by the camera lens are in green. Object 2 receives light from the source but also from Object 1, generating a second reflection, delayed with respect to the first. This unexpected second path is depicted in red. (b): Diffuse multipath. Even in absence of clear secondary paths as in the left schema, diffuse multipath arises from objects whose surfaces behave as Lambertian reflectors. This means that the light collected by the lens for a single scene point undergoes slightly different paths, with slightly different lengths. Additionally, if Object 2 is very reflective and close to the camera, most pixels of the array will be receiving a certain contribution of the light reflected by it. The strength of the effect depends on the quality of the optical system. Optical elements without anti-reflective coating might generate internal reflections (in red).

In the case of highly-reflective or specular (non-Lambertian) surfaces, several modulation frequencies are often used in order to estimate more than one depth per pixel in a closed form, without the need of scene reconstruction and ray tracing. This is the case of the method in [261] and the first method presented in [203]. Alternatively, the multiple paths can be retrieved through an optimization process, as in [153]. The methods in [261, 203, 153] cope with only two paths per pixel. The method in [153] uses only two frequencies and, therefore, solves a determined system. The closed solutions of [203] require three or four different frequencies, while the method in [261] uses five frequencies.

If more than two reflective paths occur per pixel, none of the previous approaches offer an effective solution. The simple multifrequency approach for resolving the MPI presented in [45] deals with non-unique reflections per pixel, assuming sparsity of the reflectors in depth domain, i. e., only very few return paths per pixel. The extremely sparse vector of reflected amplitudes (zero elements for all depths where no object reflects the signal) is recovered from partial Fourier measurements in a classical compressive sensing (CS) framework. The sparse signal recovery is carried out via OMP, i. e., a pseudo-l_0-minimization, with full *a priori* knowledge of the signal sparsity. Appealing due to its simplicity, the feasibility of this approach is compromised by the large number of modulation frequencies required (77 for known sparsity of 3). Additionally, if multipath arising from specular surfaces occur in combination with diffuse multipath, the assumption of exact sparsity does not hold and the number of measurements required to achieve a similar reconstruction is expected to grow. A more feasible approach is provided in [43], where k interfering paths are estimated from $2k + 1$ frequency measurements in a closed-form manner. The performance of the method, which does not require phase sampling, is evaluated with real data from an Xbox One sensor. In the experiment, 21 measurements are used to separate only two paths. The frequency domain ToF framework proposed in [249] allows for easy separation of multiple paths, at the cost of requiring large modulation bandwidth of the illumination system.

The approach presented in [189], named Sparse Reflections Analysis (SRA), is a general framework that allows both diffuse and reflective multipath estimation, exploiting the underlaying sparsity of the reflectors in depth domain. This approach discretizes the depth space and supposes that the signal received at the pixel is a sparse combination of returns from the different discrete depths. Even for low dimensionalities, the vector containing the return intensities for the different discrete depths meets the sparsity assumption, since most depths correspond to free space. If

Lambertian reflectors are considered, the vector of backscattering intensities is compressible, instead of sparse. Even in that situation, according to CS theory (Chapter 3), the sparse vector might be recovered from few measurements by l_1-minimization. This method has been demonstrated with the Xbox One sensor [471], which uses three modulation frequencies. It allows multipath removal in real time. A similar approach has been simultaneously presented in [221], where the concept of multipath is extended to scattering media, where the medium produces a number of different paths per pixel. Since scattering media are a continuum, the sparsity hypothesis in the Dirac basis does not hold and an overcomplete dictionary of exponentially-modified Gaussian functions is adopted to adapt the model and recover the sparsity of the vector of scatters. The signal received at each PMD pixel is a sparse convolution of the emitted signal convolved with some of these functions centered at different depth locations. Depending on the width of the Gaussian and the tail of the exponential, these functions can equally model specular (close-to-Dirac function) and Lambertian (non-negligible exponential decay) reflections. Similarly to [189], the signal sparsity allows solving the problem by l_1-minimization.

In general, transient imaging can be understood as way of multipath estimation. Consider, for instance, a scene that returns multiple reflections (still few) for each pixel of the ToF camera. Recovering the sparse time profile means estimating the locations of the impulses that form the scene response, that is, at which depths the reflectors are located in the scene. Consequently, one could say that multipath depth estimation is a specific case of transient imaging. For this reason, works using a PMD sensor for transient imaging, such as [248] and [220], implicitly provide a tool for multipath ToF imaging. In [248], PR codes with broad frequency spectrum and an autocorrelation function that approaches a Dirac delta function are used as modulation signal. Different measurements are acquired at different shifts of the signal, as in a conventional PMD sensor, with the peculiarity that PR codes are used as ICS and reference signal, instead of a periodic signal with close-to-zero bandwidth. The scene responses are recovered exploiting their sparsity directly in time (or, equivalently, depth) domain. This assumption might not hold if the MPI is diffuse, in which case an appropriate dictionary is to be used to sparsify the scene response functions [221]. The approach in [220] sticks to the sinusoidal modulation, originally employed in PMD sensors, instead of using PR codes. Measurements are taken at different frequencies and phase shifts. Differently from [248], where only the sparsity prior in temporal domain was used to recover the scene responses, in [220], both temporal and spatial regularization terms are included in a minimization formulation to recover the scene responses for the different pixels of the image. The spatial regularization terms are functions of the vertical and horizontal image gradients, motivated by the expectation of gradient sparsity not only in temporal domain, but also in spatial domain. One more prior related to how well the estimated time profile fits a model is also included in the cost function to minimize. The model considers Gaussian shapes for surface reflections and exponential decays for subsurface scattering. The problem is split in two simpler subproblems, namely, solving the minimization for the time profile to recover, supposing known model parameters, and estimating the model parameters supposing the time profile known. The coupled subproblems are then solved via the Chambolle and Pock primal-dual framework [98].

Differently from [248] and [220], which operate in time domain, the approach in [298] translates transient imaging to Fourier domain and obtains the time profiles by means of an inverse Fourier transform and subsequent post-processing. Similarly to [298], in [367] the measurements provided by a ToF camera are interpreted as the Fourier coefficients of the scene response. The reconstruction of the scene response in phase (or ambiguous depth or time) from the Fourier coefficients is interpreted as solving a trigonometric moment problem [252, 271]. Obviously, even for moderate time resolutions, the problem is underdetermined, unless an unfeasibly large number of frequencies are used. The authors solve this issue by looking for the solution that, fulfilling the measurements, exhibits minimal Burg entropy [64]. Using such prior allows for a closed-form solution, known as the maximum entropy spectral estimate. Alternatively, if the scene response is known to be exactly sparse, as in the case of specular multipath, the Pisarenko estimate optimally reconstructs k Dirac delta functions, i. e., $2k$ parameters, from k complex phasors (cf. [43], where $2k + 1$ measurements are required).

A single-shot transient imaging system able to capture non-repetitive events at frames rated up to 10^{11} frames per second is reported in [198]. The system operates with a single illumination pulse and, therefore, can be considered as a pulsed ToF camera. A CCD camera is used for sensing and resolution in time domain is attained by means of a DMD in combination with a streak tube. The system trades spatial resolution in the shearing dimension for temporal resolution. Such a system natively allows for resolving multiple bounces per pixel. The transient video is recovered from the compressed measurements in a CS framework.

Towards the end of the writing of this thesis, a new field has emerged that combines transient imaging with light transport analysis to achieve robust separation of the individual transport components (e. g., direct reflections, inter-reflections, caustics, etc.). While approaches oriented to determine or approximate the light transport matrix make use of different 2D patterns that are projected onto the scene, transient imaging approaches project time-coded light using different patterns. Combining both approaches means projecting 3D patterns onto the scene, being the light modulated both in spatial (e. g., using an SLM) and temporal domains simultaneously. Obviously, this would imply adding one more dimension to the light transport matrix. The core motivation seems to be getting rid of strong assumptions about the scene backscattering (e. g., sparsity or compressibility of the vector of backscattering amplitudes in depth domain), required when operating exclusively in temporal frequency domain, which might be violated in real scenes. For instance, it is known that direct transport preserves high spatial frequencies, while sub-surface scattering and diffuse inter-reflections have a low-pass filtering effect in spatial domain. In [359], sinusoidal temporal modulation is adopted and the scene is *probed* with different modulation frequencies, yielding a different complex-valued *transient frequency transport matrix* per frequency. The main hardware components are a PMD 19k-S3 sensor and a DLP LightCrafter projector using 650 nm laser diodes as illumination source. Computing the depth from direct/retroreflective images allows getting rid of the effects of diffuse MPI. The pure direct/retro-reflective image is computed subtracting the indirect-only image from a conventional PMD acquisition, containing both components. The indirect-only image is obtained by means of complementary 2D binary patterns in emission and reception of the light, so that direct paths are blocked (see [360] for the underlying concept of primal-dual coding). Spatial modulation of the emitted light is carried out by the DLP projector, while the complementary masking at reception is performed computationally, avoiding the need for a second spatial modulator. A similar idea is exposed in [209], where a phasor formulation is also adopted for the light transport model when using sinusoidally modulated illumination. The radiance at the pixel surface is the sum of direct radiance, due to direct reflection from the correspondent scene point, and global radiance, due to global light transport. The phase of the direct component contains the true depth information, while the global component contains the contributions due to MPI. Differently from [359], they exploit the fact that, for a broad range of scenes, global radiance decreases with increasing frequency. Performing the separation by temporal frequency, instead of by spatial frequency, the method in [209] avoids the need for spatial modulation/demodulation hardware. On the other hand, profiting from the vanishing effect of high modulation frequencies in the global component of the radiance means prohibitively large modulation frequencies (600 MHz to 6 GHz) when the scenes contain small geometrical features, which could be handled by MSM-PMD, but not by conventional PG-PMD technology. A direct consequence of using high modulation frequencies is the need for several frequencies to achieve an acceptable unambiguous range. In their experiments a PMD CamBoard Nano (featuring a 19k-S3 chip) with a bank of 650 nm laser diodes for illumination is used. The PMD chip limits the maximum frequency to the 100 MHz range, which, in turn, limits the effectiveness of the method to large scenes. A closed-form formula for computing MPI-compensated phase shift under single indirect bounce hypothesis is provided in [338]. The radiances due to direct and global illumination are separated using the fast method proposed in [341]. The performance of the method is demonstrated using an Xbox One sensor and a DLP Lightcommander for spatial pattern projection.

2.4.2. Lateral Resolution

The lateral resolution may be bounded by the number of pixels or by the quality of the optics. Professional photographic cameras feature sensors with often more than 20 Mpx, pushing the limitation of lateral resolution towards the optical system.

In a conventional camera setup, the limit in lateral resolution imposed by the lens can be characterized by its point spread function (PSF). If the optical system is free of imperfections, the resolution limit is given by diffraction and the system is said to be *diffraction limited.* In optics, a spot of light optimally focused by a perfect lens of circular aperture generates an Airy pattern on the surface where it is focused. The Airy pattern consists in a central bright disk, also known as Airy disk, surrounded by concentric bright rings of decaying intensity. The pattern exhibits axial symmetry and its intensity value is given by Eq. 2.33,

$$I(x) = I_0 \left(\frac{2J_1(x)}{x} \right)^2 \tag{2.33}$$

where I_0 is the maximum intensity, J_1 is the Bessel function of the first kind of order one and $x = kr_{\text{lens}} \sin\theta$, being k the wavenumber of the radiation, r_{lens} the radius of the circular aperture and θ the angle of observation. All points of the pattern at the same distance from the origin share the same observation angle and, therefore, the same intensity. For completeness, the normalized amplitude (recall that the intensity is given by the square of the amplitude) of the Airy pattern is represented in Fig. 2.8a. The resolution limit can be determined from the radius of the first Airy dark ring, given by Eq. 2.34 [465],

$$r_{\text{Airy}} = 1.22\frac{\lambda}{2NA} \tag{2.34}$$

where NA is the numerical aperture of the optical system and λ is the wavelength of the light. Clearly, larger wavelengths lead to larger Airy radius. In most cases, even using NIR light, the Airy radius lies in the micrometer range. Depending on the camera this may be higher or lower than the pixel size. In the case of a PMD camera, with a pixel size of $45 \times 45\,\mu$m (19k-S3), the lateral resolution is limited by the pixel size, unless the PSF of the optical system exhibit an abnormal width due to defocus or some system imperfection.

When the lateral resolution is given by the number of pixels (or, equivalently, by the pixel size, provided that the maximum sensor area is constant), two factors are to be taken into account: the pixel geometry and the FOV of the lens. Narrowing the FOV finer detail of the scene can be resolved with the same number of pixels. If the height and width of the pixels are different, the vertical and horizontal resolutions will also differ. The PMD pixel geometry for the specific case of the 19k-S3 chip is exposed in detail in Section 4.3.1. Here we only provide a general scheme of the two pixel geometries found in the chip, in Fig. 2.8b. The scheme has been generated from microscopic images of the pixel surface. Despite the regions do not perfectly correspond to physical boundaries, the pixel geometry and the relation between active and non-active areas is real.

Four regions per pixel are depicted in Fig. 2.8b, corresponding to the surroundings of four charge accumulation regions: two for the channel A and two for the B. In case of perfect demodulation capabilities, the location where a photon reaches the pixel should not determine in which integration channel the corresponding photogenerated charge carrier is stored. This should be only determined by the control signal. For now we adopt this hypothesis and leave further analysis for Section 4.3.1. Then, the active area of the pixel is the same for both pixel channels and equal to the colored regions in Fig. 2.8b (in red and blue in the color version, dark gray and light gray in the black and white version), while the rest of the pixel (in gray) is *blind*, i. e., non-active. The width of the active regions is calculated to be $25\,\mu$m, by simple proportionality, knowing that the size of the pixel is $45 \times 45\,\mu$m. This coarse model delivers an (optimistic) approximate fill factor of 56% and is enough to make clear that a PMD sensor will exhibit different behavior vertically and horizontally. While some scene structure that results

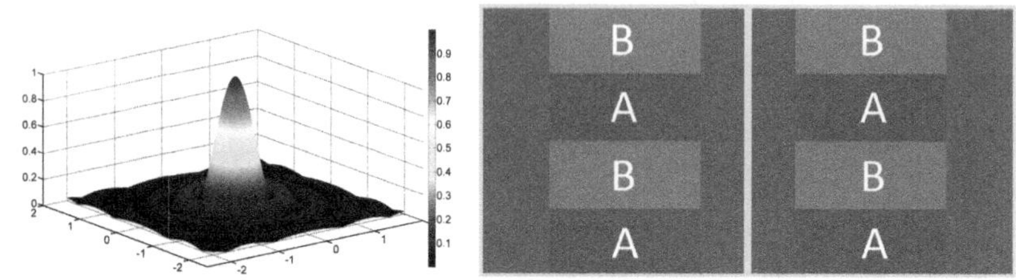

(a) Airy Pattern (b) PMD Pixels

Figure 2.8.: Normalized amplitude of an Airy diffraction pattern (a) and schematic representation of the active areas of PMD pixels in a 19k-S3 sensor (b). The axis units in (a) are arbitrary. The radius of the first Airy dark ring is, at maximum, of few micrometers, while the size of the square total pixel area (delimited in green in (b)) is $45 \times 45\,\mu$m. The gray areas in (b) are non-active areas, typically metal-shielded. For clarity, the active areas of each pixel are divided in A and B regions. A regions are centered at an A-gate readout wire, while B regions at a B-gate readout wire.

in a thin horizontal line in the image plane will be sensed—still, depending on the width, not fully resolved—, if it results in a vertical line, it might not be sensed at all, if it falls along the blind areas of the pixels. In terms of lateral resolution, if the pixel is considered as a whole, i. e., square, there is no change, but if only the active area is considered, the lateral resolution in horizontal direction is lower than in vertical direction. This means that one could either perfectly resolve or not sense at all finer structure in the horizontal direction. If the width of the non-active areas were constant and equal to the width of the active areas, splitting the image to two PMD sensors, horizontally displaced one width between each other, would be enough to obtain a PMD camera of double lateral resolution in horizontal direction. In Table 2.1 we provide values of the FOVs, angular resolution and corresponding lateral resolution at 1 m distance for some common focal lengths of the camera lens. The table columns are organized in pairs. The first pair contains the resolutions (horizontal or vertical) obtained using a pixel size of pixel is $45 \times 45\,\mu$m, while the second pair contains the horizontal resolution considering a pixel width equal to its active area ($25\,\mu$m).

In general, the lateral resolutions presented in Table 2.1 are poor, due to the low number of pixels in the chip (120×160). The lateral resolution degrades linearly with the distance to the scene. Consequently, unless

		Square Pixel Model		Rectangular Pixel Model	
f (*mm*)	FOV (°×°)	Ang. Res. (°)	Lat. Res. 1 m (*mm*)	Ang. Res. (°)	Lat. Res. 1 m (*mm*)
8.5	45.9 × 35.2	0.29	5.1	0.16	2.8
12	33.4 × 25.4	0.21	3.7	0.12	2.1
35	11.7 × 8.8	0.074	1.29	0.041	0.72

Table 2.1.: Field of View (FOV) and angular resolution of the PMD 19k-S3 sensor for different lenses of common focal lengths (f) and corresponding lateral resolution at 1 m distance. The resolutions obtained using the rectangular pixel model are theoretical horizontal resolutions, obtained using the width of the active area as pixel width.

extremely narrow FOVs are desired, the lateral resolution will easily be in the centimeter range at few meters distance. This also has an impact in the final depth accuracy, since it means that the pixel is measuring the average depth of a relatively large area of the scene. If the scene is, e. g., a plane at 45° with respect to the image plane, poor lateral resolution directly translates into an equivalent depth error, generating a staircase effect.

One could argue that the lateral resolution should be normalized by the size of the sensor, but even in that case, due to the pixel size, PMD sensors offer linear densities of 22.2 px/mm, while commercial photographic cameras range from 80 to 130 px/mm. Therefore, the low lateral resolution for the current chip size is an open issue in PMD ToF cameras.

2.4.3. Dynamic Range

The dynamic range of PMD pixels has shown to be around 70 dB (see corresponding paragraph in Section 2.3.2). Despite being relatively large, it is still below those of high-end digital cameras, close to 90 dB. The dynamic range is determined by the noise floor of the pixels and the maximum charge level they can store. Such parameters depend on the pixel design and fabrication technology and are expected to improve in new generations of PMD pixels.

We performed extensive experiments to determine the dynamic range of the pixels in a commercial PMD 19k-S3 sensor[5], which is the sensor included

[5] Note that PMD chips of newer generations, manufactured by Infineon, may exhibit significant improvements with respect to the PMD 19k-S3 we used for most experiments in this thesis, for instance, but not exclusively, in terms of dynamic range.

in the latest MultiCams and used in most experiments in this work. In our experiments, 100 different exposures were considered, ranging from 0.1 to 10 ms. For each exposure, 100 acquisitions were carried out. Since the PMD cameras operate with four phases, this means 4×100 pairs of A and B images. Two datasets with two different setups were acquired: one for determining the noise floor and another to determine the maximum value. In the first one, the camera has no optics and an opaque cap ensures that no light reaches the sensor. Additionally, the experiment was carried out in a dark room with no illumination. In the second setup, the camera is equipped with a lens and capturing a white wall, parallel to the image plane. The illumination is strong enough to drive most pixels to saturation in the cases of high exposure. It is important to note that the PMD pixels of the 19k-S3 are equipped with SBI system and, therefore, the saturation level is never achieved. Instead, the SBI does not allow the level of the pixel channels to exceed a close-to-saturation threshold (an exemplary value of $36,500$ is suggested in [284]).

All the images of all datasets are calibrated according to the per-pixel linear calibration described in Section 4.2.1. This allows averaging the results obtained for each pixel over the 120×160 pixels of the chip. The noise floor, I_{floor} in Eq. 2.17, is considered to be the standard deviation observed along the 100 acquisitions gathered per exposure time considered. The I_{floor} for a generic PMD pixel is obtained averaging the standard deviations for all pixels and exposures considered. The value of I_{ceil} in Eq. 2.17 is obtained by simply finding the maximum value for any of the pixel channels in the corresponding dataset.

Our experiments concluded that the dynamic range of a pixel channel of a PMD 19k-S3 array is 67.18 dB. This is close to the values presented in previous work, still lower, probably due to the fact that the SBI activation prevents pixels from reaching saturation level.

The SBI system can be understood as increasing the dynamic range of the PMD pixels, since it enables operation under stronger BGL conditions. In this sense, the SBI can provide an enormous improvement of the dynamic range, but it is essential to realize that the DC component of the light signals is being shrunken. If the goal is just to compute the depth using the four phases algorithm, this is irrelevant, but if we seek an *intelligent* processing of the signal that exploits eventual local correlations or data redundancy, we must be aware that information is being modified and the structure of the raw images might not meet our expectations anymore.

Another issue of capital importance is that the increment of dynamic range that the SBI can eventually provide is strongly scene-dependent, since it is bounded by the discharge process of the less-charged channel. Depending on the depth being sensed by the pixel, the charge of the less-charged channel will range from zero (e. g., at 180° total phase shift for channel A or at 0° for B) to the same value as the other channel (at 90° or 270° total phase shift). Since the SBI removes the same current from both channels, the asymmetry of charge between channels limits its correct operation. Once the less-charged channel reaches zero (or some close-to-zero value) the difference between pixel channels is no longer preserved and the depth measurement will be erroneous. For clarity, we illustrate the SBI operation in Fig. 2.9, which plots data of a real experiment. Three operation areas are depicted in different colors, from left to right: the linear region, the area of correct SBI operation and the area of incorrect operation.

Channel A (in blue) charges at a higher rate than B (in red). When channel A reaches a threshold, the SBI starts to remove current from both channels, so that the level in channel A does not increase anymore. Consequently, the level in channel B starts to decrease with a negative slope, given by the difference between the slopes of the A and B response curves during the linear region. Clearly, if both channels had the same charge level when the SBI is activated, the slope of both curves in the central area of Fig. 2.9 (correct SBI operation) would be zero. In this situation the SBI would provide a theoretical infinite dynamic range, since, regardless of the light intensity and exposure time, the difference between channels would be preserved. Conversely, if the B channel had a close-to-zero level, which might occur for certain depths in cases of low background illumination, the SBI would provide no significant increment of the dynamic range, since the level of the B channel would go to zero immediately after SBI activation.

A close observation of the response curves in Fig. 2.9 reveals that the curves in the central region are not perfect straight lines. The difference between A and B channels is being distorted due to imperfect SBI operation. Additionally, the slope of the B channel curve starts to decrease before the level reaches zero, corrupting the differential measurement (A-B). The dotted lines extend the linear regions of both curves and should continue out of the graph. In case of ideal operation, if a vertical line is traced at any exposure time, the length of the segment between the dotted lines and between the real responses should be equal, which is clearly not true for exposures in the second half of the exposure range.

Due to its scene-dependency, the activation of the SBI produces a characteristic error pattern in the depth images, where some pixels exhibit much larger depth error than others, depending on the charge asymmetry between pixel channels in the measurements, i. e., depending on the depth.

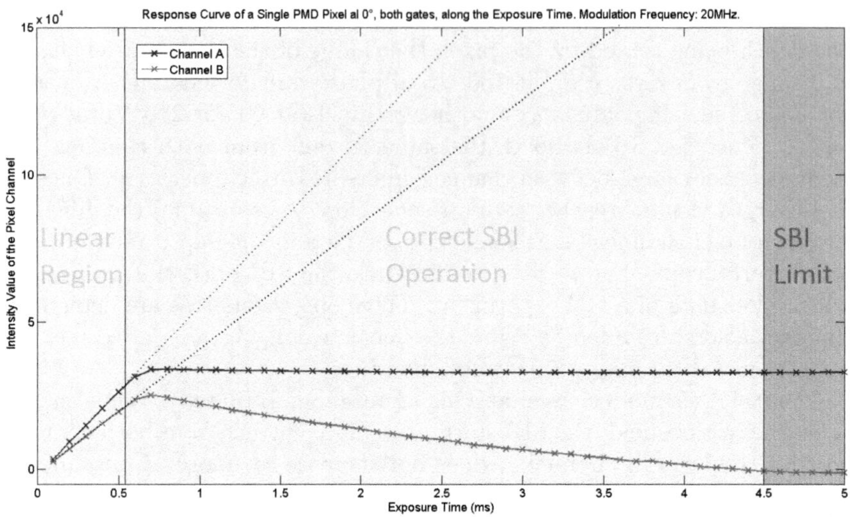

Figure 2.9.: Response curves of the A and B channels of a PMD pixel along the exposure time. The pixel corresponds to an object which is close to the camera, under strong 20 MHz-modulated IR illumination. Note the linear region of the curve, together with the good performance of the SBI, which subtracts the exact intensity to make the channel A charge remain at a constant level. The limit of the correct SBI operation is reached at almost 4.5 ms, where the charge in channel B reaches the zero level. This means that, if this close object is placed in a scene for which a higher exposure time is required, the depth measurements for the pixels of the object will be not just noisy, but wrong. ©2015 IEEE.

3. Fundamentals of Compressive Sensing

In this chapter we provide the reader with some fundamentals on the mathematical theory of Compressive Sensing, often referred as Compressed Sensing. In the following we use the abbreviation CS to refer to it. Very few research fields have undergone such a fast development as CS in the last ten years. The aim of this chapter is to present concepts and methods that will be used, implicitly or explicitly throughout the rest of the thesis. This chapter is, therefore, not a review of the state of the art in the field, but is intended to provide the reader with the necessary tools to understand the work presented in this thesis, why it has been done the way it was and why it works.

3.1. Introduction to Compressive Sensing

In this section we provide an introduction to CS theory. Basic concepts and fundamental results from reference works in CS are briefly explained.

Shannon sampling theorem and Nyquist rate It is difficult to start any kind of introduction to CS without referring to the Shannon sampling theorem and the so-called *Nyquist rate*. The Shannon sampling theorem (Theorem 1 of [413]) states that, if a function, defined in time domain over a certain interval, contains no frequencies higher than $f_{\max}$, it is completely determined by equally-spaced temporal samples at a rate $2f_{\max}$. The proof given in [413] is based on the fact that the temporal samples gathered at this rate suffice to determine the Fourier transform of the signal, provided that it is only non-zero in the frequency interval $[-f_{\max}, f_{\max}]$. Consequently, the original signal in temporal domain is also fully determined. Shannon provides a formula to reconstruct the continuous bandlimited function from its samples, which is a weighted sum of sinc functions, centered at the sample points, where the weight is the value of the function at that point. The formula appeared first in his *Mathematical Theory of Communication* [412] (Theorem 13).

For many authors, the Shannon sampling theorem had been already implicitly suggested in the works of E.T. Whittaker [463] and K. Ogura [356]. One can argue that Whittaker focused on *interpolation* theory, not on *sampling* theory. Given a set of points, interpolation means looking for an interpolant, i. e., a function of a certain class or with certain properties that passes through theses points. Differently, in sampling what is given is a function class, e. g., bandlimited functions, and what is desired is to find a representation in sampling series. Note the duality between these two concepts, nonetheless. In [463], the so-called *cardinal function* is presented as a well-behaved function (a Paley-Wiener function) that interpolates a set of equidistant sample points of the original function. Whittaker uses the Fourier integral theorem to show that all periodic constituents of frequency higher than $f_{\text{sample}}/2$ are null, being f_{sample} the rate at which the points to interpolate are acquired. This constitutes the link to the Shannon sampling theorem, since it implies that the *cardinal function* that interpolates the sample points acquired at a rate f_{sample} is bandlimited to a maximum frequency $f_{\max} = f_{\text{sample}}/2$.

Ogura also uses the term *cardinal function*, but to refer to the class of B_σ^∞ functions with $\sigma < \pi$, where B_σ^n denotes a Bernstein space. His sampling theorem [356] states that such functions can be constructed analytically from a set of values obtained at equidistant points.

For a discussion on how these early works relate to Shannon sampling theorem and posterior related research, we refer to [73]. One could classify as related even previous, purely mathematical, works. For instance, C. Carathéodory [95, 96] showed that $2k + 1$ *samples* suffice to fully determine a weighted sum of k sinusoids with positive weights. The sampling rate suggested by Shannon is often known as *Nyquist rate* and the sampling theorem often referred as *Nyquist-Shannon*. This is due to the fact that H. Nyquist had previously pointed out that approximately $2Tf_{\max}$ numbers are sufficient to represent a bandlimited function of maximum frequency $f_{\max}$ over a time interval T [353]. Equivalent results were reported by D. Gabor [194].

As a consequence of these very fundamental results on information theory, sampling at twice the bandwidth of the signal became a *rule of thumb* for engineers to avoid information loss at sensing. Many natural and man-made continuous signals are, indeed, bandlimited and can, therefore, be recovered by equally-spaced samples acquired at the Nyquist rate, at no cost in terms of information loss. One should notice that the Shannon sampling theorem was obtained under the hypothesis of regular sampling and bandlimited signal. Shannon did not explicitly discard that different sampling schemes, together with stronger hypothesis on the signal, e. g., deep *a priori* knowledge on the signal structure, could not be used to obtain sharper bounds on the minimum number of samples that allows lossless sensing.

Compression at sensing In recent times, the vertiginous development of electronics has led to an exponential growth of the volume of acquired data. Common examples are found in costumer electronic devices, such as digital photographic cameras or video cameras, where the resolution of the imaging sensor keeps growing year after year. Such devices typically perform equidistant sampling, as contemplated by the Shannon theorem, but need to perform a compression of the data right after sensing, before the data is recorded or even transmitted. This unveils a fundamental problem: on one hand, high sampling rates are required to recover high frequencies, be temporal frequency, as in the paradigmatic case of communications, or spatial frequencies, as in digital image sensors. On the other hand, the volume of data becomes excessive, both for data transmission and storage. Carrying out a compression after sensing implicitly suggests that the effort put at sensing by means of large sampling rates is partially useless, that is, the sensing scheme is not optimal. In other words, the sensor gathers redundant information.

The concept of *compressive* (or *compressed*) *sensing* denotes a novel sensing paradigm, in which the compression is performed *at* sensing and not *after* sensing. In other words, the signal is sensed in a compressed manner, so that each measurement gathers non-redundant information. Differently from the Shannon sampling theorem, that requires the signal to be bandlimited, CS theory is somewhat more general and requires the signal to be *sparse* in some known basis. Depending on the signal sparsity, the number of measurements required to achieve lossless reconstruction might be much lower than that suggested by the Nyquist rate.

Fundamental works deploying the basis of CS theory that constitute an excellent introduction to the field are those of E.J. Candès [79, 82, 90, 91, 92], R.G. Baraniuk [24] and D.L. Donoho [147]. In general, these fundamental works state that a finite-dimensional signal that admits a sparse or compressible representation in some known basis or tight frame can be exactly recovered from few non-adaptive measurements, under certain conditions related to the incoherence between the sensing process and the sparse representation. Essential works on the subject are the *Mathematical Introduction to Compressive Sensing* [184], from S. Foucart and H. Rauhut, and *Compressed Sensing* [163], from Y.C. Eldar and K. Kutyniok, [135] and references therein.

Despite CS as we know it today was born around 2006 with the aforementioned works, one can recognize previous works in the literature that already pointed in the CS direction. For brevity, we do not provide here a historical review and refer to the introduction in [163]. A paradigmatic problem in CS is, for instance, to recover of a sparse signal in time domain from partial Fourier observations. A. Beurling proposed a method to cope with this problem in 1938 [41], extrapolating the partial observations to determine the full Fourier transform. Obviously, depending on the number of measurements, this might be a massively underdetermined problem. Beurling, similarly to modern CS approaches, sought the solution that, agreeing with the measurements, had minimum l_1 norm. Different priors on the signal to recover yield different solutions, which might provide a better approximation of the real signal. An alternative approach is to look for the solution that, fulfilling the measurements, exhibits minimal Burg entropy [64], known as the maximum entropy spectral estimate. Signal recovery from partial Fourier measurements is still an actual topic nowadays, highly related to the type of sensors we deal with in this thesis. As an example, consider the recent works in transient imaging using ToF cameras [220, 298, 367] (see paragraph on multipath in Section 2.4). A recent notable precursor of CS is [446], where it is shown that signals with finite rate of innovation can be recovered from $2k$ measurements, where k is the so-called rate of innovation or degrees of freedom of the signal (e. g., number of Dirac delta functions in a stream of finite length).

There are several fundamental differences between the works on classical sampling and CS. Classical sampling theory typically deals with continuous, arbitrarily large, signals in time domain. In fact, it is common to find in the literature assumptions such as data streams of infinite length. Differently, CS theory typically considers the signal to be discrete and of finite length, e. g., a finite-dimensional vector. The concept of sampling also differs between theories. While classical sampling means gathering values of the signal at regular time intervals, in CS the *measurements* are not exactly *samples*, but the inner product between the signal to sense and some sensing function (or *sensing kernel*, as it is often named in the discrete case), defined in the same domain as the signal. While in classical sampling regularity is the rule (equidistant samples), in CS, *randomness* of the, in principle, non-adaptive, sensing functions is a key point of the approach. One last fundamental difference is how the signal is recovered. As pointed out above, Shannon recovers the signal by sinc interpolation, which is a simple step. Sparse recovery from the *compressed* measurements constitutes one of the central pillars of CS and is typically achieved by highly-nonlinear methods, iterative in most cases.

Sparsity and compressibility Let $\vec{x} \in \mathbb{R}^n$ be a vector. In most works on CS, as well as throughout this thesis, $\vec{x}$ denotes the *sparse* vector to recover. Then, the l_0 norm of $\vec{x}$ is defined as:

$$\|\vec{x}\|_0 := \lim_{p \to 0} \|\vec{x}\|_p^p = |\operatorname{supp}(\vec{x})| \tag{3.1}$$

that is, the cardinality of the support of $\vec{x}$. Then, $\vec{x}$ is called an s-sparse signal if:

$$\|\vec{x}\|_0 \leq s \tag{3.2}$$

in other words, if $\vec{x}$ has, at maximum, s non-zero elements. The sparsity s plays a key role in CS, since the minimum number of measurements required to guarantee exact recovery of the signal with high probability is typically given as a function of s and the signal dimensionality, n [79, 82, 90, 91, 92, 147, 24].

The sparsity requirement as defined above might be too strict, since it enforces $\vec{x}$ to have, at least, $n - s$ elements exactly equal to zero. Closely related to *sparsity* is the concept of *compressibility*. According to the definition given in [180] or [184], a signal $\vec{x}$ is considered *compressible* if its k-term approximation error decays rapidly in k. The k-term approximation error of $\vec{x}$ reads:

$$\sigma_k(\vec{x})_p = \min_{\hat{\vec{x}} \in \Sigma_k} \|\vec{x} - \hat{\vec{x}}\|_p \tag{3.3}$$

where Σ_k denotes the subset of $\mathbb{R}^n$ containing all k-sparse vectors. It has been shown that the best k-term approximation error can be obtained by selecting the support elements corresponding to the k highest coefficients, which is known as thresholding. k-thresholding yields the best k-term approximation, according to Eq. 3.3, $\forall p$. Defining compressibility as how *rapidly* the right hand term of Eq. 3.3 decays in k brings the need of defining what is rapid and what is not. In general, most works assume a power law decay. This translates into the requirement that the absolute values of the coefficients of $\vec{x}$ also decay according to a power law [141] when reordered from highest to lowest, specifically, in the l_2 case, we have that:

$$\sigma_k(\vec{x})_2 \leq C_1 k^{-r} \leftrightarrow |x_i| \leq C_2 i^{\frac{1}{2}-r}, \forall i \leq n \tag{3.4}$$

where C_1 and C_2 are constants and r is the decay ratio. The higher r, the more compressible the signal is, the less information it conveys and, therefore, the less sensing effort is required. In [78] it has been shown that compressible signals can also be reconstructed from a small number of random measurements.

Looking for the sparsest representation: dictionaries Until now, we have considered a discrete signal $\vec{x}$ that is itself sparse or compressible, but most natural signals are not sparse in the domain they are measured. For this reason, in most cases, a transformation is needed, which transfers the signal from its original domain to another in which it admits a sparse representation. Let $\vec{I} \in \mathbb{R}^n$ be the signal in the original domain, i. e., in a non-sparse representation or, in general, a suboptimal representation in compression terms. Then, we define a dictionary $\mathbf{\Psi} \in \mathbb{R}^{n \times n_{\text{atoms}}}$, with n_{atoms} dictionary atoms stacked by columns. If $n_{\text{atoms}} < n$ the dictionary cannot be a basis of $\mathbb{R}^n$, i. e., cannot span the entire space. If $n_{\text{atoms}} > n$, the dictionary may be overcomplete, that is, its columns may still span $\mathbb{R}^n$ if one of them is eliminated. For instance, unions of basis are overcomplete dictionaries. A square dictionary $\mathbf{\Psi} \in \mathbb{R}^{n \times n}$ is a basis of $\mathbb{R}^n$ if its columns are linearly independent, i. e., if $\text{rank}(\mathbf{\Psi}) = n$ and is an orthogonal basis if its columns (or rows) are orthogonal. In general, dictionaries might not be bases of the space, but frames.

A frame is set of vectors that allows representing any vector living in a vector space. Clearly, unless the frame is a basis of the space, it is overcomplete, i. e., redundant. Due to its similarity with the forthcoming *restricted isometry property*, of undeniable importance in CS literature, we provide here the formal definition of a frame: the columns of $\mathbf{\Psi} \in \mathbb{R}^{n \times n_{\text{atoms}}}$ define a frame if:

$$\alpha \|\vec{I}\|_2^2 \leq \|\mathbf{\Psi}^\top \vec{I}\|_2^2 \leq \beta \|\vec{I}\|_2^2, \ \forall \vec{I} \in \mathbb{R}^n \tag{3.5}$$

where $0 < \alpha \leq \beta < \infty$, being α and β the so-called lower and upper frame bounds, respectively. They are said to be optimal when α is the maximum and β the minimum possible values for which Eq. 3.5 holds. A frame is called tight if it satisfies the generalized version of the Parseval's identity, that is, if $\alpha = \beta$. Additionally, if $\alpha = \beta = 1$, it is called *Parseval frame*. Despite overcompleteness opposes to uniqueness of the representation, thanks to it, frames might be able to capture more signal structure with less number of nonzero coefficients.

Suppose that $\mathbf{\Psi}$ is a basis of $\mathbb{R}^n$, then transferring the signal $\vec{I}$ to the sparse domain does not imply any information loss. The dense signal $\vec{I}$ can be rewritten as:

$$\vec{I} = \mathbf{\Psi}\vec{x} \tag{3.6}$$

Such transformation is the basis of compression, that is, finding a basis (or a tight frame) that allows representing the signal in a sparser or more compressible way. Note that, if $\mathbf{\Psi}$ is an orthonormal basis, Eq. 3.6 can be inverted without the need for a matrix inversion and the sparse representation obtained directly as: $\vec{x} = \mathbf{\Psi}^\top \vec{I}$. If $\mathbf{\Psi}$ is not a basis, but a frame, the coefficient vector $\vec{x}$ is not uniquely determined by $\vec{I}$. The Moore-Penrose pseudoinverse, or *canonical dual frame* in frame theory nomenclature, provides the $\vec{x}$ of minimal l_2 norm: $\vec{x} = \mathbf{\Psi}^\dagger \vec{I}$.

Typically, natural signals are highly compressible in an appropriate frame. This means that, despite a number of dimensions that can be very large, the degrees of freedom of the signals are often few, that is, there are few principal *directions* or *components* that model the entire signal variability in the high-dimensional space. This is related to the obvious fact that the signals we deal with do not behave like random noise, but follow certain structure. If a dictionary provides an efficient representation of the structure of our signals of interest, high compressibility, or even exact sparsity, is attained.

Paradigmatic examples are audio or image compression. Sounds can be efficiently represented by few time-frequency atoms. Images of natural and man-made scenes are observed to have wavelet coefficients whose ordered absolute values exhibit an $1/n$ decay law [318]. It is out of the scope of this thesis to provide background on signal compression, but, in general, the power of any natural signal tends to be localized either in the original signal domain (be spatial or temporal), in its frequency domain, or, more generally, in the time-frequency plane. Confront, for instance, an image of a clear sky at night, where there is no dense texture, but many bright spots corresponding to stars, with a dense image of some repetitive man-made structure. It is difficult to estimate which image contains more information without transferring both to a compressible domain. For the first image, the original domain might be adequate, since the signal power is concentrated in specific locations. For the second image, a frequency dictionary might provide the best representation, since most of the power is expected to be concentrated at the fundamental frequency of the structure and some high-order harmonics. In natural images cases of both types and cases falling in the range between them can be found. In general, one can find multiple textured elements which cannot be optimally represented exclusively in spatial domain or frequency domain. For this reason, the most efficient compression techniques represent signals using atoms that cover different regions of the time-frequency plane, where the signal is expected to live (cf. Computational Harmonic Analysis).

Atoms representing low frequencies should have accordingly large supports, while atoms representing high frequencies should vanish much rapidly. A first implementation of this concept is the windowed Fourier transform (WFT). Unfortunately, such *windowed* Fourier atoms are not orthogonal in general and, therefore, do not define an orthogonal basis of the space. The culmination of this idea came in the shape of wavelets, which are functions able to generate orthonormal basis by means of dyadic translations and dilations. Wavelet dictionaries are, therefore, orthogonal basis whose atoms optimally cover the time-frequency plane: many short-support high-frequency atoms for accurate localization and few long-support low-frequency atoms for accurate frequency characterization.

Considering a dictionary to transfer the signal to recover to an appropriate domain, together with the fact that CS applies both to exactly sparse and compressible signals, allows considering very common families of natural signals, such as smooth signals or signals with bounded variations [92]. Smooth signals exhibit Fourier coefficients that decay with the frequency index following an exponential law [151], that is, are compressible in Fourier domain and, in general, in any frequency dictionary. The total-variation norm of a signal can be understood as an approximation of the l_1 norm of its gradient. It has been shown that the absolute value of the wavelet coefficients of a signal of bounded variation decay following an $1/n$ law, when the coefficients are rearranged in decreasing order [112]. A simple proof that Fourier coefficients also decay according the same law was previously given in [431]. For a discussion on how well natural images fit the assumption of bounded variation, we refer to [205].

Section 3.3 outlines the wide range of possibilities for the creation of CS dictionaries, be analytic (Section 3.3.1) or learned from training data (Section 3.3.2), while Section 4.4.1.1 is intended to find an appropriate 2D dictionary in the specific case of ToF imaging.

How to *sense* in a *compressed* manner: sensing matrices Until now we said that the sparse or compressible signal $\vec{x}$ can be recovered from a set of few measurements, but not how such measurements have to be carried out. In fact, this is a key element of CS and how the sensing process relates to the sparse representation determines whether a signal can be recovered or not from the measurements. As most CS-related literature, we also focus on the discrete (finite-dimensional) case, and, therefore, the measurement process is also modeled in a discrete fashion. Let $\vec{I} \in \mathbb{R}^n$ denote the signal to sense in the original domain and $\vec{y} \in \mathbb{R}^m$ the *measurement vector*, containing the m discrete measurements gathered from $\vec{I}$, where typically $m \ll n$. Then we can write:

$$\vec{y} = \mathbf{\Phi}\vec{I} \tag{3.7}$$

where $\mathbf{\Phi} \in \mathbb{R}^{m\times n}$ is the so-called *sensing matrix*, which models how each measurement is obtained. The rows of $\mathbf{\Phi}$ are the discrete *sensing functions* or *sensing kernels*. Note that in Eq. 3.7, if the coefficients of $\vec{I}$ are samples of some continuous function, they have to be according to the Nyquist rate required by the maximum frequency we want to recover. This does not mean that, in reality, a first sampling at Nyquist rate is a compulsory step before compressed sensing, since most real measurement systems will obtain the measurements as scalar product between the continuous $I(\cdot)$ and a continuous sensing function, even if the latter follows a discrete pattern. Nevertheless, the sensing model in Eq. 3.7 has to be able to capture the bandwidth of the signal we want to recover. This has an effect, not just on n, but also on the nature of the sensing functions in $\mathbf{\Phi}$. An universal rule in CS is that one cannot recover what is not being sensed. That is, the bandwidth of the measurement functions has to be, at least, as large as that of the signal we aim to recover. Otherwise, frequencies exceeding the bandwidth of the sensing functions fall into the null space of $\mathbf{\Phi}$.

This is often used as an argument against CS, claiming that the bandwidth requirement imposed by the Shannon sampling theorem cannot be overcome. At this respect, one has to realize that CS is not intended as a collision *against* Shannon, but actually a further step in the direction of Shannon. One could cast, in fact, classical sampling of continuous bandlimited signals at Nyquist rate as a special case of CS when $n \to \infty$. The sparsity is to be exploited in frequency domain, where the signal contains only the first k harmonics of the fundamental frequency. Then, according to Shannon, $2k$ measurements suffice to recover the signal from the measurements. Note that, under these conditions, CS would not do any better than classical sampling, since, in the best case, when the sparse support of the signal is known, we still would need $m = 2k$ measurements to recover the signal, provided that the sparsity of the signal in a Fourier basis is $2k$. CS is wider in its formulation, since it allows using other bases or frames that make possible to attain better signal representation (e. g., sparser) and, consequently, reducing the number of measurements needed to ensure exact recovery. Obviously, if the sensing process or the basis elements are affected by bandwidth limitations, be due to hardware constraints in the former case, or due to too aggressive discretization in the latter, the signal recovered by a CS framework will be affected by a frequency limitation due to Shannon violation.

A review of sensing matrices often used in CS is provided in Section 3.2.

Linear measurement model Combining Eq. 3.7 and Eq. 3.6, we obtain the classic CS measurement model, which is a linear system of the form:

$$\begin{aligned} \vec{y} &= \boldsymbol{A}\vec{x} \\ \boldsymbol{A} &= \boldsymbol{\Phi}\boldsymbol{\Psi} \end{aligned} \tag{3.8}$$

Clearly, the challenge is to design a *measurement matrix* $\boldsymbol{A} \in \mathbb{R}^{m \times n_{\text{atoms}}}$ that preserves the information in $\vec{x}$, while providing a substantial dimensionality reduction (in general, we have that $m \ll n \leq n_{\text{atoms}}$). Provided that the dictionary $\boldsymbol{\Psi}$ is determined by the nature of the signal $\vec{I}$, despite different options might exist for signals of a given class, it cannot be arbitrarily constructed. The sensing matrix models the sensing process, over which we suppose to have full control, and is, therefore customizable.

Solving the underdetermined system The linear system in Eq. 3.8 is typically massively underdetermined and, therefore, cannot be solved directly. The minimal-l_2-norm solution to Eq. 3.8 can be obtained by means of the Moore-Penrose pseudoinverse, but it is not an adequate solution, since our *a priori* knowledge on $\vec{x}$ is that it is sparse or, at least, compressible. The Moore-Penrose pseudoinverse would spread the power in the measurements over all coefficients of $\vec{x}$, yielding a solution that is typically dense.

In the case of exactly sparse $\vec{x}$, we look for the solution to Eq. 3.8 with minimal l_0 norm, that is:

$$\hat{\vec{x}} = \arg\min_{\vec{x}} |\text{supp}(\vec{x})| \text{ subject to } \vec{y} = \boldsymbol{A}\vec{x} \tag{3.9}$$

It has been shown that finding a solution to Eq. 3.9 is, in general, NP-hard [337]. In the literature, there are two main approaches to overcome this difficulty: either to relax the l_0 minimization into an l_1 minimization or to design a *greedy* algorithm to find the sparse support of the signal and estimate the corresponding non-zero coefficients of the sparse solution. We dedicate Section 3.4 to review some of the principal recovery methods from both categories. In CS literature, recovery guarantees are typically given under the assumption of solving Eq. 3.8 by l_1 minimization, which yields:

$$\hat{\vec{x}} = \arg\min_{\vec{x}} \|\vec{x}\|_1 \text{ subject to } \vec{y} = \boldsymbol{A}\vec{x} \tag{3.10}$$

In [108] it is shown that Eq. 3.10 can be efficiently solved as a linear program, even when dealing with overcomplete dictionaries. A central pillar of CS is to study under which conditions Eq. 3.9 and Eq. 3.10 are equivalent [149].

The null space of $\boldsymbol{A}$ The ability to recover the sparse vector $\vec{x}$ from $\vec{y}$ through Eq. 3.10 is entirely determined by the measurement matrix $\boldsymbol{A}$. A critical parameter in this regard is the *null space* or *kernel* of the matrix, defined as the set of vectors that are projected to zero by it, that is,

$$\mathcal{N}(\boldsymbol{A}) = \{\vec{x} \mid \boldsymbol{A}\vec{x} = \vec{0}\} \tag{3.11}$$

Let $\vec{p}, \vec{q} \in \mathbb{R}^{n_{\text{atoms}}}$ be two distinct s-sparse vectors and let denote by Σ_k the subset of $\mathbb{R}^{n_{\text{atoms}}}$ containing all k-sparse vectors, so that, $\vec{p}, \vec{q} \in \Sigma_s$. Then, the difference vector $\vec{\delta} = (\vec{p} - \vec{q}) \in \Sigma_{2s}$. It is clear that uniqueness can only be ensured if $\mathcal{N}(\boldsymbol{A}) \cap \Sigma_{2s} = \varnothing$. Otherwise, the same $\vec{y}$ might be obtained for different $\vec{x}$, being, therefore, impossible to uniquely recover the signal from the measurements. Note that the null vector is always a member of the null space. In the following, $\mathcal{N}(\boldsymbol{A})$ refers exclusively to the non-zero null space vectors.

A minimal set of vectors spanning the null space of a matrix can be immediately obtained from its $\boldsymbol{QR}$ factorization, simply taking the last $n_{\text{atoms}} - m$ columns of $\boldsymbol{Q}$, if $\boldsymbol{A}$ is full rank. The $\boldsymbol{QR}$ factorization can be computed using the Gram-Schmidt process or by means of Householder transformations, being the latter numerically more stable than the former. For large $\boldsymbol{A}$, explicitly calculating a basis for the null space might be unfeasible, both due to memory requirements and execution time. In order to require $\mathcal{N}(\boldsymbol{A}) \cap \Sigma_{2s} = \varnothing$ without explicitly calculating a basis for $\mathcal{N}(\boldsymbol{A})$, one can use the concept of *spark* of a matrix, defined as the smallest number of columns that are linearly dependent. Then, according to [150], given $\vec{y} \in \mathbb{R}^m$, the condition $\operatorname{spark}(\boldsymbol{A}) > 2s$ is necessary and sufficient to ensure uniqueness of the solution to Eq. 3.8, subject to $\vec{x} \in \Sigma_s$.

Provided that we suppose $m \ll n_{\text{atoms}}$, the spark is bounded by m, i. e., spark $(\boldsymbol{A}) \in [2, m+1]$. Consequently, for exactly sparse vectors, the minimum number of measurements required for recovery is

$$m \geq 2s \tag{3.12}$$

The Null Space Property The condition in Eq. 3.12 was derived for $\vec{x}$ exactly sparse, but does not take into account compressible vectors. Extending the null space analysis to compressible vectors means deriving conditions for $\boldsymbol{A}$ under which too compressible vectors do not belong to $\mathcal{N}(\boldsymbol{A})$. For notation simplicity and, in the following, let suppose $\vec{x} \in \mathbb{R}^n$, as before the introduction of overcomplete frames, and, consequently, $\boldsymbol{A} \in \mathbb{R}^{m \times n}$. An essential question was posed in [111]: what is the minimal value of m, for which there exists a pair *coder/decoder*, such that

$$\left\| \vec{x} - \hat{\vec{x}} \right\|_p \leq C_0 \sigma_k (\vec{x})_p \,, \; \forall \vec{x} \in \mathbb{R}^n, \tag{3.13}$$

where C_0 is a constant independent from s and n and $\hat{\vec{x}}$ denotes the solution to the coding in Eq. 3.8 given by the decoder? The coder is the matrix $\boldsymbol{A}$ and the decoder is the recovery method, e. g., constrained l_1-minimization, that recovers $\vec{x}$ from $\vec{y}$. A sufficient and a necessary condition for Eq. 3.13 to hold are given in Eq. 3.14 and Eq. 3.15, respectively.

$$\|\vec{x}\|_p \leq \frac{C_0}{2} \sigma_{2k} (\vec{x})_p \,, \; \vec{x} \in \mathcal{N}(\boldsymbol{A}) \,, \tag{3.14}$$

$$\|\vec{x}\|_p \leq C_0 \sigma_{2k} (\vec{x})_p \,, \; \vec{x} \in \mathcal{N}(\boldsymbol{A}) \,, \tag{3.15}$$

Provided that in l_p spaces the best k-term approximation is given by k-thresholding of the largest coefficients, the property $\|\vec{x}\|_p \leq C \sigma_k (\vec{x})_p$ can be reformulated as

$$\|\vec{x}\|_p \leq C \left\| \vec{x}_{\bar{\Omega}_k} \right\|_p \,, \; \vec{x} \in \mathcal{N}(\boldsymbol{A}) \,, \tag{3.16}$$

where Ω_k denotes any set of k support indices, so that $\vec{x}_{\Omega_k}$ denotes the vector $\vec{x}$ with support restricted to Ω_k, with $|\Omega_k| = k$. $\bar{\Omega}_k$ denotes the complement set of Ω_k. Eq. 3.16 is known as the *Null Space Property* (NSP) of order k. The answer to the aforementioned question is given in the shape of fulfillment of the NSP. More specifically, if $\boldsymbol{A}$ fulfills the NSP of order $2k$ with $C = C_0/2$, then there exists a *decoder* solving Eq. 3.10, whose estimate

$\hat{\vec{x}}$ satisfies Eq. 3.13 (cf. Eq. 3.14). Conversely, the validity of Eq. 3.13 implies that $\boldsymbol{A}$ fulfills the NSP of order $2k$ with $C = C_0$ (cf. Eq. 3.15).

Unfortunately, for the specific case of the l_2 space, meeting the recovery guarantee in Eq. 3.13 requires a large number of measurements, specifically, $m > \left(1 - \sqrt{1 - \frac{1}{C_0^2}}\right) n$, as shown in [111]. Consequently, for the l_2 case, one can relax the right hand side of Eq. 3.13 from $C_0 \sigma_k \left(\vec{x}\right)_2$ to $C_0 \sigma_k \left(\vec{x}\right)_1 / \sqrt{k}$, which holds up to a constant factor due to the Cauchy-Schwartz inequality for sparse vectors, which reads

$$\left\|\vec{x}_{\Omega_k}\right\|_2 \leq \left\|\vec{x}_{\Omega_k}\right\|_1 \leq \sqrt{k} \left\|\vec{x}_{\Omega_k}\right\|_2 , \tag{3.17}$$

Then, in the l_2 case, Eq. 3.13 boils down to

$$\left\|\vec{x} - \hat{\vec{x}}\right\|_2 \leq C_0 \frac{\sigma_k \left(\vec{x}\right)_1}{\sqrt{k}}, \ \forall \vec{x} \in \mathbb{R}^n, \tag{3.18}$$

and the NSP of order k reads

$$\left\|\vec{x}_{\Omega_k}\right\|_2 \leq \left\|\vec{x}\right\|_2 \leq C_0 \frac{\sigma_k \left(\vec{x}\right)_1}{\sqrt{k}} = C_0 \frac{\left\|\vec{x}_{\bar{\Omega}_k}\right\|_1}{\sqrt{k}}, \ \vec{x} \in \mathcal{N}\left(\boldsymbol{A}\right), \tag{3.19}$$

which is often presented as the NSP, in general. If the pair *coder/decoder*, i. e., matrix $\boldsymbol{A}$ and recovery method, satisfy Eq. 3.18, then $\boldsymbol{A}$ satisfies the NSP of order $2k$.

Note that exactly sparse vectors are just a special case of compressible vectors. Consequently, Eq. 3.16 also applies to exactly sparse vectors, in which case the NSP in l_1 space can be simplified [184] to

$$\left\|\vec{x}\right\|_1 \leq 2 \left\|\vec{x}_{\bar{\Omega}_k}\right\|_1 = 2\sigma_k \left(\vec{x}\right)_1 , \ \vec{x} \in \mathcal{N}\left(\boldsymbol{A}\right). \tag{3.20}$$

Accounting for noise Eq. 3.10 imposes that the estimate of $\vec{x}$ has to lie on the manifold of solutions to the linear system in Eq. 3.8. In practice, the measurements might be affected by noise and being that strict may yield an estimate $\hat{\vec{x}}$ that is not optimal. In the noisy case, the problem to solve is no longer that in Eq. 3.10 but the extension given by Eq. 3.21,

$$\hat{\vec{x}} = \arg\min_{\vec{x}} \left\|\vec{x}\right\|_1 \ \text{subject to} \ \left\|\boldsymbol{A}\vec{x} - \vec{y}\right\|_2 \leq \varepsilon \tag{3.21}$$

where ε is an upper bound on the l_2 norm of the noise vector in measurement domain. In [82] the stability guarantee in Eq. 3.18 is extended to a noisy measurement scenario, yielding Eq. 3.22

$$\left\| \vec{x} - \hat{\vec{x}} \right\|_2 \leq C_0 \left(\varepsilon + \frac{\sigma_k(\vec{x})_1}{\sqrt{k}} \right), \ \forall \vec{x} \in \mathbb{R}^n \tag{3.22}$$

If we deal with exactly sparse vectors and the sparse vector is to be recovered from measurements corrupted by Gaussian noise, it has been shown in [37, 38] that the corresponding constrained Cramér-Rao bound (CRB) [254] is equal to that of an estimator with knowledge of the support. This means, in turn, that in the noiseless case the constrained CRB is identical to the MSE of the oracle estimator. Also, if the SNR is high enough, a maximum likelihood estimator achieves the CRB, thus performing as good as an oracle estimator. Complementary, in cases of low SNR, some estimators may even outperform the oracle estimator in terms of MSE, e. g., the Dantzig selector [93] as shown in [39].

A *RIP-free* universal condition for recoverability Of special significance is the *uniform recoverability* analysis presented in [484], since its derivations are independent from the RIP, to be introduced in the next paragraph. This way, the author derives a sufficient condition for recoverability that does not depend on the nature of $\boldsymbol{A}$, but exclusively on its null space properties. Specifically, the estimate given by Eq. 3.10 is the unique solution to Eq. 3.8 if its sparsity satisfies

$$\sqrt{\left\| \hat{\vec{x}} \right\|_0} < \min \left\{ \frac{1}{2} \frac{\|\vec{n}\|_1}{\|\vec{n}\|_2} \mid \vec{n} \in \mathcal{N}(\boldsymbol{A}) \cap \left(\Omega - \hat{\vec{x}} \right) \setminus \vec{0} \right\} \tag{3.23}$$

where $\vec{n}$ was used to denote null space vectors, while $\vec{x}$ denotes the unknown vector to estimate. Eq. 3.23 also foresees the inclusion of custom *a priori* information on $\vec{x}$, by means of the subset $\Omega \subset \mathbb{R}^n$, where $\vec{x}$ is expected to live, i. e., $\vec{x} \in \Omega$. This might look like a trivial extension, but becomes fundamental in many real problems, where restricting $\vec{x}$ to the right Ω, be in an absolute manner (e. g., defining feasible or, conversely, forbidden regions for $\vec{x}$) or stepwise (e. g., working from coarse to fine scales in multiscale dictionaries) often determines whether $\hat{\vec{x}}$ is successfully recovered or not.

For $\vec{x}$ exactly sparse with sparsity s, the necessary and sufficient condition for recovery given in [484] for $\Omega = \mathbb{R}^n$ can be derived from the NSP in Eq. 3.19 for $k = s$ and with $C_0 = 1/2$.

The Restricted Isometry Property Until now, we have presented some recoverability considerations based on the null space of $\boldsymbol{A}$, consciously avoiding the so-called *Restricted Isometry Property* (RIP), also known as *Uniform*

Uncertainty Principle (UUP), which is a fundamental tool in recoverability analysis, profusely used in CS literature. The RIP was first introduced in [78] and further analyzed in [80]. A matrix $\boldsymbol{A}$ is said to satisfy the RIP of order k if there exists a constant $\delta_k \in (0, 1)$ such that

$$(1-\delta_k)\|\vec{x}\|_2^2 \le \|\boldsymbol{A}\vec{x}\|_2^2 \le (1+\delta_k)\|\vec{x}\|_2^2, \ \forall \vec{x} \in \Sigma_k \tag{3.24}$$

being δ_k known as the *k-restricted isometry constant.* In short terms, this property ensures that the sensing matrix is close to an isometry for k-sparse vectors, i. e., that the transformation preserves the l_2 distances between pairs of k-sparse vectors to some extent. One can establish a parallelism with frame theory and see Eq. 3.24 as the frame condition for the rows of $\boldsymbol{A}$, that is, a measure of how well the rows of $\boldsymbol{A}$ are able to capture the information in $\vec{x} \in \Sigma_k$. Under this perspective, we have that the frame bounds are $\alpha = (1-\delta_k)$ and $\beta = (1+\delta_k)$ (see Eq. 3.5). The closer δ_k is to zero, the *tighter* the frame is. In fact, one can consider arbitrary bounds, α, β, that are not symmetric about 1, as in Eq. 3.24. If $\boldsymbol{A}$ satisfies the RIP of order k with constant δ_k, then it also it also satisfies the RIP of order $k', \forall k' < k$, with $\delta_{k'} \le \delta_k$. Additionally, from [344], we have that for $k' = \alpha\lfloor\frac{k}{2}\rfloor$, $\boldsymbol{A}$ also satisfies the RIP of order k' with constant $\delta_{k'} < \alpha\delta_k$. For $\alpha \ge 3$ this property allows extending the RIP to higher orders $k' > k$, with the obvious requirement of having an initially low δ_k, e. g., $\delta_k < 1/\alpha$, in order to ensure $\delta_{k'} < 1$.

If $\boldsymbol{A}$ satisfies the RIP of order $2k$ with δ_{2k} low enough, e. g., $\delta_{2k} < \sqrt{2}-1$ [83], then successful recovery is guaranteed. Recoverability means that $\hat{\vec{x}}$ meets the error bound in Eq. 3.18 for compressible signals and implies exact recovery if $\vec{x}$ is exactly sparse with sparsity k. An obvious (and expected) implication of the RIP of order $2k$ is that vectors $\vec{x} \in \Sigma_{2k}$ cannot be in $\mathcal{N}(\boldsymbol{A})$, as introduced above. Moreover, one can show that if $\boldsymbol{A}$ satisfies the RIP of order $2k$ with $\delta_{2k} < \sqrt{2}-1$, then it also satisfies the NSP (Eq. 3.19) of order $2k$ with constant

$$C_0 = \frac{2}{1-\left(1+\sqrt{2}\right)\delta_{2k}} \tag{3.25}$$

A proof is provided in [163]. In the exactly sparse case, the value of δ_{2k} determines whether Eq. 3.9 and Eq. 3.10 are equivalent. If $\delta_{2k} < 1$, the problem in Eq. 3.9 has a unique k-sparse solution. Moreover, if $\delta_{2k} < \sqrt{2}-1$, the solution to Eq. 3.10 coincides with that of Eq. 3.9, that is, the convex relaxation is exact.

The definition in Eq. 3.24 gives rise to several fundamental questions. How well typical sensing matrices used in CS satisfy the RIP in practice? How the restricted isometry constant δ_k scales with the RIP order k? Is the RIP a too restrictive requirement in practice? In order to help the reader to address this questions, we refer to Fig. 3.2, which will be introduced later.

Stability As pointed out in [135], the RIP is just a special case of the *bi-Lipschitz* condition, which can be used, in turn, as a measure of embedding stability. In short terms, an operator (the matrix $\boldsymbol{A}$, in our case) is called *bi-Lipschitz* if it preserves distances between vectors of a certain set, so that well-separated vectors remain well-separated after projection through the operator. This property, apart from implying that the inverse transformation is feasible, ensures certain robustness to perturbations in the measurements. Consider the definition of stability given in [135], which states that a pair coder/decoder is *C-stable* if Eq. 3.26 is satisfied, where C is a constant and $\hat{\vec{x}}$ the estimate of $\vec{x}$ delivered by the decoder from perturbed measurements $\vec{y} = \boldsymbol{A}\vec{x} + \vec{e}$, being $\vec{e}$ the perturbation vector.

$$\left\| \vec{x} - \hat{\vec{x}} \right\|_2 \leq C \left\| \vec{e} \right\|_2, \quad \forall \vec{x} \in \Sigma_k, \vec{e} \in \mathbb{R}^m \tag{3.26}$$

That is, adding a certain (bounded) amount of noise in the measurements should not have an arbitrarily large effect on the recovered signal. Satisfying Eq. 3.26 requires that the lower bound of the RIP (left hand side inequality in Eq. 3.24) of order $2k$ is satisfied with a constant $\delta_{2k} = 1 - 1/C^2$. In other words, the lower bound of the RIP of order $2k$ is a necessary condition for stability.

How many measurements? Satisfying the RIP implies recoverability by l_1-minimization. The subsequent question is *how many measurements* are required to satisfy the RIP of order $2k$ with an acceptably low δ_{2k}. In other words, what is the minimum number m of rows of $\boldsymbol{A} \in \mathbb{R}^{m \times n}$ that implies satisfying the RIP and, consequently, recovering $\vec{x}$. Calculating the exact value of δ_{2k} is an NP-hard problem, since it implies, in principle, analyzing all sparse vectors $\vec{x} \in \Sigma_{2k}$. Obviously, even for large n, one can always run Monte Carlo simulations and obtain an empirical estimate of the restricted isometry constant. Despite for large numbers of Monte Carlo samples the true value might be tightly approached, one cannot conclude that it does not exist some $\vec{x} \in \Sigma_{2k}$, which was not considered in the simulation, that invalidates the empirical δ_{2k}. Consequently, it is of great interest to obtain lower bounds on m, for which the RIP of order $2k$ is satisfied, for a desired

δ_{2k}. Such lower bounds are functions of the sparsity and the dimensionality of the signal. Different tight bounds might be obtained, depending on the nature of $\boldsymbol{A}$. The most general expression for a lower bound on m, which applies to many types of random matrices [346], is of the form

$$m = \mathcal{O}\left(k \log^{\alpha}(n)\right) \tag{3.27}$$

where k matches the expected signal sparsity and α is a small integer, ideally one. An s-sparse signal is known to contain only about $s \log(n/s)$ bits of information. As a consequence, intuitively, it is expectable that $m = \mathcal{O}\left(s \log(n/s)\right)$ dimensions should suffice to embed the signal. Indeed, according to [135], if $\boldsymbol{A} \in \mathbb{R}^{m \times n}$ satisfies the RIP of order $2k$ with constant $\delta_{2k} \in (0, 1/2]$, then

$$\begin{aligned} m &\geq C_{\delta_{2k}} k \log\left(\frac{n}{k}\right) \\ C_{\delta_{2k}} &\approx \frac{0.18}{\log\left(\sqrt{\frac{1+\delta_{2k}}{1-\delta_{2k}}} + 1\right)} \end{aligned} \tag{3.28}$$

The constant $C_{\delta_{2k}}$ can be well approximated by 0.28 [163]. Interestingly, similar results can be attained by means of different derivations, for instance, examining the *Gelfand widths* of l_p balls for $0 < p \leq 1$. This allows obtaining more general results. Consider a generalization of Eq. 3.9 and Eq. 3.10 to the case of minimizing some l_p norm with $0 < p \leq 1$. Then, to allow exact recovery of s-sparse vectors, the number of measurements must be [183]

$$m \geq C_1 p s \log\left(\frac{n}{C_2 s}\right) \tag{3.29}$$

where $C_1, C_2 > 0$ are constants. Note that, for $p = 1$, i. e., the case of recovery via l_1-minimization, Eq. 3.29 is coherent with Eq. 3.28. Eq. 3.29 also shows that decreasing p while keeping n fixed makes the bound non-informative ($m \geq \varepsilon$, $\varepsilon \rightarrow 0$), specially provided that the trivial bound $m \geq 2s$ has to hold. Combining both bounds we have that s-sparse recovery by l_p-minimization requires

$$m \geq C_1 s \left(1 + p \log\left(\frac{n}{C_2 s}\right)\right) \tag{3.30}$$

where $C_1, C_2 > 0$ are some constants, different, in principle, from those in Eq. 3.29.

Returning to the specific case of recovery via l_1-minimization, different lower bounds exist for different types of random matrices. In [91], the bounds are given as upper bounds for the sparsity s of exactly sparse signals that can be recovered with *overwhelming probability* for a given pair m, n. Proofs are provided or derived from results in [92, 149, 430].

A classical case is to suppose that the entries of $\boldsymbol{A}$ are realizations of *i.i.d.* random variables following a normal distribution with zero mean and $1/m$ variance. In this case, sparse recovery is guaranteed for

$$s \leq C \frac{m}{\log\left(\frac{n}{m}\right)} \tag{3.31}$$

where $C > 0$ is, as in the following inequalities, some constant. From an implementation point of view, it is remarkable that Eq. 3.31 holds also for binary matrices with independent entries taking values $\pm 1/\sqrt{m}$ with equal probability. Due to hardware limitations, for speed or simply for the sake of simplicity, binary sensing matrices are often the preferred choice in real CS implementations, including the CS-PMD system introduced in this thesis (Chapter 5). The Hadamard-derived binary sensing matrix of which we make wide use throughout this thesis is an example of this kind.

The so-called *Fourier ensemble* has been the first measurement matrix considered in CS. A paradigmatic example is the recovery of a train of spikes in, e. g., time domain, from few Fourier measurements. The matrix is constructed by randomly selecting m rows from the $n \times n$ DFT matrix and renormalizing the columns. For such a matrix the upper bound on the sparsity reads

$$s \leq C \frac{m}{(\log n)^{\alpha}} \tag{3.32}$$

where $\alpha = 6$ was initially reported in [91] and posteriorly improved to $\alpha = 4$ in [82, 391]. The bound in Eq. 3.32 might be too restrictive. Actually, for m, it translates into $m \geq Cs\,(\log n)^{\alpha}$, while a common conjecture for Fourier ensembles is that $m \geq Cs \log n$. In fact, it is well known [79] that a sparse signal consisting on s spikes can be exactly recovered from m measurements if

$$s \leq C \frac{m}{\log n} \tag{3.33}$$

Let $\boldsymbol{A} \in \mathbb{R}^{m \times n}$ be obtained by randomly selecting m rows from an $n \times n$ orthonormal matrix $\boldsymbol{U}$ and renormalizing the columns of the resulting matrix. For this *general orthogonal measurement ensemble* we have that

$$s \leq C \frac{1}{\mu^2} \frac{m}{(\log n)^{\alpha}} \tag{3.34}$$

where $\mu := \sqrt{m} \max_{i,j} |u_{i,j}|$, being $u_{i,j}$ the coefficient of $\boldsymbol{U}$ indexed by i, j. As before, values of $\alpha = 6$ and $\alpha = 4$ were given in [91] and [82, 391], respectively. Observe that Eq. 3.32 is a special case of Eq. 3.34 ($\mu = 1$ for the DFT matrix). As in Eq. 3.32, the maximum sparsity suggested by Eq. 3.34 might be too pessimistic due to the power of α in the logarithm.

Dependency between m and the restricted isometry constant It has been shown that satisfying the RIP of order $2k$ with some sufficiently low δ_{2k} translates into a requirement in the minimum number of measurements m (recall Eq. 3.28). Nevertheless, previous equations do not characterize well the dependency of m on the restricted isometry constant. An elegant way to establish such link by means of the classic *Johnson-Lindenstrauss lemma* [244], which is related to distance-preserving embeddings of high-dimensional points into low-dimensional spaces. The lemma is, by itself, closely related to the RIP. Here we provide the formulation given in [127], which reads as follows: for any $0 < \varepsilon < 1$ and any integer p, let m be a positive integer such that

$$m \geq 4 \frac{\log p}{\left(\frac{\varepsilon^2}{2} - \frac{\varepsilon^3}{3}\right)} \tag{3.35}$$

Then, for any set V of p points in $\mathbb{R}^n$, there is a map $f : \mathbb{R}^n \mapsto \mathbb{R}^m$ such that

$$(1 - \varepsilon) \|\vec{u} - \vec{v}\|^2 \leq \|f(\vec{u}) - f(\vec{v})\|^2 \leq (1 + \varepsilon) \|\vec{u} - \vec{v}\|^2, \ \forall \vec{u}, \vec{v} \in V \tag{3.36}$$

Furthermore, $f : \mathbb{R}^n \mapsto \mathbb{R}^m$ can be found in randomized polynomial time. The original lower bound on m [244] was simply $\mathcal{O}(\log p)$. The similarity between Eq. 3.36 and Eq. 3.24 is evident, being Eq. 3.24 a special case of Eq. 3.36 in which the map $f : \mathbb{R}^n \mapsto \mathbb{R}^m$ is linear and given by the matrix $\boldsymbol{A} \in \mathbb{R}^{m \times n}$. Given a matrix $\boldsymbol{X} = [\vec{x_i}]_{1 \leq i \leq p}$, constructed stacking the p n-dimensional vectors by columns, a map f can be constructed such that the variation of distance between them is bounded by ε. As shown in [127], f can be the *Gaussian ensemble* introduced above, i. e., the entries of $\boldsymbol{A}$ are drawn from *i.i.d.* random variables following a normal distribution with zero mean and $1/m$ variance. Of special interest is the extension in [5], where the elements of the matrix $\boldsymbol{A} \in \mathbb{R}^{m \times n}$ are realizations of *i.i.d.* random variables following probability distributions different from the normal distribution, namely a binary distribution where $a_{i,j} = \pm 1$ with $\Pr(a_{i,j} = -1) = \Pr(a_{i,j} = 1) = 1/2$ and a ternary distribution where $a_{i,j} \in \sqrt{3}\{-1, 0, 1\}$ with $\Pr(a_{i,j} = 0) = 2/3$ and $\Pr\left(a_{i,j} = -\sqrt{3}\right) = \Pr\left(a_{i,j} = \sqrt{3}\right) = 1/6$. In other words, the $f : \mathbb{R}^n \mapsto \mathbb{R}^m$ can also be a binary matrix or even a *sparse* ternary matrix. This enormously eases the implementation, allowing for faster and simpler projections, be in software or in hardware. In short terms, random projectors are adequate to embed high-dimensional points into a lower-dimensional space without information loss (preserving distances between them).

In the following we adopt the lower bound on m for the Johnson-Lindenstrauss lemma given in [242], which neglects the cubic power of ε in Eq. 3.35 and reads

$$m \geq C_0 \frac{\log p}{\varepsilon^2} \tag{3.37}$$

where $C_0 > 0$ is a constant. Building upon concentration inequalities, easy to verify for probability distributions such as Gaussian and Bernoulli distributions, one can prove not only that the Johnson-Lindenstrauss lemma holds, but also that the matrix $\boldsymbol{A} \in \mathbb{R}^{m \times n}$ constructed from such distributions satisfies the RIP [26]. Conversely, matrices satisfying the RIP and with randomized column signs are shown to provide optimal Johnson-Lindenstrauss embeddings (up to logarithmic factors in n) in [270]. More specifically, if $\boldsymbol{A}$ satisfies the RIP of order $k = C_1 \log p$ with restricted isometry constant δ_k, then it can be used as distance-preserving embedding for p points with $\varepsilon = \delta_k/4$. This, in combination with the simple Johnson-Lindenstrauss lower bound on m in Eq. 3.37 yields

$$m \geq C \frac{k}{\delta_k^2}, \text{ with } C = \frac{16 C_0}{C_1} \tag{3.38}$$

which brings together both k and δ_k in a lower bound for m. Observe that, for very small δ_k, the bound given by Eq. 3.38 might be significantly higher than that of Eq. 3.28.

The *engineering workaround* to the RIP: coherence While providing sharp bounds on the restricted isometry constant for a matrix $\boldsymbol{A} \in \mathbb{R}^{m \times n}$ that satisfies the RIP is a challenging task, other measures exist that can provide a characterization of $\boldsymbol{A}$ alternative to the RIP. This is the case of the *coherence*. In the literature, two different classes of coherence are often used to characterize $\boldsymbol{A}$. The first one refers to the *mutual coherence* [82] between sensing matrix $\boldsymbol{\Phi}$ and dictionary $\boldsymbol{\Psi}$, that is how much the sensing kernels are correlated with the dictionary atoms. A general definition of mutual coherence is

$$\mu(\boldsymbol{\Phi}, \boldsymbol{\Psi}) = \max_{i,j} \frac{\left|\langle \vec{\phi_i}, \vec{\psi_j} \rangle\right|}{\left\|\vec{\phi_i}\right\|_2 \left\|\vec{\psi_j}\right\|_2} \tag{3.39}$$

where $\vec{\phi_i}$ denotes the ith row of $\boldsymbol{\Phi}$ and $\vec{\psi_j}$ the jth column of $\boldsymbol{\Psi}$. The right hand side of Eq. 3.39 is determined by the matrices $\boldsymbol{\Phi}$ and $\boldsymbol{\Psi}$ and no general lower bound can be derived without any assumption on them. Suppose that the columns of $\boldsymbol{\Psi}$ form an orthonormal basis of $\mathbb{R}^n$ and that the rows of $\boldsymbol{\Phi}$ are randomly selected vectors of another orthonormal basis of $\mathbb{R}^n$. Then, it is easy to show that

$$\frac{1}{\sqrt{n}} \leq \mu(\boldsymbol{\Phi}, \boldsymbol{\Psi}) \leq 1 \tag{3.40}$$

regardless of m. The proof of Eq. 3.40 is given in Appendix A.7.

Intuitively, one expects that the higher the incoherence, the lesser measurements needed. Indeed, the mutual coherence is to be kept as low as possible. Consider the opposite case, in which the sensing kernels are strongly correlated with the dictionary atoms. It is clear that, in the limit, if the sensing kernels *are* the dictionary atoms (in which case we talk about *basis sensing* and no longer CS) and the sparse support is unknown, a number of measurements $m = n$ (or $m = n_{\text{atoms}} \neq n$ in case of dictionaries that are not bases of $\mathbb{R}^n$) are needed. In such case, if $\boldsymbol{\Psi}$ is an orthonormal basis, $\boldsymbol{A}$ is the identity matrix and no compression takes place at sensing. Furthermore, a lower bound on the number of measurements that ensures recoverability by l_1-minimization with probability exceeding $1-\delta$ is given in [81] as a function of the mutual coherence:

$$\begin{aligned} m &\geq C s \mu^2(\boldsymbol{\Phi}, \boldsymbol{\Psi}) \log\left(\frac{n}{\delta}\right) \\ m &\geq C' \log^2\left(\frac{n}{\delta}\right) \end{aligned} \tag{3.41}$$

where $C, C' > 0$ are constants. The second type of coherence refers to the *matrix coherence* [86] of the measurement matrix $\boldsymbol{A}$ and is defined as

$$\mu(\boldsymbol{A}) = \max_{i<j} \frac{|\langle \vec{a_i}, \vec{a_j} \rangle|}{\|\vec{a_i}\|_2 \|\vec{a_j}\|_2} \tag{3.42}$$

where a_k denotes the kth column of $\boldsymbol{A}$. Low coherence of Gaussian and Bernoulli measurement matrices is pointed out in [150, 439]. The following lower bound on the matrix coherence [439] applies for both distributions with high probability:

$$\mu(\boldsymbol{A}) \geq \sqrt{\frac{C \log n}{m}} \tag{3.43}$$

where $C > 0$ is a constant. The matrix coherence can be also obtained by analyzing the Gram matrix. In general, the Gram matrix $\boldsymbol{G}$ of any set of vectors $\boldsymbol{A} = [\vec{a}_i]_{i \in [1,n]}$, stacked by columns in the matrix $\boldsymbol{A}$, is calculated as:

$$\boldsymbol{G} = \boldsymbol{A}^\top \boldsymbol{A} \tag{3.44}$$

If the columns of $\boldsymbol{A}$ are normalized, $\mu(\boldsymbol{A})$ can directly be obtained as the as the maximum of the absolute values of the off-diagonal coefficients of $\boldsymbol{G}$, that is

$$\mu(\boldsymbol{A}) = \max_{i \neq j} g_{i,j} \tag{3.45}$$

where $g_{i,j}$ denotes the i, j element of $\boldsymbol{G}$. A universal lower bound for the coherence of $\boldsymbol{A} \in \mathbb{R}^{m \times n}$ is given by the well-known Welch bound [461] (Eq. 3.46), which is zero for a square measurement matrix.

$$\mu_{\min}(\boldsymbol{A}) = \sqrt{\frac{n-m}{m(n-1)}} \tag{3.46}$$

As pointed out in [486], the Welch bound does not apply to every combination of m and n. The following necessary conditions can be derived from the absolute bounds for A-sets given in [139] (second and third rows of the Table II for the real and complex case, respectively):

$$\begin{aligned} n &\leq \frac{m(m+1)}{2}, \text{ for } \boldsymbol{A} \in \mathbb{R}^{m \times n} \\ n &\leq m^2, \text{ for } \boldsymbol{A} \in \mathbb{C}^{m \times n} \end{aligned} \tag{3.47}$$

If n is greater than the upper bounds in Eq. 3.47, the n columns of $\boldsymbol{A}$ cannot form an equiangular system [58]. For cases of high dimensionality, when the conditions in Eq. 3.47 cannot be met, the *orthoplex* or *Rankin* bound can be adopted:

$$\mu_{\min}(\boldsymbol{A}) = \frac{1}{\sqrt{m}} \tag{3.48}$$

The Rankin bound only applies up to to some level of n, specifically it can only be achieved for

$$\begin{aligned} \frac{m(m+1)}{2} < n \leq (m+1)(m+2), \text{ for } \boldsymbol{A} \in \mathbb{R}^{m\times n} \\ m^2 < n \leq 2(m^2-1), \text{ for } \boldsymbol{A} \in \mathbb{C}^{m\times n} \end{aligned} \tag{3.49}$$

If n is extremely large with respect to m, which might happen in case of sensing signals with extremely low ratios s/n, the conditions in Eq. 3.49 might not be satisfied and the lower bound in Eq. 3.48 cannot be achieved. For such cases, one can still adopt the lower bound developed by Kabatiansky and Levenshtein [247, 293] for $n \to \infty$:

$$\mu_{\min}(\boldsymbol{A}) = \begin{cases} \sqrt{\frac{3n-m^2-2m}{(m+2)(n-m)}}, & \text{for } \boldsymbol{A} \in \mathbb{R}^{m\times n} \\ \sqrt{\frac{2n-m^2-m}{(m+1)(n-m)}}, & \text{for } \boldsymbol{A} \in \mathbb{C}^{m\times n} \end{cases} \tag{3.50}$$

Alternatively, one can also make use of the lower bound given in Eq. 3.51, originally derived in [336] for the case of low-coherence beamformer codebooks and explicitly presented as a lower bound on the coherence in [472].

$$\mu_{\min}(\boldsymbol{A}) = 1 - 2n^{-\frac{1}{m-1}} \tag{3.51}$$

Now that lower bounds on the coherence of $\boldsymbol{A}$ have been given, it would be desirable to establish a link between $\mu(\boldsymbol{A})$ and $\mu(\boldsymbol{\Phi}, \boldsymbol{\Psi})$. Intuitively, one would expect that $\mu(\boldsymbol{\Phi}, \boldsymbol{\Psi})$ is somehow more informative than $\mu(\boldsymbol{A})$, provided that a pair $\boldsymbol{\Phi}, \boldsymbol{\Psi}$ uniquely define $\boldsymbol{A}$, while a given $\boldsymbol{A}$ could have been generated with different pairs $\boldsymbol{\Phi}, \boldsymbol{\Psi}$. Additionally, in many cases $\boldsymbol{\Psi}$ cannot be arbitrarily designed, since it might be given by the signal class to deal with. Therefore, the best result we might aim to derive from $\mu(\boldsymbol{\Phi}, \boldsymbol{\Psi})$ is an upper bound on $\mu(\boldsymbol{A})$. Without loss of generality, we restrict our attention to sensing matrices $\boldsymbol{\Phi}$ with normalized rows and dictionaries $\boldsymbol{\Psi}$ with normalized columns, for simplicity, but we do not make any assumption on the resulting measurement matrix $\boldsymbol{A}$. By means of a recursive ($m > 2$ summation terms) triangular inequality, we derive the following general upper bound on $\mu(\boldsymbol{A})$ given by $\mu(\boldsymbol{\Phi}, \boldsymbol{\Psi})$:

$$\mu(\boldsymbol{A}) \leq \frac{1}{C^2} m \mu^2 (\boldsymbol{\Phi}, \boldsymbol{\Psi})$$
$$C = \min_{1 \leq i \leq n} \|\vec{a}_i\|_2 \tag{3.52}$$

The proof of Eq. 3.52 is provided in Appendix A.7.

Fig. 3.1 shows the evolution of the coherence for different types of synthetic measurement matrices against the compression rate, for a fixed dimensionality $n = 19200$ (this, here arbitrary, value of n will become relevant in this thesis due to the use of PMD hardware). As a reference, the Welch bound, calculated using Eq. 3.46, is plotted in blue. The elements of the matrices are drawn from different distributions, except from the Fourier matrix, which is constructed by rows, randomly selecting m elements of a Fourier basis. Note that the coherence plots for the Gaussian matrix and for the modified Bernoulli distribution, with values of the random variable of ± 1, are almost coincident. The Fourier matrix is superior to random matrices and tends to an orthonormal system $\mu(\boldsymbol{A}) = 0$ when $m \to n$.

From the coherence of $\boldsymbol{A}$, a lower bound on the spark can be obtained, as shown in Eq. 3.53. Provided that the condition $\operatorname{spark}(\boldsymbol{A}) > 2s$ [150] is sufficient to ensure uniqueness of the solution, $\vec{x} \mid \vec{y} = \boldsymbol{A}\vec{x},\ \|\operatorname{supp}(\vec{x})| = s$, in the worst-case scenario of lowest possible spark, i. e., Eq. 3.53 holds with equality, one immediately has the (yet pessimistic) sparsity requirement in Eq. 3.54.

$$\operatorname{spark}(\boldsymbol{A}) \geq 1 + \frac{1}{\mu(\boldsymbol{A})} \tag{3.53}$$

$$s < \frac{1}{2}\left(1 + \frac{1}{\mu(\boldsymbol{A})}\right) \tag{3.54}$$

If the signal sparsity s is given and the task is to design an appropriate sensing matrix, the requirement in Eq. 3.54 immediately translates into the matrix coherence requirement in Eq. 3.55.

$$\mu(\boldsymbol{A}) < \frac{1}{2s - 1} \tag{3.55}$$

It has been shown that, for small enough coherence $\mu(\boldsymbol{A}) < c/\sqrt{m}$, the condition in Eq. 3.55, which ensures recovery via l_1 minimization or OMP, is satisfied if the number of measurements is

$$m \geq Cs^2 \tag{3.56}$$

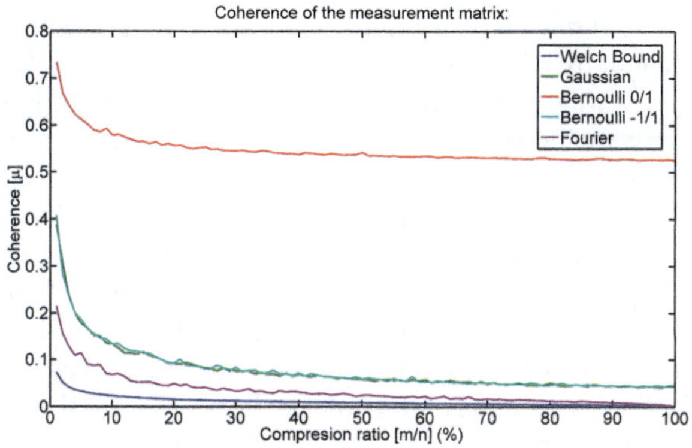

Figure 3.1.: Coherence of different measurement matrices of size $m \times n$. In all cases the matrix elements were drawn from different probability distributions, except from the Fourier matrix, whose rows are randomly selected elements of a Fourier basis. The Gaussian distribution is of zero mean and unit variance. Two Bernoulli distributions with $p = 0.5$ are presented: the real Bernoulli distribution, where the random variable takes values of 0 and 1 with equal probability, and a zero-mean distribution, with values -1 and 1. Note that the latter behaves like the Gaussian distribution, while the former is shifted vertically in approximately 0.5 due to the non-zero offset. Obviously, for $m = n$ the Fourier measurement matrix is the Fourier basis and the coherence is zero.

where $C > 0$ is a constant. The coherence can also be used to ensure that $\boldsymbol{A}$ satisfies the RIP. If the columns of $\boldsymbol{A}$ are normalized, one can affirm that $\boldsymbol{A}$ satisfies the RIP of order k with δ_k given by Eq. 3.57.

$$\delta_k = (k-1)\mu(\boldsymbol{A}), \ \forall k < 1/\mu(\boldsymbol{A}) \tag{3.57}$$

Both Eq. 3.57 and Eq. 3.53, can be derived from the Theorem 2 of [200], which states that the eigenvalues of a matrix $\boldsymbol{A} \in \mathbb{R}^{n\times n}$ lie in the union of n discs, $\bigcup_{i=1}^{n} d_i(c_i, r_i)$, centered at $c_i = a_{i,i}$ and with radius $r_i = \sum_{j\neq i} |a_{i,j}|$.

RIP Monte Carlo simulations Fig. 3.2 shows the results of intensive RIP Monte Carlo simulations carried out for different well-known types of sensing matrices. The plots show the empirical value of the restricted isometry constant δ_k obtained for matrices of different sizes, for different values of the

RIP order k. The signal dimensionality was set to 2^{10}, a value large enough to attain sufficient resolution in the ratios m/n and k/n ($9.77 \times 10^{-2}\%$ resolution in percentage). For each type of sensing matrix, different matrices constructed with ratios m/n ranging between 1% and 100% with 1% step size were evaluated. For each matrix, the restricted isometry constant δ_k was empirically obtained using populations of 10^5 k-sparse vectors. The RIP was checked for ratios k/n ranging between 1% and 100% with 1% step size. Therefore, each pixel in each plot of Fig. 3.2 is the result of checking the l_2 norm of 10^5 vectors after projecting them through the sensing matrix. For the population size of 10^5 the simulation took days in a 16-core machine with 380 Gb RAM. The binary permuted block diagonal (BPBD) matrices were created by concatenating $m \times m$ identity matrices and then randomizing by columns, so that only a non-zero element exists per column. This is the sparsest possible sensing matrix. The entries of the Gaussian matrices were drawn from a zero-mean unit-variance normal distribution. All the matrices were normalized by columns.

Observe that the restricted isometry constants derived from the minimum coherence for a given matrix size (Fig. 3.2a) are too pessimistic, unless $k \ll n$, e. g., $k/n < 5\%$. The minimum coherence comes from Eq. 3.46 and the corresponding δ_k from Eq. 3.57. Conversely, one can argue that our simulations are too optimistic due to insufficient population size. Note the diagonal transition in Fig. 3.2b. For $m \geq k$ our BPBD preserves the l_2 norm up to machine precision. Provided that each column has a single non-zero coefficient, for $m \geq k$, performs as good as having a $k \times k$ identity matrix sensing the k non-zero coefficients of the signal and the l_2 norm of k-sparse vectors is preserved. As expected, the Gaussian (Fig. 3.2c) and the binary matrix with $\{\pm 1\}$ entries (Fig. 3.2d) exhibit equivalent performance. Of special significance is the case of the binary matrix with entries in $\{0, 1\}$ (Fig. 3.2e). One would expect values of δ_k similar to those in Fig. 3.2d, but this is not the case and, for almost the complete range of k/n, δ_k is between 8 and 10, regardless of m. In other words, this matrix does not seem to satisfy Eq. 3.24 at all. Nevertheless, the experience demonstrates that both binary matrices show similar performance in practice. What happens is that the distribution of the coefficients of the second binary matrix is not centered at zero and, thus, the l_2 norm of the projected vectors does not fit the symmetric bounds of Eq. 3.24. In other words, very different δ_ks would be obtained for the lower and upper bounds in Eq. 3.24. The Fourier ensemble (Fig. 3.2f) tends to an orthonormal basis when $m \to n$ and, consequently, $\delta_k \to 0$ when $(m/n) \to 100\%$. As expected, for $m = n$, $\delta_k = 0$ up to machine precision. All the scales in Fig. 3.2 are between 0 and

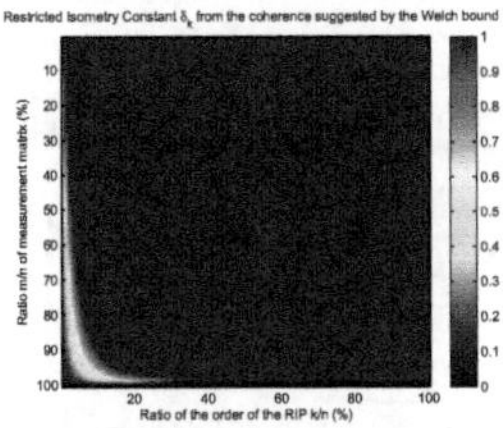

(a) From Welch-bound Coherence

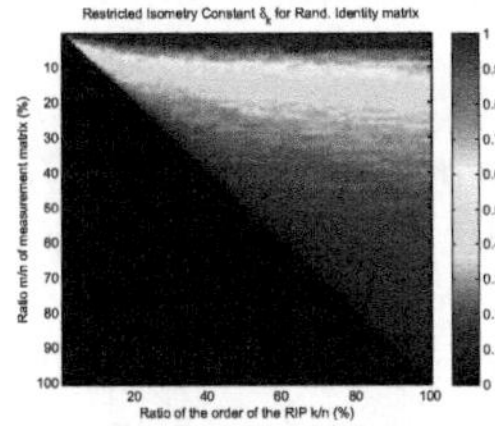

(b) BPBD

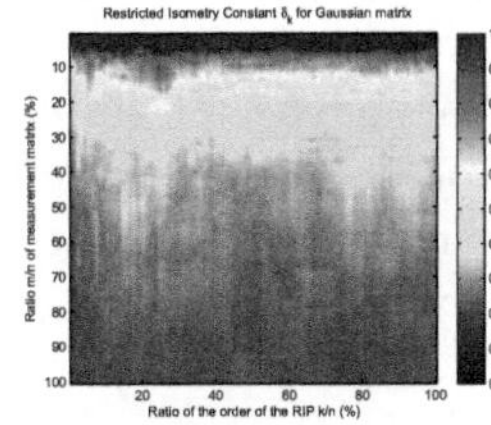

(c) Gaussian

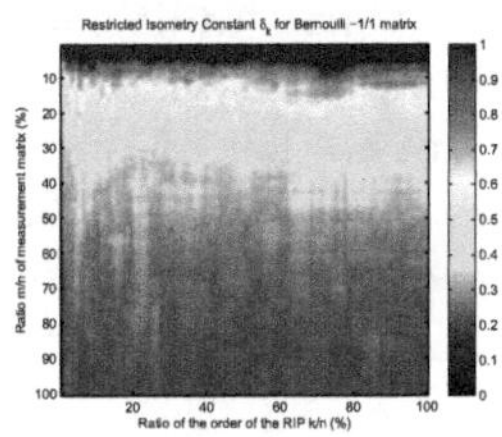

(d) Bernoulli ±1

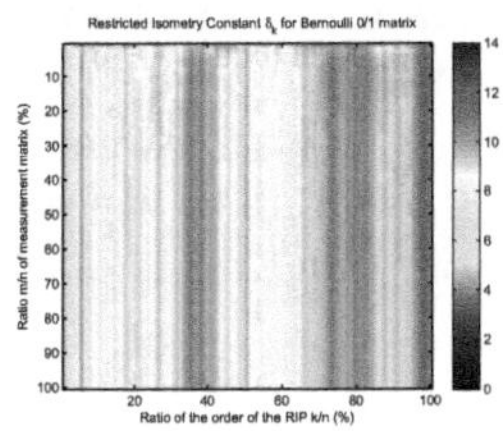

(e) Bernoulli $0/1$

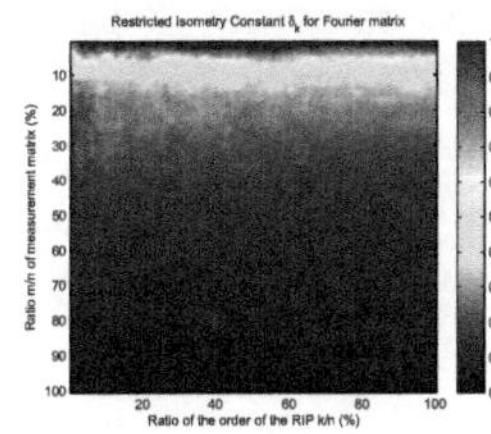

(f) Fourier

Figure 3.2.: Empirical values of the restricted isometry constant δ_k for which the RIP of order k is satisfied, for different measurement matrices. The sensing matrices were synthetically generated for different ratios m/n with $n = 2^{10}$. The types of sensing matrices considered are BPBD (b), random Gaussian (c), Bernoulli with elements $\{\pm1\}$ (d) and $\{0,1\}$ (e) and the Fourier ensemble (f). For each matrix, the RIP of order k was checked for a population of 10^5 k-sparse vectors. Both ratios m/n and k/n were sampled between 1% and 100% with 1% step size. Plot (a) shows the value of δ_k suggested by the Welch-bound coherence ($\delta_k = (k-1)\mu_{\min}(\boldsymbol{A})$).

1, except from that of Fig. 3.2e, which has been adapted to the maximum δ_k.

The unexpected results in Fig. 3.2e are an indicator that the RIP definition in Eq. 3.24 is to be handled with care. In fact, symmetry of the bounds is not required for satisfying the RIP. Consider arbitrary lower and upper bounds α and β, $0 < \alpha \leq \beta < \infty$, that substitute $(1 - \delta_k)$ and $(1 + \delta_k)$, respectively, in Eq. 3.24. If there exist such bounds, one can always scale $\boldsymbol{A}$ so that it satisfies Eq. 3.24. More specifically, multiplying $\boldsymbol{A}$ by $\sqrt{2/(\alpha + \beta)}$, Eq. 3.24 will be satisfied with $\delta_k = (\beta - \alpha)/(\alpha + \beta) < 1$. Returning to the interesting case of the binary matrix with entries in $\{0, 1\}$, we provide the empirically obtained values of α and β in Fig. 3.3a and Fig. 3.3b, respectively. Additionally, Fig. 3.3c provides the result of $(\beta - \alpha)/(\alpha + \beta)$, applied elementwise to the elements of Fig. 3.3a and Fig. 3.3b, that is, the values of δ_k that one would obtain applying an appropriate scaling to the matrices. Note that Fig. 3.3c already resembles the plots obtained for the other measurement matrices considered in Fig. 3.2.

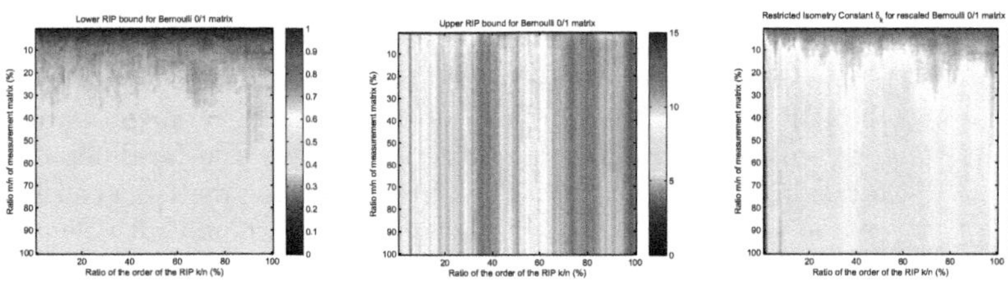

(a) RIP Lower Bound α **(b)** RIP Upper Bound β **(c)** $\delta_k = (\beta - \alpha)/(\alpha + \beta)$

Figure 3.3.: Empirical values of the RIP lower (a) and upper (b) bounds α, β for a Bernoulli matrix with elements $\{0, 1\}$ (cf. (Fig. 3.2e). The considered ratios m/n and k/n are the same as in Fig. 3.2. The dimensionality is kept to $n = 2^{10}$ and the population size to 10^5 vectors. The restricted isometry constants δ_k of the corresponding rescaled matrices are given in (c).

3.2. Sensing Matrices

In Section 3.1 sensing matrices were introduced as an element of the linear measurement model used in CS. We distinguish between sensing matrix, denoted by $\boldsymbol{\Phi}$, and measurement matrix, denoted by $\boldsymbol{A}$ and obtained as the product between sensing matrix and sparsity basis $\boldsymbol{\Psi}$. In many works it is assumed that the signal is itself sparse, thus being $\boldsymbol{\Psi}$ the identity matrix and $\boldsymbol{A} = \boldsymbol{\Phi}$. For this reason, in the literature it is common to name $\boldsymbol{\Phi}$ indistinctly as sensing or measurement matrix. In this work this equivalence is not assumed and we explicitly distinguish these matrices. The mission of the sensing matrix is to capture as much information as possible with a number of measurements that has also to be kept as low as possible. Provided that the core idea of CS is to make use of *a priori* knowledge of the signal (e. g., sparsity, joint sparsity, structured sparsity, support probabilities) to reduce the number of measurements required to achieve perfect reconstruction, the sensing matrix has to be designed observing this *a priori* knowledge. One could say that the performance of a sensing matrix is as close to optimal as the minimum number of measurements m for which successful recovery is attained approaches $\mathcal{O}(s \log n)$, where s denotes sparsity.

Each row of the matrix is a discrete sensing function or *sensing kernel*. Some orthogonality between them is desired, since it ensures that different measurements do not gather redundant information. Furthermore, the energy of all the sensing kernels has to be distributed over the dictionary atoms as uniformly as possible. This is connatural to the idea of compressing at sensing: if each measurement kernel can be represented by a single or very few dictionary atoms, many measurements will be necessary, eventually as many as the number of dictionary atoms, as in *basis sensing*. Alternatively, if each measurement kernel shows similar nonzero coherence with all dictionary atoms, it is possible that a sparse signal can be exactly recovered from few measurements, provided that they might retain all signal information. This requirement is often expressed as a minimum-coherence requirement between sensing matrix and dictionary (recall Eq. 3.39). Another indicator of the quality of a sensing matrix (or corresponding measurement matrix, in case of non-identity dictionary) is the RIP, which is intended to ensure that the transformation given by the matrix preserves distances between sparse vectors. If the RIP is satisfied, exact reovery is guaranteed. As seen before, matrix coherence can be related to satisfying the RIP.

3.2.1. Classical Sensing Matrices: Random Matrices and Fourier Ensemble

For moderate and large dimensionalities, random matrices have been shown to be excellent sensing matrices, exhibiting low coherence (Fig. 3.1) and satisfying the RIP. The entries of a random sensing matrix are realizations of *i.i.d.* random variables following certain distribution. Common distributions are the normal and the Bernoulli distributions, yielding *Gaussian* and *Bernoulli* sensing matrices. The main advantage of the latter over the former is the fact that it is binary, thus allowing simpler hardware implementations. For sub-Gaussian random matrices, such as Gaussian and Bernoulli matrices, non-asymptotic random matrix theory can be used to show that the number of measurements required to satisfy the RIP follows Eq. 3.58 (see [323]).

$$m = \mathcal{O}\left(s \log\left(\frac{n}{s}\right)\right) \tag{3.58}$$

This is due to the fact that sub-Gaussian random matrices exhibit good concentration properties [5] and from [26] we know that the RIP can be derived from concentration of measure inequalities. Unfortunately, it is widely known that the performance of random sensing matrices degrades in the presence of noise. When sensing natural images, for instance, the SNR of measurements gathered via Gaussian sensing kernels is, from [460],

$$SNR = \frac{1}{n}\frac{\pi^2}{6\sigma^2} \tag{3.59}$$

where n is the signal dimensionality (number of pixels) and σ^2 is the variance of the measurement noise. Note that, $SNR \to 0$ as $n \to \infty$. This is one of the strengths of deterministic constructions, for which the SNR can be kept constant, in comparison with random matrices. Despite the columns of a random matrix are, in general, non-orthogonal, they tend to orthogonal when $n \to \infty$. Additionally, posterior orthogonalization is always possible, e. g., by means of the Gram-Schmidt method. Optimization of sensing matrices is further commented in Section 3.2.9.

Another classical sensing matrix is the Fourier ensemble, widely used when dealing with signals that are directly sparse, e. g., spike trains, and, therefore, dense in frequency domain. It is worth mentioning that the Fourier ensemble is perfectly incoherent with the canonical basis and orthonormal wavelet bases. These common types of sensing matrices have been already introduced in Section 3.1 and are widely analyzed in the literature. In the following we present some other alternatives.

3.2.2. Scrambled Fourier Ensemble

This sensing matrix only differs from the Fourier ensemble introduced in Section 3.1 in an additional random permutation of its columns. It has been shown to yield similar performance to that of a Gaussian matrix [88], while leveraging the computational cost of the recovery procedure. Note that sensing with the scrambled Fourier ensemble is equivalent to applying the permutations to the discrete signal to measure and gather partial Fourier measurements on the permuted signal.

As noted in [218], using a Fourier ensemble as sensing matrix typically yields poor reconstructions of natural images, while after *scrambling* the columns optimal reconstruction is achieved. This is due to the fact that natural images are compressible up to a certain extent in Fourier domain and, therefore, Fourier measurements are not optimal from a CS point of view. After applying random permutations to the columns, the sensing kernels look like random noise and are actually highly incoherent with any structured dictionary, thus yielding optimal performance.

3.2.3. Hadamard-derived

A Hadamard matrix $\boldsymbol{H}$ is an orthogonal binary matrix, whose coefficients are ± 1. A Hadamard matrix is said of order n if its size is $n \times n$. This matrix is a solution to the so-called Hadamard's maximum determinant problem, i. e., a matrix (complex, in general) with elements $h_{i,j} \mid |h_{i,j}| \leq 1$ whose determinant is maximum in absolute value. Equivalently, one has that $\boldsymbol{H}\boldsymbol{H}^\top = n\boldsymbol{I}_n$, where $\boldsymbol{I}_n$ denotes the identity matrix of size n. In the different cases a Hadamard-derived sensing matrix was used within the work belonging to this thesis, regardless of whether the obtained results are presented here or not, only cases where $n = 2^r,\ r \in \mathbb{N}$ were considered. While being a sufficient condition for the existence of a Hadamard matrix, the latter is not strictly necessary. The necessary condition $n \equiv 0(\mod 4)$ was introduced by Hadamard in 1893 [211] and has been conjectured to be also sufficient [63]. Despite a fully satisfactory proof of this conjecture has not been provided yet, Hadamard matrices of size $n \times n$ fulfilling $n \equiv 0(\mod 4)$ have been constructed up to relatively large values of n (e. g., $n = 428$ in [259]). In fact, Hadamard matrices are known for all $n \equiv 0(\mod 4),\ n \leq 256$ [219]. The proof of the Hadamard conjecture given in [234] was shown wrong in [122] short after appearing. Other than $n \equiv 0(\mod 4)$, other cases such as $n \equiv 1(\mod 4)$, $n \equiv 2(\mod 4)$ and $n \equiv 3(\mod 4)$ have been studied (see [348] for a review of related literature). For arbitrary matrix sizes, we refer to the optimum binary signature set design procedure proposed in [253].

Using a Hadamard matrix as a base to generate the sensing matrix brings the following advantages:

1. **Binary**. Equivalently to Bernoulli matrices, the coefficients of a Hadamard matrix can only have two different values.

2. **Efficient construction**. Differently from random matrices, where each coefficient is a random number, generated independently from the others, there exists efficient methods for constructing Hadamard matrices. Sylvester's incremental construction generates a Hadamard matrix of size $n = 2^r$, $r \in \mathbb{N}$ with computational complexity $\mathcal{O}(r)$.

3. **Orthogonality**. Also differently from random matrices, where (approximate) orthogonality is only ensured with overwhelming probability when $n \to \infty$, Hadamard matrices are orthogonal basis. In fact, if constructed using the Sylvester's method, the Hadamard matrix is also called Walsh matrix. Regardless of the construction method, the rows (or columns) of a Hadamard matrix of size n are the Walsh functions of order n and, thus, orthogonal. Orthonormality is trivially achieved by multiplying by $1/\sqrt{n}$.

The construction of a sensing matrix of arbitrary size $\boldsymbol{\Phi} \in \mathbb{R}^{m \times n}$ from a Hadamard matrix $\boldsymbol{H} \in \mathbb{R}^{n \times n}$ involves two simple steps, namely, random permutation of its columns and random selection of m rows. The first step is just intended to destroy any eventual coherence between the rows of $\boldsymbol{H}$ and the atoms of any structured dictionary. This way, the obtained sensing kernels look like random binary patterns. Still, one can see that $\boldsymbol{\Phi}$ preserves the aforementioned properties of Hadamard matrices.

In certain applications one might be interested in binary matrices with elements $\{0, 1\}$ (e. g., the Bernoulli matrices introduced in Section 3.1) that somehow preserve the orthogonality of the Hadamard-derived sensing matrices. Unfortunately, the only possible orthogonal basis of $\mathbb{R}^n$ with elements $\{0, 1\}$ is the trivial basis of the space $\boldsymbol{I}_n$, which would be fully coherent with the dictionary in case of localized signals, such as the spike trains introduced above. Nevertheless, the Hadamard-derived sensing matrix with ± 1 elements can be rescaled to have elements in $\{0, 1\}$. Note that, while this does not yield orthogonality, the coherence between rows of the sensing matrix is exclusively given by a certain DC offset, which is common to all of them and, therefore, the redundant information gathered in every measurement is always the same, eventually null if the signal to sense has no DC component.

3.2.4. Scrambled Block Hadamard Ensemble

A matrix that derives from a Hadamard matrix as explained above is an excellent general CS sensing matrix, but when observing it an unavoidable question arises: is it really necessary that the matrix is dense?, or, equivalently, can equivalent sensing performance be attained with a sparse matrix, still preserving the highly-desirable properties of a Hadamard matrix? An answer to such a crucial question can be found in [197], where a *Scrambled Block Hadamard Ensemble* (SBHE) is proposed as sensing matrix. The SBHE is not constructed from a Hadamard matrix directly, but from a block diagonal matrix where each block is a Hadamard matrix with size submultiple of n. In the limit, if the number of Hadamard matrices is n, conventional sampling is achieved (identity sensing matrix) and no compression is possible. The construction of the sensing matrix from the block diagonal matrix is as in the Hadamard-derived case, i. e., random column permutations and random row selection.

The block diagonal structure allows parallel computation at complexity $\mathcal{O}(n \log n_{\boldsymbol{H}})$, where $n_{\boldsymbol{H}}$ denotes the size of the blocks (Hadamard matrices). The SBHE has been shown to perform as well as a Gaussian matrix if the maximum absolute value of the dictionary coefficients is bounded by $C/\sqrt{n}$, being C a constant, and $n_{\boldsymbol{H}} \to \infty \iff n \to \infty$. From [197], the number of measurements that guarantees recovery by l_1-minimization in the noiseless case is given by Eq. 3.60, which has been derived applying Eq. 3.41.

$$m \geq C \left(s \sqrt{\frac{n}{n_{\boldsymbol{H}}}} (\log n)^2 \right) \tag{3.60}$$

where $C > 0$ is a constant.

3.2.5. Binary Permuted Block Diagonal

The SBHE shows that sensing matrices can also be sparse, while performing as well as their dense counterparts. Nevertheless, a question remains unanswered: how far can we push the sparsity?, or, alternatively, what would be the simplest binary sensing matrix? The *Binary Permuted Block Diagonal* (BPBD) proposed in [218] was claimed to be the *simplest* matrix for sensing natural images. The BPBD is conceived as the maximally sparse sensing matrix that is still able to sense all the signal information. To this end, at least one element per column must be different from zero, e. g., a one. When the BPBD is constructed with only one submatrix, there is only one '1' per column and the sparsity of the matrix is maximal. In this case, the BPBD is constructed from a single block diagonal matrix where each block is a row vector of ones of length $n_b = n/m$. The BPBD is then directly obtained from this matrix after random column permutations. Observe that in this case orthogonality between rows is guaranteed, thus ensuring that each measurement gathers non-redundant information. The BPBD retains the desirable properties of the aforementioned binary sensing matrices, while providing optimal sensing efficiency.

Note that when $m \to n$ the BPBD converges to the canonical basis, i. e., the identity matrix. Nevertheless, for $m < n$ (which is the case in CS), the BPBD performs better than a partial canonical matrix, due to the fact that the former is incoherent with, e. g., wavelet and gradient basis, while the latter is not. If the dictionary is an orthogonal basis with no less than αn nonzero elements in each column and $|\Psi_{i,j}| \leq 1/\sqrt{\beta n}$, with $0 < \beta \leq \alpha \leq 1$, then the mutual coherence is bounded by:

$$\mu(\mathbf{\Phi}, \mathbf{\Psi}) \leq \sqrt{\frac{n_{\mathrm{b}}}{\beta}} \tag{3.61}$$

where the length or each block of ones is $n_{\mathrm{b}} = n/m$ in the case of a single submatrix. The proof of Eq. 3.61 is provided in [218]. One can then derive a lower bound for m to ensure recoverability substituting Eq. 3.61 in Eq. 3.41. Operating the resulting inequality for the case of a single submatrix, we have derived the bound in Eq. 3.62, which ensures recoverability with probability exceeding $1 - \delta$.

$$m \geq \sqrt{\frac{C}{\beta} sn \log\left(\frac{n}{\delta}\right)} \tag{3.62}$$

The bound in Eq. 3.62 does not seem to be tight for relatively low n, due to the n term out of the logarithm, but it tightens for high values of n due to the square root. The constants α, β seem to be well behaved, e. g., $\alpha = \beta = 1$ for the Fourier basis, yielding $\mu(\mathbf{\Phi}, \mathbf{\Psi}) = \sqrt{n_{\mathrm{b}}}$. According to the experimental evaluation of [218], the number of submatrix used to generate the BPBD matrix does not play a significant role in the reconstruction quality, thus being the single submatrix option the preferred one.

3.2.6. Noiselets

The use of noiselets as measurement functions in CS is proposed in [81], where their performance is demonstrated in the sparse recovery of a realistic synthetic image. Noiselets are presented in [113] as noise-like functions, thus incompressible by orthogonal wavelet packet methods. In particular, a complex-valued noiselet system is presented in [113] that is optimally incoherent with the Haar-Walsh wavelet basis, that is, all Haar-Walsh coefficients exhibit the same value, thus witnessing a perfect energy spread.

Consequently, using a noiselet system as sensing matrix and a Haar-Walsh wavelet basis as dictionary is optimal from a CS point of view [81], as far as the signal admits a sparse representation in such basis. The number of noiselet measurements that ensure recoverability with overwhelming probability is then $\mathcal{O}(s \log n)$. The flat Haar-Walsh and Fourier representations of noiselets are a reliable indicator of a good spread both in time (or, equivalently, space) and (spatial) frequency domains.

Similarly to an FFT or any multiscale filterbank, a full noiselet matrix can be efficiently applied with complexity $\mathcal{O}(n \log n)$ instead of the $\mathcal{O}(n^2)$ that would require a random matrix of the same size. Despite the noiselets are complex-valued, both the real and imaginary parts are binary-valued. This makes them as attractive as the aforementioned binary matrices towards a hardware implementation.

3.2.7. Best Spherical Codes

We do not adopt here the common notation used in *spherical codes* literature, where $n > 2$ denotes the dimensionality of the space and $m > n$ the number of points or codes. Seeking future coherence with our CS notation, we invert the notation, therefore being $m > 2$ the dimensionality of the space where the codes live and $n > m$ the number of codes. With the latter notation, let $\Omega_\circ\left(\vec{0}, 1\right)$ be the surface of the m-dimensional unit sphere centered at the origin of $\mathbb{R}^m$, $\vec{0}$. Then, any finite set of n points on $\Omega_\circ\left(\vec{0}, 1\right)$ is called spherical code and denoted $C_\circ(m, n)$. *Best spherical codes* (BSC) are spherical codes that maximize the minimal Euclidean distance between any two points or, equivalently, minimize the maximal inner product between them. Note that the problem is equivalent to finding the best packing of n (hyper)spheres whose centers are contained in $\Omega_\circ\left(\vec{0}, 1\right)$ [118, 172]. The best packing is that that maximizes the density, that is, the ratio between the volume occupied by spheres and the total volume of the space. This is a classical problem also known as *Tammes's problem*, or the *hard-spheres problem* [394]. Obviously, rotations of a spherical code around the origin yield the same distance distribution and the same packing density. Thus, all rotations of a BSC are regarded as the same BSC, being these often characterized by the distance distribution.

The use of BSCs as sensing matrices in CS was first suggested in [286]. For notation coherence, let's denote the code words of a given spherical code $C_\circ(m, n)$ as $\vec{\phi}_j \in \mathbb{R}^m$, $1 \leq j \leq n$. Then, the jth code word $\vec{\phi}_j$ is the jth column of $\mathbf{\Phi}$. Provided that we want to maximize the mutual distances between code words, one can define a generic energy function whose minimum is attained for BSCs:

$$E_p\left(C_\circ(m,n)\right) = \sum_{u<v}^{n} \frac{1}{\|\vec{\phi}_u - \vec{\phi}_v\|_2^p} \tag{3.63}$$

where $p \in \mathbb{N},\ p > 1$. In the case of $p = 2$, one can think of an equivalence to the Coulomb forces [40], which are also inversely proportional to the square of the distance between charges ($p = 1$ for the difference of potential between the points). That is, n charges of equal value contained in a conductive spherical shell of unit radius centered at the origin would adopt a distribution correspondent to a BSC. This leads to minimizing the potential by means of an equidistributed arrangement. When $p \to \infty$, it has been shown that the global minimum of Eq. 3.63 is attained by a BSC. Provided that minimizing Eq. 3.63 means reaching an equilibrium arrangement, one can express this equilibrium explicitly, either by expressing the location of a charge (that is, a code word) as a function of the mutual difference of locations with respect to all the rest (the so-called *equilibrium of rescaled differences of code words*, ERDW) or directly as a function of the (appropriately scaled) locations of the other charges (*equilibrium of rescaled code words*, ERW). The ERW formulation explicitly shows an important property of BSCs: the sum of all (appropriately rescaled) code words is always zero. Either using the ERDW or the ERW formulation, one gets a system of n equations with n unknowns, both m-dimensional, which can be written as

$$C_\circ(m,n) = \left[\vec{\phi}_j\right]_{1\leq j\leq n} = \left[\underline{\vec{f}}_j\left(C_\circ(m,n)\right)\right]_{1\leq j\leq n} \tag{3.64}$$

where each $\underline{\vec{f}}_j$ is the (either ERDW or ERW) normalized vectorial equation for the jth code word. Here underline denotes vector normalization. In other words, BSCs have to be fixed points of the mapping in Eq. 3.64. Approaching a solution to Eq. 3.64 iteratively fails due to the lack of convergence. Observe that the solution set can rotate endlessly around the origin, still satisfying Eq. 3.64. In order to solve this issue, one can add another term that partially *anchors* the code words to themselves, that is, a one-to-one mapping. This is the idea behind the iterative approach proposed in [286] and reproduced in Eq. 3.65, where the result of the *quasi*-one-to-one mapping at iteration k is used as BSC input of the mapping at iteration $k+1$. The first term of the sum at the right hand side of the equation is the one-to-one mapping, while the second is the mapping induced by the ERDW or ERW equations. If the *damping factor* $\alpha \in \mathbb{R}$ is low enough, the iterative process converges to the fixed points of the mapping given by the right hand side of Eq. 3.65, which coincide with those of the original mapping in Eq. 3.64, thus yielding the BSC.

$$C_\circ(m,n)^{(k+1)} = \left[\vec{\phi}_j\right]_{1\leq j\leq n}^{(k+1)} = \left[\vec{\phi}_j^{(k)} + \alpha \vec{f}_j\left(C_\circ(m,n)^{(k)}\right)\right]_{1\leq j\leq n} \tag{3.65}$$

If the matrix $\mathbf{\Phi}$ is a BSC, then the maximal inner product between any column pair is minimal. Nevertheless, this does not translate directly into minimization of the inter-column coherence, as defined in Eq. 3.42. The absolute value in Eq. 3.42 is the responsible of this mismatch. Equivalence between minimizing the maximal inner product between BSC words and minimizing the coherence holds in the case of *antipodal spherical codes*, in which, if $\vec{\phi}_j$ is a code word of the spherical code, $-\vec{\phi}_j$ is also a code word. The procedure in Eq. 3.65 can be used to generate *best antipodal spherical codes* (BASC), if the functions $\vec{f}_j$ are adapted to generate antipodal spherical codes (see [286]). An example of alternative algorithm is the hybrid algorithm proposed in [199], which combines both the aforementioned concept of energy (Eq. 3.63) minimization with maximization of the angle between code words (equivalent to minimizing the dot product).

Furthermore, it has been shown that *best complex antipodal spherical codes* (BCASC) constitute optimal sensing matrices with respect to inter-column coherence [486], that is, if there exists a set of n m-dimensional vectors attaining the minimum coherence established by the Welch bound (Eq. 3.46), then this set is given by the BCASC. If optimal coherence is attained with BCASCs, one could argue that any effort destined to look for a *better* sensing matrix is futile. Indeed, the coherence evaluation of BCASCs in [486] yields values that tightly approach or even equal the lower bounds on the coherence given in Section 3.1. Nevertheless, the fact that BCASC sensing matrices are both complex and dense, together with the high computational cost of generating them, brings up the question whether *close-to-optimal* performance, that is, *close-to-minimal* coherence could be attained with much more implementation-friendly matrices, e. g., sparse binary matrices in the direction of BPBD matrices.

3.2.8. Low-Density Parity-Check codes

BPBD matrices exemplify the fact that sparse binary matrices can be used as sensing matrices. Furthermore, if the BPBD matrix is generated from a single submatrix, it is optimal in terms of sparsity. Nevertheless, in general, close-to-optimal performance of a BPBD matrix is not guaranteed. This is mostly due to the fact that the column permutations are random. Obviously, given a certain column permutation scheme, generated at random, it seems possible, even highly probable, that there exists a different one that yields lower coherence with the dictionary and, consequently, better recovery performance. Similarly, it also seems possible that certain column permutation schemes lead to failure cases due to eventual high coherence between some sensing kernel and some dictionary atom. Despite a probabilistic analysis would assign such cases a negligible probability, the mere fact that it is not zero implies that no BPBD generated via random column permutations can be considered optimal in absolute terms. The issue is, therefore, not related to the *probable performance*, which has been shown to be good, but to its *possible instability*, which is a well-known drawback of bringing random operations into the game.

This instability issue can only be solved by means of a fully *deterministic* construction of the sensing matrix, which still has to maximize the ratio between gathered information and number of measurements. This requirement, together with the desire of a highly-sparse binary structure, naturally leads to the low-density parity-check (LDPC) codes, a widely studied area in information theory. LDPC codes were conceived by R.G. Gallager (after him, also named *Gallager codes*) [195, 196] as linear error correcting codes and are intended to enable data transmission over noisy or unreliable channels. Provided that the codes are specified by a sparse rectangular matrix, which we directly identify as our desired sensing matrix $\mathbf{\Phi} \in \mathbb{R}^{m\times n}$, each column of $\mathbf{\Phi}$ contains $d_\mathrm{s} \mid 3 \leq d_\mathrm{s} \ll m$ nonzero coefficients (ones) while each row contains $d_\mathrm{c} \mid d_\mathrm{s} < d_\mathrm{c} \ll n$ nonzero coefficients (ones). At this point, we encourage the reader to compare the first two submatrices in the Fig. 2 of [195] to the Fig. 2 of [218]. Note the equivalent goal of both constructions. Both in [195] and in [218] the number of nonzeros per column d_c equals the number of submatrices. Furthermore, despite LDPC matrices can be constructed in a deterministic manner, in [195] the construction starts with a first block diagonal submatrix and the rest are generated from it by applying different random column permutations, thus very similarly to the BPBD matrix in [218]. Interestingly, in the case of transmission through channels with memory, Gallager recommends to apply random permutations also to the first submatrix, thus yielding full equivalence between his LDPC matrix and the BPBD matrix in [218]. To the best of our knowledge, this equivalence has not been explicitly pointed out in the literature before.

The connection between channel coding and CS was formally established in [144]. The LDPC matrix is constructed in a deterministic way seeking the maximum possible *girth* of the corresponding Tanner graph [432]. Using a Tanner graph to generate the LDPC matrix translates into efficiency for large n. A LDPC matrix is uniquely defined by its Tanner graph. In graph theory, the term *girth*, denoted by g, refers to the length of the shortest path in the graph. Assuming deterministic generation of the high-girth LDPC matrix, [144] was the first work providing a deterministically-constructed binary CS sensing matrix with order-optimal number of rows. A deterministic method to construct the LDPC matrix had already been suggested in the Appendix C of [196].

As pointed out in [144], the girth can be used as an indicator of the quality of a binary sensing matrix, from a CS point of view. Indeed, if the girth is too low, the sensing might be redundant, at least locally, thus making suboptimal use of the few m measurements. A well-known method to generate Tanner graphs with large girth is the *progressive edge-growth* (PEG) construction, proposed in [231] and further developed in [232, 473]. The Tanner graph of a LDPC matrix $\mathbf{\Phi} \in \mathbb{R}^{m \times n}$ has n *symbol* nodes and m *check* nodes. An edge between the ith check node, $1 \leq i \leq m$, and the jth symbol node, $1 \leq j \leq n$, in the Tanner graph means that $\phi_{i,j} = 1$, being zero all other matrix coefficients. The number of edges starting in a symbol node (nonzero elements per column) is the *symbol node degree* d_{s}, while the number of edges reaching a check node (nonzero elements per row) is the *check node degree* d_{c}. With such graph in mind, one can define a *local girth* per symbol node as the length of the shortest cycle passing through it. Then, the girth is the minimum of such local girths. The PEG construction is *progressive* in the sense that each new edge to be added to the graph is generated depending on the previous. Specifically, suppose that all the edges of the first j symbol nodes have been established. Then, the edges of the symbol node $j+1$ are generated so that they do not impair the current temporal girth. Due to the excessive complexity of searching for the optimal set of edges at the symbol node $j+1$, the edges are selected and added one after another, so that the local girth after each addition is maximized. Note that this is a compromise solution that selects the *best* candidate between all possible edges (up to a certain depth) starting from the the symbol node $j+1$, but it is not an optimal procedure and, therefore, it does not offer any guarantee that the maximum girth is being achieved.

Despite the PEG construction is suboptimal, the resulting LDPC matrices are gifted with a lower bound on their girth, which cannot be established if the graph edges are randomly located. Lower and upper bounds on the girth are given in [231]. Another important parameter is the minimum

(Hamming) distance between codes, which directly monitors the redundancy of the sensing kernels. Thus, the minimum distance is to be kept as high as possible. Note that, between binary codes of length n, orthogonality is approached when the minimum distance approaches n. Bounds on the minimum distance of regular LDPC codes are given in [433] and extended to irregular LDPC codes in [417]. A lower bound on the minimum distance of regular PEG Tanner graphs is also given in [231]. As a direct consequence of the lower bounds on the girth and minimum distance, PEG-LDPC matrices are observed to perform better than their non-deterministic counterparts. We would like to recall here the conjecture in [231] that the decoding performance (sparse recovery in the CS counterpart) of LDPC matrices might be dominated by their *girth histogram* [321] and not exclusively by the *global* girth, especially for large n.

The study of LDPC matrices as sensing matrices in [306] showed that they exhibit higher orthogonality than equivalent randomly-generated sparse binary matrices. In fact, if any two columns cannot share more than one '1' in the same position, i. e., if $g > 4$, the coherence between them is either zero or $1/d_s$, where d_s denotes the degree of the columns. Obviously, as d_s is increased, so it does the probability of getting $1/d_s$ coherence between a randomly chosen column pair, instead of zero. A design criterion proposed in [306] to attain a tradeoff between high column degree and high girth is to choose the highest d_s for which most (eventually all) local girths are higher than four. This ensures close-to-optimal orthogonality, which cannot be attained with a random construction. As pointed out in [91], satisfying the RIP of order k is equivalent to requiring that every set of columns with cardinality equal or less than k approximately behaves like an orthonormal system. Thus, the PEG construction of LDPC matrices, which aims to maximize the girth, constitutes a deterministic method to generate sensing matrices satisfying the RIP. Apart from being binary and sparse, cyclic or quasi-cyclic structure can be attained [296], thus increasing the attractiveness of PEG-LDPC matrices towards a hardware implementation.

A near-optimal binary and sparse sensing matrix With the (recall suboptimal) girth-maximizing PEG construction procedure on one hand and the RIP on the other hand, the authors of [307] propose that the PEG-LDPC matrix constructed with the maximum column degree $d_s^{\max} < \sqrt{n} < n/2$ that still allows for $g > 4$ is a near-optimal CS binary sensing matrix. Furthermore, for k low enough, the matrix satisfies the RIP of order k with δ_k given by Eq. 3.66.

$$\delta_k = \frac{3k-2}{4d_s + k - 2} \tag{3.66}$$

When dealing with signals whose sparsity is fairly large, Eq. 3.66 is no longer valid and an alternative derivation for δ_k is provided in [307], in which the probability of getting nonzero coherence between column pairs is taken into account. PEG-LDPC matrices with appropriately-chosen d_s were shown to widely outperform both Gaussian and (sparse) binary random matrices in terms of sparse recovery performance in all experimental cases considered in [306] and [307].

Note that the near-optimality is meant in the sense of ability to satisfy the RIP, intrinsically superior than randomly-generated matrices. Nevertheless, the RIP is a sufficient, and not strictly necessary, condition for signal recovery. Thus, it might not be the optimal tool for evaluating the quality of a sensing matrix. In this regard, we would like to recall the claim in [99] that $0,1$-*matrices are bad with respect to the RIP*. This claim was motivated by the fact that such matrices were shown to need, in general, more than $\mathcal{O}(s \log n/s)$ measurements. Specifically, the binary matrices constructed in [142] require $\mathcal{O}((k/\alpha)^2)$, $/\alpha > 0$ being a constant, which is coherent with the lower bound on m given in [99]. Note that both in Theorems 3.1 and 3.4 of [142] were derived independently from the girth, offering, therefore, general results, which might be too pessimistic for PEG-LDPC matrices. The fact that PEG-LDPC matrices have been shown to outperform classical CS random sensing matrices must be interpreted as a sign in this direction. Further deterministic procedures for constructing sensing matrices exist but will not be described here. The reader is referred to [350] for a survey on this.

3.2.9. Optimization of Sensing Matrices

Random matrices exhibit low inter-column coherence, but deterministic constructions can be used to achieve (close to) minimal coherence with convenient structures, e. g., sparse binary matrices. An interesting alternative is to perform an *a posteriori* optimization of a given sensing or measurement matrix, that might have been generated randomly. Of this flavor are the *optimized projections* in [162] or posterior works [484, 477] based on the Gram matrix shrinkage introduced in [162]. These algorithms start from a fixed dictionary $\boldsymbol{\Psi}$ and a candidate sensing matrix to optimize $\boldsymbol{\Phi}$. At each iteration k, the temporal measurement matrix, $\boldsymbol{A}^{(k)} = \boldsymbol{\Phi}^{(k)}\boldsymbol{\Psi}^{(k)}$ yields a Gram matrix $\boldsymbol{G}^{(k)} = \boldsymbol{A}^{(k)\top}\boldsymbol{A}^{(k)}$. Some threshold on the inter-column coherence of $\boldsymbol{A}$ is used to *shrink* the Gram matrix by means of some shrinkage function. The rank of the *shrunk* Gram matrix is forced to be m by means of an SVD decomposition. The optimization is attained by updating $\boldsymbol{\Phi}$ so that the product with $\boldsymbol{\Psi}$ is as close as possible to the square root of the *shrunk* Gram matrix. In such an iterative framework, the optimization means progressively adapting the sensing matrix so that the inter-column coherence of $\boldsymbol{A}$ is reduced. Clearly, if $\boldsymbol{\Psi}$ is the identity matrix, that is, if we deal directly with sparse signals, these algorithms optimize $\boldsymbol{\Phi}$ in terms of its own inter-column coherence, but if this is not the case, the optimized $\boldsymbol{\Phi}$ depends on $\boldsymbol{\Psi}$.

The shrinkage function proposed in [162] contemplates three cases and was substituted by a simpler two-case function in [484]. As pointed out in [162], *hard* shrinkage functions translate into faster convergence, but increase the risk of eventual divergence. Optimizing the coherence means gathering more informative projections, thus being able to attain equivalent recovery performance with lower m when comparing to the original random $\boldsymbol{\Phi}$ or, equivalently, reducing the recovery error for equal m. It should be noted that these approaches optimize some *t-averaged coherence*, as defined in [162], and not the worst-case coherence in Eq. 3.42. As the name suggests, the t-averaged coherence is the average of the absolute values of the normalized inner products between all possible pairs of columns of $\boldsymbol{A}$ that are above the threshold t. This is one more indicator of the fact that it is not the *worst-case* coherence what determines, by itself, the quality of a measurement matrix, but its *distribution*. One can easily see that two equal columns (thus, yielding unit coherence), being the rest highly-incoherent between each other, are less harmful than having all columns highly coherent, but with coherence lower than one. Applying the optimization in [162] to a Fourier ensemble and considering an identity dictionary yields a matrix that is similar to the scrambled Fourier ensemble but that does not rely on randomness to achieve low coherence. As a result, the optimized Fourier ensemble improves the error rate by a factor of 10 when using BP and OMP as recovery methods, with respect to the non-optimized Fourier ensemble.

A different measure of coherence used when $\boldsymbol{\Psi}$ is not the identity matrix is the mutual coherence between rows of $\boldsymbol{\Phi}$ and columns of $\boldsymbol{\Psi}$, $\mu(\boldsymbol{\Phi}, \boldsymbol{\Psi})$ (Eq. 3.39). In [487] the method for constructing BASCs in [286] is adapted to obtain sensing matrices that minimize $\mu(\boldsymbol{\Phi}, \boldsymbol{\Psi})$ for a given $\boldsymbol{\Psi}$. The sparse recovery performance of the resulting sensing matrices is compared to those obtained from the optimization in [162], in terms of frequency of exact recovery. Equivalent performances are obtained using BP (Section 3.4.1.1) as recovery method, but the optimization method in [162], optimizing the inter-column coherence of $\boldsymbol{A}$, outperforms the adapted version of [286], optimizing $\mu(\mathbf{\Phi}, \mathbf{\Psi})$, when OMP (Section 3.4.2.2) is adopted as recovery method.

A different flavor of sensing matrix optimization is to *train* the sensing matrix using a training dataset, that is, to learn optimal projections from a representative set of signals, similarly to dictionary learning (Section 3.3.2). This is the idea behind the Uncertain Component Analysis (UCA) proposed in [460]. The authors show that the optimal projections are not those given by PCA or ICA (those would be the optimal representation vectors), but those that maximize the probability of the data, given the projections and the training data. UCA is closely related to the concept of *adaptive sensing*, discussed in Section 3.2.10, with the difference that in *adaptive sensing* the projections are adapted to a single signal (that being sensed) and not to a complete dataset. In the noiseless case, a random matrix is a valid solution for the UCA, since it maximizes the UCA cost function with unit probability.

3.2.10. Beyond Sensing Matrices: Adaptive Sensing

Most of the fundamental works in CS assume a fixed sensing matrix, that is, that the sensing kernels are predefined and, thus, measurements are gathered non-adaptively. This classical focus is also adopted in this thesis. Nevertheless, one should not ignore the superior results that have been reported using adaptive sensing schemes in some cases. In non-adaptive sensing, given the signal dimensionality n and the expected sparsity s, a lower bound on m can be derived and a close-to-optimal sensing matrix can be constructed, e. g., a LDPC matrix via PEG as described in [307]. Note that whatever *optimality* the sensing matrix holds, it is referred to sensing s-sparse signals, in general, but not the specific sparse signal, say $\vec{x} \in \mathbb{R}^n$, we aim to recover. For clarity, observe that an $s \times n$ sensing matrix with only a non-zero element per row at the locations given by $\text{supp}(\vec{x})$ would be optimal, superior than any other $m \times n$, $m > s$ sensing matrix in terms of sensing efficiency. Unfortunately, such a perfect adaption is only possible having oracle knowledge of $\text{supp}(\vec{x})$, which is unknown. Nonetheless, if each measurement y_i, $1 \leq i \leq m$ is taken sequentially, then the sensing kernel used to gather y_{i+1} can be designed using the information on $\vec{x}$ given by the previous i measurements.

A way of implementing adaptive sensing is making use of a Bayesian framework. Suppose that we have a prior probability distribution for $\vec{x}$, say $p(\vec{x})$. Then, after gathering each measurement, one can compute a posterior distribution via the Bayes rule. Then, the next sensing kernel is to be designed so that it maximizes the expected information gain, which is equivalent to minimizing the conditional (Shannon) entropy of the posterior distribution. Different expressions for the information gain are provided in [299].

A representative example of adaptive sensing is the *distilled sensing* (DS) procedure proposed in [217]. The algorithm is conceptually simple and operates sequentially, refining a candidate support set at each iteration, say $\Omega^{(k)}$ at iteration k. This set is originally initialized with all possible indices $i \in [1, n]$ and the corresponding i^{th} measurement kernel is a 1-sparse row vector with nonzero element at position i. That is, in the first iteration measurements are carried out according to an identity sensing matrix (conventional sampling). In each iteration, $\Omega^{(k)}$ is updated according to: $\Omega^{(k+1)} = \left\{ i \mid y_i^{(k)} > \varepsilon^{(k)} \right\}$, where $y_i^{(k)}$ denotes the measurement gathered by the ith measurement kernel at iteration k and $\varepsilon^{(k)} \geq 0$ is a noise-dependent threshold that decreases with k. In other words, the next measurements are gathered according to a sensing matrix that derives from that of the previous iteration, excluding those measurement kernels (rows) that led to (close to) zero measurements. Note that the method is designed to operate with sparse signals with coefficients $x_i \geq 0,\ \forall 1 \leq i \leq n$. In the DS implementation of [217], $|\Omega^{(k+1)}| \approx |\Omega^{(k)}|/2$, thus requiring a total of $m \approx 2n$ measurements. In other words, DS is able to estimate the support of sparse signals that are masked by noise, but at the cost of a large number of measurements $m > n$. As a consequence, the consideration of DS as a CS method is arguable.

In non-adaptive sensing, correct estimation of the support of an s-sparse signal $\vec{x} \in \mathbb{R}^n$ requires the amplitude of the nonzero components to be $x_i > C_s\sqrt{\log n}\ \forall i \in \Omega$, where $C_s > 0$ is a constant that depends on s and Ω the sparse support set. A similar result is presented in [87], where the lasso was shown to require a minimum amplitude of $8\sigma\sqrt{2\log n}$ for the nonzero components of the signal. Adaptivity allows leveraging this requirement. For instance, the aforementioned DS procedure softens it to $x_i \geq \mu(n)\ \forall i \in \Omega$, where $\mu(n) > 0$ is an arbitrarily slow strictly increasing function of n, that is, $\mu(n) \rightarrow \infty \Leftrightarrow n \rightarrow \infty$.

3.2.11. Our Proposal: Merging Adaptiveness and Determinism for Close(r)-to-Optimal Low-Density Sensing Matrices

At a first glance, *determinism* and *adaptiveness* are two fundamentally different approaches to build a sensing matrix. While a deterministic construction of a low-density binary sensing matrix distributes a limited budget of nonzero elements so that the inter-column coherence is minimal, adaptive sensing designs sensing kernels that recursively adapt to the signal support, yielding $m > n$ rows in the aggregated sensing matrix. One would wish to enjoy the benefits of a close-to-optimal construction, such as that described towards the end of Section 3.2.8, without leaving aside adaptiveness. In other words, it is desirable to find a sequential construction method of low-density binary matrices that is close-to-optimal in terms of inter-column coherence, but that also takes into account the information on the signal support contained in the measurements gathered using the previous sensing kernels, when generating the new one.

This is the idea underlying our Adaptive Progressive Edge-Growth method for constructing Low-Density Parity-Check matrices (APEG-LDPC) presented in [H9]. The authors propose a simple method for creating highly-sparse binary sensing matrices which is both *deterministic* and *adaptive* and retains the benefits of both characteristics, while overcoming the specific weaknesses of each one. More specifically, deterministic constructions of LDPC matrices have shown to provide close-to-optimal sensing matrices, but this optimality is referred to sensing s-sparse signals, in general, but not the specific sparse signal, say $\vec{x} \in \mathbb{R}^n$, we want to recover. The APEG construction focuses on this issue and progressively adapts the sensing kernels to the signal being sensed. On the other hand, the drawback of adaptiveness is an increased number of measurements, due to its iterative measurement schema. First low quality measurements are required to obtain subsequent highly-informative (adapted) measurements. The APEG construction avoids the sensing redundancy of, e. g., DS, by means of an underlying PEG construction, thus yielding a reduced number of measurements $m < n$.

In the APEG construction, adaptiveness means that the addition of new edges in the Tanner graph at the *check node* i depends not only on the existing edges in the graph, as in conventional PEG, but also on the information about $\operatorname{supp}(\vec{x})$ contained in the $i-1$ measurements gathered according to the previous $i-1$ *check codes* (recall, rows) in the temporal LDPC matrix. There are two fundamental differences between APEG and the PEG construction in [231]:

1. **Construction by check nodes**. The LDPC construction in [231] proceeds by *symbol nodes* of the Tanner graph. Provided that we use measurements obtained from previous *check codes* (LDPC matrix rows) to generate the following code, the APEG construction has to operate sequentially by *check nodes*.

2. **Forbidden support set.** In [231], each new edge added to a symbol node is selected so that it maximizes the local *girth* of the node for the given temporal Tanner graph. In our case an additional constraint is added for the creation of a new edge: a new edge originating at the check node i is to be selected so that it maximizes the local *girth* of the node and ends in a symbol node whose index is *not contained* in a temporal *forbidden support set* $\Omega_{\times}^{(i)}$.

In the following, we denote the *check* and *symbol* nodes by $c_i,\ 1 \leq i \leq m$ and $s_j,\ 1 \leq j \leq n$, respectively. Provided that the construction proceeds by check nodes, the check node degree $d_{\mathrm{c}} \geq 1$ is used to control the *density* of the matrix. The minimum degree that ensures that all n symbol nodes are connected to some check node in the Tanner graph is $d_{\mathrm{c}}^{\min} = \left\lceil \frac{n}{m} \right\rceil$. Furthermore, provided that the resulting LDPC matrix is highly regular, the minimum degree of the symbol nodes can be calculated as:

$$d_{\mathrm{s}}^{\min} = \left\lfloor \frac{m}{n} d_{\mathrm{c}} \right\rfloor \tag{3.67}$$

The edges of the Tanner graph are duplets (c_i, s_j). Let Π_i denote the set of such duplets containing c_i at the end of the construction process, that is, the edges originating in the node c_i. Note that $|\Pi_i| = d_{\mathrm{c}}\ \forall i,\ 1 \leq i \leq m$. Then the set $\Pi = \bigcup_{i=1}^{m} \Pi_i$ contains all the edges in the graph and we have that $|\Pi| = m d_{\mathrm{c}}$ is the number of nonzero elements in $\mathbf{\Phi}$. During construction, $\Pi_i^{(k)}$ denotes the temporal set of $k \leq d_{\mathrm{c}}$ edges already added to the node c_i. Similarly to [231], we define the *symbol neighborhood within depth l* of the check node c_i as the set containing all symbol nodes reached by a tree with root node c_i and l levels and is denoted as $\Omega_{c_i}^{l}$. The complementary set is denoted by $\bar{\Omega}_{c_i}^{l}$. Each *level* of the tree has two *layers*, a first layer of check nodes and a second of symbol nodes. Due to the sequential APEG construction, the measurements are also acquired sequentially: $y_i = \vec{\phi}_i^{\top} \vec{x}$, where $\vec{\phi}_i^{\top}$ denotes the ith row of the LDPC matrix under construction $\mathbf{\Phi}$. Armed with this notation, the reader can obtain a complete understanding of the operation of our APEG construction procedure by observing the pseudo-code given in Algorithm 1, where $\Omega_{c_i}^{\mathrm{cand}}$ is just an intermediate set containing the indices of candidate symbol nodes to be connected with the check node c_i and d_{s_j} denotes the temporal degree of the symbol node s_j.

The outputs of Algorithm 1 are both the LDPC sensing matrix, constructed by rows as $\mathbf{\Phi} = \left[\vec{\phi}_i^{\top} \right]_{1 \leq i \leq m}$ and the vector of measurements $\vec{y} = [y_i]_{1 \leq i \leq m}$.

Algorithm 1 Adaptive Progressive Edge-Growth (APEG)

Initialize: $\Omega_{\times}^{(1)} = \emptyset,\ \Pi_i^{(0)} = \emptyset\ \forall i,\ 1 \le i \le m$
for $i = 1; i := i + 1$ **to** $i = m$ **do**
  for $k = 1; k := k + 1$ **to** $k = d_{\mathrm{c}}$ **do**
    if $k = 1$ **then**
      Candidate set: $\Omega_{c_i}^{\mathrm{cand}} = \bar{\Omega}_{\times}^{(i)}$
    else
      Expand tree up to depth $l \mid \bar{\Omega}_{c_i}^{l} \neq \emptyset$ **and** $\bar{\Omega}_{c_i}^{l+1} = \emptyset$
      Candidate set: $\Omega_{c_i}^{\mathrm{cand}} = \bar{\Omega}_{c_i}^{l} \cap \bar{\Omega}_{\times}^{(i)}$
    end if
    Select symbol node index: $j = \underset{j \in \Omega_{c_i}^{\mathrm{cand}}}{\arg\min}\, d_{s_j}$
    Add new edge: $\Pi_i^{(k)} = \Pi_i^{(k-1)} \cup (c_i, s_j)$
  end for
  Sensing kernel: $\vec{\phi}_i \in \mathbb{R}^n \mid \phi_{i,j} = 1 \Leftrightarrow (c_i, s_j) \in \Pi_i$
  Measure: $y_i = \vec{\phi}_i^{\top} \vec{x}$
  if $y_i < \varepsilon$ **then**
    Update forbidden support: $\Omega_{\times}^{(i+1)} = \Omega_{\times}^{(i)} \cup \mathrm{supp}\left(\vec{\phi}_i\right)$
  else
    Preserve forbidden support: $\Omega_{\times}^{(i+1)} = \Omega_{\times}^{(i)}$
  end if
end for

Algorithm 1 ensures that each new edge (c_i, s_j) added to the temporal Tanner graph has the lowest possible impact on g, by selecting the symbol node s_j that appears for the first time in the deepest level of the subtree expanded from c_i. Then, the length of the shortest cycle passing through the new edge (c_i, s_j) is guaranteed to be at least $2l$, $l \ge 1$. Additionally, the information given by each new measurement y_i is used to update the *forbidden support set* for the next sensing kernel $\Omega_{\times}^{(i+1)}$. The symbol nodes whose indices are contained in $\Omega_{\times}^{(i+1)}$ are not considered when generating new edges in the Tanner graph. Consequently, the resulting LDPC codes used as sensing kernels progressively focus on the signal support and avoid sensing the zero locations. As in [217], we consider nonnegative sparse signals, which is the case, for instance, in pulsed ToF depth sensing. The noise dependent threshold ε can be calculated as $\varepsilon = d_{\mathrm{c}}\sigma / c_{\sigma}$, where σ is the expected standard deviation of some additive noise on the entries of $\vec{x}$

and $c_\sigma \geq 1$ is just an user-defined safety factor. High c_σ yields robustness to noise, but precludes the adaptation process, while too low values (e. g., $c_\sigma < 1$) might lead to failure due to a wrong adaptation.

For comparison purposes, we consider also a baseline for PEG construction without adaptation, which follows Algorithm 1 with the only difference of having a fixed $\Omega_{\times}^{(i)} = \emptyset \; \forall i, \; 1 \leq i \leq m$, thus allowing precalculating $\mathbf{\Phi}$ independently from the measurements, as in conventional PEG. Fig. 3.4 depicts two LDPC matrices of size 250×1024 , constructed using the PEG baseline (Fig. 3.4a) and our APEG algorithm (Fig. 3.4b). In both cases $d_\mathrm{c} = 20$, that is, there are exactly 20 nonzero elements per row. The signal $\vec{x}$ to measure was 10-sparse. Note the homogeneous distribution of the nonzero elements in Fig. 3.4a, visually equivalent to a random distribution, still being fully deterministic and maximizing g. In Fig. 3.4b the first rows of the matrix look like in Fig. 3.4a, but, as the adaptation continues, the measurement kernels progressively focus on a smaller set of support elements.

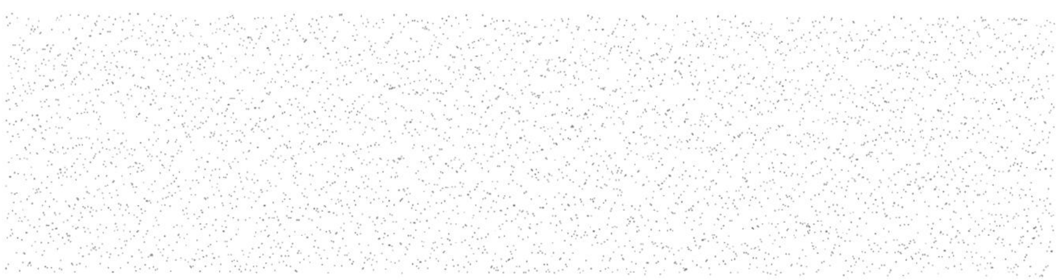

(a) PEG

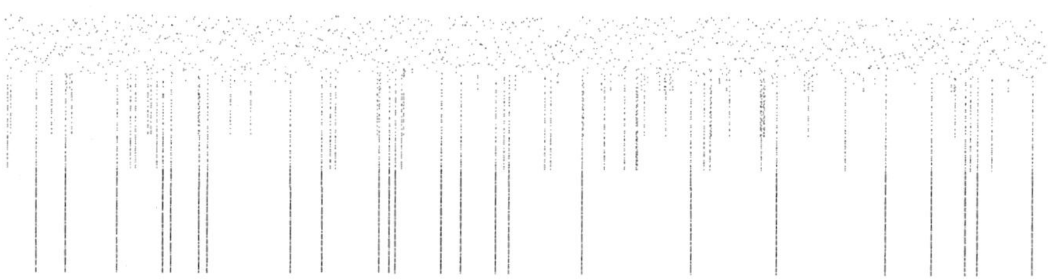

(b) APEG

Figure 3.4.: Graphical representations of two LDPC matrices of size 250×1024 and equal check node degree $d_\mathrm{c} = 20$. Black squares symbolize nonzero elements. The matrix in (a) was obtained via non-adaptive PEG construction, while (b) was generated via our APEG algorithm, therefore progressively adapting the LDPC codes (binary sensing kernels) to a sparse signal of s=10. ©2016 IEEE.

In the following we compare the inter-column coherence and the recovery performance of LDPC sensing matrices constructed via our APEG algorithm to those of equivalent matrices constructed using the PEG baseline and random matrices generated using a zero-mean and unit-variance normal distribution and Bernoulli distributions with elements ± 1 and $\{0, 1\}$. In all the experiments the signal dimensionality is $n = 1024$.

Inter-column Coherence If d_c is chosen to be sufficiently low, our PEG baseline yields a Tanner graph with $g > 4$, as well as the PEG procedure in [231] does for d_s low enough. As pointed out in [307], this yields a maximum inter-column coherence of $\mu^{\max} = 1/d_s$. Using Eq. 3.67 to express this upper bound in terms of our input parameter d_c yields:

$$\mu \leq \frac{1}{\left\lfloor \frac{m}{n} d_c \right\rfloor} \tag{3.68}$$

In our experiments we consider relatively low values of d_c, since further increments do not yield to improvements in the sparse recovery. For these low values, $g > 4$ and Eq. 3.68 was observed to hold in the non-adaptive case. Fig. 3.5 shows the inter-column coherence obtained for Gaussian matrices (Fig. 3.5a), Bernoulli matrices with ± 1 elements (Fig. 3.5b), Bernoulli matrices with $\{0, 1\}$ elements (Fig. 3.5c) and matrices constructed using our PEG baseline with $d_c = 60$ (Fig. 3.5d), for different values of m.

Gaussian and Bernoulli matrices with ± 1 elements exhibit the lowest values of coherence. Bernoulli matrices with $\{0, 1\}$ elements have slightly worse coherence due to the fact that the originating distribution is not centered at zero. This translates into a DC offset that rises the coherence. The PEG-LDPC matrices exhibit good coherence properties for s and m/s high enough, that is, the bottom-right area of Fig. 3.5d, where the coherences are almost equal to those in Fig. 3.5c. This is possible thanks to the deterministic PEG construction and constitutes a remarkable fact, provided that the density of the sparse Bernoulli matrices is more than 8.5 times greater than that of the PEG-LDPC matrices with $d_c = 60$. The bound in Eq. 3.68 was observed to hold for the results in Fig. 3.5d, even without the floor operator. Our APEG-LDPC matrices have not been included in the coherence comparison, provided that the adaptation process blows up the inter-column coherence, making any comparison meaningless. Nevertheless, note that the APEG goes from a close-to-PEG construction for the first rows, where minimum coherence is pursued, to a close-to-oracle-driven sparse sensing for the last rows. Therefore, the high coherence of the APEG matrices is *good*, in the sense that is due to sensing at the *right* locations.

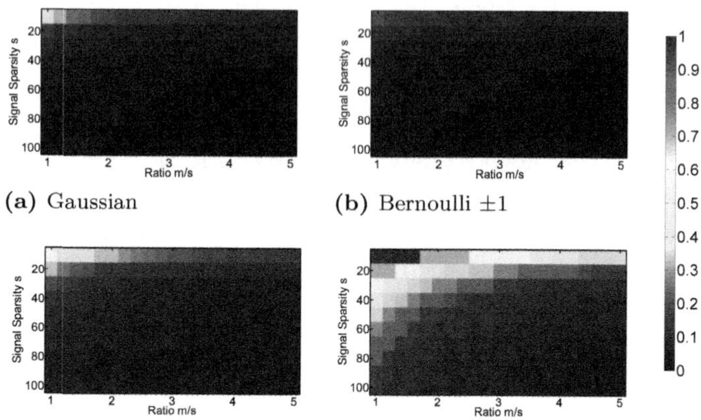

(a) Gaussian (b) Bernoulli ± 1

(c) Bernoulli $\{0, 1\}$ (d) PEG-LDPC, $d_c = 60$

Figure 3.5.: Coherence comparison between random matrices and deterministic LDPC matrices constructed via the PEG baseline. The random matrices are Gaussian (a) and Bernoulli matrices with ± 1 elements (b) and $\{0, 1\}$ elements (c). Signal sparsities between 10 and 100 are considered. For consistency with further experiments, m is given as a function of the sparsity s. Ratios of m/s between 1 and 5 are considered for each s.

As d_c increases, the advantage of the APEG construction over the PEG is expected to vanish, since the room for adaptivity decreases with increasing density of the sensing kernels. Consequently, the coherence plots obtained for APEG should show a progressive transition from almost-unit coherence for all s and m considered to plots equal to those obtained for PEG, with increasing d_c. Such evolution with excessively high values of d_c is illustrated for PEG and APEG LDPC matrices in Figs. 3.6 and 3.7, respectively. Note that in these cases $g = 4$ and the coherence is some entire multiple of the bound in Eq. 3.68. Observe the clear superiority of the PEG construction (Fig. 3.6) over a random construction (Fig. 3.5c). For values of $d_s \approx m/2$ (50% density), that is, $d_c \approx n/2$ (from Eq. 3.67), the coherence of LDPC matrices constructed using PEG (Fig. 3.6e) and APEG (Fig. 3.7e) are close to each other, due to the fact that the high density of the sensing kernels obtained for $d_c = 500$ precludes adaptivity. Additionally, the plots are also close, still superior, to that for equivalent Bernoulli matrices with $\{0, 1\}$ elements (Fig. 3.5c).

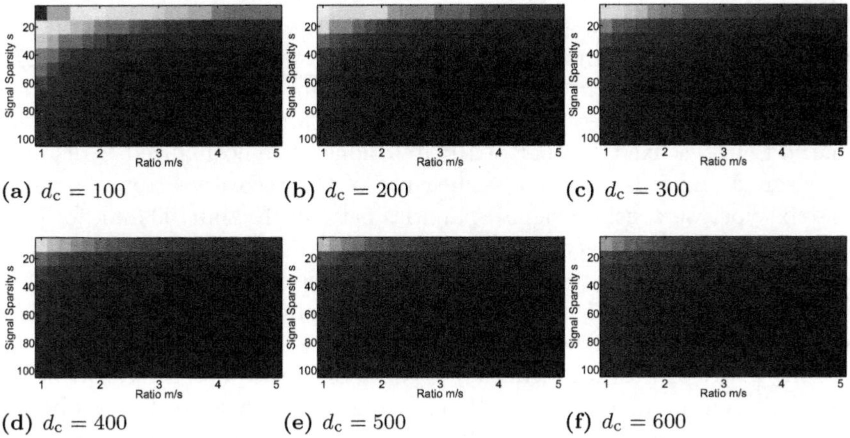

(a) $d_c = 100$ (b) $d_c = 200$ (c) $d_c = 300$

(d) $d_c = 400$ (e) $d_c = 500$ (f) $d_c = 600$

Figure 3.6.: Coherence of non-adaptive PEG sensing matrices for excessively high values of check node degree $d_c \in \{100, 200, 300, 400, 500, 600\}$. For each d_c, the same cases for s and the ratio m/s as in Fig. 3.5 are considered. Scale bar in Fig. 3.5-right.

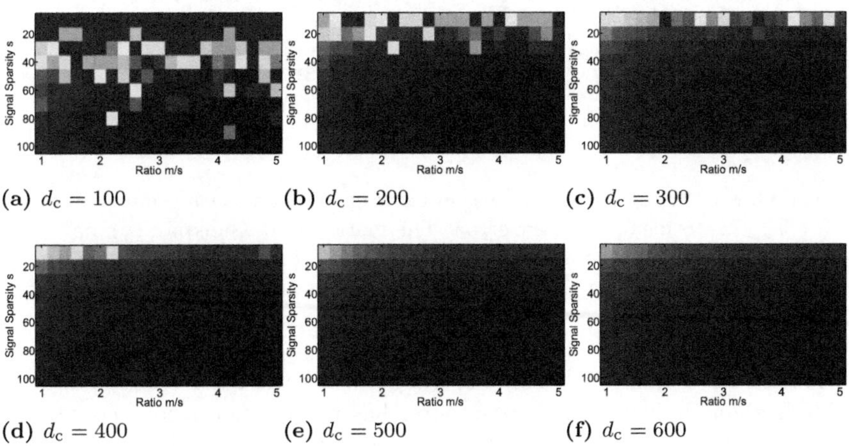

(a) $d_c = 100$ (b) $d_c = 200$ (c) $d_c = 300$

(d) $d_c = 400$ (e) $d_c = 500$ (f) $d_c = 600$

Figure 3.7.: Coherence of our APEG sensing matrices for excessively high values of check node degree $d_c \in \{100, 200, 300, 400, 500, 600\}$. For each d_c, the same cases for s and the ratio m/s as in Fig. 3.5 are considered. Scale bar in Fig. 3.5-right.

Sparse Recovery Performance In the following, we present some experimental results on the sparse recovery performance. The goal of the experiments was to recover an s-sparse n-dimensional signal $\vec{x}$ from $m \ll n$ linear measurements $\vec{y} = \mathbf{\Phi}\vec{x}$. The efficient Chambolle and Pock's algorithm for l_1-minimization (see Section 3.4.1.3 for details) is adopted as recovery method, with a fixed number of 10^3 iterations. The normalized l_2 distance between $\hat{\vec{x}}$ and $\vec{x}$ is adopted as measure of the recovery error. For each matrix type, we consider signal sparsities between 10 and 100 and, for each of them, ratios m/s between 1 and 5. This way we avoid using too many measurements to sense too sparse signals or gathering too less measurements when s is high. Fig. 3.8 shows the recovery errors obtained for the random sensing matrices. Observe the good performance of Gaussian and Bernoulli sensing matrices with ± 1 elements, in contrast to the poorer performance of Bernoulli sensing matrices with $\{0, 1\}$ elements, which seems to corroborate the aforementioned claim in [99]. In Fig. 3.8a and Fig. 3.8b a sharp transition between exact recovery and fail is observed, unveiling the minimum ratio m/s for which exact recovery is achieved.

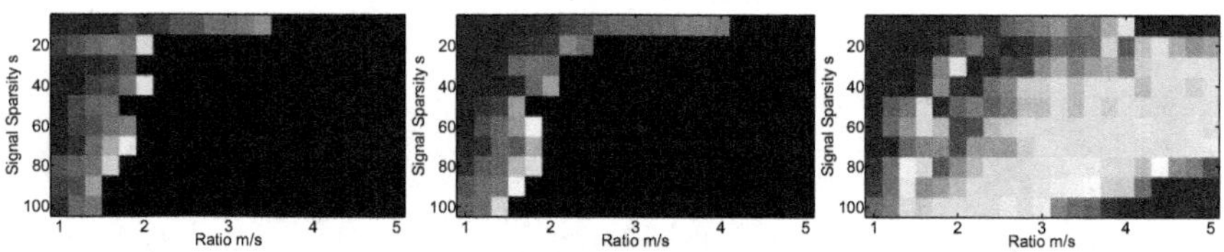

(a) Gaussian **(b)** Bernoulli ± 1 **(c)** Bernoulli $\{0, 1\}$

Figure 3.8.: Performance comparison between random sensing matrices in terms of recovery error. The matrices are Gaussian (a) and Bernoulli matrices with ± 1 elements (b) and $\{0, 1\}$ elements (c). For the considered combinations of s and the ratio m/s and the scale bar, see Fig. 3.5. ©2016 IEEE.

For LDPC sensing matrices constructed using the PEG baseline and the APEG algorithm, one more parameter has to be taken into account: the check node degree d_c. For both constructions, $d_c \in \{10, 20, 30, 40, 50, 60\}$ are considered. Figs. 3.9 and 3.10 provide the results obtained for PEG and APEG LDPC matrices, respectively. Note that all plots in Fig.3.10 show a superior recovery performance than the corresponding plots in Fig.3.9, that is, our APEG construction generates sensing matrices that allow for successful recovery of an s-sparse signal from typically less measurements than those required when using matrices constructed using the non-adaptive PEG baseline. A visual indicator of this superior performance is the area in dark blue, which means exact recovery (normalized errors ranging between 10^{-7} and 10^{-14}). Conversely, experimental cases in colors different than dark blue are almost always failure cases. The superiority of APEG over PEG is clearer for low d_c, e. g., $d_c = 10, 20$, when the sensing kernels are so sparse that a *guidance* on where to place the nonzero elements turns a failure case into exact recovery. The performance of the APEG-LDPC matrices seems to improve with d_c, specially when it is low, but no significant improvement was observed for $d_c > 60$, while the advantage over the PEG baseline decreases when increasing d_c. When compared to the performance of random matrices (Fig. 3.8), both PEG and APEG LDPC matrices widely outperform the binary $\{0, 1\}$ Bernoulli matrix. Additionally, if d_c is conveniently chosen, APEG-LDPC matrices outperform all random matrices considered, regardless of s, but especially for very low s.

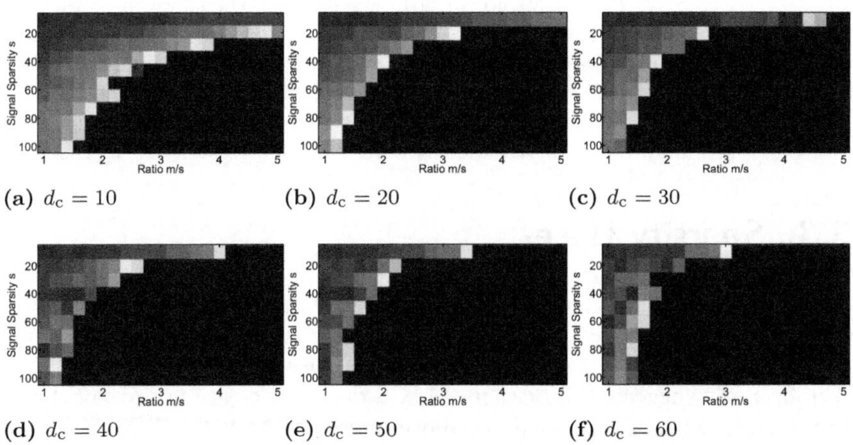

(a) $d_c = 10$ (b) $d_c = 20$ (c) $d_c = 30$

(d) $d_c = 40$ (e) $d_c = 50$ (f) $d_c = 60$

Figure 3.9.: Recovery errors obtained using non-adaptive PEG sensing matrices for $d_c \in \{10, 20, 30, 40, 50, 60\}$. For each d_c, the same cases for s and the ratio m/s as in Fig. 3.5 are considered. Scale bar in Fig. 3.5-right. ©2016 IEEE.

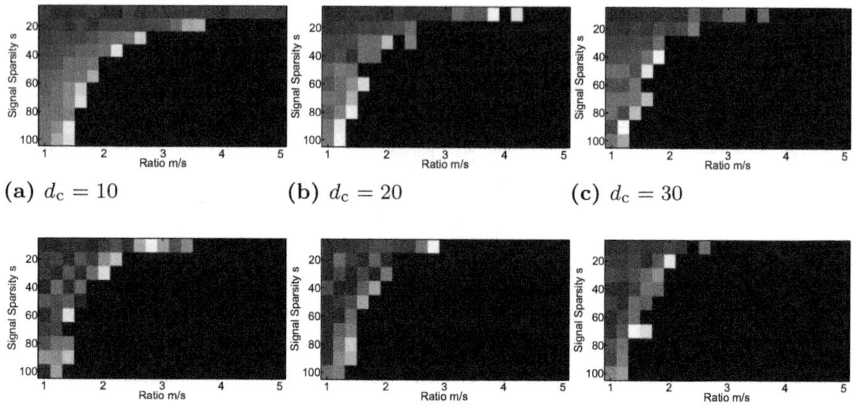

(a) $d_c = 10$ (b) $d_c = 20$ (c) $d_c = 30$

(d) $d_c = 40$ (e) $d_c = 50$ (f) $d_c = 60$

Figure 3.10.: Recovery errors obtained using our APEG sensing matrices for $d_c \in \{10, 20, 30, 40, 50, 60\}$. For each d_c, the same cases for s and the ratio m/s as in Fig. 3.5 are considered. Differently from the results in Fig. 3.8 and Fig. 3.9, where s only influences the values of m given by the m/s ratios we consider, in the APEG case, the specific realization of the s-sparse vector influences the final LDPC matrix. Scale bar in Fig. 3.5-right. ©2016 IEEE.

3.3. Sparsity Bases

Given a signal $\vec{I} \in \mathbb{R}^n$ not necessarily sparse, we call *sparsity basis* or *dictionary* the set of n linearly independent vectors, grouped by columns in a matrix $\mathbf{\Psi} \in \mathbb{R}^{n \times n}$, such that the representation $\vec{x} \mid \vec{I} = \mathbf{\Psi}\vec{x}$ is sparse (Eq. 3.2). As suggested in Section 3.1, CS dictionaries are not always bases, but frames with some degree of redundancy [267, 268]. A quantitative notion of redundancy for infinite frames is given in [17]. In this thesis, as in most CS applications, we deal with discrete dictionaries, that is, with finite-dimensional frames. Faithful to the nomenclature introduced in Section 3.1, let $\mathbf{\Psi} \in \mathbb{R}^{n \times n_{\text{atoms}}}$ be a frame, defined by n_{atoms} n-dimensional column vectors. Then, the classical notion of redundancy is given by the quotient

$$\rho_{\mathbf{\Psi}} = \frac{n_{\text{atoms}}}{n} \tag{3.69}$$

In a tight frame of unit norm, this quotient coincides with the frame bounds $\alpha = \beta$ and is indeed informative of the redundancy. Nevertheless, one can easily think of fundamentally different cases of redundant frames yielding the same redundancy just for having equal n and n_{atoms}, as pointed out in [59]. One of the aspects that cannot be captured by Eq. 3.69 is how *localized* the redundancy is. Repeating the same element n_{atoms} times in the frame yields the same redundancy as repeating every element only once, while the structure of the two resulting frames is fundamentally different. Deleting one element from them would have an effect in the first case (with probability 0.5) and no effect in the second (with probability 1). More interesting is the *redundancy function* defined in [59], provided here adapted to our matrix notation:

$$\begin{aligned}\rho_{\mathbf{\Psi}}\left(\vec{I}\right) &= \sum_{i=1}^{n_{\text{atoms}}} \frac{\left|\left\langle \vec{\psi_i}^{\top}, \vec{I} \right\rangle\right|^2}{\left\|\vec{\psi_i}\right\|_2^2} \\ &= \left\|\mathbf{\Psi}^{\top}\vec{I}\right\|_2^2 \leftrightarrow \left\|\vec{\psi_i}\right\|_2^2 = 1,\ \forall i \leq n_{\text{atoms}}\end{aligned} \tag{3.70}$$

where the second equality only holds if the columns of $\mathbf{\Psi}$ are normalized. From Eq. 3.70, the concepts of *lower redundancy* ($\rho_{\mathbf{\Psi}}^{-}$) and *upper redundancy* ($\rho_{\mathbf{\Psi}}^{+}$) are defined as:

$$\rho_{\mathbf{\Psi}}^{-} = \min_{\vec{I} \in \mathbb{R}^n} \rho_{\mathbf{\Psi}}\left(\vec{I}\right),\ \rho_{\mathbf{\Psi}}^{+} = \max_{\vec{I} \in \mathbb{R}^n} \rho_{\mathbf{\Psi}}\left(\vec{I}\right) \tag{3.71}$$

As proven in [59], in an equal-norm frame, the previous redundancies can be immediately computed from the frame bounds:

$$\rho_{\mathbf{\Psi}}^{-} = \frac{\alpha}{c},\ \rho_{\mathbf{\Psi}}^{+} = \frac{\beta}{c} \tag{3.72}$$

where $c = \left\|\vec{\psi_i}\right\|_2$, $\forall i \leq n_{\text{atoms}}$. By comparing the values of $\rho_{\mathbf{\Psi}}^{-}$ and $\rho_{\mathbf{\Psi}}^{+}$ one can determine how uniform or, conversely, how localized, the redundancy is. Additionally, for equal-norm Parseval frames, both redundancies boil down to Eq. 3.69.

Highly redundant frames allow sparse representation of signals that might not admit sparse representation in a single basis. For instance, signals that are the superposition of two or more signal types that admit, in turn, sparse representations in known basis, might not be sparse in each of those basis separately, while they are indeed sparse in the overcomplete ($n_{\text{atoms}} > n$) dictionary generated as the union of the individual sparsity basis. For illustration, consider the case of an electrical network where active elements, e. g., MOSFETs or IGBTs, are used for fast switching of some highly-inductive charge, e. g., a motor. The large inductance of the charge does not allow for violent changes of the electrical current flowing through it. Consequently, when switching, violent peaks will appear in the network voltage. As a result, if the network is monitored with a device with enough bandwidth, the waveform of the voltage will be the superposition of a sinusoid at the network frequency, typically low (50 or 60 Hz) and spikes at the switching instants. A train of spikes is sparse by itself, that is, in the identity dictionary or trivial basis, while the network signal is very close to a perfect sinusoid and, therefore, extremely sparse in a frequency dictionary, e. g. the Fourier basis. Despite the great sparsity of both components in their respective basis, the resulting signal is not sparse in any of both separately.

Using a union of basis or, in general, any kind of overcomplete dictionary to attain a sparse representation harms the coherence of the resulting measurement matrix, which might be very close to one. As introduced in Section 3.1, incoherence between the columns of the measurement matrix can be posed as a requirement to ensure sparse recovery. Even the RIP is related to the notion of matrix incoherence. Clearly, uniqueness of the solution cannot be guaranteed if the dictionary is constructed as the union of two or more sparsity basis. Nevertheless, we might not be interested in obtaining a unique sparse representation $\vec{x}$, but the *sparsest* possible $\vec{x}$ that explains the signal $\vec{I} = \mathbf{\Psi}\vec{x}$.

In order to show that sparse recovery with redundant dictionaries is feasible, the RIP is extended in [86] to the case of redundant dictionaries in a natural manner. Instead of considering for the RIP of order k all the k-sparse vectors, all vectors lying in the union of all subspaces spanned by all subsets of k columns of the dictionary are considered. The extended RIP is named *D-RIP*, where D stands for dictionary. Adapting the results in [86] to our notation, let $\mathbf{\Psi} \in \mathbb{R}^{n \times n_{\text{atoms}}}$ be an arbitrary tight frame and let $\mathbf{\Phi} \in \mathbb{R}^{m \times n}$ be a sensing matrix satisfying the D-RIP with $\delta_{2k} < 0.08$, then the solution to the l_1-*analysis* problem (cf. the classical l_1-*synthesis* formulation in Eq. 3.10 and 3.21) given in Eq. 3.73 satisfies Eq. 3.74.

$$\hat{\vec{I}} = \underset{\vec{I} \in \mathbb{R}^n}{\arg\min} \left\| \mathbf{\Psi}^\top \vec{I} \right\|_1 \text{ subject to } \left\| \mathbf{\Phi}\vec{I} - \vec{y} \right\|_2 \leq \varepsilon, \tag{3.73}$$

where ε is, as in Eq. 3.21, the expected l_2 norm of the noise vector in measurement domain.

$$\left\|\vec{I}-\hat{\vec{I}}\right\|_2 \leq C_0\varepsilon + C_1 \frac{\left\|\boldsymbol{\Psi}^\top \vec{I} - \left(\boldsymbol{\Psi}^\top \vec{I}\right)_k\right\|_1}{\sqrt{k}}, \ \forall \vec{x} \in \mathbb{R}^n, \tag{3.74}$$

where $C_0, C_1 > 0$ may only depend on δ_{2k} and $(\cdot)_k$ denotes k-thresholding. Observe that the l_1 norm term in the right hand side of Eq. 3.74 corresponds to the k-term approximation error in terms of l_1 norm (cf. Eq. 3.22). In case of dealing with complex matrices, the transposes in Eq. 3.73 and 3.74 ($\boldsymbol{\Psi}^\top$) should be conjugate (Hermitian) transposes.

From now on we use the general term *dictionary*, which may denote both a basis or a highly-redundant frame, depending on the ratio between n_{atoms} and n. As pointed out in [389], there are two fundamentally different ways of generating a dictionary. The first, somewhat classical, approach is to generate the dictionary building upon a mathematical model of the signals to represent. This requires the ability of expressing in mathematical terms some *a priori* knowledge of the signal. We refer to this kind of dictionaries as *analytic dictionaries* and a brief survey is given in Section 3.3.1. The other approach is to *learn* the dictionary from a set of *training* data. Simple signal properties, such as smoothness or periodicity can be easily modeled mathematically, thus adequate dictionaries can be constructed via the first approach. Nevertheless, providing a mathematical model to represent complex knowledge on the signal might not be an easy task, if feasible at all. The optimal representation of periodic signals is achieved by means of a frequency dictionary, but what is the best dictionary for representing, say, images of *horses* or *cars* or any other class of real objects that have very tight restrictions in their shape and structure but also an unmodelable variability? Such cases are examples for which the second approach provides a mean of overcoming the lack of a satisfactory mathematical model. Some of the most common methods for learning dictionaries from training data are reviewed in Section 3.3.2.

3.3.1. Analytic Dictionaries

Some of the analytic dictionaries described below have been tested with real PMD raw images in order to find an appropriate CS dictionary in spatial domain. We refer to Section 4.4.1.1 for some experimental results (see Fig. 4.37 and Fig. 4.38).

3.3.1.1. Frequency Dictionaries

The main drawback of pure frequency dictionaries is the lack of localization. A direct consequence is the inability to sparsely represent discontinuities. For this reason most natural images do not exhibit an exactly-sparse Fourier representation, but *compressible.* The Fourier basis, either complete or selecting the atoms corresponding to the frequency range where the signal is expected to live, e. g., the k-lowest frequencies with $k < n$, is arguably the most representative frequency dictionary. In the continuous case, the (infinite) set of Fourier functions can be expressed as

$$\mathcal{F} = \{\psi_k(t)\}_{k\in\mathbb{Z}} = \left\{e^{\frac{2\pi k t}{T} i}\right\}_{k\in\mathbb{Z}} \tag{3.75}$$

where T is the period corresponding to the fundamental frequency. Alternatively, one can explicitly formulate the basis functions as sines and cosines, thus avoiding the need for complex coefficients when representing real signals, at the cost of two atoms per frequency:

$$\mathcal{F}_{\text{real}} = \left\{\begin{matrix}\psi_{2k-1}(t)\\ \psi_{2k}(t)\end{matrix}\right\}_{k\in\mathbb{Z}} = \left\{\begin{matrix}\sin\left(\frac{2\pi k t}{T}\right)\\ \cos\left(\frac{2\pi k t}{T}\right)\end{matrix}\right\}_{k\in\mathbb{Z}} \tag{3.76}$$

In the discrete case, the continuous functions translate into vectors of finite dimensionality, that are the dictionary atoms. The discrete version of t can only take values $t_j = j(T/n)$, with $1 \leq j \leq n$. The discrete basis functions are directly obtained from Eq. 3.75 and yield the Fourier dictionary:

$$\mathbf{\Psi} = [\psi_{j,k}]_{\substack{1\leq j\leq n\\ 1\leq k\leq n_{\text{atoms}}}} = \left[e^{\frac{2\pi k t_j}{T} i}\right]_{\substack{1\leq j\leq n\\ 0\leq k\leq n_{\text{atoms}}-1}} = \left[e^{\frac{2\pi k j}{n} i}\right]_{\substack{1\leq j\leq n\\ 0\leq k\leq n_{\text{atoms}}-1}} \tag{3.77}$$

Note that we use j, k as row and column indices, respectively, of the elements of $\mathbf{\Psi}$, since i already denotes the imaginary unit. The popularization of the Fourier representation was boosted by the advent of the Fast Fourier Transform (FFT) [119], a method that allows computing the complex Fourier coefficients of an n-dimensional signal in $\mathcal{O}(n \log n)$, rather than $\mathcal{O}(n^2)$, as expected for trivial implementations. The FFT yields a massive speedup for large n. The CS linear model requires an explicit dictionary, $\mathbf{\Psi} \in \mathbb{R}^{n\times n}$ in the case of a basis, thus forcing the complexity $\mathcal{O}(n^2)$. Nevertheless, one could act as if $\mathbf{\Psi}$ were known and simply apply an FFT (or an *inverse* FFT) whenever a left-product by $\mathbf{\Psi}$ (or by $\mathbf{\Psi}^{-1} = \mathbf{\Psi}^\top$) is required in the CS recovery pipeline.

The Discrete Cosine Transform (DCT) [8] can be seen as a simplification of the Discrete Fourier Transform (DFT) which uses exclusively cosines as basis functions. Consequently, for real signals with even symmetry, the DCT is equivalent to a DFT of twice the length (signals with even symmetry have null sine coefficients in the basis given by Eq. 3.76). Conversely, a Discrete Sine Transform (DST) would be equivalent to a DFT for odd signals. All periodic bases imply a periodic extension of the signal, outside its initial domain, provided that the basis functions are periodic. The DCT basis implies an even extension, while the DST basis implies an odd extension. Similarly to the Fourier transform, the DCT requires, in principle, $\mathcal{O}(n^2)$ operations. Transferring the concept of FFT to the DCT, one can implement Fast Cosine Transform (FCT) algorithms running in $\mathcal{O}(n \log n)$ operations. There are eight types of DCTs and another eight of DSTs. The different types differ from each other in around which point the even or odd symmetry of the signal is defined. The most common DCT is the DCT-II, which is, in fact, simply referred to as DCT. The DCT basis functions are given in Eq. 3.78. Note that the first atom ($k = 0$ in the most-right hand of the equation) is a vector of ones and requires further normalization. The block-definition in Eq. 3.78 is just for simplicity.

$$\mathbf{\Psi} = [\psi_{j,k}]_{\substack{1 \leq j \leq n \\ 1 \leq k \leq n_{\text{atoms}}}} = \left[\cos \left[\frac{\pi k \left(j + \frac{1}{2} \right)}{n} \right] \right]_{\substack{0 \leq j \leq n-1 \\ 0 \leq k \leq n_{\text{atoms}}-1}} \tag{3.78}$$

A paradigmatic use case of the DCT is in JPEG compression. For images, the basis elements are two-dimensional, but can be easily obtained from the one-dimensional formulation, by elementwise multiplication of basis elements corresponding to each orthogonal direction. After each 2D basis element is generated, it can be packed in a one-dimensional container using any vectorization scheme. Then, these vectors can be stacked by columns yielding the dictionary $\mathbf{\Psi}$, as presented above. The only detail to take into account is to apply the same vectorization scheme to the image to be sensed. Regarding localization, DCT or DST, as well as DFT, offer null temporal (or spatial) frequency localization. This becomes an important drawback when dealing with large natural images, not just for computational reasons, but due to the fact that the (eventually very few) frequencies present in some area of the image will be missing in others, thus harming the ability of the transforms to yield sparse representations. Nevertheless, in JPEG compression, a DCT transform of size 8×8 pixels (plus border) is used. This allows massive parallel processing of image patches with a GPU-friendly size (a common power of two), thus increasing the speed, at the time that provides a trivial way of localization, at the cost of limiting the maximum period of the low frequencies to the size of the patch. That is, a constant image of size $8k \times 8k$ would require k^2 nonzero DCT coefficients instead of only one. On the other hand, if it contains only relatively high frequencies in the range represented by the 8×8 DCT patches, very localized in space, higher sparsity is attained.

3.3.1.2. Localized-Frequency Dictionaries

As pointed out above, a major disadvantage of pure frequency dictionaries is the lack of localization. Therefore, a logic step is to try to concentrate the support of the dictionary atoms in different small regions, enabling sparse representations of signals that are *locally* very sparse in frequency but not so much globally. This is the idea behind the Windowed Fourier Transform (WFT), also known as Short-Time Fourier Transform (STFT) [12]. In STFT, some kind of *window function* is used to restrict the domain where the signal is *seen* by the transform, provided that the function vanishes out of that region. The 1991 JPEG compression standard [365] makes use of a simple type of STFT (8×8 DCT patches). Considering different types of window functions, e. g., step, triangular or Gaussian functions, different families of STFTs are obtained. One can consider that window functions have two fundamental parameters, one adjusting its position and another its width, thus providing two more degrees of freedom when generating the dictionary atoms. Using large widths allows covering the whole space with a low number of atoms, at the cost of limited localization. Conversely, excessively short widths allow accurate localization at the cost of quality loss in the frequency representation, e. g., inability to represent low frequencies. If the number of considered locations is high and the supports of the window functions overlap, the STFT might yield redundant dictionaries. The STFT covers the general *time-frequency* analysis introduced by D. Gabor [194], from which *time* and *frequency* analysis are just special cases. For this reason, dictionaries constructed from the STFT are also known as *Gabor* dictionaries. An efficient algorithm for producing the Discrete Gabor Transform (DGT) of a signal is given in [381]. Eq. 3.79 provides an example of Gabor dictionary construction, where k_s, $0 \leq k_s \leq n_s - 1$ is the index corresponding to temporal or spatial location and k_f, $0 \leq k_f \leq n_f - 1$ is the frequency index. Therefore, the dictionary considers n_s locations for each of the n_f frequencies, yielding $n_{atoms} = n_s n_f$ atoms. The window function $w(\cdot)$ is a low-pass function, e. g., Gaussian, centered at zero. The parameters α and β adjust the temporal/spatial and frequency resolution, respectively. Low values of α yield better temporal/spatial resolution, but require larger n_s. Equivalently, low values of β yield better frequency resolution, but require larger n_f.

$$\begin{aligned}\boldsymbol{\Psi} = [\psi_{j,k}]_{\substack{1\leq j\leq n\\ 1\leq k\leq n_{\mathrm{atoms}}}} &= \left[w(j-\alpha k_\mathrm{s})e^{\frac{2\pi\beta k_\mathrm{f} j}{n}i}\right]_{\substack{1\leq j\leq n\\ 0\leq k_\mathrm{s}\leq n_\mathrm{s}-1\\ 0\leq k_\mathrm{f}\leq n_\mathrm{f}-1}}\\ \text{with: } k_\mathrm{s} &= k(\mod n_\mathrm{s})-1\\ k_\mathrm{f} &= \lfloor k/n_\mathrm{s}\rfloor\end{aligned} \tag{3.79}$$

As well as in Eq. 3.77 and Eq. 3.78, the frequency index k_f starts at 0 in the right side of Eq. 3.79, in order to include the zero frequency. Also the temporal/spatial index k_s starts at zero. This should not be interpreted as a conflict with respect to the indices of the left sides of the equations, which refer to matrix elements and start always at 1. In the multidimensional case, Gabor atoms can have an orientation, thus yielding more degrees of freedom during construction, which should translate into a better adaptation to the target signals. It is natural to see a connection between the two-dimensional Gabor atoms and the cortical receptive field profiles (RFPs) proposed by J.D. Daugman [133]. In fact, Daugman himself showed in [134] that the 2D RFPs of simple cells in mammalian visual cortex can be well represented by Gabor functions. We refer to the Fig. 5 of [132] for an illustration of the astonishing similarity between empirical RFPs and Gabor functions.

We would like to finish this short reference to Gabor functions with the original formulation by Gabor of a function that *occupies the minimum area* $\Delta t\Delta f=\frac{1}{2}$, which is the *modulation product of a harmonic oscillation of any frequency with a pulse of the form of a probability function*, that is, a general Gabor function with Gaussian window $w(t)=e^{-\frac{(t-t_0)^2}{\sigma_w^2}}$. Such function, in combination with the complex Fourier formulation yields the elegant exponential formulation in Eq. 3.80 for the continuous case. The location and the frequency of the Gabor function are determined by t_0 and f_0, respectively, and σ_w adjusts the sharpness of the Gaussian modulation.

$$\begin{aligned}\psi_{t_0,f_0}(t) &= e^{-\frac{(t-t_0)^2}{\sigma_w^2}+2\pi f_0 t i}\\ \Delta t &= \sqrt{\frac{\pi}{2}}\sigma_w,\ \Delta f=\frac{1}{\sqrt{2\pi}\sigma_w}\end{aligned} \tag{3.80}$$

3.3.1.3. Localization plus Multiple Scales: Wavelet Dictionaries

Gabor dictionaries provide an appropriate sparsifying basis for any signal whose power is concentrated in few or a single region of the time-frequency plane. Not in vain Gabor named his time-frequency diagrams *information diagrams*, since they show exactly where the information of the signal is contained. Nevertheless, Gabor frames will not constitute, in general, an orthogonal basis and the Gabor coefficients cannot be directly obtained via scalar product of the signal with the Gabor atoms. In [132] a three-layer neural network (NN) is proposed to find the best Gabor coefficients in terms of MSE of the reconstructed image. Note that this might not be necessarily the sparsest solution. Furthermore, if α and β in Eq. 3.79 are chosen to be too small, the computational cost of any sparse recovery algorithm derived from the resulting n_{atoms} might be unaffordable. Also, recall that for $n_{\text{atoms}} > n$ uniqueness of the sparse representation cannot be guaranteed. In other words, we seek a way to choose temporal/spatial and frequency resolutions so that the Gabor dictionary remains an excellent representation matrix, while having as less redundancy as possible.

The previous issue was solved with the advent of *wavelets*. In short terms, a function is called a wavelet if it can be used to define a Hilbert basis, i. e., a complete orthonormal basis, by means of dyadic translations and dilations. Such a function is often called *mother* wavelet, being her children called simply wavelets. Note that wavelets can be seen as special cases of Gabor frames, where both the window function and the locations and frequencies are chosen in such a way that the set of Gabor functions is a complete orthonormal basis of the space. In fact, the wavelets introduced by Morlet in the early works [333, 334], dedicated to the analysis of seismic signals, were Gaussian-modulated, as in Eq. 3.80 (thus, nonorthogonal) and were called Gabor wavelets. We refer to the works of I. Daubechies [129], S. Mallat [315, 318] and Y. Meyer [325] for complete first-hand introductions into wavelets. A very general formal definition of wavelets in the continuous case is given in Eq. 3.81, where $\psi(t)$ denotes the basic wavelet, centered at zero, and $\hat{\psi}(\omega)$ its Fourier transform.

$$\begin{aligned} \psi_{\alpha,\beta}(t) &= \alpha^{-\frac{1}{2}}\psi\left(\frac{t-\beta}{\alpha}\right) \\ \text{with: } \int \psi(t)dt &= 0 \Leftrightarrow \hat{\psi}(\omega)|_{\omega=0} = 0 \end{aligned} \tag{3.81}$$

Clearly, the parameter β provides localization, while α *scales* the mother wavelet, thus providing control over its frequency. Large values of the scaling parameter α correspond to low frequencies and vice-versa. The zero-mean condition in Eq. 3.81 derives from the so-called *admissibility condition*, given in Eq. 3.82, which is more general, in combination with the assumption that $\psi(t)$ decays rapidly.

$$\int \frac{\left|\hat{\psi}(\omega)\right|^2}{\omega} d\omega < \infty \tag{3.82}$$

Then a fundamental question is the choice of $\psi(t)$. A common function that satisfies the second equality in Eq. 3.81 is the so-called *Mexican hat* function, which is the second derivative of a Gaussian function: $\psi(t) = (1 - t^2)e^{\frac{-t^2}{2}}$. The key difference with respect to a WFT is the use of *scaling* instead of a window function of fixed size, that is, the width of wavelets is adapted to its frequency, so that high frequency wavelets are narrower and low frequency wavelets are wider. This is indeed the idea behind the initial formulation of the *constant shape ratio* Gabor wavelets in [334]. For a formal comparison between WFT and wavelets, we refer to [128] Eq. 3.81 can also be reformulated for discrete values of α and β, e. g., $\alpha = \alpha_0^m$ and $\beta = n\left(\beta_0 \alpha_0^m\right)$ with $\alpha_0 > 1$, $\beta_0 > 0$. Note that the position parameter also scales accordingly with α, that is, lower frequency atoms (thus wider temporal support) exhibit higher separation between each other than higher frequency atoms. Note the natural way this three degrees of freedom of the WFT, namely, position, frequency and width, are tied together in the wavelet transform. The discrete version of Eq. 3.81 can then be used to construct a dictionary as follows:

$$\boldsymbol{\Psi} = \left[\psi_{j,k}\right]_{\substack{1 \leq j \leq n \\ 1 \leq k \leq n_{\mathrm{atoms}}}} = \left[\alpha_0^{-\frac{k_s}{2}} \psi\left(\alpha_0^{-k_s} j - k_{\mathrm{p}} \beta_0\right)\right]_{\substack{1 \leq j \leq n \\ 0 \leq k_{\mathrm{s}} \leq n_{\mathrm{s}}-1 \\ 0 \leq k_{\mathrm{p}} \leq n_{\mathrm{p}}(k_{\mathrm{s}})-1}} \tag{3.83}$$

where n_{s} denotes the number of scales considered in the dictionary and $n_{\mathrm{p}}\left(k_{\mathrm{s}}\right)$ the number of locations considered at scale k_{s}. Differently from WFT, different number of locations are considered for each scale, since coarser scales require less temporal/spatial resolution than finer scales. If the number of locations at the first scale $n_{\mathrm{p}}(0) = n_{\mathrm{p}}^0$ is a power of α_0, then one can define $n_{\mathrm{p}}\left(k_{\mathrm{s}}\right) = n_{\mathrm{p}}^0 / \left\lceil (\alpha_0)^{k_{\mathrm{s}}} \right\rceil$. The index k in the left hand side of

Eq. 3.83 is some bijective mapping of the indices k_s, k_p in the right hand side, needed to get the 2D matrix $\mathbf{\Psi}$. Obviously, $n_{atoms} = \sum_{k_s=0}^{n_s-1} n_p(k_s)$.

Orthogonality was brought to wavelets by Meyer in [324], who proposed to select $\alpha_0 = 2$, thus inducing a dyadic system. In the continuous case, the wavelet system proposed by Meyer takes the form:

$$\psi_{k_s,k_p}(t) = 2^{\frac{k_s}{2}} \psi\left(2^{k_s} t - k_p\right), \; k_s, k_p \in \mathbb{Z} \tag{3.84}$$

Bases constructed from Eq. 3.84 where the first analytically-constructed orthogonal bases having good localization properties both in temporal/spatial and frequency domains. The reader might have noticed a mismatch in the sign of the exponents in Eq. 3.84 with respect to the generic wavelet equations. In practice, this only determines whether the construction goes from fine to coarse scales or from coarse to fine. Provided that the base is 2, a coarser scale has always half the resolution of the previous or, conversely, a finer scale doubles the resolution of the previous. It should be noted that the approximation of a signal at a scale of resolution 2^{k_s+1} suffices to compute its approximation in the next scale of smaller resolution 2^{k_s}, thus allowing for a sequential *multiresolution* analysis [315] that can be efficiently implemented. Mallat constructs his multiscale wavelets from a a pair of localized functions, namely, the *scaling function* (often called *father function*) and the *mother wavelet*. The latter has the effect of a bandpass filter, whose bandwidth is halved at each coarser scale. The former is used to filter the lowest (coarsest) level of the transform. A framework for computing the Discrete Wavelet Transform (DWT) is provided in [416]. The DWT is obtained by means of a filter bank structure with low-pass and bandpass filters. The low-pass filters are applied prior to decimation, while the bandpass filters yield the wavelet coefficients at each scale. Matrix forms of the 2D DWT are provided in [455] as an explicit dictionary for CS applications.

Similarly to frequency dictionaries,wavelets can also be extended to multiple dimensions. Two-dimensional wavelet bases yield efficient representations of natural images and the JPEG 2000 compression standard [9] uses a DWT instead of a DCT. However, in the multidimensional case, the wavelets are generated from their one-dimensional versions, thus yielding more degrees of freedom in the dictionary design. Consider, for instance the two-dimensional case for images, with *horizontal* and *vertical* dimensions. If the 2D wavelets are generated from 1D wavelets of equal frequency, the dictionary will only be effective for signal with *point* singularities, but will fail when representing *edge* singularities. If a wavelet of high frequency along one dimension is combined with another of lower frequency in the other dimension, edges along the latter dimension might be sparsely represented, but such versatility of the dictionary comes again at the cost of massive redundancy. Still, edges with arbitrary orientations cannot be appropriately handled unless some mean to *steer* the wavelets is introduced.

A crucial observation regarding 2D wavelet representations of natural images is its sparse tree structure, due to the fact that natural images show a decaying frequency spectrum. Consequently, when representing natural images, the energy contained in the different subbands decreases for finer scales. The values of wavelet coefficients at a certain scale are lower than those of the corresponding *parents* in the immediately coarser scale. For this reason, zero *parents* cannot have nonzero *children*, thus yielding a sparse tree, where nodes in the deepest levels (finer scales) are only due to (few) localized singularities in the image. In a 2D wavelet decomposition, each coefficient has *four* descendants (note that each dyadic square obviously splits into four dyadic square children), thus yielding a so-called *quad-tree.* Quad-trees for which all nodes are equal or smaller than the root are called *zerotrees.* Fast wavelet encoding strategies using a zerotree, such as the embedded zerotree wavelet algorithm (EZW) [414] exploit the fact that the wavelet coefficients of natural images decrease with scale. Similarly to the NN for estimating Gabor coefficients in [132], NNs can also be used to obtain optimal wavelet decompositions, e. g., the *wavelet networks* (WNN) introduced in [483].

We would like to finish this section with brief notes on some of the most important wavelet classes from which orthonormal bases can be constructed, namely the Haar, Meyer, Daubechies and Coiflet families. All the formulas of these and other wavelets can be found in the excellent compilation by I. Daubechies [129], to which we also refer for orthonormality proofs. The suitability of the 2D extensions of some of these families as sparsity bases for our signals in spatial domain is analyzed in Section 4.4.1.1.

Haar wavelets Provided that Haar wavelets date back to 1909, when Alfréd Haar introduced the after him named *Haar sequence* [210], they can be considered the earliest wavelets. The Haar functions were intended to be an orthonormal system for the space of square-integrable functions on $[0,1]$. What makes Haar wavelets of special interest is the fact that the mother wavelet takes values ± 1 on $[0,1)$ and is zero everywhere else, while the scaling function is 1 on $[0,1)$ and also zero everywhere else. For completeness, the formal definitions of both functions are given in Eq. 3.85 and Eq. 3.86, respectively. Similarly to the case of binary (or ternary) sensing matrices, Haar dictionaries may be used as a faster alternative to more complex wavelets in a hardware implementation. If Gabor wavelets are the counterpart of the Fourier basis, endowed with localization capabilities, Haar wavelets are the localized counterpart of the Walsh basis. Recall that Walsh functions form an orthogonal binary basis for periodic signals. The rows (or columns) of the Hadamard matrices used in Section 3.2.3 for generating pseudorandom binary sensing matrices are sets of discrete Walsh functions. The Haar wavelet family can also be seen as a special case of a Daubechies wavelet, with only one vanishing moment (DB1 in our notation). The lack of continuity of Haar wavelets may be an advantage when representing signals with sharp transitions.

$$\psi(t) := \begin{cases} 1, \ 0 \le t < \frac{1}{2} \\ -1, \ \frac{1}{2} \le t < 1 \\ 0, \ \text{elsewhere} \end{cases} \tag{3.85}$$

$$\phi(t) := \begin{cases} 1, \ 0 \le t < 1 \\ 0, \ \text{elsewhere} \end{cases} \tag{3.86}$$

Meyer wavelets Meyer wavelets are also defined by a mother wavelet and a scaling function. These functions are both defined in frequency domain as given in Eq. 3.87 and Eq. 3.88, respectively. A common choice for the (smooth) function $\nu(x)$ is $\nu(x) = x, \ 0 < x < 1$, satisfying the conditions in Eq. 3.87 elsewhere.

$$\begin{aligned} \psi(\omega) &:= \begin{cases} \frac{1}{\sqrt{2\pi}} \sin\left[\frac{\pi}{2}\nu\left(\frac{3|\omega|}{2\pi} - 1\right)\right] e^{\frac{\omega}{2}i}, \ \frac{2\pi}{3} \le |\omega| \le \frac{4\pi}{3} \\ \frac{1}{\sqrt{2\pi}} \cos\left[\frac{\pi}{2}\nu\left(\frac{3|\omega|}{4\pi} - 1\right)\right] e^{\frac{\omega}{2}i}, \ \frac{4\pi}{3} \le |\omega| \le \frac{8\pi}{3} \\ 0, \ \text{elsewhere} \end{cases} \\ \nu(x) &:= \begin{cases} 0, \ x \le 0 \\ 1, \ x \ge 1 \end{cases}, \ \text{satisfying: } \nu(x) + \nu(1-x) = 1 \end{aligned} \tag{3.87}$$

$$\phi(\omega) := \begin{cases} \frac{1}{\sqrt{2\pi}}, \ |\omega| \le \frac{2\pi}{3} \\ \frac{1}{\sqrt{2\pi}} \cos\left[\frac{\pi}{2}\nu\left(\frac{3|\omega|}{2\pi} - 1\right)\right], \ \frac{2\pi}{3} \le |\omega| \le \frac{4\pi}{3} \\ 0, \ \text{elsewhere} \end{cases} \tag{3.88}$$

Daubechies wavelets Differently from the previous wavelets, Daubechies wavelets cannot be given as a combination of a mother wavelet and a scaling function in a closed form. Daubechies wavelets [130] are defined via the desired number of *vanishing moments.* Given the constraints due to the moment and orthogonality conditions, the wavelet is designed so that the scaling filter has extremal phase. There are two different notations for designing a Daubechies family, namely 'Dn_{taps}' and 'dbn_{moms}', where n_{taps} denotes the number of taps and $n_{\text{moms}} = n_{\text{taps}}/2$ the number of vanishing moments. In this thesis we adopt the notation 'DBn_{moms}' to refer to Daubechies wavelets with n_{moms} vanishing moments. Daubechies wavelets are referred in [130, 129] as *compactly supported wavelets* and they exhibit the highest possible number of vanishing moments that is compatible with their support width. Wavelets with n_{moms} vanishing moments have a support of $2n_{\text{moms}} - 1$. Having a high number of vanishing moments leads to high compressibility because the wavelet coefficients of fine scales are zero in the smooth regions of the signal. Complete symmetry between mother wavelet and scaling function cannot be achieved except for the case DB1 (Haar wavelets). The extremal phase requirement is achieved by selecting always the roots inside (or outside) the unit circle. If the selected roots are those inside the unit circle, minimum phase is attained and maximum phase when chosen outside. In the minimum and maximum phase cases, the obtained basis functions are mirror images of each other.

Coiflets I. Daubechies gave the name *coiflets* to this wavelet class because it was R. Coifman who suggested to construct orthonormal wavelet bases with vanishing moments not only for the mother wavelet, as DB wavelets, but also for the scaling function, which is the core idea of this family. Consequently, coiflets have an *order*, say L, which is the number of vanishing moments of the wavelet, i. e., $n^{\psi}_{\text{moms}} = L$. The scaling function has then $n^{\phi}_{\text{moms}} = L - 1$ vanishing moments, provided that the zero-order moment has to be unit, thus nonzero. As a result, more regular and symmetric basis functions are obtained.

3.3.1.4. Overcomplete Wavelet Dictionaries

Wavelets can be used to build orthogonal bases that offer an optimal tradeoff between temporal/spatial localization and frequency localization. If the goal is to represent signals that are confined to small regions of the time-frequency plane, wavelet bases seem to be the best option. Indeed, as pointed out in [334] in a brief remark on the absence of contradiction with respect to the classic results by Shannon, wavelet representations are able to retain more information in fewer bits than the standard sampling method. Nevertheless, there are two major weaknesses of wavelets, namely, *translation* sensitivity and *rotation* sensitivity in the multidimensional case. The first was pointed out in [42] as lack of *shift invariance* of the wavelet coefficients. Indeed, as mentioned in [419], *no subsampled subband transform can be translation invariant.* The solution is to eliminate the subsampling step associated with each coarser scale, thus yielding an *undecimated* or *non-subsampled* wavelet transform, known as *stationary wavelet transform*, among other names [185]. Eliminating the subsampling suggests the need of $\mathcal{O}(n^2)$ operations to compute the coefficients, but the algorithm in [42] computes the coefficients of all n circulant shifts of the n-dimensional signal in $\mathcal{O}(n \log n)$ operations. The second issue relates to the multidimensional case and demands some way of rotating the wavelets around a number of orthogonal rotation axes that is given by the number of dimensions of the space.

A 2D wavelet transform that is *jointly shiftable* both in position and orientation is proposed in [419] and referred to as *steerable pyramid.* They use the frequency transformation method [297] to generate the 2D bandpass filters with quasi-circular symmetry from the original 1D radial filters. For constructing the set of oriented filters, the annular bandpass spectrum is divided into orientation subbands and then the filters are obtained by means of the frequency sampling method. The obtained basis functions are translations, dilations and rotations of a single kernel, thus yielding a representation in which the signal power contained in each subband is invariant to translations and to the relative alignment between the input and the transform sampling lattices. Overcoming translation and rotation sensitivity translates into larger n_{atoms} of the dictionary and the subsequent loss of orthogonality, yielding overcomplete dictionaries.

3.3.1.5. Going Multidimensional: Wedgelet and Derivatives

The translation to higher dimensions comes at the cost of massively overcomplete dictionaries and, for achieving good translation and rotation invariance of the representation quality, prohibitive values of n_{atoms} might be required. For this reason, multidimensional dictionaries have to be carefully designed, so that a good tradeoff between redundancy and representation efficiency is achieved. *Wedgelet* dictionaries [148] are 2D overcomplete dictionaries whose atoms are squares of constant value aligned with the axes and bisected by straight lines. The basic function is repeated for different scales and locations in a wavelet fashion. Wedgelet dictionaries are optimal for representing piecewise constant images. One of the merits of wedgelets is the moderate number of dictionary atoms, $n_{\text{atoms}} = \mathcal{O}(\log_2 nn^2)$, for square $n \times n$ images. A wedgelet construction method and a fast algorithm for wedgelet decomposition are also provided in [148].

For smooth images, wavelet dictionaries yield sparse decompositions. If the image contains isolated point singularities, any tree-based decomposition will retain most information in the coarser scales, being the coefficients of fine scales nonzero only in the locations of the singularities. The situation changes when dealing with *piecewise smooth* images. Wavelets cannot provide sparse representations of edges and enforcing sparsity translates into the aforementioned *ringing* artifacts. In [454] the authors approach a solution to this issue by combining wavelets, efficient in the smooth areas, with their *wedgeprints*, orthogonal projections of *wedgelets* onto a wavelet subspace, that is, efficient in representing the edges. Consequently, wedgelets are *mounted* on an underlying wavelet tree structure and entire subtrees of wavelet coefficients, necessary to represent an edge in a conventional wavelet representation, can be estimated from a single wedgeprint coefficient.

Another derivative of wedgelets are *platelets* [466], which were conceived for application in photon-limited imaging, where the data is severely corrupted by photon shot noise (listed as one of the factors affecting the depth measurement uncertainty in Section 2.3.3). Except from the Haar case (recall, DB1), wavelets are not suitable for denoising Poisson-corrupted data, mostly because the kind of *salt-and-pepper* noise that appears in dark areas is eventually highly-correlated with some of the well-localized fine-scale (high frequency) wavelets. Similarly to *steerable* wavelets, platelets are localized atoms generated at different locations, scales and orientations. They were shown to outperform conventional wavelet and wedgelet approximations in terms of representation efficiency and, differently from these and most previous representations, are suitable for the analysis of Poisson data. For each dyadic square corresponding to a terminal node of a recursive partition, the platelet function is defined within the dyadic square as an oriented plane, thus having only three parameters, namely, two slopes, corresponding to the two spatial dimensions and a constant offset. Out of its dyadic square the platelet function is zero. Such functions provide accurate sparse representations of smooth images containing smooth boundaries.

Surflets [100, 101] are an extension of wedgelets to the high-dimensional case, say $N \gg 2$ dimensions. Each surflet atom is then an N-dimensional piecewise constant function with an $(N-1)$-dimensional polynomial discontinuity. Consequently, wedgelets are just a special case of surflets with $N = 2$. Similarly to wedgelets, surflets are intended to provide efficient representations of piecewise constant or smooth signals, containing discontinuities of arbitrary smoothness. A cornerstone of surflets is its method to quantize and encode higher order polynomial coefficients, without which the number of polynomial parameters and thus the size of the dictionary would become intractable. In the same way *wedgeprints* where derived from *wedgelets*, in [101], *surflets* serve as a basis to define *surfprints*. Equivalently to the wedgeprints-wedgelets pair, surfprints are linear combinations of wavelet basis functions, which are obtained by projecting the piecewise polynomial surflets onto a wavelet subspace.

Wedgelets, in the 2D case, and surflets for higher dimensionality, are to be confronted to *ridgelets* [94, 85], which were also created to deal with $(N-1)$-dimensional polynomial discontinuities in N-dimensional signals (lines for $N = 2$, planes for $N = 3$, etc.). The *ridge* functions are N-dimensional functions that are constant along hyperplanes, thus looking like a *ridge* in the 2D case. For its attractive combination of formulation simplicity and effectiveness, we provide the definition of a ridgelet function in the 2D case,

$$\begin{aligned} \psi_{\alpha,\beta,\theta}(\vec{x}) &= \frac{1}{\sqrt{\alpha}}\psi\left(\frac{x_1\cos\theta + x_2\sin\theta - \beta}{\alpha}\right) \\ \int \psi(x)dx &= 0 \end{aligned} \tag{3.89}$$

where $\psi : \mathbb{R} \to \mathbb{R}$ is a smooth univariate function with sufficient decay and vanishing mean, $\alpha > 0$ is the scaling parameter, $\beta \in \mathbb{R}$ is the location parameter and $\theta \in [0, 2\pi)$ is the orientation parameter. The function is constant along the *ridges* $x_1\cos\theta + x_2\sin\theta = C$, being C a constant. The suffix *-let* comes from the fact that, traverse to these *ridges*, the function is a wavelet. The ridgelet transform can be seen as applying a one-dimensional wavelet transform to the slices of the Radon transform [382] computed over hyperplanes $\vec{a}'\vec{x} = c$, with $\vec{a}$ constant and c varying. Similarly to the NN approach for Gabor decomposition in [132] and the WNN in [483], also ridgelet decompositions were implemented in a NN fashion. More specifically, a three-layer *feed-forward* NN (FNN) architecture is proposed in [478].

3.3.1.6. Beyond Straight Edges: Curvelets

Until the advent of *curvelets* the only way to efficiently represent edges of arbitrary shape was making use of adaptive techniques, able to track the shape of the discontinuity. Therefore, curvelet frames were the first allowing for stable, efficient and near-optimal representations of smooth objects having discontinuities along smooth *curves*. Curvelets were first introduced in [85] and further refined in [84], where two different discrete implementations of the *second generation* curvelet transform are provided, for two and three dimensions. The fast discrete curvelet transforms (FDCTs) in [84] exhibit $\mathcal{O}(n^2 \log n)$ complexities, both for the forward and the inverse transform, for arrays of size $n \times n$. One of the novelties of curvelets with respect to previous representations in 2D is the parabolic scaling relation, which ties two of the free parameters to the scale. At the jth scale each atom has an envelope aligned along a *ridge* of length $2^{-j/2}$ and width 2^{-j}. Apart from forming an efficient dictionary, curvelets have been shown to be near-eigenfunctions of the solution operator to a large class of hyperbolic differential equations.

The first 2D FDCT proposed in [84] is via *unequispaced* FFTs (USFFT). The USFFT first applies the conventional 2D FFT to the image, thus obtaining the discrete 2D Fourier coefficients on a regular square grid. For each scale-angle pair, the result of the FFT must be interpolated to obtain the values for the points in the corresponding grid, which is no longer square and different for each pair. The interpolated (sheared) result has to be multiplied with the parabolic window corresponding to the scale, yielding localization. Then an inverse 2D FFT applied to the previous results yield the discrete curvelet coefficients.

The second FDCT in [84] is via *wrapping*. Similarly to USFFT, the first step is a 2D FFT. Instead resampling according to a tilted grid, as in USFFT, a rectangular grid with parabolic ratio is used. The problem is that the parabolic window used for restricting the support does not fit in a rectangle of size $2^j \times 2^{j/2}$ (for the jth scale) aligned with the axes. Thus, the 2D IFFT cannot be directly applied as before. For this reason, after the product with the window, some *wrapping* around the origin is required before applying the 2D IFFT. This transform exhibits a computational complexity that is between 6 and 10 times that of an FFT operating on an array of the same size.

We refer to [479] for the extension of the discrete curvelet transform to 3D, where the curvelets are plate-like objects, preserving the parabolic scaling between the long axes and the short one, oscillating only along the latter. An alternative to curvelets are the *shearlets* [274], which are obtained by dilating, *shearing* and translating a fixed function.

3.3.1.7. Curvelets Simplified: Contourlets

Provided that curvelet dictionaries solve the only remaining issue of efficiently representing curved edges, the FDCT methods in [84] could be seen as the definitive transforms for achieving sparse representations, making any further research senseless. Nevertheless, the complexity of these FDCTs, arising from appropriately handling the discretization, motivated further research towards simpler alternatives. The *contourlet* transform [146] constitutes such an alternative, that avoids complex discretization schemes by adopting a discrete domain construction and then studying an eventual extension to a continuous domain. Contourlets preserve the localization, orientation and parabolic scaling, characteristic of curvelets. The first-generation contourlet transform is implemented via a *double filter bank* structure, which combines a Laplacian pyramid with directional filter banks (DFB) [18]. At each subband of the pyramid, the DFBs are intended to link point discontinuities into linear structures.

The second-generation contourlet transform [308] substitutes the Laplacian pyramid by another multiscale decomposition defined in frequency domain. In each step of the new pyramid there is a low-pass filter prior to downsampling and a high-pass filter prior to the DFB. The downsampled image is the input of the next pyramid level. This change was motivated by an issue regarding frequency-localization in the first-generation transform, due to aliasing effects. Second-generation contourlets are also less redundant than second-generation curvelets.

3.3.1.8. An Adaptive Representation: Bandelets

The analytic dictionaries presented until now are adapted, by construction, to a certain type of signals, but not to the specific signal that they should represent. Once the dictionary is created, it should represent well all signals of the expected class. *Signal-adaptive transforms* provide a representation that is adapted, not just to the signal class, but to the signal itself. Note that despite such a transform might yield higher representation efficiency than a non-adaptive counterpart, labeling the set of atoms as a CS dictionary is not acceptable, in principle. Note that an adaptive representation needs the signal to generate the dictionary and in CS, differently from the classical image compression scenario, we do not have access to the image, but to a reduced set of $m < n$ measurements, being n the signal dimensionality. One could argue that the dictionary could be generated using some previous signal for adaptation and used to represent the next signals in continuous operation, under the hypothesis that the adaptation for the previous signal remains valid for the next. Despite such considerations already step on slippery ground, we provide here an example of an adaptive transform, namely the *bandelet* transform, because the resulting adapted atoms could be eventually used as a CS dictionary under certain conditions. There exist other adaptive dictionaries, such as *directionlet* [444] and *grouplet* [317] dictionaries, not analyzed here for brevity.

The *bandelet* transform [287] is a 2D multiscale transform intended to exploit the geometric regularity of the image for designing a set of atoms that are *adapted* to it. The image is first characterized by a geometric flow of vectors, which give the local directions in which the image has regular variations. Then the bandelet bases are constructed by dividing the image in regions where the geometric flow is parallel. If there is no geometric flow over some region (isotropic region), the wavelet basis is used for representation. Each region of parallel flow is divided in subregions, where the flow vectors are parallel either horizontally or vertically. For each case, a family of *warped wavelets* is constructed, integrating the flow. The warped wavelets undergo then a process of *bandeletization*, which is intended to correct the fact that horizontally-oriented wavelets (separable wavelets with scaling function in the vertical direction and mother wavelet in the horizontal direction) do not have vanishing moments in the vertical direction and, equivalently, vertically-oriented wavelets (separable wavelets with scaling function in the horizontal direction and mother wavelet in the vertical direction) do not have vanishing moments in the horizontal direction. The bandeletization consists in substituting the family of scaling functions causing the issue (those acting in the vertical dimension for the case of vertically-parallel flow and those acting in the horizontal dimension for the case of horizontally-parallel flow) by a family of orthonormal wavelets spanning the same space. The final result is an orthonormal bandelet basis for each subregion. The union of them forms a block orthonormal basis for the region. The geometric flow can be computed using a fast algorithm that requires $\mathcal{O}(n^2(\log_2 n)^2)$ operations for an arrays of size $n \times n$.

Unfortunately, the resulting bandelet transform is not orthogonal and the warping introduces boundary artifacts. The second-generation bandelet bases proposed in [369] is intended to solve these issues by means of a construction over a standard orthogonal wavelet transform. The 2D wavelet coefficients are reordered and then a 1D wavelet transform is applied. The resulting set of bandelet atoms is orthogonal.

3.3.2. Dictionary Learning

If large datasets of the signals for which we seek a sparse representation are available, one can directly use the implicit information on the signal structure contained in the dataset as a substitute of a more rigid analytic formulation. This is the idea behind *dictionary learning* or *training*: using a large dataset to obtain a set of atoms that optimally represent our signals. This allows obtaining good dictionaries for complex signal classes that might not fit a simple mathematical model. The clearest drawback of dictionary learning is the training or learning process itself, which requires a large dataset. In some applications large datasets of the signals might not be available. Additionally, the datasets have to be representative of the signal variability, that is, should not be biased. This has to be taken into consideration when creating or acquiring the dataset. If some of the degrees of freedom of the signal are not represented by elements in the dataset, the obtained dictionary will not be able to efficiently represent eventual variability of the signal along them. Furthermore, for signals of high dimensionality, a large number of elements in the dataset might lead to excessively large memory requirements during training, as well as to unacceptably long training times. In the following we briefly introduce some of the classical methods for dictionary learning and state-of-the-art approaches.

3.3.2.1. Method of Optimal Directions

The formulation of the problem for which the Method of Optimal Directions (MOD) [169] was proposed is simple: given a dataset of n_{samples} signal realizations, stacked by columns in the matrix $\boldsymbol{Y} = [\vec{y}_i]_{1 \leq i \leq n_{\text{samples}}}$, $\vec{y}_i \in \mathbb{R}^n$, the goal is to find a dictionary $\boldsymbol{\Psi} \in \mathbb{R}^{n \times n_{\text{atoms}}}$ and the corresponding sparse matrix of coefficients $\boldsymbol{X} = [\vec{x}_i]_{1 \leq i \leq n_{\text{samples}}}$, $\vec{x}_i \in \mathbb{R}^{n_{\text{atoms}}}$, which minimize the l_2 approximation error:

$$\hat{\boldsymbol{\Psi}}, \hat{\boldsymbol{X}} = \underset{\boldsymbol{\Psi}, \boldsymbol{X}}{\arg\min} \left\| \boldsymbol{Y} - \boldsymbol{\Psi}\boldsymbol{X} \right\|_F^2, \text{ subject to: } \|\vec{x}_i\|_0 \leq s_{\max}, \ \forall i \tag{3.90}$$

One could establish some parallelism between Eq. 3.90 and the MMV CS formulation (Eq. 3.128). In the MMV recovery the goal is, given $\boldsymbol{Y}$ and $\boldsymbol{\Psi}$, determine the sparsest representation, $\boldsymbol{X}$, given some additional *a priori* knowledge on the supports of the constituent $\vec{x}_i$, namely, joint sparsity. In MOD the goal is more ambitious, since $\boldsymbol{\Psi}$ is also to be determined, but the absence of compression at sensing (identity sensing matrix), together with the typically large number of samples ($n_{\text{samples}} \gg n$), make possible approaching a solution. Note that the condition on the supports of the vectors $\vec{x}_i$ in Eq. 3.90 is softer than in Eq. 3.4.2.8. Instead of joint sparsity, the requirement is that their sparsity is upper bounded by $s_{\max}$.

MOD operates in an iterative fashion, starting from an initial guess for $\boldsymbol{\Psi}^{(0)}$. The columns of $\boldsymbol{X}^{(0)}$ (sparse representations) are calculated using a vector selection algorithm, enforcing $\|\vec{x}_i\|_0 = s_{\max}$. Given the temporal approximation $\boldsymbol{Y}^{(k)} = \boldsymbol{\Psi}^{(k)}\boldsymbol{X}^{(k)}$ ($k = 0$ for the first iteration) and the residual, the dictionary vectors are adjusted. After adjustment, the new approximations and residuals are computed. At this point some stopping criterion is checked and, unless it is met, the previous two steps are repeated. The crucial aspect is how the dictionary elements are *adjusted*. Each atom $\vec{\psi}_i$ (a point in $\mathbb{R}^n$) is displaced along a vector $\vec{\delta}_i \in \mathbb{R}^n$, being the set of optimal directions $\boldsymbol{\Delta} = \left[\vec{\delta}_i\right]_{1 \le i \le n_{\text{samples}}}$ chosen such that the total MSE, given by the sum of squared l_2 norms of the n_{samples} residuals is reduced as much as possible. Obviously, this calculus has to be done supposing constancy of the (temporal) sparse coefficients $\boldsymbol{X}^{(k)}$, which are recomputed afterwards with the adjusted dictionary. For clarity, the dictionary update reads $\boldsymbol{\Psi}^{(k+1)} = \boldsymbol{\Psi}^{(k)} + \boldsymbol{\Delta}$. When the algorithm ends, after some K iterations, we have that $\hat{\boldsymbol{\Psi}} = \boldsymbol{\Psi}^{(K)}$ and $\hat{\boldsymbol{X}} = \boldsymbol{X}^{(K)}$.

We have provided the above formulation mostly because the set of optimal directions $\boldsymbol{\Delta}$ gives name to the method, but clearly, if we seek the maximum error reduction in terms of l_2 norm, the Moore-Penrose pseudo-inverse can be used directly for the dictionary update as follows:

$$\boldsymbol{\Psi}^{(k+1)} = \boldsymbol{Y}^{(k)}\boldsymbol{X}^{(k)\dagger} \tag{3.91}$$

3.3.2.2. Union of Orthonormal Bases

The overcomplete dictionary generated via MOD is adapted to the dataset, but is not *structured*. One could attempt to train an overcomplete dictionary defined as a *union of orthonormal bases* (ONBs), so that, while being adapted to the signals of interest, the dictionary holds some internal structure. Learning such a dictionary was proposed in [292], where the general formulation in Eq. 3.90 was slightly modified and split into the two update steps that are actually performed in their alternating strategy, namely:

$$
\begin{aligned}
\hat{\boldsymbol{X}} &= \arg\min_{\boldsymbol{X}} \|\boldsymbol{Y} - \boldsymbol{\Psi}\boldsymbol{X}\|_2^2 + \lambda \|\boldsymbol{X}\|_1 \\
\hat{\boldsymbol{\Psi}} &= \arg\min_{\boldsymbol{\Psi}} \|\boldsymbol{Y} - \boldsymbol{\Psi}\boldsymbol{X}\|_2^2, \text{ subject to: } \boldsymbol{\Psi} = [\boldsymbol{\Psi}_i]_{1\leq i\leq n_{\mathrm{ONB}}} \\
\boldsymbol{\Psi}_i &\text{ orthonormal } \forall i
\end{aligned} \tag{3.92}
$$

where $\boldsymbol{\Psi}_i$ denotes each of the n_{ONB} orthobases in the union. Note that the restriction on the cardinality of the sparse supports of the vectors $\vec{x}_i$ has been substituted by the more general condition of minimizing the l_0 norm, which, in turn, has been relaxed to an l_1 norm regularization term in the first equality of Eq. 3.92. While in the case of $\boldsymbol{\Psi}$ being an orthonormal basis the solution to this minimization problem can be obtained by soft thresholding, when $\boldsymbol{\Psi}$ is a union of ONBs, $\hat{\boldsymbol{X}}$ cannot be directly obtained and the Block Coordinate Relaxation (BCR) algorithm [429] is used to approach a solution in an iterative fashion. The dictionary update step is an iterative process that makes use of SVD and takes into account that $\boldsymbol{\Psi}$ is a union of n_{ONB} ONBs. Specifically, at each iteration of the learning process, for one of the basis in $\boldsymbol{\Psi}$, say $\boldsymbol{\Psi}_i$, the dictionary update follows the steps below:

$$
\begin{aligned}
\boldsymbol{Y}_i &= \boldsymbol{Y} - \sum_{j\neq i} \boldsymbol{\Psi}_j \boldsymbol{X}_j \\
\text{SVD: } \boldsymbol{Y}_i \boldsymbol{X}_i^\top &= \boldsymbol{U}\boldsymbol{\Sigma}\boldsymbol{V}^\top \\
\text{Update: } \boldsymbol{\Psi}_i &= \boldsymbol{U}\boldsymbol{V}^\top
\end{aligned} \tag{3.93}
$$

3.3.2.3. Karhunen-Loève Transform or Principal Components Analysis

The Karhunen-Loève Transform (KLT) [318], first suggested in [266], is a linear transform that, as well as the Fourier transform, provides a set of orthogonal functions to represent a given function on a bounded interval in a somewhat efficient manner. The peculiarity of the KLT is that it is adapted to the statistics of the signal. Consequently, it yields an optimal basis if the distribution from which the signals come is known. The idea is to find a low-dimensional subspace to represent the data, which minimizes the l_2 approximation error. Given the covariance matrix of the data $\boldsymbol{\Sigma} \in \mathbb{R}^{n\times n}$, the basis vectors of this optimal k-dimensional subspace, $k < n$, are given by the first k eigenvectors of the SVD decomposition of $\boldsymbol{\Sigma}$. Provided that $\boldsymbol{\Sigma}$ is symmetric, one can write $\boldsymbol{\Sigma} = \boldsymbol{U}\boldsymbol{\Lambda}\boldsymbol{U}^\top$ and the basis vectors are the first k columns of $\mathbf{U}$. As the reader might have already noticed, this is exactly the widely-spread Principal Component Analysis (PCA), originally formulated in [191]. The DCT is known to be a good non-adaptive approximation of the KLT for natural images.

Generalized Principal Components Analysis Generalized Principal Components Analysis (GPCA) [448] is an extension of PCA. While PCA seeks the optimal basis for the signals in the dataset, the GPCA formulation is wider and does not force the signal to live in a single subspace but in a *union* of low-dimensional subspaces. The number of subspaces might be unknown and the dimensionality of each one might be different. The subspaces might exhibit different degrees of overlap between each other. Differently from the union of ONBs method, where the different bases of the dictionary are jointly used to sparsely represent the signals, in GPCA each signal lives in one of the subspaces of the union and is to be represented exclusively with the basis vectors spanning that subspace. The algebraic-geometric GPCA algorithm simultaneously determines these subspaces and fits orthogonal bases to them. Suppose that the number of subspaces N is known. Then GPCA states that the union of N subspaces of $\mathbb{R}^n$ can be represented with a set of homogeneous polynomials of degree N in n variables. If enough sample points are given, the coefficients of the polynomials can be linearly determined. Differentiating the polynomials in the different subspaces yields a basis for the complement of each subspace. The points where the derivatives are to be computed can be recursively selected via polynomial division. A recursive extension of GPCA for the case of an unknown number of underlying subspaces is also included in [448]. The main drawback of this extension is the excessive computational cost for large dimensionality and large number of subspaces.

3.3.2.4. K-SVD

The K-SVD algorithm [6] is a dictionary learning framework that aims to solve the minimization problem in Eq. 3.90 by means of a block relaxation approach similar to that of MOD. At each iteration a step of sparse coding of the sample data using the current dictionary is followed by a dictionary update to improve the fitting. The K-SVD algorithm is a generalization of the K-means clustering algorithm. Despite clustering and sparse representation are, in principle, different tasks, once can think of clustering as a 1-sparse representation of the data in a dictionary whose atoms are the centers of the clusters and the nonzero coefficient must be one. Adopting this line of thinking, the problem of sparse representation can be seen as an extension of clustering. Similarly to dictionary learning, the K-means algorithm, also known as generalized Lloyd's algorithm (GLA) [301], iterates between one step of assigning some set of data points to their *nearest neighbors* and a subsequent step of updating the cluster centers. One of the strong points of K-SVD is its flexibility, since it can use any pursuit algorithm for the sparse coding step, e. g., basis pursuit, FOCUSS, MP, OMP, etc. Regarding the dictionary update, K-SVD differs from MOD and other learning approaches in the fact that the columns of the dictionary are updated sequentially, as in K-means, and not in a single block operation. For the ith dictionary atom, the general minimization problem in Eq. 3.90 is accordingly reformulated as:

$$\hat{\vec{\psi}}_i, \hat{\vec{x}}_i^\top = \underset{\vec{\psi}_i, \vec{x}_i^\top}{\arg\min} \left\| \boldsymbol{Y} - \sum_{j \neq i} \vec{\psi}_j \vec{x}_j^\top - \vec{\psi}_i \vec{x}_i^\top \right\|_F^2 = \left\| \boldsymbol{R}_i - \vec{\psi}_i \vec{x}_i^\top \right\|_F^2 \quad (3.94)$$

where $\vec{x}_i^\top$ denotes the ith row of $\boldsymbol{X}$ and $\boldsymbol{R}_i$ is the residual matrix. A simple solution to Eq. 3.94 is obtained by a rank one approximation of $\boldsymbol{R}_i$, that can be computed via SVD. Note that SVD cannot be applied directly to $\boldsymbol{R}_i$, since this would result in a non-sparse $\vec{x}_i^\top$. Consequently, one has to first restrict $\boldsymbol{R}_i$ to the columns corresponding to nonzeros in $\vec{x}_i^\top$, $\Omega_i = \{j \mid 1 \leq j \leq n_{\text{samples}},\ \vec{x}_i^\top(j) \neq 0\}$. We denote restriction to Ω_i as $(\cdot)_{\Omega_i}$. Then, if $(\boldsymbol{R}_i)_{\Omega_i} = \boldsymbol{U\Sigma V}^\top$ is the SVD of $\boldsymbol{R}_i$, the solutions to Eq. 3.94 are $\hat{\vec{\psi}}_i = \vec{u}_1$ and $\left(\hat{\vec{x}}_i^\top\right)_{\Omega_i} = \sigma_{1,1}\vec{v}_1$. Note that $\hat{\vec{x}}_i^\top$ is immediately obtained from $\left(\hat{\vec{x}}_i^\top\right)_{\Omega_i}$ by filling up with zeros those coefficients not indexed in Ω_i. Adjusting the sparse coefficients $\vec{x}_i^\top$ at the same time the ith dictionary atom is updated accelerates convergence (cf. Gauss-Seidel methods). The dictionary atom update in Eq. 3.94 does not substitute the sparse coding step, which is still necessary in order to allow for changes in the support of the representations.

Provided that, at each atom update, either additional reduction or no change of the MSE is guaranteed, the algorithm ensures monotonic MSE reduction, thus guaranteeing convergence to a local minimum. This convergence is subject to the performance of the method used for sparse coding. If the sparse representation of the training vectors is required to be one-sparse, K-SVD converges to the *gain-shape* Vector Quantization (VQ) [288]. If, additionally, the sparse coefficients are forced to have unit value, it converges to the standard K-means algorithm.

3.3.2.5. Parametric Training

Parametric training methods constitute a bridge between pure dictionary learning, which does not require any signal modeling, and analytic dictionaries, which are generated according to some model of the signal. In the case of adaptive analytic dictionaries, the signal itself is considered in the design of the optimal dictionary atoms, which still are constructed according to some model (localization in the space-frequency-orientation domain). Similarly, parametric training proposes combining dictionary learning from a set of sample signals and some model of the signal that reduces the complexity to a reduced number of parameters, typically much lower than the signal dimensionality. The advantages of an appropriate parametrization are many, e. g., reducing the number of samples, accelerating convergence, avoiding local minima and guiding towards a *better* (say, sparser or more realistic) representation. In the limit, if the number of parameters does not grow with the signal dimensionality, one can efficiently deal with arbitrary-sized signals, eventually infinite-dimensional. Following the classification in [389], we distinguish here between *translation-invariant*, *multiscale* and *sparse* parametric trained dictionaries.

Translation-invariant A translation-invariant dictionary is able to represent signals of arbitrarily high dimensionality and can be derived from a given dictionary for a low-size signal patch, simply by collecting all possible translations of the dictionary atoms in the extended domain. Shift-invariant sparse coding was proposed in [357] to efficiently represent time-varying natural images, by means of time-varying basis functions. The results led to interpret neural spike trains as a sparse code in time, strengthening the parallelism between neuroscience and sparse coding. However, the explicit shift-invariant sparse coding formulation in [357] does not scale with the problem size, making it intractable for many real world problems. An alternative to time-domain approaches is to use a phase-blind spectral model. We refer to [373] for a comparison between phase-blind spectral methods and the shift-invariant time-domain approach. A method to avoid the course of dimensionality when training translation-invariant dictionaries is proposed in [53]. The authors propose a subset selection step to select a small subset of features depending on their correlation with the observation.

The Matching of Time Invariant Filters (MoTIF) method [246] is a method for sequential computation of the generating functions. The training signals are allowed to have infinite size but are nonzero only on their support, of limited size. The size of the support of the generating functions is upper-bounded by the latter. MoTIF learns each translation invariant filter iteratively, using two fundamental steps: in the first step, given the generating function, the best translation is found. In the second step, for that translation, the generating function is optimized so that it maximizes the correlation with the training signals, while minimizing the correlation with previously generated functions. The explicit translation-invariant dictionary can be created by applying all possible translations to the obtained set of generating functions.

The MOD is extended to the translation-invariant case in [170], in which case, $\boldsymbol{X}$ (whose pseudoinverse is required for the dictionary update, cf. Eq. 3.91) is a Toeplitz matrix. Note that translation-invariant dictionaries are just a particular case of their *unrestricted overlapping dictionary* class, with a shift of one row between consecutive submatrices. The formulation is also extended to the case of including additional constraints.

The *image-signature-dictionary* (ISD) proposed in [7] for 2D signals is simply a small image, such that every patch in it, with different location and size, is a different dictionary atom. Obviously, the main advantage of the ISD is its compact structure. For instance, n_{atoms} 2D dictionary atoms can be implicitly contained in an ISD of size $\sqrt{n_{\text{atoms}}} \times \sqrt{n_{\text{atoms}}}$. Note that, if the atoms are all of equal size, say $\sqrt{n} \times \sqrt{n}$, the size of a conventional explicit dictionary would be $n_{\text{atoms}} \times n$. This approach builds upon the idea of *epitome* [245, 110] or smallest image that can be used to represent a certain class of images of larger size. This is a way to implement the idea of shift invariance, since two patches (atoms) extracted from nearby locations of the ISD will resemble each other. Extracting patches of different sizes allows exploring different scales, but if this option is to be exploited, it has to be taken into account during training of the ISD. The difference between the epitome concept and the ISD is that, in the epitome case, each image patch is represented as a one-sparse combination of epitome patches, that is, by a single patch of the epitome, while in the ISD case, it is represented as an s-sparse combination, where the sparsity $s \leq s_{\max}$ is a small number. Note the parallelism between the pair of K-means clustering and the K-SVD algorithm. In the conventional K-means each point is represented by its *nearest neighbor* between the cluster centers, thus equivalent to a one-sparse approximation with unit sparse coefficient, while the K-SVD algorithm allows for general s-sparse representations. As well as the K-SVD algorithm converges to K-means when $s = 1$ and the sparse coefficient is unit, so it does the ISD with respect to the epitome. As usually, the learning process iterates between one step of updating the ISD and a subsequent update of the sparse representations of the training patches. If the sparse coding step is solved via conventional techniques (e. g., OMP) using the $n \times n_{\text{atoms}}$ explicit dictionary, $\mathcal{O}(n \times n_{\text{atoms}})$ operations are required, thus loosing the advantage provided by the compact ISD structure. Alternatively, one can exploit the fact that a convolution in the spatial domain is equivalent to an inner product in frequency domain and reduce the complexity to $\mathcal{O}(n_{\text{atoms}} \log n_{\text{atoms}})$, using forward and inverse Fourier transforms of the signals. The dictionary update is performed in a way similar to that of the MOD.

Multiscale The multiscale dictionary learning approach in [395] adapts an overcomplete wavelet basis to the statistics of the training dataset. The dictionary atoms are, therefore, dilations and translations of a small set of trained patches, which act as *non-separable* mother wavelet functions. A prior distribution for the sparse coefficients is modeled as a superposition of a Gaussian and a (infinitely narrower) Dirac delta function, in order to encourage sparsity. Inspired by the K-SVD algorithm, the authors of [313] propose learning a multiscale representation for images, obtained via quadtree decomposition of the learned dictionary, using overlapping image patches. At each finer scale, the support of the dictionary atoms is halved in each dimension, with respect to those of previous scale, thus yielding four possible translations per atom. During training of the multiscale dictionary, each atom at each scale is updated independently. Extending the K-SVD to multiple scales avoids the risks derived from selecting a single patch size for the dictionary atoms. Too small patch sizes lead to an inefficient representation of image structures whose size is larger than the patch. Conversely, too large patches might be unnecessary when representing most of the fine details in the image. One could think that the K-SVD algorithm with large patch sizes should automatically learn the underlying multiscale structure in the image, but this does not seem to be the case and oversmoothing is observed when the patches are too large.

Sparse Sparse parametric dictionaries can be seen as a generalization of both previous classes, since they build upon some analytic dictionary admitting calculus of the coefficients via a fast transform. Two of the most desired characteristics of analytic dictionaries are certain translation invariance and multiscale representation. Sparse dictionary learning [390] proposes to express the dictionary to learn as the product of a fixed basic dictionary $\boldsymbol{\Psi}^0$ and a sparse selection matrix $\boldsymbol{S}$, i. e., $\boldsymbol{\Psi} = \boldsymbol{\Psi}^0 \boldsymbol{S}$. Including this dictionary model into the general minimization in Eq. 3.90 yields

$$\hat{\boldsymbol{S}}, \hat{\boldsymbol{X}} = \underset{\boldsymbol{S},\boldsymbol{X}}{\arg\min} \left\| \boldsymbol{Y} - \boldsymbol{\Psi}^0 \boldsymbol{S}\boldsymbol{X} \right\|_F^2$$
$$\text{subject to: } \begin{cases} \|\vec{x}_i\|_0 \leq s_{\max}^{\vec{x}}, \ \forall i \\ \|\vec{s}_j\|_0 \leq s_{\max}^{\vec{s}}, \ \left\|\boldsymbol{\Psi}^0 \vec{s}_j\right\|_2 = 1, \ \forall j \end{cases} \tag{3.95}$$

where $s_{\max}^{\vec{x}}$ and $s_{\max}^{\vec{s}}$ are the maximum number of nonzero elements allowed in the columns of $\boldsymbol{X}$ and $\boldsymbol{S}$, respectively. The problem can be attacked by means of the K-SVD algorithm, restricted to atoms of the form $\vec{\psi}_i = \boldsymbol{\Psi}^0 \vec{s}_i$ with $\operatorname{supp}(\vec{s}_i) \leq s_{\max}^{\vec{s}}$. The modified K-SVD atom update for the ith atom (Eq. 3.94) reads

$$\hat{\vec{s}}_i, \hat{\vec{x}}_i^\top = \underset{\vec{s}_i, \vec{x}_i^\top}{\arg\min} \left\| \boldsymbol{Y} - \sum_{j \neq i} \boldsymbol{\Psi}^0 \vec{s}_j \vec{x}_j^\top - \boldsymbol{\Psi}^0 \vec{s}_i \vec{x}_i^\top \right\|_F^2 = \left\| \boldsymbol{R}_i - \boldsymbol{\Psi}^0 \vec{s}_i \vec{x}_i^\top \right\|_F^2$$
$$\text{subject to: } \|\vec{s}_i\|_0 \leq s_{\max}^{\vec{s}}, \ \left\|\boldsymbol{\Psi}^0 \vec{s}_i\right\|_2 = 1, \ \forall i \tag{3.96}$$

After some derivations, it is shown that Eq. 3.96 can be solved via the two-step approach given in Eq. 3.97. Prior to solving Eq. 3.97, $\vec{x}_i$ has to be normalized to unit length. After solving the minimization, the obtained vector of coefficients $\hat{\vec{s}}_i$ has also to be normalized so that $\left\|\boldsymbol{\Psi}^0 \hat{\vec{s}}_i\right\|_2 = 1$. After calculating $\hat{\vec{x}}_i$ via the second equality of Eq. 3.97, the process can be repeated (still for the same dictionary atom) to refine the solution, if necessary.

$$\hat{\vec{s}}_i = \underset{\vec{s}_i}{\arg\min} \left\| \boldsymbol{R}_i \vec{x}_i - \boldsymbol{\Psi}^0 \vec{s}_i \right\|_2^2 \text{ subject to: } \|\vec{s}_i\|_0 \leq s_{\max}^{\vec{s}}$$
$$\hat{\vec{x}}_i = \boldsymbol{R}_i^\top \boldsymbol{\Psi}^0 \vec{s}_i \tag{3.97}$$

The dictionary learning approach in [109] uses *steerable* Riesz-wavelets as starting point for training. Consequently, the dictionary atoms are (possibly sparse) linear combinations of Riesz-wavelet vectors.

3.3.2.6. Simultaneous Sensing Matrix and Dictionary Optimization

All the aforementioned dictionary learning methods use a set of representative signals to generate a set of dictionary atoms that allows optimally representing them. In a CS framework, we have no direct access to the signals. Instead, one has vectors of reduced dimensionality, which are obtained projecting the signals by a *sensing matrix* $\boldsymbol{\Phi} \in \mathbb{R}^{m \times n}$, where n is the signal dimensionality and $m < n$ is the dimensionality of the measurement space. In Section 3.2.9, methods for optimizing a given (often randomly-generated) $\boldsymbol{\Phi}$ were presented. Such methods typically seek the lowest possible coherence between columns of $\boldsymbol{\Phi}$ in an attempt to boost the sensing efficiency. At the end of this section on dictionary learning a question that arises is whether it is possible to optimize both the sensing matrix and the dictionary in a joint manner. A framework for design and optimization of both matrices from a set of representative signals is proposed in [157]. For learning the sensing matrix, supposing the dictionary known, the authors seek the $\boldsymbol{\Phi}$, such that the Gram matrix of the corresponding measurement matrix $\boldsymbol{A} = \boldsymbol{\Phi\Psi}$ (Eq. 3.44) is as close as possible to an identity matrix. The core of the dictionary training stage is the K-SVD algorithm, appropriately coupled with the sensing matrix optimization. We omit here the details of the method, but give the general formulation of the minimization problem to solve in [157].

$$\begin{aligned} \hat{\boldsymbol{\Phi}}, \hat{\boldsymbol{\Psi}}, \hat{\boldsymbol{X}} &= \underset{\boldsymbol{\Phi},\boldsymbol{\Psi},\boldsymbol{X}}{\arg\min} \left\{ \alpha \left\| \boldsymbol{I} - \boldsymbol{\Psi X} \right\|_F^2 + \left\| \boldsymbol{Y} - \boldsymbol{\Phi\Psi X} \right\|_F^2 \right\} \\ &\text{subject to: } \left\| \vec{x}_i \right\|_0 \leq s_{\max},\ \forall i \end{aligned} \tag{3.98}$$

where $\boldsymbol{I}$ is the set of training data, which should admit a sparse representation in the dictionary $\boldsymbol{\Psi}$, that is, $\boldsymbol{I} = \boldsymbol{\Psi X}$ with $\|\vec{x}_i\|_0 \leq s_{\max}$, $\forall i$. The matrix $\boldsymbol{Y}$ contains the vectors of measurements by columns, i. e., $\boldsymbol{Y} = \boldsymbol{\Phi I} + \boldsymbol{\nu}$, where $\boldsymbol{\nu}$ simply models eventual measurement noise. The scalar $0 < \alpha \leq 1$ is a weighting factor, aimed to compensate for the typically larger value of the reconstruction error. The proposed coupled-K-SVD algorithm iterates between a step of optimizing $\boldsymbol{\Phi}$, given $\boldsymbol{\Psi}$, a step of sparse coding, i. e., obtaining $\boldsymbol{X}$, given $\boldsymbol{\Psi}$ (e. g., via OMP), and a step of dictionary update (atom by atom, as in K-SVD).

An extension of the coupled-K-SVD algorithm to block-sparse signals is given in [295].

3.4. Recovery Methods

Given the sensing matrix, $\boldsymbol{\Phi} \in \mathbb{R}^{m \times n}$, $m < n$, which models the sensing process, and the dictionary $\boldsymbol{\Psi} \in \mathbb{R}^{n \times n_{\text{atoms}}}$, which is a basis ($n_{\text{atoms}} = n$) or a frame where the signal $\vec{I} \in \mathbb{R}^n$ admits a sparse representation $\vec{I} = \boldsymbol{\Psi}\vec{x}$, the main challenge of any CS framework is to recover the s-sparse coefficient vector $\vec{x} \in \mathbb{R}^{n_{\text{atoms}}}$ from the measurement vector $\vec{y} \in \mathbb{R}^m$, that is, to solve the minimization problem in Eq. 3.9, where the measurement matrix is $\boldsymbol{A} = \boldsymbol{\Phi}\boldsymbol{\Psi}$. As already pointed out in Section 3.1, solving Eq. 3.9 is NP-hard and a common workaround is to *convexify* it and solve Eq. 3.10 instead, where the l_0 minimization has been substituted by l_1. Methods in the literature for recovering $\vec{x}$ from $\vec{y}$ can, therefore, be divided between those trying to give an approximate solution to Eq. 3.9 and those solving Eq. 3.10 instead. The first are *greedy* algorithms that operate sequentially, taking at each step the *best* decision according to some criterion, subject to those taken in the previous steps. This sequential operation prevents greedy algorithms from offering optimality guarantees, in principle. The second are methods for convex optimization. While convergence guarantees can be derived for convex optimization methods, the simplicity of some greedy approaches, which often translates into lower memory footprint and lower computational cost, enhances their attractiveness with respect to optimal convex optimizations. In practice, greedy algorithms often show equivalent or even better performance than convex-optimization-based methods.

3.4.1. l_1-Minimization Algorithms

3.4.1.1. Basis Pursuit

Basis Pursuit (BP) [108] was born as a method for finding the *optimal* representation of a signal in an overcomplete dictionary. Optimality is meant here in terms of representation efficiency, that is, BP searches for the decomposition having the lowest l_1 norm, between all possible decompositions of the signal in the dictionary. As already pointed out in [108], BP itself is not an algorithm, but an optimization principle that seeks a sparse representation by minimizing the l_1 norm of the coefficients. Consequently, all algorithms described in the following within Section 3.4.1 can be thought as different implementations of the BP concept, that is, different ways of solving the convex problem in Eq. 3.10.

Here we only make a short reference to the two basic algorithms for BP suggested in [108]. The first algorithm is the *simplex* method [125, 275], which requires selecting an initial support set Ω_0, for which the corresponding solution, $\vec{x}_0 = \boldsymbol{A}_{\Omega_0}^{-1}\vec{y}$, is feasible, i. e., has no negative elements, where $\boldsymbol{A}_{\Omega_0}$ denotes the matrix obtained selecting the columns of $\boldsymbol{A}$ indexed by Ω_0. Note that the simplex algorithm does not deal natively with negative coefficients. Then, the algorithm operates iteratively, swapping one support element contained in the temporal support by another not contained in it in each iteration, using the swap that best improves the objective function. When the objective function does not improve anymore, convergence has been achieved. *Anti-cyclic* rules (e. g., Bland's rule [51]) can be applied to ensure convergence of the algorithm and avoid getting stuck in some cycle of swaps. Additionally, the algorithm for selecting the *best basis* proposed in [114], which seeks the orthogonal basis (from an overcomplete dictionary) yielding the minimum Shannon entropy of the signal expansion, can be used to obtain an appropriate initialization of Ω_0.

The second algorithm is an *interior point method.* While the simplex method makes the temporal solution *walk along* the edges of the convex polytope that defines the feasible region, interior point methods, also called *barrier methods*, make the temporal solution traverse the interior of the polytope, thus starting from a point $\vec{x}_0$ that is well inside the interior. In consecutive iterations the temporal solution approaches the boundary. In the BP case, at each iteration the coefficients of $\vec{x}$ are modified maintaining feasibility ($\boldsymbol{A}\vec{x} = \vec{y}$) and then a transformation that sparsifies $\vec{x}$ is applied (e. g., a thresholding operation).

3.4.1.2. The Homotopy Method

A *homotopy* between two continuous functions, say f and g, from a topological space X to a topological space Y is defined as a continuous function $H : X \times [0,1] \to Y$, where the origin space is the product of the space X with the closed unit interval and the destination space coincides with Y. The defining property of the homotopy is that $H(x,0) = f(x)$, $\forall x \in X$ and $H(x,1) = g(x)$. That is, the second parameter, say λ, describes a *continuous* deformation of the function f into g. Now consider the minimization problem in Eq. 3.10 and the following l_1-regularized least squares functional:

$$F_\lambda(\vec{x}) = \frac{1}{2}\left\|\boldsymbol{A}\vec{x} - \vec{y}\right\|_2^2 + \lambda\left\|\vec{x}\right\|_1, \ \vec{x} \in \mathbb{R}^n \tag{3.99}$$

We know from Section 3.3 that, if the dictionary is overcomplete, we have a dimensionality of $\vec{x}$ equal to some $n_{\text{atoms}} > n$, nevertheless, for notation simplicity and without loss of generality, we adopt here $\vec{x} \in \mathbb{R}^n$. Let $\vec{x}_\lambda$ be a minimizer of Eq. 3.99, then, it can be shown that $\hat{\vec{x}} = \lim_{\lambda \to 0^+} \vec{x}_\lambda$ is a solution to Eq. 3.10. In short terms, the homotopy method proposes to trace a solution path from $\vec{x}_{\lambda^{(0)}} = \vec{0}$ to the solution $\vec{x}_{\lambda^{(K)}} = \hat{\vec{x}}$, attained for a vanishing $\lambda = \lambda^{(K)}$. Therefore, the method minimizes Eq. 3.99 for a vanishing sequence of values of λ, say $\left\{\lambda^{(0)}, \lambda^{(1)}, ..., \lambda^{(K)}\right\}$, thus obtaining a sequence of minimizers, $\{\vec{x}^{(0)}, \vec{x}^{(1)}, ..., \vec{x}^{(K)}\}$ (omitting from now on the subindex λ) that converges to the solution to Eq. 3.10, i. e., $\hat{\vec{x}} = \vec{x}^{(K)}$.

Due to space constraints, the derivations of the homotopy method equations are omitted here. Instead, the method is presented in a compact algorithmic form in Algorithm 2, which has been directly derived from the equations in Section 15.1 of [184]. The reader is referred to this work for derivations and further details.

The residual vector at iteration k is defined in terms of the previous estimate $\vec{x}^{(k-1)}$ as $\vec{c}^{(k)} = \boldsymbol{A}^* \left(\boldsymbol{A}\vec{x}^{(k-1)} - \vec{y}\right)$. The algorithm ensures that, at each iteration, $\lambda^{(k)} = \left\|\vec{c}^{(k+1)}\right\|_\infty$ and, therefore, a threshold on $\lambda^{(k)}$ can be used directly as stopping criterion. $\boldsymbol{A}^*$ denotes the adjoint of the matrix $\boldsymbol{A}$ and $(\cdot)_i$ applied to a matrix simply selects the ith column. The symbol $+$ over the minimizations denotes that only the positive values are considered. The first element to be added to the temporal support set i^1 is the index corresponding to the greatest residual coefficient in absolute value (line 4). The second element to be added to the temporal support set, denoted by i^2 in line 9, is simply the index that minimizes the right hand side of line 7.

Note that a division by zero in line 14 does not occur, since the minimization is restricted to the indices contained in Ω^k, for which $d_i^{(k)} \neq 0$ by construction (line 12). The indices i_+^k and i_-^k, to be added or removed, respectively, from Ω^k (lines 17 and 20, respectively) are just the indices that minimize the right hand sides of lines 13 and 14, respectively.

In general, the fundamental tasks of the algorithm at each iteration k are the calculus of the direction, along which the temporal solution should evolve, $\vec{d}^{(k)}$ (see line 12 for $k \geq 2$) and the step size $\gamma^{(k)}$, that is, how much the temporal solution should move in that direction. Then, the temporal estimate of $\vec{x}$ is updated, as shown in line 22. The parameter λ (Eq. 3.99) is decreased by $\gamma^{(k)}$ in each iteration (line 23). Both $\gamma^{(k)}$ and $\lambda^{(k)}$ vanish as the algorithm advances ($k \uparrow$) and $\vec{x}^{(k)}$ approaches $\hat{\vec{x}}$.

3.4.1.3. Primal-Dual Interior Point Methods

Several primal-dual algorithms have been proposed to solve the convex problem in Eq. 3.10. One of the most well-known libraries for sparse recovery from compressed measurements is the l_1-magic library [89]. The library provides programs for solving, not only Eq. 3.10, but also other versions of the l_1 minimization problem, such as the minimum l_1 error approximation, the minimum l_1 with quadratic constraints (Eq. 3.21), the *Dantzig Selector* and the minimum TV (that is, sparsity in gradient domain), both with

Algorithm 2 Homotopy Method for l_1 Minimization

Initialize: $\vec{x}^{(0)} = \vec{0},\ \lambda^{(0)} = \|\boldsymbol{A}^*\vec{y}\|_\infty\ \ k = 0$

while $\left(\lambda^{(k)} > \varepsilon\right)$ **do**

  $k := k + 1$

  if $(k = 1)$ **then**

    $i^1 = \arg\max\limits_{1 \le i \le n} |(\boldsymbol{A}^*\vec{y})_i| = \arg\max\limits_{1 \le i \le n} \left|c_i^{(1)}\right|$

    $\Omega^1 = \left\{i^1\right\}$

    Calc. direction: $d_{i^1}^{(1)} = \frac{\operatorname{sgn}\left((\boldsymbol{A}^*\vec{y})_{i^1}\right)}{\|\vec{a}_{i^1}\|_2^2} = \frac{\operatorname{sgn}\left(c_{i^1}^{(1)}\right)}{\|\vec{a}_{i^1}\|_2^2},\ d_i = 0\ \forall i \neq i^1$

    Calculate step: $\gamma^{(1)} = \overset{+}{\min\limits_{i \neq i^1}} \left\{ \frac{\lambda^{(0)} + c_i^{(1)}}{1 - \left(\boldsymbol{A}^*\boldsymbol{A}\vec{d}^{(1)}\right)_i}, \frac{\lambda^{(0)} + c_i^{(1)}}{1 + \left(\boldsymbol{A}^*\boldsymbol{A}\vec{d}^{(1)}\right)_i} \right\}$

    Update estimate: $\vec{x}^{(1)} = \vec{x}^{(0)} + \gamma^{(1)}\vec{d}^{(1)} = \gamma^{(1)}\vec{d}^{(1)}$

    $i^2 = \arg\overset{+}{\min\limits_{i \neq i^1}} \left\{ \frac{\lambda^{(0)} + c_i^{(1)}}{1 - \left(\boldsymbol{A}^*\boldsymbol{A}\vec{d}^{(1)}\right)_i}, \frac{\lambda^{(0)} + c_i^{(1)}}{1 + \left(\boldsymbol{A}^*\boldsymbol{A}\vec{d}^{(1)}\right)_i} \right\}$

    Update support: $\Omega^2 = \Omega^1 \cup i^2 = \left\{i^1, i^2\right\}$

  else

    Calc. direction: $\vec{d}^{(k)} : \boldsymbol{A}_{\Omega^k}^*\boldsymbol{A}_{\Omega^k}\vec{d}_{\Omega^k}^{(k)} = -\operatorname{sgn}\left(\vec{c}_{\Omega^k}^{(k)}\right),\ d_i = 0\ \forall i \notin \Omega^k$

    Step candidate: $\gamma_+^{(k)} = \overset{+}{\min\limits_{i \notin \Omega^k}} \left\{ \frac{\lambda^{(k-1)} + c_i^{(k)}}{1 - \left(\boldsymbol{A}^*\boldsymbol{A}\vec{d}^{(k)}\right)_i}, \frac{\lambda^{(k-1)} + c_i^{(k)}}{1 + \left(\boldsymbol{A}^*\boldsymbol{A}\vec{d}^{(k)}\right)_i} \right\}$

    Step candidate: $\gamma_-^{(k)} = \overset{+}{\min\limits_{i \in \Omega^k}} \left\{ \frac{-x_i^{(k-1)}}{d_i^{(k)}} \right\}$

    if $\left(\gamma_+^{(k)} < \gamma_-^{(k)}\right)$ **then**

      Set step: $\gamma^{(k)} = \gamma_+^{(k)}$

      Update support: $\Omega^{k+1} = \Omega^k \cup i_+^k$

    else if $\left(\gamma_-^{(k)} < \gamma_+^{(k)}\right)$ **then**

      Set step: $\gamma^{(k)} = \gamma_-^{(k)}$

      Update support: $\Omega^{k+1} = \Omega^k \setminus i_-^k$

    end if

    Update estimate: $\vec{x}^{(k)} = \vec{x}^{(k-1)} + \gamma^{(k)}\vec{d}^{(k)}$

    Decrease λ: $\lambda^{(k)} = \lambda^{(k-1)} - \gamma^{(k)}$

  end if

end while

equality and quadratic constraints. The problem in Eq. 3.10 can be directly expressed as a linear program, which is then solved by means of the classical Newton method. The peculiarity is that the step towards the new point is to be adjusted so that it remains in the interior of the polytope. In practice, a parameter is used to bias the solution towards the interior, so that, in successive iterations, the solution is allowed to approach more and more the boundary, where the solution lies. The primal-dual algorithm repeats the Newton method until the duality gap is lower than some user-defined tolerance.

Chambolle and Pock's primal-dual algorithm [98] is often referred as the fastest method for performing constrained l_1 minimization. The algorithm is a first-order primal-dual method for convex optimization problems with saddle-point structure and it converges to such a saddle point with rate $\mathcal{O}(1/n)$ for the n-dimensional case, which is, indeed, an optimal rate for non-smooth problems. In general terms, if the primal problem is given by Eq. 3.100, then the corresponding dual problem is given by Eq. 3.101.

$$\hat{\vec{x}} = \arg\min_{\vec{x} \in X} F\left(\boldsymbol{A}\vec{x}\right) + G\left(\vec{x}\right) \tag{3.100}$$

$$\hat{\vec{\xi}} = \arg\max_{\vec{\xi} \in \Xi} - \left(G^*\left(-\boldsymbol{A}^*\vec{\xi}\right) + F^*\left(\vec{\xi}\right)\right) \tag{3.101}$$

Here X and Ξ denote the subspaces where the primal and the dual variables, $\vec{x}$ and $\vec{\xi}$, live, following the notation in [98]. For the type of problems we consider in this thesis $X = \mathbb{R}^n$, $\Xi = \mathbb{R}^m$ and $\boldsymbol{A} \in \mathbb{R}^{m \times n}$. The functions $F : X \to [0, +\infty)$ and $G : \Xi \to [0, +\infty)$ are proper, convex, lower-semicontinuous functions and F^*, G^* denote their respective convex conjugate. We omit here the derivations for brevity and directly present the general structure of the Chambolle and Pock's primal-dual algorithm in Algorithm 3. For details we refer to [98] and Section 15.2 of [184]. The algorithm uses the so-called *proximal mappings* of F^* and G, which are denoted by P_{F^*} and P_G, respectively.

The stopping criterion in Algorithm 3 is a threshold on the primal-dual gap $E\left(\vec{x}, \vec{\xi}\right) = F\left(\boldsymbol{A}\vec{x}\right) + G\left(\vec{x}\right) + F^*\left(\vec{\xi}\right) + G^*\left(-\boldsymbol{A}^*\vec{\xi}\right)$, $\nu > 0$. The outputs of the algorithm are the estimates of the primal and dual variables $\hat{\vec{x}} = \vec{x}^{(K)}$ and $\hat{\vec{\xi}} = \vec{\xi}^{(K)}$, respectively, being K the index of the last iteration. For $\theta = 0$, thus $\bar{\vec{x}}^{(k)} = \vec{x}^{(k)}$, Algorithm 3 is equivalent to the classical Arrow-Hurwicz method [14].

Algorithm 3 Primal-Dual Algorithm

Initialize: $\theta \in [0,1]$, $\tau, \sigma > 0$ such that: $\tau\sigma \|\boldsymbol{A}\|_2^2 < 1$, $\vec{\bar{x}}^{(0)} = \vec{x}^{(0)}$, $k = 0$

while $\left(E\left(\vec{x}^{(k)}, \vec{\xi}^{(k)}\right) > \nu\right)$ **do**

  $k := k + 1$

  $\vec{\xi}^{(k)} := P_{F^*}\left(\sigma; \vec{\xi}^{(k-1)} + \sigma \boldsymbol{A}\vec{\bar{x}}^{(k-1)}\right)$

  $\vec{x}^{(k)} := P_G\left(\tau; \vec{x}^{(k-1)} - \tau \boldsymbol{A}^* \vec{\xi}^{(k)}\right)$

  $\vec{\bar{x}}^{(k)} := \vec{x}^{(k)} + \theta\left(\vec{x}^{(k)} - \vec{x}^{(k-1)}\right)$

end while

In Section 15.2 of [184] Algorithm 3 is adapted to the specific case of solving Eq. 3.10, in which case the functions F and G read

$$F(\vec{z}) = \chi_{\vec{y}}(\vec{z}) = \begin{cases} 0 \text{ if } \vec{z} = \vec{y} \\ \infty \text{ if } \vec{z} \neq \vec{y} \end{cases} \tag{3.102}$$

$$G(\vec{x}) = \|\vec{x}\|_1 \tag{3.103}$$

The corresponding convex conjugates are given by

$$F^*\left(\vec{\xi}\right) = \Re\left\langle \vec{y}, \vec{\xi} \right\rangle \tag{3.104}$$

$$G^*\left(\vec{\zeta}\right) = \begin{cases} 0 \text{ if } \|\vec{\zeta}\|_\infty \leq 1 \\ \infty \text{ otherwise} \end{cases} \tag{3.105}$$

Then it can be shown that the corresponding proximal mappings of F^* and G are given by

$$P_{F^*}\left(\sigma; \vec{\xi}\right) = \vec{\xi} - \sigma \vec{y} \tag{3.106}$$

$$P_G(\tau; \vec{z}) = S_\tau(\vec{z}) \tag{3.107}$$

where $S_\tau(\vec{z})$ is a (complex) soft-thresholding operator with threshold value τ. Taking into account Eq. 3.106 and 3.107, one can reformulate Algorithm 3 for the specific case of constrained l_1 minimization, yielding the well-known formulation given in Algorithm 4.

For further details and a similar derivation of the primal-dual algorithm for the case of the quadratically-constrained l_1 minimization (Eq. 3.21) we refer again to Section 15.2 of [184].

Algorithm 4 Primal-Dual Algorithm for l_1 Minimization

Initialize: $\theta \in [0,1]$, $\tau, \sigma > 0$ such that: $\tau\sigma \|\boldsymbol{A}\|_2^2 < 1$, $\bar{\vec{x}}^{(0)} = \vec{x}^{(0)}$, $k = 0$

while $\left(E\left(\vec{x}^{(k)}, \vec{\xi}^{(k)}\right) > \nu\right)$ **do**

$k := k + 1$

$\vec{\xi}^{(k)} = \vec{\xi}^{(k-1)} + \sigma\left(\boldsymbol{A}\bar{\vec{x}}^{(k-1)} - \vec{y}\right)$

$\vec{x}^{(k)} = S_\tau\left(\vec{x}^{(k-1)} - \tau\boldsymbol{A}^*\vec{\xi}^{(k)}\right)$

$\bar{\vec{x}}^{(k)} = \vec{x}^{(k)} + \theta\left(\vec{x}^{(k)} - \vec{x}^{(k-1)}\right)$

end while

3.4.1.4. Iteratively Reweighted l_1 Minimization

Consider the following *weighted* extension of the l_0 minimization in Eq. 3.9:

$$\hat{\vec{x}} = \arg\min_{\vec{x}} |\mathrm{supp}(\boldsymbol{W}\vec{x})| \text{ subject to } \vec{y} = \boldsymbol{A}\vec{x} \tag{3.108}$$

where $\boldsymbol{W}$ is a diagonal matrix containing the weights corresponding to each coefficient of $\vec{x}$. Clearly, as far as the weights are nonzero, if the solution to Eq. 3.9 is unique, so it is the solution to Eq. 3.108. Similarly to the relaxation of Eq. 3.9 into Eq. 3.10, one can also relax the l_0 minimization in Eq. 3.108 into an l_1 minimization, obtaining

$$\hat{\vec{x}} = \arg\min_{\vec{x}} \|\boldsymbol{W}\vec{x}\|_1 = \arg\min_{\vec{x}} \sum_{i=1}^{n} w_i |x_i| \text{ subject to } \vec{y} = \boldsymbol{A}\vec{x} \tag{3.109}$$

where w_i denotes the ith diagonal element of $\boldsymbol{W}$. Note that the solutions to Eq. 3.10 and Eq. 3.109 will be different, in general, since the solution of Eq. 3.109 is under the influence of the values of the weights w_i, $1 \leq i \leq n$. A large weight w_i will penalize the ith coefficient in the minimization process, while a low value will favor adopting that coefficient as part of the signal support, since it has a negligible influence on the cost function. In other words, an appropriate weighting can *discourage* some entries of $\vec{x}$, which are unlikely to belong to the signal support and, conversely, *encourage* some other entries that are likely to belong to the signal support. The challenge is, therefore, the selection of the weights, so that the solution of Eq. 3.109 is superior to that of its unweighted counterpart (Eq. 3.10).

If the sparse signal $\vec{x}$ is known beforehand, the weights can be chosen to be inverse of the absolute value of each coefficient, that is,

$$w_i = \begin{cases} \frac{1}{|x_i|} & \text{if } x_i \neq 0 \\ \infty & \text{if } x_i = 0 \end{cases} \tag{3.110}$$

This weight selection is optimal, in the sense that it makes Eq. 3.109 equivalent to Eq. 3.9. Unfortunately, full *a priory* knowledge of $\vec{x}$ is an unrealistic hypothesis, since it is the unknown of the problem. This difficulty is overcome by means of an iterative formulation in [166]. The authors solve Eq. 3.109 in each iteration, where the weights are calculated from the previous estimate of $\vec{x}$ via Eq. 3.110.

The pseudocode of this simple approach is given in Algorithm 5. The l_1 minimization in line 3 can be solved by means of any conventional, i. e., non-reweighted, BP method. The formula for the weights update in line 4 includes a small constant $\varepsilon \to 0^+$, intended to avoid an explicit division by zero when computing the weight for a zero-valued coefficient of $\vec{x}$. Algorithm 5 is a piecewise linear approach for minimizing a cost function of the form $\sum_{i=1}^{n} \log(|x_i| + \varepsilon)$, which is concave. Provided that the original log-sum penalty function is concave, the initialization of the algorithm is crucial to attain convergence. Both in [174] and [166], the authors suggest using the solution to the unweighted l_1 minimization as starting point for the weighted version. In Algorithm 5 we account for this via the unit initialization of the weights prior to the first iteration.

Algorithm 5 Iteratively Reweighted l_1 Minimization

Initialize: $w_i^{(0)} = 1,\ \forall 1 \leq i \leq n,\ k = 0$

1: **while not**(Stopping conditions) **do**

2: $\quad k := k + 1$

3: $\quad$ Update estimate: $\vec{x}^{(k)} = \arg\min_{\vec{x}} \left\| \boldsymbol{W}^{(k-1)} \vec{x} \right\|_1 = \arg\min_{\vec{x}} \sum_{i=1}^{n} w_i^{(k-1)} |x_i|$ subject to $\vec{y} = \boldsymbol{A}\vec{x}$

4: $\quad$ Update weights: $w_i^{(k)} = \frac{1}{\left|x_i^{(k)}\right| + \varepsilon}\ \forall 1 \leq i \leq n$

5: **end while**

Reweighted extensions of other l_1 minimization problems, such as the minimum l_1 error approximation, the minimum l_1 with quadratic constraints (Eq. 3.21), the *Dantzig Selector* and the minimum TV are also provided in [166]. The authors use the primal-dual-based solvers provided by the l_1-magic library [89].

FOCUSS The idea of a reweighted iterative approach for l_1 minimization reminds and, in fact, is tightly connected to the FOcal Underdetermined System Solver (FOCUSS) [204]. At each iteration, FOCUSS solves a reweighted l_2 minimization of the form given in Eq. 3.112 for the more general IRLS framework, which will be introduced later, but substituting $\boldsymbol{W}$ by $\boldsymbol{W}^2$. The weights are directly given by the inverse of the coefficients of the previous estimate of $\vec{x}$. A generalization of FOCUSS, including two additional parameters is also suggested in [204]. The first parameter is an exponent l for the coefficients of $\vec{x}$ when computing the corresponding weights. Note that this constitutes an early bridge between the reweighted l_1 formulation presented above and the, intrinsically more general, IRLS which will be introduced later. The other parameter is a fixed weighting matrix $\boldsymbol{W}_0$, which is intended to integrate some *a priori* information on $\vec{x}$ that might be available, e. g., which entries are surely nonzero and which ones are unlikely to belong to the signal support. Then, the resulting weighting matrix is $\boldsymbol{W} = \boldsymbol{W}_0\boldsymbol{W}_{\vec{x}}$. The simplified pseudocode of the FOCUSS is provided in Algorithm 6.

Algorithm 6 General FOcal Underdetermined System Solver (FOCUSS)

Initialize: $w_{x_i}^{(0)} = 1,\ \forall 1 \le i \le n,\ \boldsymbol{W}^{(0)} = \boldsymbol{W}_0\boldsymbol{W}_{\vec{x}}^{(0)},\ k = 0$
1: **while not**(Stopping conditions) **do**
2: $k := k + 1$
3: Update estimate: $\vec{x}^{(k)} = \boldsymbol{W}^{(k-1)^{-1}} \left(\boldsymbol{A}\boldsymbol{W}^{(k-1)^{-1}}\right)^{\dagger} \vec{y}$
4: Update $\boldsymbol{W}$: $\boldsymbol{W}^{(k)} = \boldsymbol{W}_0\boldsymbol{W}_{\vec{x}}^{(k)},\ w_{x_i}^{(k)} = \frac{1}{\left(x_i^{(k)}\right)^l}\ \forall 1 \le i \le n$
5: **end while**

Note that we use here an inverse notation for the weighting matrices, with respect to that used in the original paper. The reader should notice that both notations differ only in a matrix inversion. Here we restrict our attention to diagonal weighting matrices $\boldsymbol{W}_0$ and $\boldsymbol{W}_{\vec{x}}$, thus yielding a diagonal $\boldsymbol{W}$. Consequently, the inverses are obtained by inverting each of the diagonal elements. The estimate in line 3 is the minimum l_2 norm solution subject to the constraints, obtained by means of a Moore-Penrose pseudounverse. The derivations imply a change of variable and are equivalent to those for the more general IRLS case, thus omitted here. Also note that, for the first iteration, and if $\boldsymbol{W}_0$ is the identity matrix, the first estimate is the unweighted LS solution, given by $\tilde{\vec{x}} = \boldsymbol{A}^{\dagger}\vec{y}$, which is an appropriate starting point for the algorithm.

IRLS A conceptually similar reweighted scheme that can be used for l_1 minimization is the Iteratively Reweighted Least Squares approach (IRLS) [105, 131]. In the most general setting, IRLS aims to solve a minimization problem of the form

$$\hat{\vec{x}} = \arg\min_{\vec{x}} \|\vec{x}\|_p^p \text{ subject to } \vec{y} = \boldsymbol{A}\vec{x} \tag{3.111}$$

by replacing the l_p norm by a weighted l_2 norm. The weighted version of Eq. 3.111 then reads

$$\hat{\vec{x}} = \arg\min_{\vec{x}} \vec{x}^{\top}\boldsymbol{W}\vec{x} = \arg\min_{\vec{x}} \sum_{i=1}^{n} w_i x_i^2 \text{ subject to } \vec{y} = \boldsymbol{A}\vec{x} \tag{3.112}$$

where $\boldsymbol{W}$ is the diagonal weighting matrix with diagonal elements w_i, $1 \leq i \leq n$. In order to look for a solution to Eq. 3.112, consider the change of variable $\vec{z} = \boldsymbol{W}^{\frac{1}{2}}\vec{x}$. Then the weighted least squares problem in Eq. 3.112 can be rewritten as

$$\begin{aligned} \hat{\vec{z}} = \arg\min_{\vec{z}} \sqrt{\vec{z}^{\top}\vec{z}} = \arg\min_{\vec{z}} \sqrt{\sum_{i=1}^{n} z_i^2} = \arg\min_{\vec{z}} \|\vec{z}\|_2 \\ \text{subject to } \vec{y} = \boldsymbol{A}\boldsymbol{W}^{-\frac{1}{2}}\vec{z} \end{aligned} \tag{3.113}$$

where the equality constraint follows directly from that in Eq. 3.112 inverting the change of variable. Then, a closed form solution to Eq. 3.113 can be obtained via the Moore-Penrose pseudoinverse: $\hat{\vec{z}} = \left(\boldsymbol{A}\boldsymbol{W}^{-\frac{1}{2}}\right)^{\dagger}\vec{y}$. Undoing the change of variable yields a solution to the weighted least squares problem in Eq. 3.112:

$$\hat{\vec{x}} = \boldsymbol{W}^{-\frac{1}{2}}\left(\boldsymbol{A}\boldsymbol{W}^{-\frac{1}{2}}\right)^{\dagger}\vec{y} = \boldsymbol{W}^{-1}\boldsymbol{A}^{\top}\left(\boldsymbol{A}\boldsymbol{W}^{-1}\boldsymbol{A}^{\top}\right)^{-1}\vec{y} \tag{3.114}$$

As before, the optimal weights w_i yielding the best approximation to the problem in Eq. 3.111 are determined by the signal coefficients x_i, inversely, as in Eq. 3.110, but this time taking into account the specific l_p norm to minimize and the fact that the cost function is a weighted sum of *squared* coefficients. Then, the optimal weights are given by

$$w_i = \begin{cases} \frac{1}{|x_i|^{2-p}} \text{ if } x_i \neq 0 \\ \infty \text{ if } x_i = 0 \end{cases} \tag{3.115}$$

where the interesting cases correspond to $0 \leq p \leq 1$. For $p = 0$, this framework is equivalent to the previous, since the squared denominator for the weights for nonzero entries of $\vec{x}$ compensates the fact that the cost function in Eq. 3.112 is a weighted sum of squares. Also as before, $\vec{x}$ is unknown *a priori* and the weighted least squares problem has to be solved in an iterative fashion, thus yielding an algorithm similar to Algorithm 5 but for a generic l_p norm, which we describe in Algorithm 7.

Algorithm 7 Iteratively Reweighted Least Squares (IRLS)

Initialize: $w_i^{(0)} = 1,\ \forall 1 \leq i \leq n,\ k = 0$

1: **while not**(Stopping conditions) **do**

2: $\quad k := k + 1$

3: $\quad$ Update estimate: $\vec{x}^{(k)} = \boldsymbol{W}^{(k-1)^{-1}} \boldsymbol{A}^\top \left(\boldsymbol{A} \boldsymbol{W}^{(k-1)^{-1}} \boldsymbol{A}^\top \right)^{-1} \vec{y}$

4: $\quad$ Update $\boldsymbol{W}$: $w_i^{(k)} = \frac{1}{\left(\left| x_i^{(k)} \right|^2 + \varepsilon^2 \right)^{1-\frac{p}{2}}} \quad \forall 1 \leq i \leq n$

5: **end while**

The small constant $\varepsilon \to 0^+$ introduced in the weight update (line 4) fulfills the same damping function as in Algorithm 5. In practice, the selection of ε does influence the result of the algorithm. Using a moderate value of ε and then rerunning Algorithm 7 for a finer ε after convergence has been shown to yield better results than using a single value [105].

Note that the IRLS framework allows approximating solutions to *quasi-norm* minimizations (l_p norms with $0 < p < 1$). Good approximations to such optimization problems, often characterized as intractable in the literature, can be achieved with the appropriate initialization. For instance, the iterative procedure proposed in [104], which weights each signal coefficient using weights calculated as in line 4 to calculate the direction along which the temporal solution should move, showed better reconstruction results for some quasinorms (e. g., $p = 0.5$) than for l_1 minimization when the number of measurements is low.

3.4.1.5. From IRLS to an l_1-Minimizing Kalman Filter

For a linear sensing model and variables following Gaussian probability density functions, the Kalman filter offers an optimal estimate of the time-varying state, for system states that propagate also according to a linear model. In [L1], Loffeld *et al.* proposed to understand the l_1 norm as a non-linear observation of the (sparse) state vector $\vec{x}$ and use an iterative linearized Kalman filter framework to approach an estimate $\hat{\vec{x}}$ in a recursive manner. This technique is further developed and analyzed in [L2].

The algorithm starts from the LS estimate of $\vec{x}$, given by the Moore-Penrose pseudoinverse $\tilde{\vec{x}} = \boldsymbol{A}^{\dagger}\vec{y}$ and estimates, in each iteration, a difference vector contained in the nullspace of $\boldsymbol{A}$, such that, added to the LS solution $\tilde{\vec{x}}$, yields an estimate of $\vec{x}$ with reduced l_1 norm, while fulfilling the constraints $\boldsymbol{A}\vec{x} = \vec{y}$. In the following we explain briefly how the algorithm, whose pseudocode for the real case is provided in Algorithm 8, operates.

Algorithm 8 l_1-Minimizing Kalman Filter

Initialize: $\tilde{\vec{x}} = \boldsymbol{A}^{\dagger}\vec{y}$, $\boldsymbol{E}_{\mathcal{N}(\boldsymbol{A})} \in \mathbb{R}^{n\times(n-m)}$ basis of $\mathcal{N}(\boldsymbol{A})$, $\vec{x}_{\lambda}^{(0)} = \vec{0}$, $\vec{x}^{(0)} = \tilde{\vec{x}}$, $\boldsymbol{P}_{\lambda}^{+\,(0)} = \boldsymbol{P}_0$, $k = 0$

while $\left(\Delta \left\|\vec{x}^{(k)}\right\|_1 > \varepsilon\right)$ **do**

  $k := k + 1$

 Prediction:

  Propagate state: $\vec{x}_{\lambda}^{-\,(k)} := \vec{x}_{\lambda}^{+\,(k-1)}$

  Propagate covariance: $\boldsymbol{P}_{\lambda}^{-\,(k)} := \boldsymbol{P}_{\lambda}^{+\,(k-1)} + \boldsymbol{Q}$

 Measurement update:

  Measure: $y_{\lambda}^{(k)} = \gamma_k \left\|\vec{x}^{(k-1)}\right\|_1$

  Calculate Jacobian: $\boldsymbol{C}_{\lambda}^{(k)} = \left[\frac{x_i^{(k-1)}}{\left|x_i^{(k-1)}\right|}\right]_{1\le i\le n}^{\top} \boldsymbol{E}_{\mathcal{N}(\boldsymbol{A})}$

  Kalman gain: $\boldsymbol{K}^{(k)} := \boldsymbol{P}_{\lambda}^{-\,(k)}\boldsymbol{C}_{\lambda}^{\top\,(k)}\left(\boldsymbol{C}_{\lambda}^{(k)}\boldsymbol{P}_{\lambda}^{-\,(k)}\boldsymbol{C}_{\lambda}^{\top\,(k)} + R\right)^{-1}$

  Update state: $\vec{x}_{\lambda}^{+\,(k)} = \vec{x}_{\lambda}^{-\,(k)} + \boldsymbol{K}^{(k)}\left(y_{\lambda}^{(k)} - \left\|\vec{x}^{(k-1)}\right\|_1\right)$

  Update estimate: $\vec{x}^{(k)} = \tilde{\vec{x}} + \boldsymbol{E}_{\mathcal{N}(\boldsymbol{A})}\vec{x}_{\lambda}^{(k)}$

  Update covariance: $\boldsymbol{P}_{\lambda}^{+\,(k)} := \boldsymbol{P}_{\lambda}^{-\,(k)} - \boldsymbol{K}^{(k)}\boldsymbol{C}_{\lambda}^{(k)}\boldsymbol{P}_{\lambda}^{-\,(k)}$

end while

First of all, a basis for the null space of $\boldsymbol{A}$ is calculated and denoted $\boldsymbol{E}_{\mathcal{N}(\boldsymbol{A})}$ (by columns). Note that $\mathcal{N}(\boldsymbol{A})$ is a $(n-m)$-dimensional subspace of $\mathbb{R}^n$. The difference vector to be added to the LS solution in each iteration is calculated via its coefficients in the null space basis $\vec{x}_\lambda \in \mathbb{R}^{(n-m)}$. The initial estimate of the sparse vector is the non-sparse LS estimate $\vec{x}^{(0)} = \tilde{\vec{x}}$ and, thus, $\vec{x}_\lambda^{(0)} = \vec{0}$.

The state propagation model corresponds to a constant state, provided that the sparse vector to estimate does not vary along iterations. The model in state space (null space coefficients) reads

$$\vec{x}_\lambda^{(k+1)} = \vec{x}_\lambda^{(k)} + \vec{w}_\lambda^{(k)} \tag{3.116}$$

where $\vec{w}_\lambda^{(k)}$ is the realization of some Gaussian driving noise with covariance matrix $\boldsymbol{Q} \in \mathbb{R}^{(n-m)\times(n-m)}$ at the iteration k and has nothing to do with the weighting factors introduced in Section 3.4.1.4.

The nonlinear measurement model is

$$\begin{aligned} y_\lambda^{(k)} &= h\left(\vec{x}^{(k)}\right) + \nu^{(k)} = h\left(\tilde{\vec{x}} + \boldsymbol{E}_{\mathcal{N}(\boldsymbol{A})}\vec{x}_\lambda^{(k)}\right) + \nu^{(k)} \\ &= \left\|\vec{x}^{(k)}\right\|_1 + \nu^{(k)} = \left\|\tilde{\vec{x}} + \boldsymbol{E}_{\mathcal{N}(\boldsymbol{A})}\vec{x}_\lambda^{(k)}\right\|_1 + \nu^{(k)} \end{aligned} \tag{3.117}$$

where $h(\cdot) = \|\cdot\|_1$ is the nonlinear measurement function, which generates the observation $y^{(k)}$ from the current estimate of the state vector $\vec{x}_\lambda^{(k)}$. Note that, provided that the CS measurement matrix $\boldsymbol{A}$ remains constant, so it does $\mathcal{N}(\boldsymbol{A})$, its basis $\boldsymbol{E}_{\mathcal{N}(\boldsymbol{A})}$, and also the LS solution $\tilde{\vec{x}}$. As a consequence, the kth estimate of the state vector $\vec{x}_\lambda^{(k)}$ fully determines the kth estimate of the sparse vector $\vec{x}^{(k)} = \tilde{\vec{x}} + \boldsymbol{E}_{\mathcal{N}(\boldsymbol{A})}\vec{x}_\lambda^{(k)}$ and, thus, $y_\lambda^{(k)}$. The observation $y_\lambda^{(k)}$ is not to be confused with the CS vector of measurements $\vec{y}$, which is already taken into account in the algorithm as a static constraint by means of the LS estimate $\tilde{\vec{x}}$. The kth realization of the Gaussian observation noise, with variance R, is denoted by $\nu^{(k)}$ in Eq. 3.117.

A linearization of $h(\cdot)$ is required in order to provide the Kalman filter with a linear measurement model. To that end, the partial derivative of $h(\cdot)$ with respect to the state vector $\vec{x}_\lambda$ is computed, yielding the Jacobian matrix

$$\begin{aligned} \boldsymbol{C}_\lambda &= \frac{\delta}{\delta\vec{x}_\lambda} h\left(\tilde{\vec{x}} + \boldsymbol{E}_{\mathcal{N}(\boldsymbol{A})}\vec{x}_\lambda\right) = \frac{\delta h(\vec{x})}{\delta\vec{x}}\frac{\delta\vec{x}}{\delta\vec{x}_\lambda} \\ &= \left[\frac{x_i}{|x_i|}\right]_{1\le i\le n}^\top \boldsymbol{E}_{\mathcal{N}(\boldsymbol{A})} \end{aligned} \tag{3.118}$$

which, in this case, is a matrix $\boldsymbol{C}_\lambda \in \mathbb{R}^{1\times(n-m)}$, that is, a row vector of length $(n-m)$. This is because the dimensionality of the measurement space of the filter is one, but $\boldsymbol{C}_\lambda$ is, in general, a matrix. The derivative of the absolute value function is the signature function when working in $\mathbb{R}$, while the derivative of the sparse vector $\vec{x}$ with respect to the state vector $\vec{x}_\lambda^{(k)}$ is, by construction, the matrix $\boldsymbol{E}_{\mathcal{N}(\boldsymbol{A})}$.

The Kalman filter in Algorithm 8 has, as any Kalman filter, two main stages: one of prediction of the next state vector and corresponding covariance matrix, using the updated state from the previous iteration and the propagation model in Eq. 3.116, and another of correction or update of the predicted state integrating the information given by a new measurement. The prediction of the new state vector with measurements until the previous time instant, denoted by $\vec{x}_\lambda^{-\,(k)}$ is done by adopting the updated estimate for the previous instant, $\vec{x}_\lambda^{+\,(k-1)}$ (line 3). Similarly, the predicted covariance matrix $\boldsymbol{P}_\lambda^{-\,(k)}$ is the sum of the updated covariance matrix from the previous state $\boldsymbol{P}_\lambda^{+\,(k-1)}$ plus the covariance matrix of the driving noise $\boldsymbol{Q}$ (line 4).

The key step of the algorithm is the *simulated* measurement operation in line 5. Provided that there is no real measurement process, this measurement is an illusion, which, in short terms, tells the filter that the real sparse vector $\vec{x}$ has an l_1 norm slightly lower than that that would be obtained from the previous estimate of the filter. This is done by means of the parameter $0 < \gamma_k < 1$, which has to be $\gamma_k \to 1^-$. The parameter can be different for each iteration and it has been shown that the performance of the algorithm profits from γ_k increasing asymptotically to one [L2], with respect to a constant value.

The Jacobian matrix can be computed as in line 6, evaluating the partial derivatives at the previous estimate of the sparse vector $\vec{x}$. Then, the Kalman gain can be calculated from the predicted covariance and the current Jacobian, integrating also the variance of the measurement noise R (line 7). Note that, in general, when the dimensionality of the measurement space is greater than one, we deal with a *covariance* matrix $\boldsymbol{R}$ and not a scalar variance. Nevertheless, in our case the dimensionality is one, R is scalar and the resulting Kalman gain $\boldsymbol{K}$ is a column vector of length n. The Kalman gain is then used in line 8 to update the estimate of the state vector $\vec{x}_\lambda$ according to the the measurement error, over which we have full control via the parameter γ_k (recall line 5). The updated estimate of $\vec{x}_\lambda$ fully determines the corresponding estimate of $\vec{x}$ (line 9). The algorithm ends when convergence is attained or, in practice, when the l_1 norm of $\vec{x}$ does not change more than some threshold $\varepsilon > 0$, close to zero, between two consecutive iterations.

3.4.2. Greedy Algorithms

Greedy algorithms can be classified in two broad categories: purely *greedy pursuits* and *thresholding* algorithms. The general structure of a greedy pursuit is given in Algorithm 9 and implies sequential discovery of the nonzero coefficients of $\vec{x}$. Each algorithm iteration consists of two fundamental steps. In the first one, an element of the support (necessarily new or eventually already selected, depending on the algorithm) is selected, while in the second, the temporal estimate of $\vec{x}$ is updated. These two steps correspond to lines 3 and 4 in Algorithm 9, where the updates are given by the arguments that optimize some general *support selection* function f_{sel} and *cost* function f_{cost}, respectively. The first function is responsible for picking a support element, i. e., selecting a column index of $\boldsymbol{A}$, given the current residual $\vec{r}^{(k-1)}$. The second is a cost function intended to deliver the estimate of $\vec{x}$ that best fits the measurements $\vec{y}$, given the temporal support Ω^k. Note the vast generality of this framework. The use of the temporal support Ω^k means that the algorithm could have memory about previously selected support elements, but also includes approaches that consider a single coefficient update per iteration, thus ignoring (and possibly repeating) previous selections.

Algorithm 9 Greedy Pursuit (General)

Initialize: $\vec{r}^{(0)} = \vec{y},\ \vec{x}^{(0)} = \vec{0},\ \Omega^0 = \emptyset,\ k = 0$

1: **while not**(Stopping conditions) **do**
2: $\quad k := k + 1$
3: $\quad$ Update support: $\Omega^k = \Omega^{k-1} \cup i^k,\ i^k = \underset{1 \leq i \leq n_{\text{atoms}}}{\arg\max}\ f_{\text{sel}}\left(\vec{r}^{(k-1)}, \boldsymbol{A}, i\right)$
4: $\quad$ Update estimate: $\vec{x}^{(k)} = \underset{\vec{x}}{\arg\max}\, f_{\text{cost}}\left(\vec{x}, \boldsymbol{A}, \Omega^k, \vec{y}\right)$
5: $\quad$ Update residual: $\vec{r}^{(k)} = \vec{y} - \boldsymbol{A}\vec{x}^{(k)}$
6: **end while**

Algorithms following this structure can eventually attain close-to-optimal performance. Consider, for instance, the greedy algorithm proposed in [201], which finds the principal k Fourier coefficients of a signal, from which only m non-adaptive samples are given. This algorithm, despite its *greediness*, is *near-optimal*, in the sense that the l_2 error is within $(1 + \varepsilon)$ of the optimal error (best k-term approximation). The computational cost is polynomial in

$m \log k \log r/\varepsilon$, where r is the ratio between largest and smallest numerical quantities.

Thresholding algorithms iterate between a step of selection of a support element, similarly to greedy pursuits, and a second step of support *pruning*, which christens the category. Thresholding algorithms are conceptually very simple and very fast in their implementations. Additionally, recovery guarantees similar to those for convex relaxation methods can be obtained for algorithms of this class.

In the following we briefly review some of the most well-known greedy algorithms, both pure greedy pursuits, such as MP, OMP, ROMP, ORMP, and thresholding algorithms, namely IHT, CoSaMP and SP.

3.4.2.1. Matching Pursuit

Matching Pursuit (MP) [316] was conceived as a mean to find the best representation of a signal in an arbitrary redundant dictionary, such as Gabor and wavelet dictionaries. MP is attractive due to its conceptual simplicity and iterative structure, allowing for fast implementations, when compared to BP. The algorithm, given in Algorithm 10, operates in an incremental manner. At each iteration, one column of $\boldsymbol{A}$ is selected and only the corresponding coefficient of $\vec{x}$ is updated. Nothing prevents the algorithm from updating the same coefficient several times. Nevertheless, the residual norm $\|\vec{r}\|_2$ converges linearly to zero if the columns of $\boldsymbol{A}$ span $\mathbb{R}^m$. Consequently, the algorithm stops in a finite number of iterations due to the stopping condition on $\|\vec{r}\|_2$, being ε_{tol} a tolerance based on the expected signal noise or the desired accuracy. The other stopping condition simply offers the possibility of integrating some *a priori* information on the maximum cardinality of the signal support set Ω, that is, it forces $\vec{x}$ to have at maximum $s_{\max}$ non-zero elements. If there is no cue about $s_{\max}$, this stopping condition can be ignored and line 5 can be omitted, since there is no need for keeping track of the temporal support set. These two stopping conditions will be used in other algorithms described in next sections, but they are not mandatory and can be substituted or complemented by some other custom conditions. The update step in line 6 minimizes the square approximation error $\left\|\vec{y} - \boldsymbol{A}\vec{x}^{(k)}\right\|_2^2$ with respect to the coefficient selected in line 4, i^k.

Note that an orthonormal decomposition is a particular case of MP where the dictionary is an orthonormal basis. All algorithm operations are vector or scalar operations except from the matrix-vector product in line 3, which, consequently, dominates the computational cost. If the matrix product can

Algorithm 10 Matching Pursuit (MP)

Initialize: $\vec{r}^{(0)} = \vec{y},\ \vec{x}^{(0)} = \vec{0},\ \Omega^0 = \emptyset,\ k = 0$

while $\left(\left\|\vec{r}^{(k)}\right\|_2 > \varepsilon_{\text{tol}}\right)$ **and** $\left(\left|\Omega^k\right| < s_{\max}\right)$ **do**

  $k := k + 1$

  $\vec{g}^{(k)} = \boldsymbol{A}^\top \vec{r}^{(k-1)}$

  $i^k = \underset{1 \leq i \leq n_{\text{atoms}}}{\arg\max} \dfrac{\left|g_i^{(k)}\right|}{\left\|\vec{a}_i\right\|_2}$

  Update support: $\Omega^k = \Omega^{k-1} \cup i^k$

  Update coefficient: $x_{i^k}^{(k)} = x_{i^k}^{(k-1)} + \dfrac{g_{i^k}^{(k)}}{\left\|\vec{a}_{i^k}\right\|_2^2}$

  Update residual: $\vec{r}^{(k)} = \vec{r}^{(k-1)} - \vec{a}_{i^k} \dfrac{g_{i^k}^{(k)}}{\left\|\vec{a}_{i^k}\right\|_2^2}$

end while

be implemented as a fast transform, e. g., a Fourier dictionary via FFT, we know from Section 3.3.1.1 that the complexity of the operation will be reduced from $\mathcal{O}(n^2)$ to $\mathcal{O}(n \log n)$ in the square dictionary case ($n_{\text{atoms}} = n$). Additionally, if the dictionary has a structure similar to a multiscale Gabor or wavelet dictionary, that is, if the dictionary atoms have small support and the support of each atom intersects those of few other atoms, then an extremely fast implementation of BP with complexity of each iteration of only $\mathcal{O}(\log n)$ is presented in [272] (cf. to that of naïve $[\mathcal{O}(n^2)]$ and 'fast' $[\mathcal{O}(n)]$ implementations).

3.4.2.2. Orthogonal Matching Pursuit

In MP only one coefficient of $\vec{x}$ is updated per iteration and no memory of previously updated coefficients is needed. An evolution of MP is the so-called Orthogonal Matching Pursuit (OMP) [361], which necessarily keeps track of the temporal support Ω and updates all the coefficients of $\vec{x}$ indexed by Ω at each iteration. The OMP algorithm is reproduced in Algorithm 11. The update of $\vec{x}$ consists in an orthogonal projection of $\vec{y}$ onto the columns of $\boldsymbol{A}$ indexed by Ω, which we denote by $\boldsymbol{A}_\Omega$, for which the Moore-Penrose pseudoinverse is used (line 6). This yields the optimal $\vec{x}^{(k)} = \underset{\vec{x} | \operatorname{supp}(\vec{x}) = \Omega^k}{\arg\min} \left\|\vec{y} - \boldsymbol{A}\vec{x}\right\|_2^2$ at each iteration k.

Differently from MP, OMP selects each support element only once. OMP typically performs better than MP, but at the cost of higher computational

Algorithm 11 Orthogonal Matching Pursuit (OMP)

Initialize: $\vec{r}^{(0)} = \vec{y},\ \vec{x}^{(0)} = \vec{0},\ \Omega^0 = \emptyset,\ k = 0$

while $\left(\left\|\vec{r}^{(k)}\right\|_2 > \varepsilon_{\text{tol}}\right)$ **and** $\left(|\Omega^k| < s_{\max}\right)$ **do**

  $k := k + 1$

  $\vec{g}^{(k)} = \boldsymbol{A}^\top \vec{r}^{(k-1)}$

  $i^k = \underset{1 \le i \le n_{\text{atoms}}}{\arg\max} \dfrac{\left|g_i^{(k)}\right|}{\|\vec{a}_i\|_2}$

  Update support: $\Omega^k = \Omega^{k-1} \cup i^k$

  Update estimate: $\vec{x}^{(k)}_{\Omega^k} = \boldsymbol{A}^\dagger_{\Omega^k} \vec{y}$

  Update residual: $\vec{r}^{(k)} = \vec{y} - \boldsymbol{A}_{\Omega^k} \vec{x}^{(k)}_{\Omega^k}$

end while

complexity. This might become an issue when n is large. Due to the sequential addition of support elements, OMP requires s iterations to deliver the s-sparse $\vec{x}$. In [436] it is shown that the exact recovery condition is met if $s < \left(\mu^{-1} + 1\right)/2$, or, put in terms of the maximum allowable intercolumn coherence:

$$\mu < \frac{1}{2s - 1} \tag{3.119}$$

The condition in Eq. 3.119 is a sufficient condition for both OMP and BP to resolve sparse signals. Additionally, also from [436], we have that, if the *cumulative coherence* (also called Babel function, see definition in [436, 437]) is $\mu_1(s) \le 1/3$, then OMP provides an s-term approximant whose approximation error is, at most, a factor $\sqrt{1 + 6s}$ times the best s-term approximation error, in terms of l_2 norm. An upper bound for the cumulative coherence can be given in terms of the intercolumn coherence: $\mu_1(s) \le s\mu,\ \forall s \in \mathbb{N}$. Previous results in [202] showed that, for $s < \frac{1}{8\sqrt{2}}\frac{1}{\mu} - 1$, OMP generates an s-term approximant whose approximation error is, at most, $8\sqrt{s}$ times the best s-term approximation error.

Both MP and OMP fit into the wider class of *general MP* algorithms. In fact, MP and OMP can be considered special cases of *directional pursuit* [54] with update directions δ_{i^k} and $\boldsymbol{A}^\dagger_{\Omega^k} \vec{g}^{(k)}$, respectively. In [439] it has been shown that OMP can reliably recover the s-sparse $\vec{x}$ from $m = \mathcal{O}(s \log n)$ measurements gathered via an *admissible* measurement matrix (e. g., Gaussian or Bernoulli), similarly to BP, thus offering a fast alternative to the latter. Additionally, they found that in practice successful recovery is attained with fewer measurements than those theoretically required. Nevertheless, the recovery guarantees for OMP are not uniform, as pointed out in [384], that is, despite they apply for *each* fixed s-sparse signal $\vec{x}$, they do not apply for *all* s-sparse signals.

3.4.2.3. Regularized Orthogonal Matching Pursuit

The Regularized Orthogonal Matching Pursuit (ROMP) [347, 345] is an evolution of OMP intended to achieve uniform recoverability for all measurement matrices satisfying the RIP. To achieve this, ROMP proposes selecting more than one support element per iteration, specifically it suggests considering, at each iteration k, a set of support elements for which the corresponding coefficients in the correlation vector $\vec{g}^{(k)}$ have similar value. Indeed, in the OMP structure it might be possible that more than one support elements are appropriate solutions of the maximization in line 4 (Algorithm 11). Selecting only one when differences between the best candidates are negligible is an obvious instability. The ROMP algorithm is provided in Algorithm 12.

Algorithm 12 Regularized Orthogonal Matching Pursuit (ROMP)

Initialize: $\vec{r}^{(0)} = \vec{y},\ \vec{x}^{(0)} = \vec{0},\ \Omega^0 = \emptyset,\ k = 0$

1: **while** $\left(\left\|\vec{r}^{(k)}\right\|_2 > \varepsilon_{\text{tol}}\right)$ **and** $\left(\left|\Omega^k\right| < s_{\max}\right)$ **do**

2: $k := k + 1$

3: $\vec{g}^{(k)} = \boldsymbol{A}^\top \vec{r}^{(k-1)}$

4: Candidate support:

$$\Omega_{\text{c}}^k = \left\{ i \;\middle|\; \frac{\left|g_i^{(k)}\right|}{\|\vec{a}_i\|_2} > \frac{\left|g_j^{(k)}\right|}{\|\vec{a}_j\|_2},\ \left|\Omega_{\text{c}}^k\right| = \min\left(s_{\max}, \left\|\vec{g}^{(k)}\right\|_0\right) \right\},\ \forall j \notin \Omega_{\text{c}}^k$$

5: Regularize:

$$\Omega_{\text{r}}^k = \underset{\Omega \subset \Omega_{\text{c}}^k}{\arg\max} \sum_{i \in \Omega} \left(\frac{\left|g_i^{(k)}\right|}{\|\vec{a}_i\|_2} \right)^2,\ \Omega \;\middle|\; \frac{\left|g_i^{(k)}\right|}{\|\vec{a}_i\|_2} \leq \frac{1}{r} \frac{\left|g_j^{(k)}\right|}{\|\vec{a}_j\|_2},\ \forall i, j \in \Omega$$

6: Update support: $\Omega^k = \Omega^{k-1} \cup \Omega_{\text{r}}^k$

7: Update estimate: $\vec{x}_{\Omega^k}^{(k)} = \boldsymbol{A}_{\Omega^k}^\dagger \vec{y}$

8: Update residual: $\vec{r}^{(k)} = \vec{y} - \boldsymbol{A}_{\Omega^k} \vec{x}_{\Omega^k}^{(k)}$

9: **end while**

After computing the vector of scalar products between columns of $\boldsymbol{A}$ and residual $\vec{r}$, denoted by $\vec{g}$ (line 3), the ROMP first selects a *candidate* support set Ω_{c}, containing the indices corresponding to the entries of $\vec{g}$ with highest absolute value after appropriate normalization by the l_2 norm of the corresponding column of $\boldsymbol{A}$ (line 4). The cardinality of Ω_{c} is the minimum between the maximum sparsity $s_{\max}$ and the number of nonzeros in $\vec{g}$. Sorting algorithms exist that can perform this step in $\mathcal{O}(n \log n)$ operations (recall n is to be substituted by n_{atoms} if the dictionary is overcomplete). The crucial step of ROMP is the regularization step in line 5, which substitutes the single-coefficient selection of OMP. In this step one considers all possible subsets $\Omega \subset \Omega_{\text{c}}$ that group the (normalized) coefficients of $\vec{g}$ indexed by Ω_{c} into sets with comparable values. To this end, a regularization parameter $0 < r \leq 1$ is introduced. Note that, if $r = 1$, unless some of the normalized coefficients exhibit exactly equal values, only one set with a unique coefficient can be formed, thus converging to OMP. In [347, 345] it is set to $r = 1/2$. The regularization step can be done in $\mathcal{O}(\log s)$ operations (here $s = \|\Omega_{\text{c}}\| \leq s_{\max}$).

If $\boldsymbol{A}$ satisfies the RIP of order $k = 2s$ with $\delta_k = 0.03/\sqrt{\log s}$, the authors proved that ROMP outputs, in at most s iterations, a support set Ω such that $\operatorname{supp}(\vec{x}) \subset \Omega$, $|\Omega| \leq 2s$, for every possible s-sparse vector $\vec{x}$ (*uniform* recoverability). Note that this result permits using ROMP with *RIP-friendly* matrices, such as Gaussian and Bernoulli random matrices or even partial Fourier matrices, for which OMP could not provide (even non-uniform) recovery guarantees.

3.4.2.4. Order Recursive Matching Pursuit

The Order Recursive Matching Pursuit (ORMP) [339] is a greedy pursuit that aims to improve the quality of the solution over that of OMP, at the cost of higher computational complexity. The algorithm is found in different forms and under different names in the literature, for instance, as *Stepwise Projection* or *Orthogonal Least Squares* [107]. This second name points the main difference between ORMP and OMP: while OMP selects, at each iteration, the dictionary atom that is most correlated with the current residual, ORMP selects the atom that minimizes the residual in the least squares sense. At a first glance, one might expect that selecting the atom that is best correlated with the residual should lead to the maximum error reduction. This might be the case if $\boldsymbol{A}$ satisfies the RIP with a sufficiently small constant and the intercolumn coherence is low enough. Nevertheless, in general, if this is not the case, since OMP ignores the correlations between the remaining candidate atoms and those already selected in previous iterations, selecting the atom that is best correlated with the residual does not necessarily lead to the best residual reduction. ORMP selects the atom that will produce the maximum residual reduction in the subsequent update of $\vec{x}$ by means of an *orthogonal projection.* We provide the ORMP pseudocode in Algorithm 13.

Algorithm 13 Order Recursive Matching Pursuit (ORMP)

Initialize: $\vec{r}^{(0)} = \vec{y}, \vec{x}^{(0)} = \vec{0}, \Omega^0 = \emptyset, k = 0$
1: **while** $\left(\left\|\vec{r}^{(k)}\right\|_2 > \varepsilon_{\text{tol}}\right)$ **and** $\left(\left|\Omega^k\right| < s_{\max}\right)$ **do**
2: $k := k + 1$
3: $\vec{g}^{(k)} = \boldsymbol{A}^\top \vec{r}^{(k-1)}$
4: $i^k = \underset{i \notin \Omega^{k-1}}{\arg\max} \dfrac{\left|g_i^{(k)}\right|}{\left\|\boldsymbol{P}_{\Omega^{k-1}}^{\perp} \vec{a}_i\right\|_2}$
5: Update support: $\Omega^k = \Omega^{k-1} \cup i^k$
6: Update estimate: $\vec{x}_{\Omega^k}^{(k)} = \boldsymbol{A}_{\Omega^k}^{\dagger} \vec{y}$
7: Calculate orthogonal projector: $\boldsymbol{P}_{\Omega^k}^{\perp} := \left(\boldsymbol{I}_m - \boldsymbol{A}_{\Omega^k} \boldsymbol{A}_{\Omega^k}^{\dagger}\right)$
8: Update residual: $\vec{r}^{(k)} = \boldsymbol{P}_{\Omega^k}^{\perp} \vec{y} = \vec{y} - \boldsymbol{A}_{\Omega^k} \vec{x}_{\Omega^k}^{(k)}$
9: **end while**

Note that the main difference between Algorithm 13 and Algorithm 11 is in line 4 of Algorithm 13. This selection of the support element by maximization is intended to minimize the residual after coefficient update, namely,

$$\begin{aligned} i^k &= \underset{1 \le i \le n_{\text{atoms}}}{\arg\min} \left\|\vec{r}^{(k)}\right\|_2^2 \\ &= \underset{1 \le i \le n_{\text{atoms}}}{\arg\min} \min_{x_i} \left\|\vec{r}^{(k-1)} - \vec{a}_i x_i\right\|_2^2 \\ &= \underset{1 \le i \le n_{\text{atoms}}}{\arg\min} \min_{x_i} \left\|\vec{y} - \boldsymbol{A}_{\Omega^{k-1}} \vec{x}_{\Omega^{k-1}} - \vec{a}_i x_i\right\|_2^2 \end{aligned} \tag{3.120}$$

where $\vec{a}_i$ denotes the ith column of $\boldsymbol{A}$. The reader might expect that the projection operator $\boldsymbol{P}_{\Omega}^{\perp}$ also premultiplies $\boldsymbol{A}$ in line 3, coherently with the l_2 normalization in line 4. While this is, in principle, true, note that it is completely unnecessary, provided that, by construction, the residual from the previous iteration, say $k-1$, $\vec{r}^{(k-1)}$ is orthogonal to the subspace spanned by the columns of $\boldsymbol{A}$ indexed by Ω^{k-1}. Therefore, despite such projection would, indeed, modify the columns of $\boldsymbol{A}$, it would not lead to changes in the vector of scalar products, $\vec{g}^{(k)}$ at iteration k. The projection step is the main source of computational cost with respect to OMP. Complexity of $\mathcal{O}(mn)$

per iteration was reported in [339] for his *Algorithm Greedy*, which is an early ORMP formulation. The advantage of ORMP over OMP vanishes as the orthogonality between columns of $\boldsymbol{A}$ increases.

3.4.2.5. Iterative Hard Thresholding

As introduced before, thresholding algorithms are known to be simple and fast and to exhibit similar recovery guarantees to those of convex relaxation. It should be clarified, nevertheless, that such recovery guarantees for thresholding are not uniform [384], as those for OMP. Thresholding algorithms work iteratively, correcting the current estimate of $\vec{x}$ according to the current residual and then *thresholding* the coefficients of the updated $\vec{x}$ to generate a new sparse estimate. The simplest algorithm implementing this scheme is the Iterative Hard Thresholding (IHT), proposed independently in [57] and [375], under the names of IHT and Iterated Marginal Minimization, respectively. We provide the basic structure of IHT in Algorithm 14. Note the simplicity of the algorithm, which could be further reduced to a single iterative step, if line 4 is included in line 3. The step size α in line 3 only regulates the aggressiveness of the update step. Too low values of α require an excessively large number of iterations, while, if α is too large, convergence might be compromised.

Algorithm 14 Iterative Hard Thresholding (IHT)

Initialize: $\vec{r}^{(0)} = \vec{y}, \vec{x}^{(0)} = \vec{0}$
1: **while** $\left(\left\|\vec{r}^{(k)}\right\|_2 > \varepsilon_{\text{tol}}\right)$ **do**
2: $\quad k := k + 1$
3: $\quad$ Update estimate: $\vec{x}^{(k)} = H_s\left(\vec{x}^{(k-1)} + \alpha \boldsymbol{A}^\top \vec{r}^{(k-1)}\right)$
4: $\quad$ Update residual: $\vec{r}^{(k)} = \vec{y} - \boldsymbol{A}\vec{x}^{(k)}$
5: **end while**

The core of Algorithm 14 is the thresholding operation in line 3, expressed by means of the hard thresholding operator H_s, where s is the (enforced) sparsity of the result it delivers. In other words, H_s preserves the largest s entries of the input vector and set the rest to zero. If $\|\boldsymbol{A}\|_2 < 1$ and assuming that its columns contain a basis for the signal space and that $\|\vec{a}_i\|_2 > c > 0$, $\forall i$, then IHT converges to a local minimum. In case $\|\vec{a}_i\|_2 = c > 0$, $\forall i$ (columns of equal norm) and if $c^2 > \mu_1 (s-1)$, then at the fixed point $\hat{\vec{x}}$ the residual norm is bounded by

$$\|\vec{r}\|_2 = \left\|\vec{y} - \boldsymbol{A}\hat{\vec{x}}\right\|_2 \leq \frac{c\,\|\vec{y}\|_2}{c^2 - \mu_1\,(s-1)} \tag{3.121}$$

where $\mu_1(s),\ s \in \mathbb{N}$ denotes cumulative coherence. The right hand side of Eq. 3.121 is also an upper bound for the actual threshold value that is implicitly implemented by selecting the s largest coefficients (to ensure coherence with the following notation, one could denote it by ε_s). Under the hypothesis of fixed support, the optimal step size α can be calculated at each iteration by means of Eq. 3.122, which seeks the step size that maximally reduces the error at each iteration k.

$$\alpha_{\text{opt}}^{(k)} = \frac{\left\|\vec{g}_{\Omega^{k-1}}^{(k-1)}\right\|_2^2}{\left\|\boldsymbol{A}_{\Omega^{k-1}}\vec{g}_{\Omega^{k-1}}^{(k-1)}\right\|_2^2} \tag{3.122}$$

where $\vec{g}_{\Omega^k}^{(k)}$ denotes the vector $\vec{g}^{(k)}$, as defined in previous algorithms (e. g., Algorithm 11), restricted to the entries contained in the temporal support set Ω^k. If $\operatorname{supp}\left(\vec{x}^{(k)}\right) = \operatorname{supp}\left(\vec{x}^{(k-1)}\right)$, then using the step size $\alpha_{\text{opt}}^{(k)}$ yields maximal error reduction, which ensures stability. Nevertheless, if this is not the case and the signal support exhibits a non-negligible variation between consecutive iterations, the optimality of $\alpha_{\text{opt}}^{(k)}$ is to be put into question. In this case, a sufficient condition for convergence is given by Eq. 3.123, from [56].

$$\alpha^{(k)} \leq (1-c)\frac{\left\|\vec{x}^{(k)} - \vec{x}^{(k-1)}\right\|_2^2}{\left\|\boldsymbol{A}\left(\vec{x}^{(k)} - \vec{x}^{(k-1)}\right)\right\|_2^2} \tag{3.123}$$

where $c > 0$ is a small constant. Note that evaluating the right hand side of Eq. 3.123 requires computing an estimate $\vec{x}^{(k)}$. One could first suppose that $\operatorname{supp}\left(\vec{x}^{(k)}\right) = \operatorname{supp}\left(\vec{x}^{(k-1)}\right)$ and adopt the optimal step size suggested by Eq. 3.122 and then, once $\vec{x}^{(k)}$ is obtained, check whether the support was actually preserved. If this is the case, one can continue as in Algorithm 14. Otherwise, one can use Eq. 3.123 to generate a shrunk step size that ensures convergence. This is the scheme adopted in the *Normalized* IHT algorithm (NIHT) proposed in [56]. We omit further details for brevity.

Different thresholding operators can be defined. For instance, if the signal sparsity is unknown, the thresholding operator might actually threshold according to the coefficient values, as in Eq. 3.124, and not to some expected sparsity.

$$H_\varepsilon(\vec{x}) = [h_i]_{1\le i\le n} \mid h_i = \begin{cases} 0 \text{ if } |x_i| \le \varepsilon \\ x_i \text{ if } |x_i| > \varepsilon \end{cases} \tag{3.124}$$

where $\vec{x} \in \mathbb{R}^n$ and ε is the threshold value. In [57] it is set to $\varepsilon = \sqrt{\lambda}$, where λ is the relative weight given to the l_0 cost term in the generic minimization in Eq. 3.125.

$$\hat{\vec{x}} = \arg\min_{\vec{x}} \|\vec{y} - \boldsymbol{A}\vec{x}\|_2^2 + \lambda \|\vec{x}\|_0 \tag{3.125}$$

The only computationally-relevant operations performed by Algorithm 14 are two matrix-vector multiplications, by $\boldsymbol{A}$ and $\boldsymbol{A}^\top$, and the coefficient sorting required for thresholding, thus yielding outstanding speed.

A modification of IHT which is defined as a combination of IHT and CoSaMP is provided in [182] and named Hard Thresholding Pursuit (HTP). In short terms, the core difference between HTP and IHT is how the s-sparse estimate of $\vec{x}$ is obtained. While in IHT this estimate is obtained directly by thresholding the coefficients of the corrected version of $\vec{x}$ (line 3), HTP first estimates the support as the IHT thresholding operator would do, that is, identifying the s largest coefficients of the corrected version of $\vec{x}$, but then the s-sparse estimate of $\vec{x}$ is not obtained by setting to zero the coefficients not contained in the support, but solving the minimization problem in Eq. 3.126, as in the intermediate estimate of CoSaMP.

$$\vec{x}^{(k)} = \arg\min_{\vec{x}} \|\vec{y} - \boldsymbol{A}\vec{x}\|_2, \ \operatorname{supp}(\vec{x}) \subseteq \Omega^k \tag{3.126}$$

The authors show that, if $\boldsymbol{A}$ satisfies the $3s$-RIP with constant $\delta_{3s} < 1/\sqrt{3}$, HTP recovers all s-sparse solutions. Additionally, this condition also guarantees stability with respect to sparsity defect and robustness to measurement error. The sufficient condition $\delta_{3s} < 1/\sqrt{3}$ also applies to IHT. Simulations showed that the recovery times of HTP are lower than those of IHT and NIHT, both in case of successful and unsuccessful recovery. Exactly as in the IHT case, it is also possible to conceive a *normalized* version of HTP, which is denoted by NHTP.

3.4.2.6. Compressive Sampling Matching Pursuit

Differently from IHT or HTP, the Compressive Sampling Matching Pursuit algorithm (CoSaMP) [344] keeps track of the temporal support set Ω along the iterations. The CoSaMP algorithm is summarized in Algorithm 15.

Algorithm 15 Compressive Sampling Matching Pursuit (CoSaMP)

Initialize: $\vec{r}^{(0)} = \vec{y}, \vec{x}^{(0)} = \vec{0}, \Omega^0 = \text{supp}\left(H_s\left(\boldsymbol{A}^\top \vec{y}\right)\right), k = 0$

1: **while** $\left(\left\|\vec{r}^{(k)}\right\|_2 > \varepsilon_{\text{tol}}\right)$ **do**

2: $k := k + 1$

3: $\vec{g}^{(k)} = \boldsymbol{A}^\top \vec{r}^{(k-1)}$

4: Candidate support:

$$\Omega_{\text{c}}^k = \left\{ i \,\middle|\, \frac{\left|g_i^{(k)}\right|}{\|\vec{a}_i\|_2} > \frac{\left|g_j^{(k)}\right|}{\|\vec{a}_j\|_2},\ \left|\Omega_{\text{c}}^k\right| = \min\left(2s, \left\|\vec{g}^{(k)}\right\|_0\right) \right\},\ \forall j \notin \Omega_{\text{c}}^k$$

5: Temporal support (merging): $\Omega_{\text{temp}}^k = \Omega^{k-1} \cup \Omega_{\text{c}}^k$

6: Temporal estimate: $\vec{x}^{(k)}_{\Omega^k_{\text{temp}}} = \boldsymbol{A}^\dagger_{\Omega^k_{\text{temp}}} \vec{y},\ \vec{x}^{(k)}_{\overline{\Omega^k_{\text{temp}}}} = 0$

7: Estimate support: $\Omega^k = \text{supp}\left(H_s\left(\vec{x}^{(k)}_{\Omega^k_{\text{temp}}}\right)\right)$

8: Update estimate: $\vec{x}^{(k)}_{\Omega^k} = \left[\vec{x}^{(k)}_{\Omega^k_{\text{temp}}}\right]_{\Omega^k},\ \vec{x}^{(k)}_{\overline{\Omega^k}} = 0$

9: Update residual: $\vec{r}^{(k)} = \vec{y} - \boldsymbol{A}\vec{x}^{(k)}$

10: **end while**

The main peculiarity of CoSaMP is the existence of an intermediate estimate of $\vec{x}$ of sparsity at most $3s$ (or $3s_{\text{max}}$ if the real sparsity is unknown and the upper bound s_{max} is used instead). The support of this intermediate estimate Ω_{temp} is the union of the signal support Ω from the previous estimate and the set of support indices corresponding to the $2s$ greatest coefficients of $\vec{g}$, after eventual normalization, if the columns of $\boldsymbol{A}$ are not of equal norm. This intermediate estimate is the least-squares solution, restricted to that temporal support, computed by means of the Moore-Penrose pseudoinverse in line 6. Then, this temporal $3s$-sparse solution is used to estimate the support by means of a hard thresholding operator that selects the s largest coefficients (line 7). The estimate of $\vec{x}$ at each iteration is given by that hard thresholding operator, that is, those coefficients indexed by the support set are equal to those of the temporal solution and the rest are set to zero (line 8).

3.4.2.7. Subspace Pursuit

Subspace Pursuit (SP) [124] is very close to CoSaMP. In fact, similarly to CoSaMP, SP also keeps track of the temporal support set and also calculates an intermediate estimate of $\vec{x}$ with a temporal augmented support. The main differences between SP and CoSaMP are the cardinality of that augmented support set and the way the estimate of $\vec{x}$ is calculated once the support is estimated from the intermediate estimate. For clarity, these details will be indicated in the SP pseudocode, given in Algorithm 16.

Algorithm 16 Subspace Pursuit (SP)

Initialize: $\vec{r}^{(0)} = \vec{y}, \Omega^0 = \operatorname{supp}\left(H_s\left(\boldsymbol{A}^\top \vec{y}\right)\right), \vec{x}^{(0)} = \boldsymbol{A}_{\Omega^0}^{\dagger} \vec{y}, k = 0$

1: **while** $\left(\left\|\vec{r}^{(k)}\right\|_2 > \varepsilon_{\text{tol}}\right)$ **do**

2: $k := k + 1$

3: $\vec{g}^{(k)} = \boldsymbol{A}^\top \vec{r}^{(k-1)}$

4: Candidate support:

$$\Omega_{\text{c}}^k = \left\{ i \;\middle|\; \frac{\left|g_i^{(k)}\right|}{\|\vec{a}_i\|_2} > \frac{\left|g_j^{(k)}\right|}{\|\vec{a}_j\|_2}, \; \left|\Omega_{\text{c}}^k\right| = \min\left(s, \left\|\vec{g}^{(k)}\right\|_0\right) \right\}, \quad \forall j \notin \Omega_{\text{c}}^k$$

5: Temporal support (merging): $\Omega_{\text{temp}}^k = \Omega^{k-1} \cup \Omega_{\text{c}}^k$

6: Temporal estimate: $\vec{x}_{\Omega_{\text{temp}}^k}^{(k)} = \boldsymbol{A}_{\Omega_{\text{temp}}^k}^{\dagger} \vec{y}, \; \vec{x}_{\overline{\Omega_{\text{temp}}^k}}^{(k)} = 0$

7: Estimate support: $\Omega^k = \operatorname{supp}\left(H_s\left(\vec{x}_{\Omega_{\text{temp}}^k}^{(k)}\right)\right)$

8: Update estimate: $\vec{x}_{\Omega^k}^{(k)} = \boldsymbol{A}_{\Omega^k}^{\dagger} \vec{y}$

9: Update residual: $\vec{r}^{(k)} = \vec{y} - \boldsymbol{A}\vec{x}^{(k)}$

10: **end while**

The first difference is to be observed in line 4, where the cardinality of the *candidate* support set Ω_{c}, to be merged with the previous support set, is upper-bounded by s, that is, Ω_{c} contains the indices corresponding to the s largest coefficients of $\vec{g}$ (recall, after normalization, if needed), unless the number of nonzero elements of $\vec{g}$ is lower than s, of course. Consequently, the temporal support set obtained in line 5, Ω_{temp} has, at most, cardinality $2s$, differently from CoSaMP, where its cardinality was, at most $3s$. The other fundamental difference with respect to CoSaMP is the final estimation of $\vec{x}$, once the support of cardinality s is estimated. While CoSaMP calculates the s-sparse approximation of $\vec{x}$ by s-hard-thresholding of the intermediate solution, SP computes the least-squares solution, restricted to the estimated support, as shown in line 8, using the Moore-Penrose pseudoinverse.

Exact recovery via SP is guaranteed if $\boldsymbol{A}$ satisfies the $3s$-RIP with $\delta_{3s} < 0.165$. The authors of [124] also suggest using a threshold on the the variation of residual norm between consecutive iterations as stopping criterion, instead of a threshold on the residual norm directly, as we have suggested for all greedy algorithms. In that case, the condition $\delta_{3s} < 0.165$ can be relaxed to $\delta_{3s} < 0.205$. The complexity of the SP algorithm is upper-bounded by

$\mathcal{O}(m(n+s^2)s)$ for sparse signals, which, in the common case of $s^2 < n$, simplifies to $\mathcal{O}(mns)$.

3.4.2.8. Multiple Measurement Vector Extensions

Until now all recovery methods considered are for a *single measurement vector*, thus named SMV recovery methods, that compute an estimate of the sparse vector $\vec{x}$ that explains the measurements $\vec{y}$. Now consider a system of N measurement vectors, for convenience stacked by columns in a matrix of measurements $\boldsymbol{Y} = [\vec{y}_i]_{1 \leq i \leq N}$, $\vec{y}_i \in \mathbb{R}^m\ \forall i$. Each measurement vector is gathered by means of the same sensing matrix $\boldsymbol{\Phi}$ and each signal being measured admits a sparse representation in the same dictionary $\boldsymbol{\Psi}$. For convenience, let the sparse representations be stacked by columns in a matrix $\boldsymbol{X} = [\vec{x}_i]_{1 \leq i \leq N}$, $\vec{x}_i \in \mathbb{R}^{n_{\text{atoms}}}\ \forall i$, exactly as their corresponding vectors of measurements. Suppose, for simplicity, that all the vectors $\vec{x}_i$ are s-sparse, that is, $|\operatorname{supp}(\vec{x}_i)| \leq s,\ \forall i$. Obviously, the problem of recovering the matrix of sparse vectors $\boldsymbol{X}$ from the matrix of measurement vectors $\boldsymbol{Y}$ can be solved sequentially, that is, solving Eq. 3.9 or, equivalently, Eq. 3.10, for each pair $\vec{y}_i, \vec{x}_i$, $1 \leq i \leq N$. Nevertheless, one can intuit that a sequential SMV scheme might not be the best way to approach this problem. In order to find a better alternative, consider the notion of *joint sparsity* as defined in [137]: a matrix $\boldsymbol{X} \in \mathbb{R}^{n \times N}$ is called k jointly sparse, or k joint sparse, if $|\operatorname{supp}(\boldsymbol{X})| \leq k$, being

$$\operatorname{supp}(\boldsymbol{X}) := \bigcup_{1 \leq i \leq N} \operatorname{supp}(\vec{x}_i) \tag{3.127}$$

We adopt the term *joint sparse* for being that originally used in [137]. According to this definition, our matrix $\boldsymbol{X}$ of sparse vectors defined above is k joint sparse with $k \leq Ns$. Of special interest is the case when all signals share approximately the same support and, consequently, $k \approx s$. In such case one can estimate the support of the signals in a joint manner, solving one single *multiple measurement vector* (MMV) problem instead of N sequential SMV recoveries. The MMV recovery can be formulated in a general manner as the minimization of the cardinality of the joint support subject to the measurements:

$$\hat{\boldsymbol{X}} = \underset{\boldsymbol{X} \in \mathbb{R}^{n \times N}}{\arg\min} |\operatorname{supp}(\boldsymbol{X})| \text{ subject to: } \boldsymbol{AX} = \boldsymbol{Y} \tag{3.128}$$

for a generic signal dimensionality n, to be substituted by n_{atoms} if an overcomplete dictionary is used. As usual, $\boldsymbol{A} = \boldsymbol{\Phi\Psi}$. Note that the SMV problem (Eq. 3.9) is just a special case of Eq. 3.128 for $N = 1$. Solving the MMV problem in Eq. 3.128 requires less computational cost than solving the equivalent sequence of SMV problems, despite the required memory increases. Additionally, estimating the support in a joint manner yields better robustness to noise and incompleteness of the measurements. Disturbances will be different for each vector of measurements, thus they will cancel each other up to some extent if the support estimation is carried out in a joint manner. The sparse matrix $\boldsymbol{X}$ can be uniquely determined from the matrix of measurements $\boldsymbol{Y}$ if the following sufficient condition [106] is satisfied:

$$|\mathrm{supp}\,(\boldsymbol{X})| < \frac{\mathrm{spark}\,(\boldsymbol{A}) - 1 + \mathrm{rank}\,(\boldsymbol{Y})}{2} \tag{3.129}$$

Eq. 3.129 is to be compared to the condition $\mathrm{spark}\,(\boldsymbol{A}) > 2s$, which was shown to be sufficient to ensure uniqueness of the solution to the SMV case [150]. Indeed, for $\mathrm{rank}\,(\boldsymbol{Y}) = 1$ (as in the SMV case), Eq. 3.129 boils down to the latter SMV uniqueness condition with $k = s$, as expected. A direct consequence of the sufficient condition in Eq. 3.129 is that, in the MMV framework, the sparsity requirement can be leveraged and less-sparse signals can be recovered from the same number of measurements. Conversely, at constant sparsity, the MMV framework allows reducing the number of measurements that ensure successful recovery. Consider the following optimal scenario: $\boldsymbol{\Phi}$ is appropriately constructed so that the spark is maximized: $\mathrm{spark}\,(\boldsymbol{A}) = m + 1$ (recall $m \ll n$). The number of vectors N is sufficiently large, so that $\mathrm{rank}\,(\boldsymbol{X})$ is only bounded by the joint sparsity k, that is, the dimensionality of the subspace where all sparse columns of $\boldsymbol{X}$ live. Obviously, this bound on the rank immediately translates to $\boldsymbol{Y}$: $\mathrm{rank}\,(\boldsymbol{Y}) \leq \mathrm{rank}\,(\boldsymbol{X}) \leq k$. Then, the condition in Eq. 3.129 translates into $m \geq k + 1$. In other words, in an optimal scenario, a MMV framework with enough number of vectors only requires one more measurement than an *oracle-guided* sensing to estimate the sparse coefficients, that is, $s+1$ if $k = s$. Note that the inequality in Eq. 3.129 also holds if $\mathrm{rank}\,(\boldsymbol{Y})$ is substituted by $\mathrm{rank}\,(\boldsymbol{X}) \geq \mathrm{rank}\,(\boldsymbol{Y})$. In the following we enumerate and briefly comment some well-known methods for solving Eq. 3.128, mostly extensions of SMV greedy algorithms to the MMV case.

Full-rank MMV: MUSIC The maximum profit of an MMV framework is taken when $\mathrm{rank}\,(\boldsymbol{Y}) = k$. In this case, a solution to Eq. 3.128 can be obtained using a discrete version of the Multiple Signal Classification (MUSIC) algorithm, first introduced in [399] and popular in array signal processing. Discrete MUSIC was first applied to the MMV sparse recovery in [175], in the context of reconstruction of multiband signals without prior knowledge of the spectral support. Let $\mathcal{R}(\boldsymbol{M})$ denote the *image* or *range* of the matrix $\boldsymbol{M}$, that is, the space spanned by its columns. Then, if $\operatorname{rank}(\boldsymbol{Y}) = k$, it follows that $\mathcal{R}(\boldsymbol{Y}) = \mathcal{R}(\boldsymbol{A}_\Omega)$, that is, every column of $\boldsymbol{A}$ indexed by the joint support set $\Omega = \operatorname{supp}(\boldsymbol{X})$ lives in the subspace spanned by the columns of $\boldsymbol{Y}$. In order to exploit this fact, MUSIC constructs first an orthonormal basis of this subspace from $\boldsymbol{Y}$, say $\boldsymbol{U}$, e. g., via SVD. Then estimating the support means finding the columns of $\boldsymbol{A}$ that are best correlated with the basis vectors. In theory we have that

$$i \in \Omega \Leftrightarrow \frac{\left\|\vec{a}_i^\top \boldsymbol{U}\right\|_2}{\left\|\vec{a}_i\right\|_2} = 1 \tag{3.130}$$

but, in practice, due to the presence of noise, one simply looks for the k support indices for which the right hand side of Eq. 3.130 is the largest. Once Ω is determined, the problem becomes overdetermined (typically $m > k$) and $\boldsymbol{X}$ can be robustly estimated from $\boldsymbol{Y}$ via the Moore-Penrose pseudoinverse of $\boldsymbol{A}_\Omega$. If necessary, MUSIC can attain improved robustness to noise via eigenvalue thresholding. The simplified pseudocode is given in Algorithm 17.

Algorithm 17 Discrete MUSIC

1: Orthonormal basis for $\mathcal{R}(\boldsymbol{Y})$: $U = \operatorname{orth}(\boldsymbol{Y})$

2: Support set: $\Omega = \left\{ i \,\middle|\, \frac{\left\|\vec{a}_i^\top \boldsymbol{U}\right\|_2}{\left\|\vec{a}_i\right\|_2} > \frac{\left\|\vec{a}_j^\top \boldsymbol{U}\right\|_2}{\left\|\vec{a}_j\right\|_2},\ |\Omega| = k \right\},\ \forall j \notin \Omega$

3: Calculate estimate: $\hat{\boldsymbol{X}} = \boldsymbol{A}_\Omega^\dagger \boldsymbol{Y}$

MMV-OMP MMV extensions of MP and OMP are studied in [311], under the names of Vector Weak Greedy Algorithm (VWGA) and Vector Weak Orthogonal Greedy Algorithm (VWOGA), respectively. The MMV extension of OMP has also been studied in [294] for the general setting of uniformly smooth Banach spaces.

OMP for MMV was studied in [106] as a natural extension of the SMV OMP [361] to the MMV problem. The MMV-OMP approximately follows the structure outlined in Algorithm 11, with the corresponding adaption of the vectors $\vec{x}$, $\vec{y}$ and $\vec{r}$ into the corresponding MMV matrices $\boldsymbol{X}, \boldsymbol{Y}$ and $\boldsymbol{R}$. A fundamental difference is that the scalar products between each column of $\boldsymbol{A}$ and the residual vectors (columns of $\boldsymbol{R}$) yield N-dimensional vectors and no longer a single scalar, as in the SMV case. Consequently, the support element selection step (line 4 in Algorithm 11) has to be modified accordingly and the term to maximize changes from $|g_i|/\|\vec{a}_i\|_2$ to $\|\vec{g}_i\|_p/\|\vec{a}_i\|_2$, where $\vec{g}_i \in \mathbb{R}^N$ is the vector containing the scalar products between the ith column of $\boldsymbol{A}$ and each of the N columns of $\boldsymbol{R}$. Different selections of p in the norm $\|\cdot\|_p$ yield different versions of the MMV-OMP algorithm. In [438], $p = 1$ is chosen and the resulting algorithm is named *Simultaneous Orthogonal Matching Pursuit* (SOMP, see [440] for details and comparisons). In [120] $p = 2$ is chosen for both MMV-MP and MMV-OMP direct extensions. In [106] a general algorithm with arbitrary $p \geq 1$ is considered. The authors show that, when $\boldsymbol{X}$ is very sparse, the selection of p does not play an important role and the MMV-OMP recovers the sparsest representation, regardless of the chosen l_p norm. Apart from these variations in the support selection step, the different MMV-OMP algorithms exhibit equivalent structure, close to that of SMV-OMP.

MMV-Thresholding MMV extensions of well-known thresholding algorithms, such as IHT, NIHT, HTP, NHTP and CoSaMP, are given in [50] and named adding an 'S' (from *simultaneous*) before their abbreviations, as in SOMP. All these MMV extensions implement a step of joint support detection that is carried out on the current estimate of $\boldsymbol{X}$, which identifies the k rows of $\boldsymbol{X}$ with largest l_2 norm. In the MMV extensions of IHT, NIHT, HTP and NHTP, that estimate of $\boldsymbol{X}$ is after residual-driven correction, while in the case of the MMV-CoSaMP, the joint support detection is carried out on the intermediate estimate of $\boldsymbol{X}$ (see the SMV CoSaMP in Algorithm 15 for reference). This estimate of the joint support is then used in a subsequent step of hard thresholding, which sets to zero all rows of $\boldsymbol{X}$ not indexed in that support estimate, thus equivalent to the support *pruning* step in the SMV formulation.

Similarly, the MMV extension of the Alternating Direction Method (MMV-ADM) proposed in [305] makes use of a shrinkage operator that operates jointly over each of the rows of $\boldsymbol{X}$, thus enforcing the joint sparsity of the columns.

MMV-ORMP An MMV extension of ORMP is proposed in [120], together with a review of other MMV extensions, such as MMV-MP and MMV-OMP. The extension is, as previous MMV extensions, trivial and, in fact, the MMV version of ORMP can be immediately derived from Algorithm 13, just by substituting the vectors $\vec{x}$, $\vec{y}$ and $\vec{r}$ by the corresponding MMV matrices $\boldsymbol{X}$, $\boldsymbol{Y}$ and $\boldsymbol{R}$ and the absolute value of g_i in line 4 by the l_2 norm of the corresponding vector $\vec{g}_i$, as explained when introducing MMV-OMP. Apart from that, MMV-ORMP operates according to Algorithm 13 and the estimate of $\boldsymbol{X}$ at each iteration is obtained from $\boldsymbol{Y}$ by means of the Moore-Penrose pseudoinverse of $\boldsymbol{A}$ restricted to the temporal support set.

RA-ORMP All the MMV extensions presented so far, except from MUSIC, which operates in the full-rank case, are *rank blind*, since they do not explicitly handle the rank of $\boldsymbol{X}$. Consequently, the worst-case recovery performance of the MMV extensions cannot be shown to improve that of their SMV counterparts. Indeed, one can easily imagine, as an example of worst-case scenario, the case of all columns of $\boldsymbol{X}$ being identical, which is equivalent to an SMV. In order to attain improved worst-case guarantees, an MMV recovery method is needed that *sees* the rank and integrates this valuable information into the joint sparse recovery pipeline. Such algorithms are referred to as *rank aware* (RA) algorithms. A paradigmatic example is the Rank Aware Order Recursive Matching Pursuit (RA-ORMP) [137], which is, as the name indicates, an RA MMV extension of ORMP. The pseudocode of RA-ORMP is provided in Algorithm 18.

Algorithm 18 Rank Aware Order Recursive Matching Pursuit (RA-ORMP)

Initialize: $\boldsymbol{R}^{(0)} = \boldsymbol{Y}, \boldsymbol{X}^{(0)} = \boldsymbol{0}, \Omega^0 = \emptyset, k = 0$

1: **while** $\left(\left\|\boldsymbol{R}^{(k)}\right\|_2 > \varepsilon_{\text{tol}}\right)$ **and** $\left(\left|\Omega^k\right| < s_{max}\right)$ **do**

2: $\quad k := k + 1$

3: $\quad \boldsymbol{U}^{(k-1)} = \text{orth}\left(\boldsymbol{R}^{(k-1)}\right)$

4: $\quad i^k = \underset{i \notin \Omega^{k-1}}{\arg\max} \dfrac{\left\|\vec{a}_i^\top \boldsymbol{U}^{(k-1)}\right\|_2}{\left\|\boldsymbol{P}_{\Omega^{k-1}}^\perp \vec{a}_i\right\|_2}$

5: $\quad$ Update support: $\Omega^k = \Omega^{k-1} \cup i^k$

6: $\quad$ Update estimate: $\boldsymbol{X}_{\Omega^k}^{(k)} = \boldsymbol{A}_{\Omega^k}^\dagger \boldsymbol{Y}$

7: $\quad$ Calculate orthogonal projector: $\boldsymbol{P}_{\Omega^k}^\perp := \left(\boldsymbol{I}_m - \boldsymbol{A}_{\Omega^k} \boldsymbol{A}_{\Omega^k}^\dagger\right)$

8: $\quad$ Update residual: $\boldsymbol{R}^{(k)} = \boldsymbol{P}_{\Omega^k}^\perp \boldsymbol{Y} = \boldsymbol{Y} - \boldsymbol{A}_{\Omega^k} \boldsymbol{X}_{\Omega^k}^{(k)}$

9: **end while**

Observe that the fundamental difference with respect to a naïve (rank blind) MMV extension of the ORMP algorithm (see Algorithm 13 for reference) is in line 3, where an orthonormal basis for the previous residual is calculated and denoted by $\boldsymbol{U}^{(k-1)}$, at iteration k. Then, the next support element i^k is chosen to be that for which the corresponding column of $\boldsymbol{A}$, projected by $\boldsymbol{P}_{\Omega^{k-1}}^\perp$, is best-correlated with the basis vectors (columns) of

$\boldsymbol{U}^{(k-1)}$, as indicated in line 4. As already pointed out in Section 3.4.2.4 for the SMV-ORMP, it is not necessary to explicitly project each column of $\boldsymbol{A}$ in the numerator of the quotient to maximize in line 4, since the residual, lives in a subspace that is already orthogonal to that spanned by the columns of $\boldsymbol{A}$ indexed by Ω^{k-1}, by construction.

Similarly to MUSIC, RA-ORMP enjoys guaranteed recovery in the full-rank case. In the rank one case, RA-ORMP reduces to ORMP.

3.4.3. Recovery Using *a priori* Knowledge: Model-Based Compressive Sensing

CS assumes that the only knowledge we have about the signal to recover $\vec{x} \in \mathbb{R}^n$, apart from a low-dimensional vector of measurements $\vec{y} \in \mathbb{R}^m$, $m < n$, is the fact that it admits a sparse or, at least, compressible representation in some dictionary, also known. With this knowledge, the recovery methods presented in this section aim to reconstruct $\vec{x}$ from $\vec{y}$. Successful recovery is only guaranteed if m is greater than some lower bound that depends on n and the sparsity of $\vec{x}$, s (see Section 3.1). In practice, different algorithms exhibit different minimum values of m for which $\vec{x}$ is successfully recovered, despite all look for the sparsest solution that is coherent with the measurements. In many cases sparsity is not the only knowledge on $\vec{x}$ available. Further knowledge on the *structure* of the sparse support might also be available. Such knowledge can be more or less complex and can range, for instance, from full knowledge of the signal support (*oracle-given* support) to some relations of mutual exclusion or mutual inclusion between support elements or groups of them. One can think of different cases where such *a priori* knowledge determines or, at least, influences, the structure of the support. Intuitively, one expects that considering this knowledge in the recovery algorithm improves its performance, i. e., that the same reconstruction error is achieved with lower m or, conversely, that better reconstruction is attained using the same number of measurements.

This idea is formally studied as *model-based* CS in [25], where the concept of *structured sparsity* is introduced to characterize sparse signals where interdependencies among coefficients can be modeled, thus defining some structure in the support. Modeling this underlying structure reduces the degrees of freedom of $\vec{x}$ by permitting only certain support configurations. The authors provide *structured sparsity* extensions of the IHT and CoSaMP algorithms. *Block sparsity* refers to the case when the nonzero coefficients appear grouped around very few locations. In this case, not only the coefficients are sparse, but blocks of coefficients are even sparser. If some estimate of the block size is known, accounting for block sparsity during recovery is an easy task, while being a valuable guide for finding the signal support. Reconstruction of block-sparse signals from a reduced number of measurements have been studied in [426]. A different case is the *sparse tree* structure induced, for instance, by 2D wavelet dictionaries (see the probabilistic tree formulation of [123]). Similarly to wavelet encoding, the recovery method can profit from a multiscale operation, where the membership of a signal coefficient in the support is conditioned by that of *parent* coefficients. The tree structure of wavelet dictionaries has been exploited in [273, H7] within a CS framework to attain better signal reconstruction. The *block sparse* CoSaMP extension in [25] needs $m = \mathcal{O}\left(n_{\text{block}} s + s \log\left(\frac{n}{s}\right)\right)$ measurements (cf. Eq. 3.28) to attain robust recovery. Note that, as the block size n_{block} grows, the dependency on n is reduced. The *tree structure* CoSaMP extension requires only $m = \mathcal{O}(s)$ measurements, thus eliminating the dependency on n.

A model-based RIP is derived in [55] for the union of subspaces model. Indeed, structured sparsity can be modeled as a union of subspaces, where each subspace corresponds to a different s-sparse support configuration. It should be noted that the model-based RIP *extends* the conventional RIP, in the sense that the latter can be retrieved from the former when the number of subspaces in the union is the unrestricted combination $\binom{n}{s}$. For an arbitrary support size, k, which might or not correspond to the sparsity of the signals to recover, s, we denote the model-based RIP by $\mathcal{M}_k$-RIP. Restrictions in the support structure translate into a lower number of measurements needed to guarantee successful recovery. Specifically, any subgaussian measurement matrix $\boldsymbol{A} \in \mathbb{R}^{m \times n}$ satisfies the $\mathcal{M}_k$-RIP with constant $\delta_{\mathcal{M}_k}$ with probability at least $1 - e^{-t}$ if the number of measurements is

$$m \geq \frac{2}{C\delta_{\mathcal{M}_k}^2} \left(\ln(2n_k) + k \ln \frac{12}{\delta_{\mathcal{M}_k}} + t \right) \tag{3.131}$$

where n_k is the number of subspaces in the union where the k-sparse signals live and $C > 0$ is a constant. Parallel to structured sparsity, the authors of [25] also introduce the concept of *structured compressibility*, for signals whose coefficients decay according to a *structured* power law. In order to study the recovery stability for such signals, the authors consider the effect of a k-term approximation, say $\vec{x}_k$, which translates into an augmented noise in the measurement side: $\vec{n}_\text{a} = \vec{n} + \boldsymbol{A}(\vec{x} - \vec{x}_k)$, being $\vec{n} \in \mathbb{R}^m$ the original measurement noise vector. Then, $\boldsymbol{A}$ should not *amplify* much the l_2 norm of vectors living in the k-term approximation residual subspaces. This immediately leads to the formulation of a *Restricted Amplification Property* (RAmP), which is to be fulfilled to ensure recoverability. The detailed formulation, together with a lower bound on m that ensures satisfaction of the RAmP with high probability are also provided in [25].

4. Compressive Sensing for the Photonic Mixer Device

At the light of the fundamentals of Compressive Sensing (CS) presented in Chapter 3, it is clear that the phase-shift-based ToF imaging systems described in Chapter 2 offer an appropriate application area of CS. The signals to be sensed are typically not random, but follow a specific structure, which allows them to be sparsely represented in an appropriate dictionary and, therefore, eventually recovered from few measurements. In this chapter we study the most appealing ways of applying CS to PMD-based ToF imaging systems, with the main goal of overcoming the current limits of the technology, introduced in Section 2.4, mostly the poor lateral resolution, derived from the low number of pixels, and the low depth accuracy.

4.1. Introduction and Application Domains

CS theory states that a signal can be recovered from few measurements, fewer than those suggested by the Shannon criterion, if it admits a sparse or compressible representation in an appropriate dictionary or basis. In order to achieve exact signal recovery, a certain minimum number of measurements is required, which depends on the signal sparsity, and are to be gathered incoherently with the representation matrix or dictionary. Consequently, the first task to identify appropriate application domains is to look for sparsity or compressibility of the different signals being sensed by a PMD-based ToF imaging system.

Another requirement of CS is the linearity of the sensing model, i.e., the measurements can be modeled as a scalar product between the signal to recover and different sensing functions. In the discrete case, the signal to recover is a vector and the sensing kernels are the rows of the sensing matrix. At the light of Eq. 2.19, the process of depth calculus from the raw measurements is highly non-linear and, therefore, CS cannot be applied to the full ToF imaging pipeline, as a whole. Clearly, one could linearize the process around a certain operation point using, e. g., the first order Taylor series expansion, which, in the multivariable case, means using the Jacobian matrix in the linear model. This requires having an initial guess of the depth and reflectance maps and any eventual DC offset, since the measurements depend on the offset, amplitude and phase of the light signal received at the pixel. In principle, we face another *chicken and egg* problem. If the camera provides a video stream of depth images at high frame rate, the previous result can be used as initial guess for the next. A Kalman filter can be used to keep track of the current state for each pixel. The prediction step may use information from neighboring pixels and additional state variables (e. g., *speed* of the pixel in the scene or camera motion) to predict the reflectance and depth in the next instant. While this is theoretically possible, the strong non-linearity of Eq. 2.19 makes any approach based on linearization risky. The filter might easily get lost and diverge from the real state. This is specially applicable to the PMD case, where insufficient illumination often leads to measurements corrupted by strong Poisson noise. For this reason and in order to provide general methods working in a per-depth-image basis, not requiring a continuous stream, we look for our application domains in the raw data level, before the calculus in Eq. 2.19 is carried out.

Provided that the system is an imaging system, i.e., a camera, the data it generates are images of natural or man-made scenes. Both natural images and images of man-made structure are known to exhibit a compressible representation [318], if not sparse, in an appropriate dictionary. This is the key point underlying the widely-spread lossy compression formats JPEG and JPEG2000, based on the discrete cosine transform (DCT) basis and the wavelet basis, respectively. As well as images acquired with conventional photographic cameras, the raw images delivered by PMD sensors are compressible and can, therefore be recovered from few stochastic measurements. That is, the image *spatial* domain (pixel domain) is suitable for the application of CS.

PMD cameras normally operate with a continuous wave (CW) periodic modulation of the emitted light. Provided that the light itself is an electromagnetic wave serving as carrier, this modulation is often called continuous wave amplitude modulation (CWAM). Clearly, if the modulation waveform is a periodic signal, it admits a sparse representation in any frequency-based dictionary or basis, such as the Walsh basis, the DCT basis or the Fourier basis, which is known to be the optimal basis for periodic signals. Conversely, if the modulation signal is a pulsed signal, i.e., light is emitted in few or a single light pulse, then the signal is sparse directly in time domain. In other words, the modulation signal is sparse in time or frequency domain or, more generally, in the time-frequency plane, unless PN modulation is used, in which case the signal is densely distributed in both domains, since ideal noise does not admit a sparse representation.

4.1.1. Spatial Domain

As far as the raw data are images, compressibility in, e. g., wavelet domain can be assumed and exploited in a CS framework. There are two fundamental directions of applying CS to optimize the operation of a sensor. One is oriented to recover the same information with less measurements, while the other pursues recovering more information with a number of measurements increased in a lower amount than that of the recovered information.

In spatial domain, adopting the first focus would mean attempting to recover the raw images from a number of measurements that is lower than the number of pixels. Some results on sparse recovery of PMD raw images from few measurements are given in Section 4.4.1. The requirement of a set of several raw images per depth image immediately brings the idea of exploiting the common structure in the raw images, e. g., in a MMV approach. This idea is further developed in Section 4.4.3, where a *greedy* method is presented that performs a synergistic joint recovery of the raw images from a reduced number of measurements.

The second focus would be oriented to obtain a super-resolved image, i.e., to recover images with a higher number of pixels than that of the PMD sensor from a number of measurements that is lower than the number of pixels in the super-resolved image. One way to implement this concept would be to superimpose custom patterns on the image section projected onto a single pixel, in a similar way to the well-known *single pixel camera* [156]. This way a small image patch can be recovered per pixel of the PMD sensor, being the superresolution factor given by the size of the (square) patch side. Such a system requires a specific hardware development, able to superimpose the patterns directly on the pixel surface. Alternatively, one could place some sort of external spatial light modulator (SLM) at the image plane and refocus later the result on a PMD chip, avoiding the need for a custom chip development. This hardware is implemented as an external addition to the first prototype of a CS-based PMD camera presented in Section 5.2.1. An alternative simple way of obtaining the required stochastic measurements without an SLM would be to displace the PMD sensor to different positions on the image plane, with a step size lower than the pixel size. This way, a pseudorandom sensing kernel can be achieved by combining the measurements obtained for different displacements. This method is further developed in Section 4.4.2 and used to recover superresolved FLIM images. Note that using chip displacements to gather the spatial frequencies that are lost due to averaging on the pixel surface in conventional imaging allows eliminating the *blind* areas of the PMD chip.

Any superresolution method requires a model of the pass from a high-resolution scene—infinite, in the continuous case—to a low-resolution image. The complex structure of the PMD pixels makes conventional downsampling approaches (e. g., Gaussian blur plus regular decimation) completely invalid for modeling the PMD sensing process, since they rely on implicit hypotheses such as a high fill factor and uniformity of the pixel response. A fast look at the simple schema of the PMD 19k-S3 pixels depicted in Fig. 2.8b reveals that approximately half of the pixel is non-sensitive, while the sensitive area is structured in *fingers* around several readout wires (4 per pixel). Consequently, the pixel response in spatial domain is expected to be different for the A and B channels and probably also not uniform over the A and B areas depicted in Fig. 2.8b. An accurate characterization of the PMD pixel response is a *conditio sine qua non* for modeling the resolution loss due to the non-infinitesimal pixel area or, equivalently, to develop a superresolution framework, which is based on the inverse model. In a CS framework, this model has to be taken into account in the construction of the sensing matrix. For this reason, Section 4.3 provides a thorough characterization of the PMD pixel response in spatial domain.

4.1.2. Time-Frequency Domain

As introduced above, any ToF imaging system based on phase-shift emits a signal (and receives the corresponding echo), which is highly sparse in the time-frequency plane, unless PN modulation is adopted. Being faithful to the conventional PMD camera design, we adopt the hypothesis of a quasi-sinusoidal modulation, that is, an illumination signal that is highly sparse in frequency domain. This does not imply a loss of generality, since any method developed under this hypothesis can be trivially adapted to the case of sparsity in time domain by simply substituting the frequency dictionary by an identity matrix or equivalent spikes dictionary. Other signal shapes, such as Gaussian functions, exponentially-modified Gaussian functions or sinc functions, could also be considered, as far as they are equally sparse in the corresponding dictionary and sensing kernels equally incoherent with that dictionary are provided. In general, random sensing kernels are always incoherent with time-frequency dictionaries, if the noise bandwidth is large enough.

The real shape of the illumination signal depends both on the ICS used to drive the light modules and the nature of the light sources. Clearly, if the light source is a femtosecond laser, one can consider that it does not impose any limitation and the shape of the illumination signal replicates that of the ICS. This is not possible in practice, since it supposes rising the cost of the ToF system from the 100 $ range to the 50-100 k$ range. A common option that minimizes the cost while maximizing the efficiency of the electro-optical conversion is the use of LEDs as light sources. As pointed out in Section 2.3.3, the rising and falling times in the nanosecond range produce a low-pass filtering effect in the conversion of electrical signals to optical signals when the fundamental frequency is of tens of megahertz or higher. That is, the higher the frequency, the higher the attenuation and, consequently, harmonics of very high frequency may be completely eliminated in the conversion. The frequency response of the LEDs of the illumination system used in this work (and analyzed in Appendix A.6) has been evaluated and the result is shown in Fig. 4.1.

Differently from a simple one-dimensional analysis of the frequency response, in which the ratio between output and input power is measured for different frequencies supposing sinusoidal signals, the 2D analysis in Fig. 4.1 allows observing how the transferred power is redistributed in frequency. By means of another dimension in the plots we avoid subjecting the analysis to an unrealistic hypothesis of sinusoidality. The ICS is, in fact, a square signal. 15 experimental cases are considered, with ICS fundamental frequencies equally spaced between 2 and 30 MHz. For each case, the power distributions in frequency of the ICS (Fig. 4.1a) and the light signal emitted by the LED (Fig. 4.1b) are obtained, with 1 MHz resolution. Observe the change in the signal shape: in Fig. 4.1a only the odd harmonics of the fundamental frequencies exhibit non-zero power (as it corresponds to a square wave), while in Fig. 4.1b both even and odd harmonics of the fundamental frequencies have non-zero power. Despite the output signals preserve the power peaks at the fundamental frequency present in the corresponding inputs and exhibit remarkable attenuation of high-order harmonics, it is clear that the output signals are not square, neither sinusoidal. Despite the sinusoidal approximation might be valid for relatively low frequencies, the large attenuation of the fundamental frequency component for high frequencies rises the THD to unacceptable levels for such hypothesis to hold. Consider, for instance, the cases of 20 or 22 MHz, where the power of the first harmonic is of the same order of magnitude as that of the fundamental component.

The normalized frequency response in Fig. 4.1c was obtained dividing the

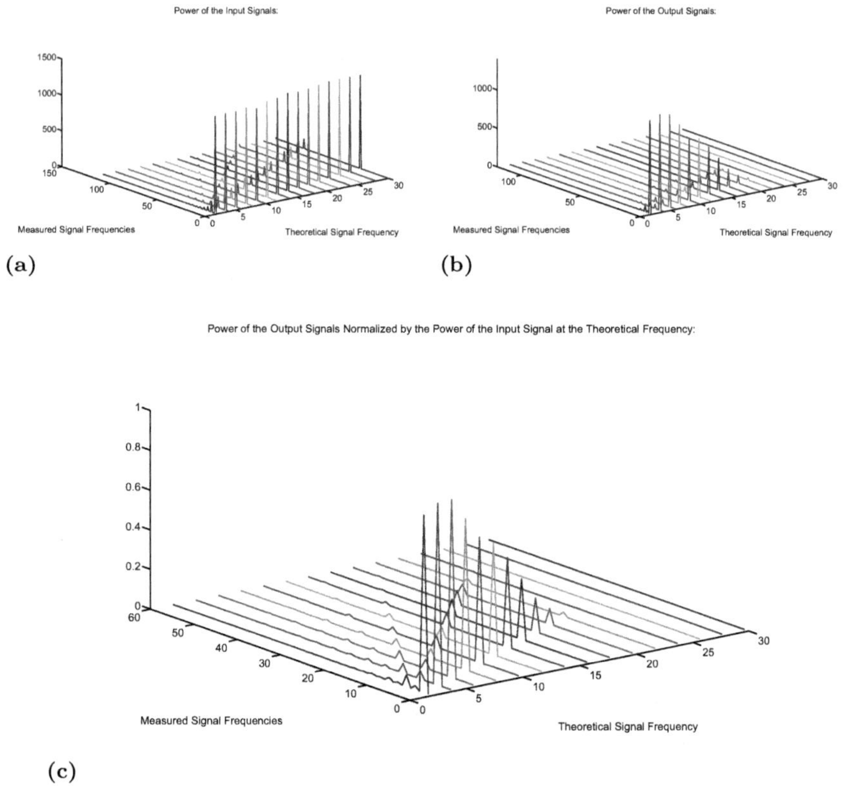

Figure 4.1.: Frequency response of the NIR LEDs used in our illumination systems. The ICS is a square signal, whose fundamental frequency was varied from 2 to 30 MHz with 2 MHz step size. Observe the characteristic power distribution in frequency of the square inputs (a) with only odd harmonic frequencies. The corresponding output signal power distribution in frequency (b) clearly shows a low-pass filtering effect. For fundamental frequencies higher than 24 MHz the transferred power is negligible. A normalized frequency response can be obtained by dividing the output powers by that of the input signal at the fundamental frequency (c). The scale of measured and theoretical frequencies have resolutions of 1 MHz and 2 MHz, respectively. The scale of measured frequencies has been restricted to 60 MHz in (c) for visibility.

output powers in Fig. 4.1b by the power of the input signals at the fundamental frequencies, for each fundamental frequency considered. Both the attenuation of high-frequency harmonics and the decay of the diagonal peaks at the fundamental frequencies with increasing frequency in Fig. 4.1c witness that the LED has a remarkable low-pass filtering effect. The relatively low cutoff frequency is due to the fact that the LED was analyzed in conjunction with our driving modules and additional wiring, which significantly reduce the operative bandwidth. Consequently, the results are not to be interpreted as the frequency characteristics of the LED itself. Improvements in the signal routing and the drivers might lead to bandwidth extension.

The low-pass effect illustrated in Fig. 4.1 favors the compressibility of the optical signal in frequency domain. In order to illustrate the effect on a real PMD ICS, in Fig. 4.2 we plot the output of an LED of our illumination system (in blue) when driven with a perfectly square ICS. The relatively slow rising and falling characteristics deform the square shape into quasi-trapezoidal. The capacitive behavior of the LED filters out the highest frequencies, being the small high-frequency oscillations due to measurement noise. The red line in Fig. 4.2 shows the normalized DCT coefficients of the original signal. Note the fast decay of their value, indicator of great compressibility, and the clear sparsity of the signal in DCT domain. This means that the optical signal can be accurately recovered from very few DCT coefficients. To show this, the 10% highest DCT coefficients, in absolute value, are selected and the rest are set to zero. The corresponding reconstructed signal is plot in green and is hardly distinguishable from the original signal. The reconstruction is so good that one could say it is exact, since the reconstruction error is within the measurement noise.

Equivalent results are observed using a Fourier representation. The power of the signal is always concentrated in few coefficients, quite regardless of the sampling rate, provided that it is high enough. Consequently, the profile of the illumination signal can be recovered from few stochastic measurements using CS. Recovering all the frequencies in the optical signal allows depth calculus substituting the unrealistic hypothesis of sinusoidal illumination required by Eq. 2.19 by the real condition of compressibility in frequency domain.

A basic approach for recovering the illumination signal from few measurements using a PMD camera that does not require any hardware modification is presented in Section 4.5.1. A more general framework, gathering measurements according to pseudorandom binary codes is demonstrated in Section 4.5.2. This framework requires modifying the PMD frontend to allow using custom codes as patterns for correlating the light signal. Consequently,

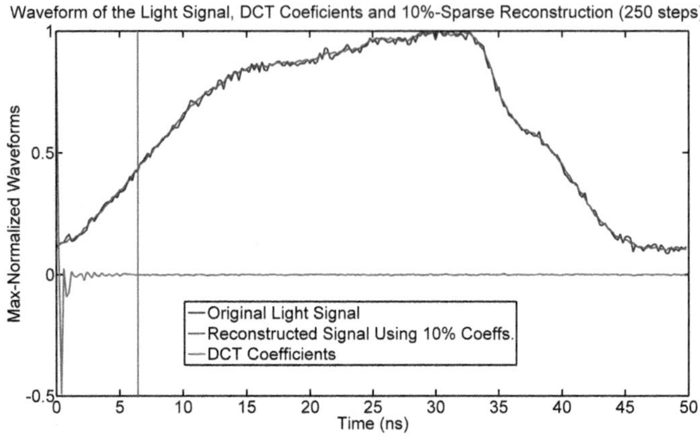

Figure 4.2.: Compressibility of the optical signal in frequency domain. The optical signal (in blue in the color version) is a real illumination waveform from one of our NIR-LEDs, driven with a square ICS of 20 MHz, gathered with a fast photodiode. The corresponding DCT coefficients unveil the great sparsity of the signal. The reconstructed waveform (in green in the color version) makes use of only the 10% highest DCT coefficients (those at the left of the vertical line). All the curves are maximum-normalized. ©2015 IEEE.

the results presented in Section 4.5.2 come from simulations. A real implementation of this framework is subject of Chapter 5, where a prototype of a CS-PMD camera is presented.

4.2. Solving Preliminary Issues

As introduced in Section 4.1, the PMD sensing process seems to be appropriate to implement a compressed sensing framework and posterior sparse reconstruction of the signals, at least in spatial domain, exploiting compressibility of the raw images, and in frequency domain, exploiting the periodicity of the CW illumination signal. Unfortunately, despite its theoretical feasibility, CS does not work as expected when applied directly without carefully taking into account some preliminary issues, which are endemic to the PMD technology. From the high-level, one could sum up these issues as a large amount of *noise* in the raw data. Since random noise does not admit a sparse representation in any structured dictionary, the signal sparsity and compressibility are compromised. If the issue is to be corrected, a more accurate characterization of the noise sources is required. In real operation of PMD cameras, three issues were found to be responsible of the corruption of raw data: the activation of the SBI to avoid saturation of one of the pixel channels, the so-called fixed-pattern noise (FPN), due to the ADCs in the PMD frontend, and the shot noise when operating in low-light conditions.

Fixed-Pattern Noise The fixed-pattern noise (FPN) is an effect of the analog-to-digital conversion that takes place in the PMD frontend. In the case of the frontend for the PMD 19k-S3, the conversion is carried out outside the PMD chip, by means of three ADCs. The fact that each ADC exhibits different offsets and gains results in a fixed pattern in the raw images. This pattern is given by a constant part, due to the ADC offsets, and a signal-dependent part, given by the differences between ACD gains. Consequently, a linear calibration is an effective method to compensate this effect, as shown in Section 4.2.1.

Raw Data Corruption due to SBI Activation The SBI operation is illustrated in Fig. 2.9. The pixel response is linear for both channels until one of them reaches the SBI activation threshold. After this point the responses cannot be considered linear anymore, breaking one of the basic requirements of the CS measurement model. As a result, the structure in the raw images is often washed out by the variation of signal level produced by the SBI, for those pixels where it is active. If the SBI is operating in the central region of Fig. 2.9 (correct operation), the structure would be recovered using differential (A-B) images, but the DC component could not be recovered. The problem of raw data corruption by SBI activation is motivated by the limited dynamic range of the PMD pixels (see Section 2.4.3) and is illustrated in Fig. 4.3.

The scene in Fig. 4.3 is a long corridor with a computer monitor on a table, a cabinet on the wall, a metallic U-shaped pipe with a valve at the bottom and some other structure. Such a scene is as common as challenging, due to the high illumination power required to get some echo from the end of the corridor, which requires, in turn, a high dynamic range to avoid saturation for the structure close to the camera. In practice, one has to choose between sensing the close objects and reducing the operative range or sensing most of the corridor at the cost of low depth accuracy in the close objects due to SBI activation.

(a) Raw Img. Exp.: 2 ms (b) Raw Img. Exp.: 6 ms (c) Raw Img. Exp.: 10 ms

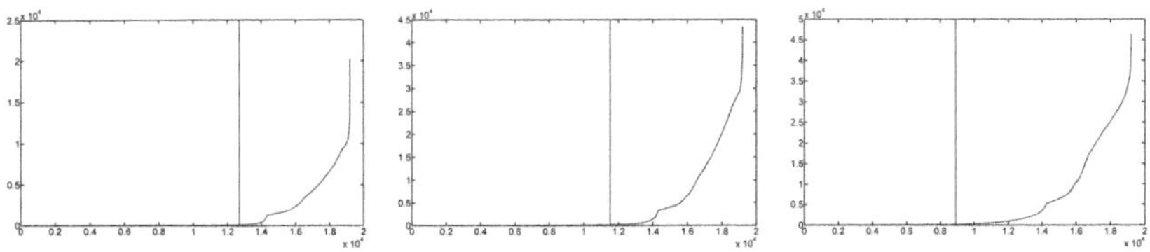

(d) DB4 Coef. Exp.: 2 ms (e) DB4 Coef. Exp.: 6 ms (f) DB4 Coef. Exp.: 10 ms

Figure 4.3.: First raw image delivered by a PMD camera (A channel, 0°) for three different exposure times: 2, 6 and 10 ms. The plots in the second row are intended to illustrate the degradation of signal sparsity due to SBI operation and show the absolute values of the 2D Daubechies 4 (DB4) coefficients of the images in the first row, ordered from lowest to highest.

Note the low intensity in Fig. 4.3a, approaching to zero for medium and far areas of the corridor, together with spots of high-intensity at the metallic corners of the table and the metallic pipe, which are acting as specular reflectors. In Fig. 4.3b the exposure is high enough to achieve good intensity on the wall and the screen, at the cost of triggering the SBI in shiny surfaces frontal to the camera, such as the lateral of the table and part of the cabinet and the wall. For 10 ms exposure (Fig. 4.3c) a second door and some farther structure in the ceiling and wall are resolved that were not visible for lower exposures. As a drawback, the SBI reaches its limit for many pixels of Fig. 4.3c (dark blue spots), corrupting even the differential (A-B) measurements.

Apart from invalidating simple linear sensing models, the SBI activation breaks the smoothness of the raw images and eclipses the real structure, damaging the signal sparsity in wavelet domain. To illustrate this effect, the absolute values of the 2D Daubechies 4 (DB4) coefficients of the images in the first row of Fig. 4.3 are plotted in ascendant order under the corresponding images. The vertical line is located at approximately the same close-to-zero value for all three plots. That is, the number of DB4 coefficients whose absolute value is higher than a close-to-zero threshold increases with the number of pixels for which the SBI is active.

Exact recovery of the raw images in wavelet domain from few measurements is not possible when they are affected by SBI-related noise, due to the non-structured nature of this noise. In the best case, the rest of the image is accurately recovered, while the noise is not. This means loosing the depth information for that area, since the differential (A-B) signal will also be removed. A method to artificially extend the dynamic range of PMD pixels while avoiding SBI-related noise is, therefore, necessary for the application of CS to recover the raw images. This is the subject of Section 4.2.2.

Low-Light Conditions Challenging scenes such as the corridor of Fig. 4.3 might lead to saturation (or SBI activation) in the raw images for some objects close to the camera while far areas are still poorly illuminated. While the effect of saturation can be avoided with additional acquisitions at shorter exposure times (see Section 4.2.2), gathering information from far or poorly illuminated scene points requires larger exposure times. In most cases this is simply not possible due to frame rate requirements. In such situation, the raw images are affected by severe shot noise in those areas. The uncertainty in raw data affects the quality of the depth estimation, as described by Eq. 2.28. If shot noise is dominant, the pixel values of the raw images follow a Poisson distribution and the induced depth uncertainty is given by Eq. 2.29. One can easily observe that the depth uncertainty is inversely proportional to the amplitude of the light signal. Consequently, in poorly-illuminated areas, depth variabilities of the same order of magnitude as the unambiguous range might be achieved and the depth image looks like random noise. For this reason, we provide a method for raw image denoising in Section 4.2.3 that exploits the similarity between raw images to filter out the noise.

4.2.1. Per-Pixel Linear Calibration of Raw Data

The fixed-pattern noise (FPN) is produced by the differences between pixels and the different offsets and gains of the ADCs used for digitalization. Since different pixels undergo different transformations, the pattern looks like noise and the compressibility of the raw images is degraded. Provided that such transformations are linear, a per-pixel linear calibration suffices to remove this effect. In this section we expose the details of the calibration method we use to remove the FPN of PMD raw images, in all experiments involving real acquisitions in this work. We also show the effect of the calibration in the sparsity of the images.

The method we adopt has already been described in Section 4.1 of [283]. We rewrite here the main equations, trying to fuse our notation with that of [283]. Let $\Gamma \in \{A, B\}$ denote the pixel channel and $I_i^\Gamma(\vec{x})$ the intensity value of the pixel indexed by $\vec{x} \in [1, n_{\text{rows}}] \times [1, n_{\text{cols}}]$ in the raw image acquired with an exposure time t_i^{exp}, $i \in [1, n]$, being $n_{\text{rows}} \times n_{\text{cols}}$ the image size and n the total number of acquisitions. Consider the average values for intensity and exposure time given by Eq. 4.1.

$$\overline{I^\Gamma}(\vec{x}) = \frac{1}{n}\sum_{i=1}^{n} I_i^\Gamma(\vec{x}), \ \overline{t_{\text{exp}}} = \frac{1}{n}\sum_{i=1}^{n} t_i^{\text{exp}} \tag{4.1}$$

We define the difference of intensity and exposure time with respect to the mean values given in Eq. 4.1 as:

$$\Delta I_i^\Gamma(\vec{x}) = I_i^\Gamma(\vec{x}) - \overline{I^\Gamma}(\vec{x}), \ \Delta t_i^{\text{exp}} = t_i^{\text{exp}} - \overline{t_{\text{exp}}} \tag{4.2}$$

For each pixel, the slope of the line that best fits the response curve described by the different sampling points is given by Eq. 4.3.

$$\alpha^\Gamma(\vec{x}) = \frac{\sum_{i=1}^{n} \Delta t_i^{\text{exp}} \Delta I_i^\Gamma(\vec{x})}{\sum_{i=1}^{n} (\Delta t_i^{\text{exp}})^2} \tag{4.3}$$

The offset of each pixel can be calculated subtracting the level predicted by the slope from Eq. 4.3 for the average exposure time from the average intensity, as in Eq. 4.4.

$$\beta^\Gamma(\vec{x}) = \overline{I^\Gamma}(\vec{x}) - \alpha^\Gamma(\vec{x})\overline{t_{\text{exp}}} \tag{4.4}$$

The slopes will be slightly different for each pixel, depending on the gain of the ADC used to digitize the intensity value. A mean slope is defined in Eq. 4.5.

$$\overline{\alpha^\Gamma} = \frac{1}{n_{\text{rows}} n_{\text{cols}}} \sum_{\substack{1 \le x \le n_{\text{rows}} \\ 1 \le y \le n_{\text{cols}}}} \alpha^\Gamma(\vec{x}) \tag{4.5}$$

Then the per-pixel linear calibration simply consists in subtracting the offset given by Eq. 4.4 and rescaling the intensity so that all the pixels show a slope equal to that given by Eq. 4.5 after calibration. That is, the value of each pixel of the calibrated image is calculated as:

$$\hat{I}^{\Gamma}(\vec{x}) = \left(I^{\Gamma}(\vec{x}) - \beta^{\Gamma}(\vec{x})\right) \frac{\overline{\alpha^{\Gamma}}}{\alpha^{\Gamma}(\vec{x})} \tag{4.6}$$

From Eq. 4.6 it is clear that only two parameters are to be stored per pixel and channel: the offset $\beta^{\Gamma}(\vec{x})$ and the scaling factor $\overline{\alpha^{\Gamma}}/\alpha^{\Gamma}(\vec{x})$. These parameters are to be computed only once for a given PMD frontend. Then, the calibration operation consists in a subtraction plus a product.

Since the calculus of the calibration parameters has to be performed only once and the accuracy of these parameters will affect all the calibrated raw data afterwards, we take a conservative position and gather an extensive dataset of 10,000 PMD acquisitions. The exposure time is varied from 0.1 ms to 10 ms with a step size of 0.1 ms. For each of the 100 exposure steps, 100 acquisitions are gathered, at the same exposure. Gathering many measurements for each exposure further reduces the effect of random noise in the measurements. Provided that each acquisition of a PMD camera consists of four pairs of raw images, we have $n = 40,000$ sample points per pixel and channel to compute the calibration parameters. The experiment is set in such way that saturation is avoided for all pixels, even for the maximum exposure, i. e., all sampling points are usable.

The camera used in the experiments is our ZESS MultiCam, which features a 19k-S3 PMD chip. This MultiCam is the sensor of choice along this work for gathering PMD raw data. The experimental setup consists of the camera, without optics and equipped with a thick layer of translucent diffusive material in front of the aperture. The illumination system is the medium-range NIR LED system analyzed in Appendix A.6. The diffusive material makes that each pixel receives the same amount of light, coming from all directions. The phase information is totally lost and the four acquisitions gathered per depth image can be used as independent measurements at the same exposure time. We further avoid eventual influence of the scene by placing the camera at a few meters from a plain and white wall, which reflects the light of the LEDs. The experiment is carried out in a dark room without people inside, in order to avoid the undesired effect of a non-constant background illumination.

The calibration parameters obtained from this high-quality dataset provide an excellent intensity calibration, able to totally remove the FPN. To illustrate the performance of the calibration in real data acquired with the camera, in Fig. 4.4 we provide all the raw images of a PMD acquisition of the challenging scene in Fig. 4.3, before and after calibration. Despite the large depth range of the scene, the data was acquired with 2 ms exposure time to avoid saturation of close objects. Images in Fig. 4.3 are after calibration, in order to separate the SBI-related noise from the intense FPN. Consequently, Fig. 4.3a is the top-left raw image in Fig. 4.4b. Fig. 4.4a shows a positive average offset in the raw images, which is totally eliminated in Fig. 4.4b.

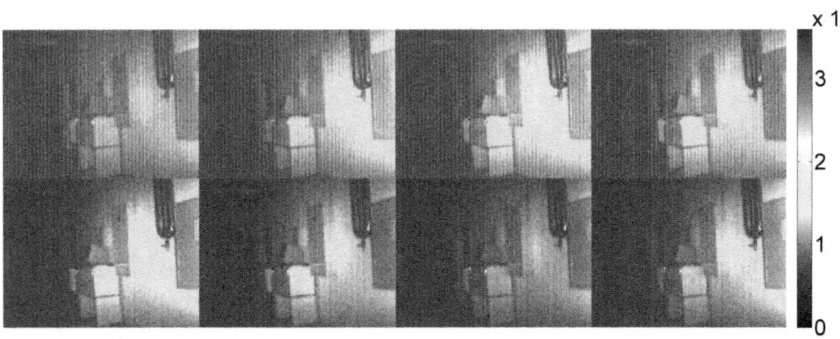

(a) Non-Calibrated Raw Images

(b) Calibrated Raw Images

Figure 4.4.: Raw images delivered by a 19k-S3 sensor before (a) and after (b) intensity calibration. For each set of raw images, the first row is for the A channel and the second for the B. The images are ordered by columns according to the phase: 0°, 90°, 180° and 270°, from left to right.

Effect of the Calibration in the Sparsity and Compressibility From the CS point of view, we are interested to check the gain in signal sparsity or compressibility. To this end, we capture a real office scene and perform a DB4 wavelet decomposition of the first raw image delivered by the PMD camera (A channel, 0°), before and after applying the calibration. The results are given in Fig. 4.5. Equivalent results are obtained for the other channel and for any of the four phases.

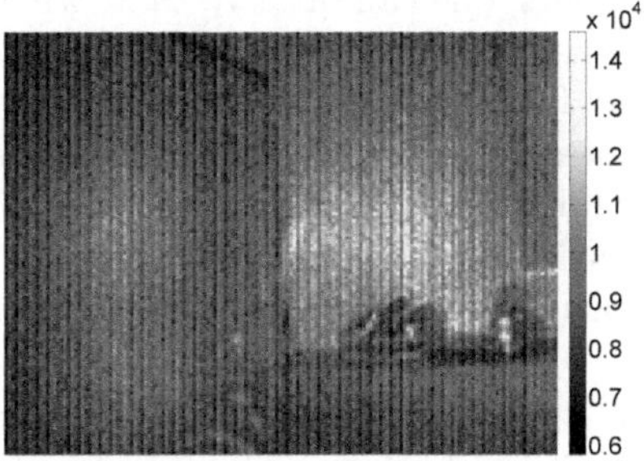

(a) Non-Calibrated Raw Image

(b) Calibrated Raw Image

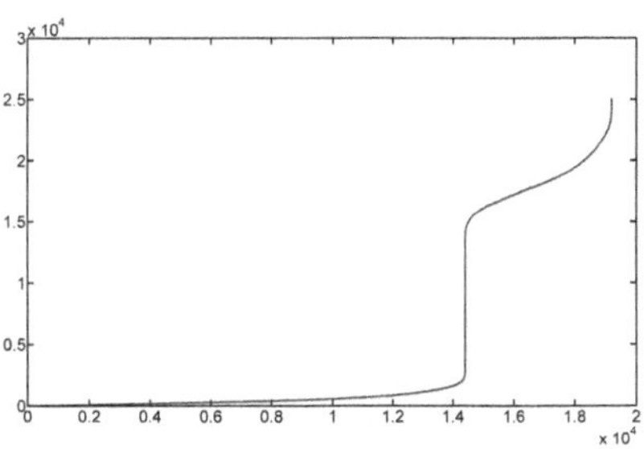

(c) DB4 Coef. of Non-Calibrated Raw Image

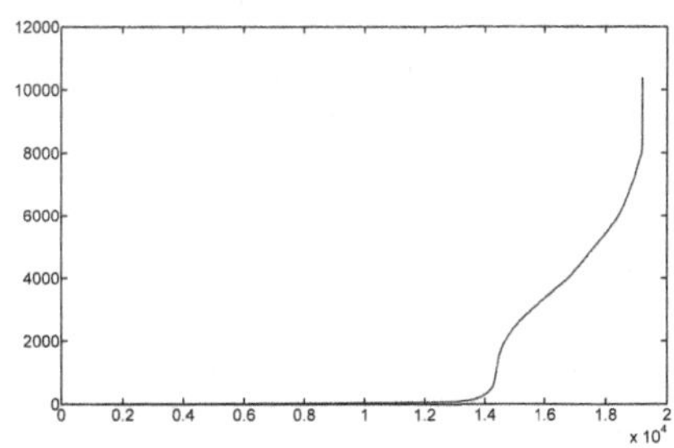

(d) DB4 Coef. of Calibrated Raw Image

Figure 4.5.: Effect of the per-pixel linear calibration of raw images on the signal sparsity and compressibility. Images (a) and (b) show the raw image corresponding to the A channel of a PMD sensor for the first acquisition (0° phase shift), before and after applying the calibration. Plots (c) and (d) show the absolute values of the Daubechies-4 (DB4) coefficients for the images (a) and (b), respectively, ordered from lowest to highest.

The scene in Fig. 4.5 is an office environment with a cabinet at the left, a poster roll behind it, several objects on a table and part of a whiteboard at the right. Additionally, a holder with a very thin steel wire, ending in a small square white surface, parallel to the image plane, is placed in the center of the scene, close to the camera. The wire cannot be resolved due to the limited lateral resolution. The small white square, which is masked by the FPN in Fig. 4.5a, is clearly visible in Fig. 4.5b. Also the poster roll and the objects on the table are hardly distinguishable in Fig. 4.5a. Comparing Fig. 4.5a and Fig. 4.5b one can clearly observe that most of the power of Fig. 4.5a is due to the FPN, which cannot be sparsely represented.

Looking at the plots of the DB4 coefficients, one can observe that, while Fig. 4.5d represents a sparse signal (most values are close to zero), the signal in Fig. 4.5c is not sparse (the values decay following a power law, but they are not zero) and its classification as compressible is also doubtful, since the power of the FPN pushes the coefficients up, breaking the smooth decay of Fig. 4.5d. Therefore, the per-pixel calibration can be considered a crucial step towards the application of CS, since it boosts the sparsity of the signal.

Effect of the Calibration in the Sparse Signal Recovery Provided the considerable improvement in terms of sparsity and compressibility achieved by linear calibration of the raw images, it is expectable that this has a direct impact on the quality of the recovered images, for a fixed number of measurements. Equivalently, one could hope to attain equivalent reconstruction quality with a lower number of measurements, provided that the minimum number of measurements required to achieve exact recovery in a noiseless scenario grows linearly with the sparsity of the signal.

In our experiments, we perform the sparse recovery of the PMD raw images in DB4 domain from a reduced number of stochastic measurements. As introduced in Section 3.1, $\mathbf{\Psi} \in \mathbb{R}^{n \times n}$ denotes the dictionary, whose atoms are stored by columns, and $\mathbf{\Phi} \in \mathbb{R}^{m \times n}$ denotes the sensing matrix, whose sensing kernels are stored by rows. Consequently, n is the dimensionality of the signal to recover and m the number of measurements. $\mathbf{\Psi}$ is an orthonormal DB4 wavelet dictionary, while $\mathbf{\Phi}$ is a pseudorandom binary matrix, whose elements are ± 1. $\mathbf{\Phi}$ is derived from a Hadamard matrix, randomized by columns, from which m rows are randomly selected. Recall that the rows of a Hadamard matrix are the basis vectors of the orthogonal Walsh basis. This way, we assure that each measurement gathers information that is independent from the previous, while preserving the random nature of the sensing process.

Let $\vec{i}^{\theta,\Gamma} \in \mathbb{R}^n$ denote a (now vectorized) PMD raw image (then $n = n_{\text{rows}} \times n_{\text{cols}}$) from the pixel channel Γ at a phase θ. If the image admits an exactly-sparse representation in the DB4 dictionary, then one can write $\vec{i}^{\theta,\Gamma} = \mathbf{\Psi}\vec{x}^{\theta,\Gamma}$, where the vector $\vec{x}^{\theta,\Gamma}$ is exactly sparse. If the image is compressible, but not exactly sparse, then the elements of $\vec{x}^{\theta,\Gamma}$ decay according to a power law. In both cases CS theory states that $\vec{x}^{\theta,\Gamma}$ can be recovered from a number of measurements $m < n$ that depends linearly on the sparsity and logarithmically on the signal dimensionality. Let $\vec{y}^{\theta,\Gamma} = \boldsymbol{\Phi}\vec{i}^{\theta,\Gamma}$ be the measurement vector obtained projecting $\vec{i}^{\theta,\Gamma}$ through $\boldsymbol{\Phi}$. The measurements impose an underdetermined system of linear equations $\vec{y}^{\theta,\Gamma} = \boldsymbol{A}\vec{x}^{\theta,\Gamma}$, as in Eq. 3.8, where $\boldsymbol{A} = \boldsymbol{\Phi}\boldsymbol{\Psi}$ is the measurement matrix. Then $\vec{x}^{\theta,\Gamma}$ is recovered by l_1-minimization, subject to these constrains, as in Eq. 3.10.

In this section we provide the recovered images obtained from measurements gathered from a non-calibrated and a calibrated PMD raw image. The original raw image was obtained from a real scene using our MultiCam. The original images, before and after calibration, are those in Fig. 4.5 and the corresponding recovered images are given in Fig. 4.6. In both cases $m = 0.5n$ measurements were considered. While Fig. 4.6b shows a fairly good recovery (cf. Fig. 4.5b), Fig. 4.6a shows that the number of measurements is not enough to accurately reconstruct the FPN. Nevertheless, the energy of the FPN captured by the measurements appear in the shape of an unstructured noise in Fig. 4.6a. The high intensity of the noise suggests not considering the result in Fig. 4.6a as a successful reconstruction of Fig. 4.5a. In fact, no depth image can be computed from raw images like that in Fig. 4.6a. While the four phases algorithm (Eq. 2.19) is natively immune to the FPN, raw images affected by strong noise due to a non-exact recovery would lead to random depth values. The reconstruction error with respect to the original is, in terms of RMSE, 797 for the non-calibrated case and 53.7 for the calibrated case, both in PMD pixel units. In other words, the linear calibration allows reducing the reconstruction error in a 93%, enabling successful sparse recovery from as few as $m = 0.5n$.

4.2.2. Increasing the Dynamic Range: Adaptive High Dynamic Range

Fig. 4.3 illustrates the effect of the SBI activation in the PMD raw images and its effect on the sparsity in wavelet domain. A method for artificially increasing the dynamic range of PMD pixels is highly desired. High Dynamic Range (HDR) approaches extend the dynamic range of digital cameras by combining several images acquired at different exposure times. Sections 4.2.2.1 and 4.2.2.2 provide an overview of HDR imaging and a simple HDR framework for the case of PMD depth imaging. Section 4.2.2.3 presents the adaptive HDR for ToF cameras of [H4] and demonstrates its superiority over conventional HDR on real PMD data.

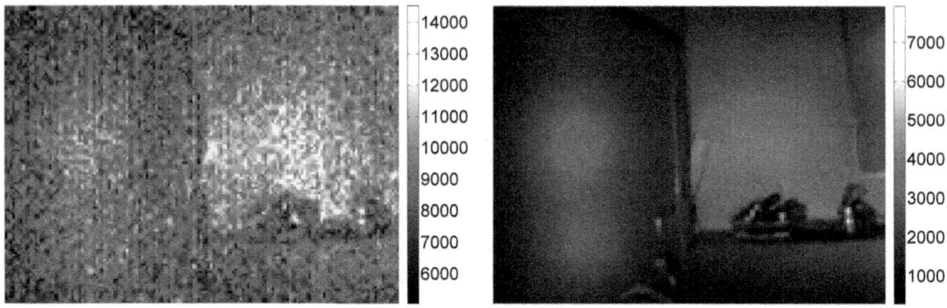

(a) From Non-Calibrated Data **(b)** From Calibrated Data

Figure 4.6.: Effect of the per-pixel linear calibration of raw images on the signal recoverability. The images are generated from the sparse vectors of DB4 coefficients, obtained by l_1-minimization. The result obtained when measuring on non-calibrated data (a) is clearly of lower quality than that obtained using calibrated data (b).

4.2.2.1. High Dynamic Range

Real-world scenes may exhibit a huge dynamic range. For instance, the illuminance of a clear night sky is around 10^{-3} lx, while being 10^4 lx for daylight or 10^5 lx for direct sunlight [225]. With this in mind, one can show that any kind of camera facing a scene with an outdoor part under twilight illumination (say 10 lx) plus some indoor scene, e. g., looking through a window into a well-illuminated room (suppose daylight level, 10^4 lx), would need a dynamic range of 100 dB to capture the whole scene without saturation. The human eye is able to capture such scenes due to its large dynamic range, known to be around six orders of magnitude, that is, 120 dB. This exceeds that of conventional cameras and also of the PMD pixels (Section 2.4.3). Nowadays, some HDR video cameras exist that achieve greater dynamic range than that of the human vision system (HVS), which make use of especial pixels showing a logarithmic response inspired by the HVS. For instance, the HDRC VGAx sensors [410] have been reported to have a dynamic range of 160 dB with a single acquisition [409].

In general, camera pixels show a response that is approximately linear along most of their operating range, leading to a moderate dynamic range, lower tan that of most challenging real-world scenes. This is not an exclusive problem of digital photography. In fact, analog photography is also affected by dynamic range limitations and, consequently, it is in this area where we find the first approaches to extend the dynamic range that could be captured by means of a photosensitive film. One of the first advances in this field that is often referred as a primitive HDR method is the extended response film presented by Charles W. Wyckoff [470]. His film consists of three layers with different light sensitivities and, therefore, the dynamic range of the film is the composition of those of the different layers. This is equivalent to taking several images at different exposure times but at the same instant and from the same point. If we will to directly translate this idea to digital imaging, several sensors working with several exposure times will be required, together with an appropriate beam-splitting setup after a single lens, so that the same image is formed in all sensors. Alternatively, an array of cameras [464] or a multiaperture camera might be used to gather simultaneous images with different exposure times and slightly different points of view. Clearly, this requires an additional step of registration and cropping of the common area. In case we have no access to a multiaperture camera, have no possibility of building an array of cameras or use a special camera whose peculiarities do not allow for such setups, there are two alternatives: either taking multiple images from different points or from the same point at consecutive instants, using different exposure times.

Most HDR approaches [33] make use of a dataset of low dynamic range images of a scene, taken from the same camera position with different exposure times. The main issue is then to optimally combine the images to get a single HDR result, referenced to an arbitrary exposure time. This process cannot be carried out directly in *pixel-charge* space, but in the original *irradiance* space. Consequently, most works focus on finding the pixel response function, under certain hypothesis, e. g., that the irradiance remains constant during the image acquisition. A simple, yet realistic, approach to HDR imaging is given in [312], where a linear response function is assumed for CCD pixels and the gains, i. e., the slopes of the functions, are trivially deduced from 2D-histograms of corresponding pixel values between two images of differing sensitivity. In [320] the authors present a recursive algorithm that generates a plot of the response curve from only two images of the scene gathered using different exposure times. In [138] the estimation of the response function is recast as a minimization problem, which can be solved through least squares estimation. In our case the response functions of our pixel channels do not follow the typical pixel saturation curve, since the channel with less charge start loosing it when the exposure time exceeds that for which the charge in the other channel reaches the SBI activation threshold. This means that the response function is no longer monotonic (cf. Fig. 2.9), as assumed in most related work.

Another important point is how the different images are combined. For any pixel, measurements taken in the linear region of the response curve (if such region exists) should contribute more to the result than measurements taken close to the complete saturation of the pixel. A certainty function was computed in [320] by differentiating the logarithmic response curves in order to perform such weighting. A simple hat function is used in [138] as weighting function to avoid saturation disturbance. Our approach applies a simple thresholding to exclude the non-linear region and considers a linear response curve.

4.2.2.2. HDR for the Photonic Mixer Device

A naive HDR approach for PMD ToF imaging is proposed in [255], where two exposure times are manually selected and the final depth values are those obtained with short exposure for saturated pixels in the long-exposed acquisition and those obtained with long exposure for the rest. Here we propose a more general framework, which combines any number of raw images, acquired using different exposure times, into a single HDR image. The method automatically handles SBI activation and ensures that only measurements in the linear region of the pixel response (Fig. 2.9) are considered.

A priori knowledge of the imaging system is crucial to bound the sources of non-linearities in the pixel response function and design an efficient and accurate method to compute it and compose the HDR image. Several sources of non-linearities are present in conventional imaging devices [138], such as digital cameras, being sometimes due to software modifications for image enhancement. In the case of ToF imaging systems, the response function of the pixel itself is also not the only transformation that the irradiance undergoes until we get the final intensity values for each pixel channel in the raw images. The ADC's of the readout circuit add an additional linear transformation. We use the calibration method described in Section 4.2.1 to remove the effect of the different ADC offsets and scalings. After calibration, all pixels can be considered to follow the same response curve for certain illumination, environmental conditions and phase. The linear region of the pixel response curve (green area in Fig. 2.9), where the pixel charge is proportional to the observed irradiance for a given exposure time, is uniquely defined by a straight line. The SBI effect is the only source of non-linearity. The activation threshold of the SBI can be robustly and uniquely determined in an empirical way analyzing the response curve of the PMD pixels after ADC-calibration.

The input of our algorithm is a dataset of raw PMD image sets, taken using different exposure times. Two consecutive exposure times differ in a constant time value, which has to be low enough to ensure certain overlap between captured dynamic ranges (no information loss), as well as a good estimation (improved accuracy through redundant measures). Let $t_i, i \in [1, n_{\text{exp}}]$ be the exposure time used to acquire the ith raw image set, being n_{exp} the total number of exposure times in the dataset. Each image of the ith set is denoted as $I_i^{\theta,\Gamma}$, where $\theta \in \{0°, 90°, 180°, 270°\}$ denotes the phase and $\Gamma \in \{A, B\}$ the pixel channel. A pixel of $I_i^{\theta,\Gamma}$, indexed by the coordinates vector $\vec{x} \in [1, n_{\text{rows}}] \times [1, n_{\text{cols}}]$, is denoted by $I_i^{\theta,\Gamma}(\vec{x})$. Our algorithm works in a per-pixel basis and each pair of raw channel images is processed independently for each phase. Each pixel channel pair is, nevertheless, processed in a coupled way, since the activation of the SBI by one channel breaks the linearity of the response function of both channels. In order to select only points from the linear region, we introduce a membership function, rather than a weighting function. Since our response curve is perfectly linear in that region, all sample points should contribute equally to the estimation. The membership function is given by Eq. 4.7 and establishes that a certain measurement $I_i^{\theta,\Gamma}(\vec{x})$ was performed in the linear region (therefore valid for the estimation) if *both* pixel channels are below the activation threshold of the SBI, I_{SBI}. $\mathcal{L}_\theta^\Gamma(\vec{x})$ represents the subset of all such measurements for a certain phase, θ.

$$\mathcal{L}_\theta^\Gamma(\vec{x}) = \{I_i^{\theta,\Gamma}(\vec{x}) \mid I_i^{\theta,\Psi}(\vec{x}) \leq I_{\text{SBI}}, \forall \Psi \in \{A, B\}\},\ \Gamma \in \{A, B\} \tag{4.7}$$

The linear dependency between pixel charge and exposure time yields:

$$I_i^{\theta,\Gamma}(\vec{x}) = a_1^{\theta,\Gamma}(\vec{x}) t_i + a_0^{\theta,\Gamma}(\vec{x}), \forall I_i^{\theta,\Gamma} \in \mathcal{L}_\theta^\Gamma \tag{4.8}$$

where $a_1^{\theta,\Gamma}(\vec{x})$ and $a_0^{\theta,\Gamma}(\vec{x})$ are constant values for each pixel channel measurement along exposure time. $a_1^{\theta,\Gamma}(\vec{x})$ is an estimate of the real HDR signal, that we will denote with $\hat{I}_i^{\theta,\Gamma}$, for unit exposure time ($t_i = 1$). Since we deal with calibrated data, $a_0^{\theta,\Gamma}(\vec{x}) \approx 0, \forall \vec{x}$. Provided that all data in $\mathcal{L}_\theta^\Gamma(\vec{x})$ satisfy Eq. 4.8, we can obtain $a_1^{\theta,\Gamma}(\vec{x})$ and $a_0^{\theta,\Gamma}(\vec{x})$ solving an overdetermined system of linear equations:

$$\begin{aligned} I^{\theta,\Gamma}(\vec{x}) &= \boldsymbol{T}_{\text{exp}} \vec{a}^{\theta,\Gamma}(\vec{x}), \text{ where:} \\ I^{\theta,\Gamma}(\vec{x}) &= \{I_i^{\theta,\Gamma}(\vec{x}) \mid I_i^{\theta,\Gamma}(\vec{x}) \in \mathcal{L}_\theta^\Gamma(\vec{x})\} \\ \vec{t}_{\text{exp}}^{\,t} &= (t_i\ 1), \vec{a}^{\theta,\Gamma}(\vec{x}) = \begin{pmatrix} a_1^{\theta,\Gamma}(\vec{x}) \\ a_0^{\theta,\Gamma}(\vec{x}) \end{pmatrix} \end{aligned} \tag{4.9}$$

where $I^{\theta,\Gamma}(\vec{x}) \in \mathbb{R}^n$, $n = |\mathcal{L}_\theta^\Gamma(\vec{x})|$ and $\vec{t}_{\text{exp}}^{\,i}$ is the ith row of $\boldsymbol{T}_{\text{exp}}$. Since there are no outliers among the measures we select to build the system, we solve Equation 4.9 through simple least squares. The final HDR value for each pixel channel is then computed using Equation 4.8 for an arbitrary reference exposure time.

Note the simplicity of our method, which relies on the linearity of the PMD pixel response curve to get rid of non-linear mapping estimations, as well as of complex weighting functions. Note that this implies taking full profit from those measurements obtained close to saturation, which exhibit higher SNR (due to the lower shot noise) and are typically neglected by severe weighting functions, e. g., the hat function used in [138].

We test the effectiveness of our approach in the long corridor scene depicted in Fig. 4.3. A MultiCam with our medium-range NIR LED illumination system (Appendix A.6) is used to gather the corresponding dataset, with exposure times ranging from 0.1 to 10 ms, with step size of 0.1 ms. For each exposure time, 100 images were acquired, yielding a dataset of 10,000 acquisitions. The lens mounted in this experiment and in posterior ones, unless indication in contrary, had a focal length of 8.5 mm, offering a wide FOV of 46° × 35°. After calibration, the HDR method presented in this section was applied to the full dataset and HDR raw images were generated for an arbitrary exposure time, e. g., $t_{\text{exp}}^{\text{ref}} = 1\,\text{ms}$, that is, $\hat{I}^{\theta,\Gamma} = a_1^{\theta,\Gamma}$. This set of HDR images is then used to compute the depth image. Fig. 4.7 shows the depth image obtained from HDR data and compares it with the depth image obtained from unprocessed raw data acquired with 1 ms exposure time. The sharp cut at around 7.5 m in the images is due to the depth unambiguous range, provided that the PMD chip was working at 20 MHz modulation frequency. For a better understanding of the scene, we refer to the raw images in Fig. 4.4. The exposure time of 1 ms was chosen to obtain an good compromise between background quality and non-saturated foreground in the single-exposure case. Still, the farthest part of the corridor is lost in Fig. 4.7a. There are two doors in the closer part of the corridor, which are perfectly reconstructed in Fig. 4.7b, while the second door cannot be distinguished in Fig. 4.7a. A third door in the far area (not visible in the images of Fig. 4.4 due to insufficient exposure time) is distinguishable in Fig. 4.7b, while completely masked by noise in Fig. 4.7a. Two illumination modules (horizontal rectangles) on the ceiling are also visible in Fig. 4.7b, while only one is in Fig. 4.7a.

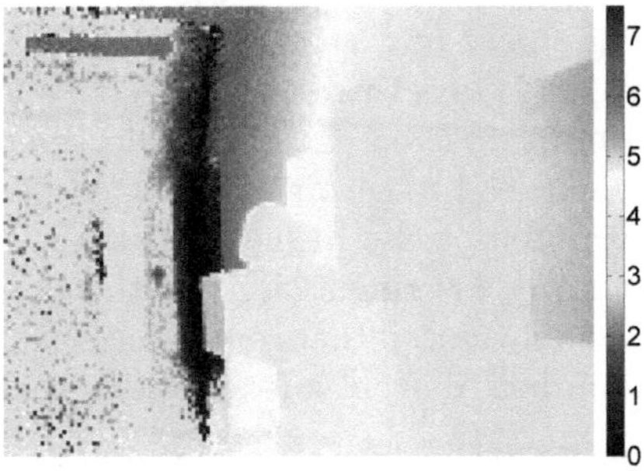

(a) Depth from Raw Data

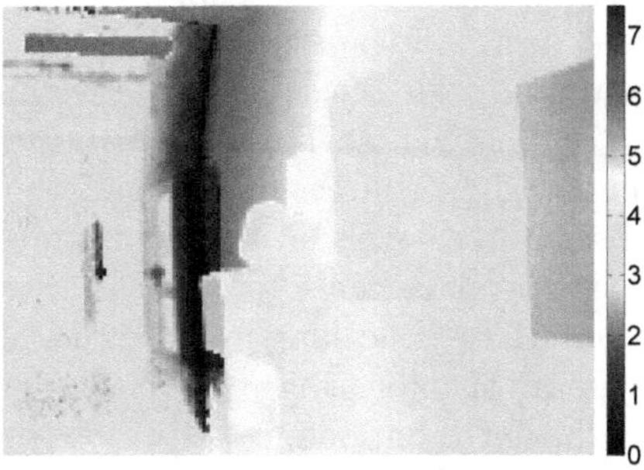

(b) Depth from HDR Raw Data

Figure 4.7.: Depth result from HDR raw data. Compare the depth image computed from a single PMD acquisition at 1 ms (a) and the depth image computed from HDR raw data (b), for the same reference exposure time. The HDR raw data was obtained using the fast method presented in this section and an extensive dataset with exposure times from 0.1 to 10 ms. All scales are in meters.

4.2.2.3. Adaptive High Dynamic Range (AHDR)

Using our medium-range illumination system, ranges over a hundred meters can be achieved. In such cases, if a far scene is being observed, while some objects are very close to the camera, the dynamic range to capture is huge. The simple HDR method presented in Section 4.2.2.2 requires several acquisitions at different exposure times, so that the operative ranges of each of them partially overlap each other and cover the entire range of the scene. This eventually leads to an excessively large number of acquisitions, which reduce, in turn, the frame rate of the ToF camera and harden the requirement of a static scene. The previous algorithm remains an attractive HDR framework for offline enhancement of ToF images, making use of a relatively large dataset of raw images of the scene, but it is not suitable for online operation. In order to cover this gap, we present here an Adaptive High Dynamic Range (AHDR) approach [H4], which deals with common HDR situations in medium- and large-range ToF imaging, adapting its behavior to changes in the scene and operating in real time.

The case of study contemplated in the AHDR method is the following: the ToF camera is fixed, sensing a distant and static scene, using a fixed exposure time, which has been set according to an area of interest. During operation, unexpected saturation occurs for some pixels, often grouped forming a saturated region in the raw images. This is a very common case that occurs, for instance, when a moving object (e. g., a person, a vehicle) comes into the scene, close to the camera, and stays there for a certain time before leaving it. This case also includes unexpected punctual reflections, intense light sources or sunlight, that might lead to several disjoint groups of saturated pixels or even to disperse saturated pixels. In such situations, it does not make sense to apply HDR constantly covering a large depth range where there is nothing in the scene, for the only reason that something might appear, or expecting punctual saturation due to eventual reflections that might never happen.

In the following we distinguish two different kinds of pixels in the raw images. Those pixels belonging to the scene we are expected to sense, that will not get saturated at the current illumination conditions and exposure time, are called *harmless* pixels and they form a *harmless* region. Of the other kind are those pixels that would achieve saturation for the exposure time selected to sense the scene, either because they correspond to very close objects that were not expected beforehand, either because they are strong direct reflections, light sources, etc. We call this second type *harmful* or saturating pixels and they might form a *saturated* region if the default exposure time is used.

The block diagram of the AHDR algorithm is given in Fig. 4.8. The process is triggered by each raw data acquisition (marker [M1] in the block diagram) of the ToF camera. Raw images at different exposure times are acquired. Typically there is one reference exposure time (related to the *harmless* static region of interest in the scene), which is always acquired [M2], and a small set of short exposure times (related to eventual saturating objects) [M3]. The set of short exposure times is what is adapted according to the scene. Originally the complete set of exposure times to be acquired might contain exclusively the reference exposure time and so will continue until a change in the acquired raw data is registered.

In order to detect scene changes [M4], we compare the current acquisition to the previous using a distance between images. We consider a kind of Hamming distance between raw data, namely eight distances, one for each one of the eight PMD raw images (two per phase). The distance is defined as the count of pixels showing a variation higher than a threshold. In order to avoid instability due to noise effects that would occur for pixels at boundaries of objects, the variation is weighted using information of the pixel neighborhood to penalize pixels in unstable regions, that are prone to be noisy along time. Given two correspondent raw images, $I_{k-1}^{\theta,\Gamma}$, $I_{k}^{\theta,\Gamma}$, gathered at the instants $k-1$ and k, respectively, the distance between them

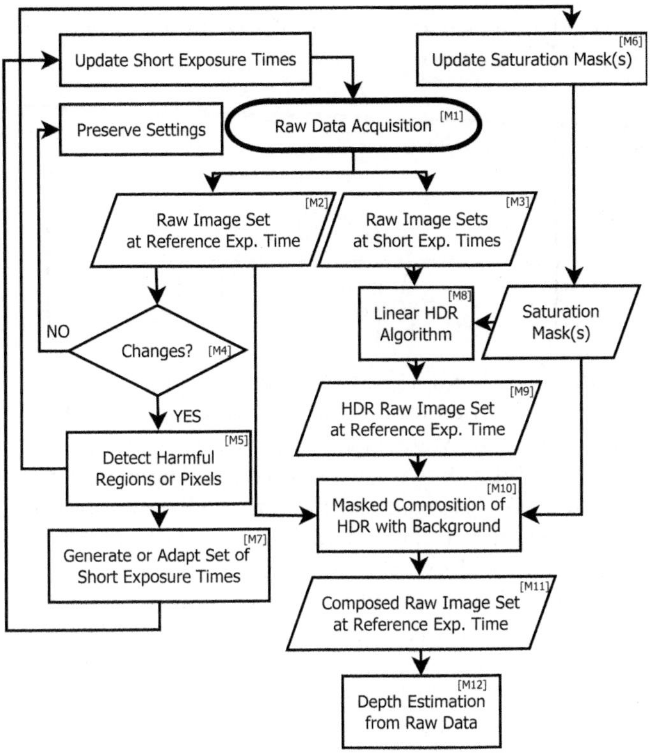

Figure 4.8.: Block diagram of our Adaptive High Dynamic Range (AHDR) algorithm. A change in the scene triggers a process of adaptation to the new scene. Eventual close saturating objects are detected and segmented. The exposure times to be acquired are adapted to these objects. An efficient HDR algorithm is applied over raw images gathered at those exposure times to accurately recover the close objects. The final depth image is computed combining the HDR result for the *saturable* areas and a background depth image obtained from long-exposed raw images. ©2015 IEEE.

is given by Eq. 4.10 as the cardinality of the subset of pixels fulfilling our weighted variation condition.

$$\Delta I_k^{\theta,\Gamma} = \left| V_k^{\theta,\Gamma} \right|, \text{ where:}$$
$$V_k^{\theta,\Gamma} = \left\{ I_k^{\theta,\Gamma}(\vec{x}) \mid \frac{\left| I_k^{\theta,\Gamma}(\vec{x}) - I_{k-1}^{\theta,\Gamma}(\vec{x}) \right|}{\max\limits_{\substack{\vec{x}' \in \Omega_r(\vec{x}) \\ \vec{x}' \neq \vec{x}}} \left| I_k^{\theta,\Gamma}(\vec{x}) - I_k^{\theta,\Gamma}(\vec{x}') \right|} \geq v_{\text{th}} \right\} \tag{4.10}$$

where $\vec{x} \in [1, n_{\text{rows}}] \times [1, n_{\text{cols}}]$ is a vector indexing a pixel, $\Omega_r(\vec{x})$ is a certain neighborhood of $\vec{x}$, defined by its radius r, and v_{th} is the threshold on the weighted variation. This value is to be adjusted experimentally: too high values increase reliability but reduce sensitivity. The parameter r has to be adjusted according to the expected motion or vibration of the camera. In static conditions values of one or two pixels are sufficient. If the distance $\Delta I_k^{\theta,\Gamma}$ is higher than a threshold, ΔI_{th}, the algorithm interprets that the scene has changed. This threshold is adjusted depending on the application and in a coherent way with v_{th}. e. g., if we want to be sensitive to eventual punctual saturating reflections, we will set v_{th} to a high value and a value of only a few pixels for ΔI_{th}. Otherwise, if we just want to detect unexpected foreground objects, v_{th} can be relaxed, while ΔI_{th}, must be drastically increased.

If the scene does not change ($\Delta I_k^{\theta,\Gamma} \leq \Delta I_{\text{th}}$), no adaptation process takes place and the algorithm uses the previous exposure parameters for next acquisition. If $\Delta I_k^{\theta,\Gamma} > \Delta I_{\text{th}}$, the algorithm proceeds to detect the *harmful* pixels [M5], creating a mask [M6] (or several masks, if several saturating levels are to be considered). A set of short exposure times—needed to sense the saturating object in the next acquisition—will be computed [M7] from an analysis of the detected *harmful* pixels. The next acquisition will be performed using exposure times that are already adapted to the scene and the raw images taken at short exposure times will be inputs of the HDR algorithm.

In the implementation of [H4], the step of detection and segmentation of saturating objects requires a dataset of raw images taken at different exposure times, as for the non-adaptive HDR presented before. This implies that, always a change in the scene is detected, a relatively slow acquisition process takes place. Nevertheless, note that, when the AHDR is already running, the set of raw images gathered at short exposure times, together with the raw image gathered at reference time, might be enough data for the *harmful* pixel detection module. The process of detection computes the linear response function for each pixel channel in a similar way to the basic HDR algorithm. A pixel is considered a *harmful* pixel (typically part of a saturating object) if the slope of the response function is higher than a threshold for any of the channels, at any of the four double raw images. Note the wideness of this definition, which includes both pixels belonging to excessively close objects, high-reflective scene objects and eventual intense light sources. One single saturating mask is simultaneously computed, where *harmful* pixels are assigned the value 1, being 0 for the rest. Note that, if different slope thresholds are considered, several levels of saturation sensitivity can be detected, leading to several masks. In such case, an accurate sensing of the scene covered by each mask would be provided by acquiring different ranges of exposure times for each level. HDR would be applied separately to each of these datasets. We do not present this case because it was observed that one single slope threshold is enough in most real cases. Very important is to remark that, if no *harmful* pixels are detected, the algorithm performs a conventional acquisition and no HDR takes place, being the final raw data the original one, taken at the reference exposure time.

The process of creating or adapting the set of short exposure times uses information generated in the previous step of *harmful* pixel detection. In that step, the exposure range where the pixel channels offer a linear response is registered, more specifically, the maximum exposure time before saturation is stored for each pixel. Then, the minimum of these maximum exposure times is found between the pixels belonging to the saturating mask. That is, we compute the maximum exposure time for which no pixel in the mask would be saturated. The adapted exposures are calculated in a linear way from this value. In [H4] this process is adjusted through two parameters: one that scales the minimum step of the exposure range and another that influences the width of the range.

Once the adaptation steps have taken place, the algorithm performs the basic HDR process [M8] described before, for each acquisition, over few short-exposure raw images. The use of the saturating mask allows performing the process only for those pixels in the mask, typically very few. The result of the HDR algorithm is a set of HDR raw images at an arbitrary virtual exposure time [M9], which is set to the reference exposure time. The HDR result is then combined with the set of raw images gathered at the reference exposure time according to the saturating mask [M10]: values for pixels in the mask are taken from the HDR result and the rest from the single-exposure raw image set. Note that the set is composed by four double-channel images, but one single mask is used for all of them. This is necessary because the final composed raw image set [M11] is the input for the depth estimation process [M12], from which a single depth image is computed from these eight using the four phases algorithm (Eq. 2.19).

4.2.2.4. Experimental Evaluation

We make use of the same hardware (the MultiCam with a 8.5 mm lens and medium-range illumination system) to gather two datasets with two different large real-world scenes, in which an unexpected object is introduced very close to the camera. The results presented in this section are to be expanded with those presented in Appendix A.8. The main scene of the first experimental setup (in the following, Exp. 1) is a corridor (Fig. 4.9a) and is intended to be an easily-reproducible case allowing quantitative evaluation of improvements. The second case (in the following, Exp. 2) takes place in a large laboratory (Fig. 4.9b) and is intended to provide qualitative evidence of good performance in high-complexity and large-range scenes. In both cases the exposure time is adjusted to sense the scene. Consequently, only a HDR system will provide accurate results for both the original scene and the intrusive object.

(a) Experimental Setup 1: Corridor **(b)** Experimental Setup 2: Laboratory

Figure 4.9.: Two real scenes with large depth ranges are used to test the performance of our AHDR algorithm. The first one is a corridor, where a panel with two Böhler stars is placed close to the sensor. The second is a large laboratory, requiring a very high dynamic range. In the latter case, an apple is used as close object. ©2015 IEEE.

Böhler Stars Experiment The scene in Fig. 4.9a is a corridor, observed in longitudinal direction, with a width variation in the middle and ending in a wall with a closed door at approximately 14.5 m from the camera. The exposure time of the PMD chip is set to 2 ms, sensing the scene without pixel saturation. The experiment consists in placing a panel containing two Böhler stars [60], which are the 3D counterpart of the widely-used Siemens stars, at approximately one meter from the camera. The Böhler stars are both of 20 cm diameter. The upper star features 12 fields, and the lower one 24. The Böhler stars are so close to the camera that require an exposure time as low as 0.1 ms to get an accurate depth image. In Fig. 4.10, the depth images obtained from the outputs of our non-adaptive HDR and the AHDR algorithms are put in contrast with those obtained using single exposure times of 2 and 0.1 ms. The figure is to be complemented with Fig. A.7.

The depth result for 2 ms exposure time (Fig. 4.10b) shows very low noise in the middle and far areas of the scene, while being very noisy in the Böhler stars. The combination of close range and intense illumination forces the SBI to its limit, producing an intense noise pattern on the surface of the stars' panel. Complementary, for 0.1 ms (Fig. 4.10c), the plain surface of the stars is captured more accurately but the depth estimation for the rest of the scene is seriously deteriorated. The strong *salt and pepper* noise in Fig. 4.10c is due to the short exposure: pixels deliver values within their noise floor and, therefore, the depth estimation produces a random value. The result of the exhaustive HDR algorithm (Fig. 4.10a) is free of the artifacts observed in the single-exposure-time cases. Fig. 4.10d confirms that the AHDR algorithm delivers the highest-quality depth images, while minimizing the number of acquisitions in an adaptive way. Note the superior reconstruction of the lower Böhler star and the significant blur reduction on the edges of the stars' panel. Even the horizontal border between the two star panels becomes visible.

Fig. 4.11 offers further comparison of the four depth results in Fig. 4.10 by means of two horizontal cuts in the depth images: one at the height of a plain area of the Böhler stars' panel (marked as cut C in Fig. 4.10) and another one crossing four fields of the upper part of the lower star (cut D in Fig. 4.10). The left plots depict the depth values and the right plots the standard deviations, obtained over 100 consecutive experiments. Note the plain and noiseless depth estimation of the AHDR approach for the pixels of cut C within the panel. Additionally, out of the panel, the depth and standard deviation graphs of the AHDR approach cover the equivalent graphs of the single acquisition at 2 ms. The depth plot obtained for cut D explicitly shows that the AHDR method brings an axial resolution

(a) HDR

(b) Single Exposure: 2 ms

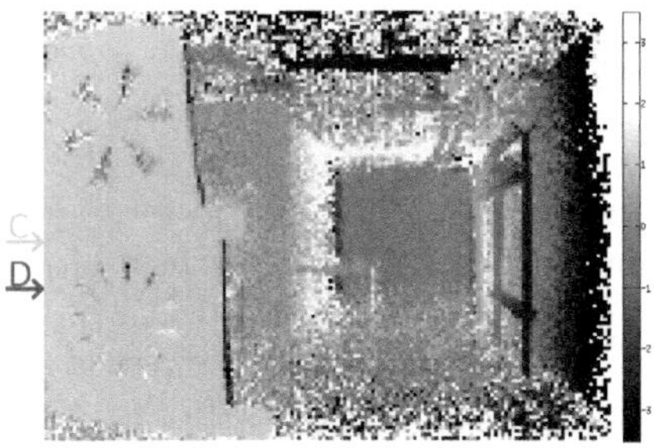

(c) Single Exposure: 0.1 ms

(d) AHDR

Figure 4.10.: Depth maps obtained for the Exp. 1. The corresponding Multi-Cam color image is given in Fig. 4.9a. The depth result obtained from the intensive HDR algorithm is shown in (a). Single exposure times of 2 ms, set to sense the entire corridor, and 0.1 ms, adjusted to optimally capture the Böhler stars, were used in (b) and (c), respectively. Finally, (d) shows the result of our AHDR approach. ©2015 IEEE.

enhancement, allowing clear distinction of the four star fields traversed by cut D. In general, Fig. 4.11 confirms that the AHDR approach offers the best depth estimation, while keeping a very high stability—low standard deviation—along the entire depth range.

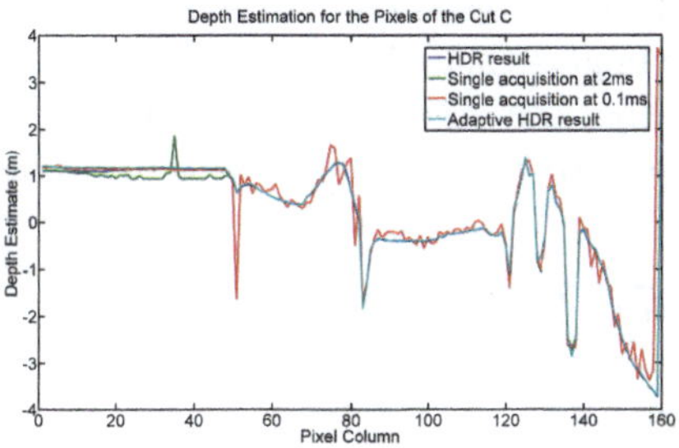

(a) Cut C: Depth Profile

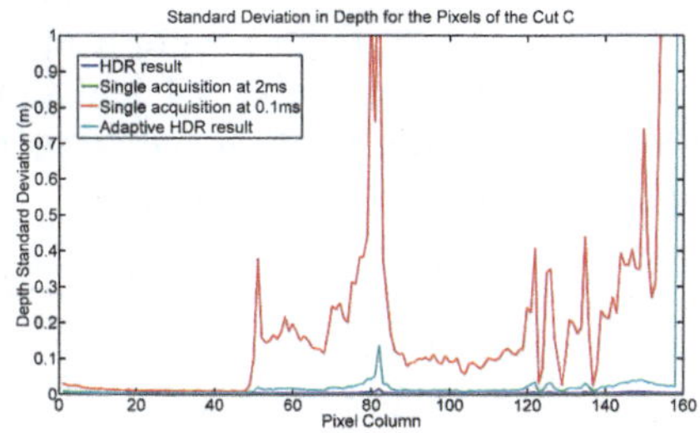

(b) Cut C: Depth Standard Deviation

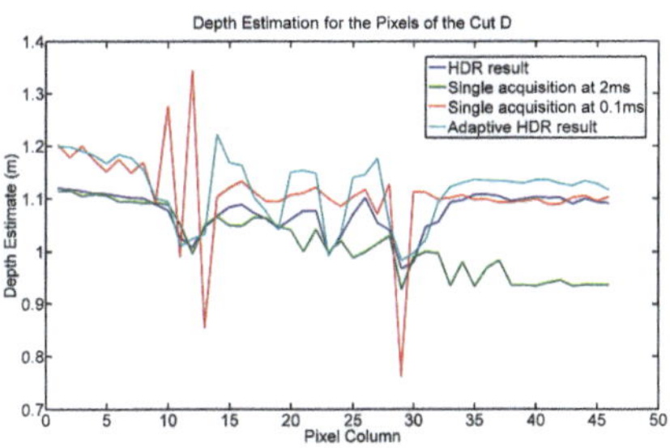

(c) Cut D: Depth Profile

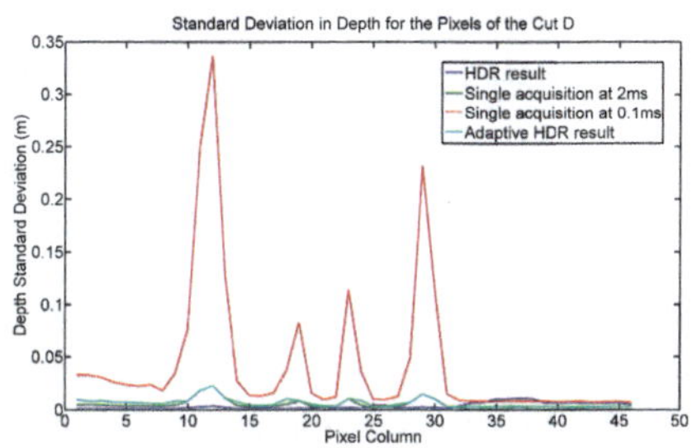

(d) Cut D: Depth Standard Deviation

Figure 4.11.: The plots at the left depict the depth profiles of cuts C (a) and D (c) in Fig. 4.10, while the plots (b) and (d), at the right, show the corresponding standard deviation. Four experimental cases are considered: depth estimation using our reference HDR algorithm (blue), from single raw data acquisitions at 2 ms (green) and at 0.1 ms (red) exposure time and using our AHDR algorithm (cyan). ©2015 IEEE.

Lateral Resolution The use of Böhler stars allows us to give an experimental measure of the angular resolution, one of the most critical parameters of ToF cameras. The effective lateral resolution of a depth imaging device at some distance from the camera can be computed from the depth image of a Böhler star placed at that distance using Eq. 4.11:

$$r = \frac{\pi d_{\text{int}}}{n} \tag{4.11}$$

where d_{int} is the diameter of the incorrectly measured interior circle in the middle of the star and n is the number of fields of the star. The previous formula is the quotient between longitude of the interior circle and number of fields, and, for n high enough, it approximates the axial resolution. Results are summarized in Table 4.1 and show that our AHDR approach improves the axial resolution by a factor of three, with respect to the resolutions obtained for single acquisitions.

	Böhler Lateral Resolution (pixels)			
Böhler star	HDR	2 ms Exp.	0.1 ms Exp.	AHDR
12-Fields Star	1.57	2.36	2.36	0.79
24-Fields Star	1.57	2.36	2.36	0.92
Average	1.57	2.36	2.36	0.85

Table 4.1.: Lateral resolution in depth domain obtained using Böhler stars. The result of our Adaptive HDR (AHDR) approach is compared to those obtained using our reference HDR and two single exposure times. ©2015 IEEE.

Depth Error The planar structure of part of the scene allows reliably fitting a plane model to the depth data in some regions, which can be used to generate depth ground truth (GT). Therefore, for those regions, depth errors can be computed. Two regions of interest (ROI) are chosen: one in the Böhler stars' panel, on a plain area, at the up-right corner of the lower Böhler star, namely, a foreground region, and the other one covering the wall at the end of the corridor, excluding the door, i. e., a background region, far from the camera. For both cases, the GT distance from the camera to the planar region is obtained by fitting a plane measurement model to the data. This measurement model depends on the FOV of the camera and on the target parameter d_{plane}, the distance from camera to plane, measured orthogonally to the plane. The model is, for a given parameter set, a simple continuous mapping from 2D pixel index domain to 1D depth domain, as expressed by Eq. 4.12,

$$\begin{aligned} M(\text{FOV}, d_{\text{plane}}): \quad \mathbb{N}^2 &\rightarrow \mathbb{R} \\ \vec{x} &\rightarrow d_{\vec{x}}^{M(\text{FOV}, d_{\text{plane}})} \end{aligned} \tag{4.12}$$

where $\vec{x} \in [1, n_{\text{rows}}] \times [1, n_{\text{cols}}]$ is the vector of pixel indexes, being $n_{\text{rows}} \times n_{\text{cols}}$ the size of the depth image. Therefore, fitting the model to some data means finding the value of d_{plane} that minimizes the distance between the measurements predicted by the model ($d_{\vec{x}}^{M(\text{FOV}, d_{\text{plane}})}$) and those observed through real measurements ($d_{\vec{x}}^{O}$), as given by Equation 4.13. The instantiation of the model gives a reliable estimate of the GT distance between camera and plane area, $\hat{d}_{GT}$, together with an error bound, ε.

$$\hat{d}_{GT} = \underset{d_{\text{plane}} \in \mathbb{R}}{\arg\min}\, \varepsilon = \sqrt{\frac{1}{n_{\text{rows}} n_{\text{cols}}} \sum_{\vec{x} \in \Omega_{\text{ROI}}} \left| d_{\vec{x}}^{M(\text{FOV}, d_{\text{plane}})} - d_{\vec{x}}^{O} \right|^2} \tag{4.13}$$

where Ω_{ROI} is a set containing the index vectors of the pixels in the ROI. For the first ROI the model is instantiated using data acquired at short exposure time (0.1 ms), while for the second data acquired at long exposure time (2 ms) is used. The minimization problem is convex for both cases and converges to a distance value of 1.070 m for the first ROI and 14.59 m for the second, with residuals of 7.40 mm and 3.88 cm, respectively. Using these distances and the plane model, GT depths are obtained for the pixels of each ROI. Table 4.2 summarizes the errors with respect to GT obtained for the four results presented in Figure 4.10.

	Depth RMSE (m)			
ROI Location	HDR	2 ms Exp.	0.1 ms Exp.	AHDR
Stars Panels	2.26×10^{-2}	1.46×10^{-1}	7.40×10^{-3}	3.85×10^{-2}
Frontal Wall	3.98×10^{-2}	3.88×10^{-2}	6.10×10^{-2}	3.88×10^{-2}

Table 4.2.: Depth error with respect to ground truth for two planar ROIs in the Böhler stars experiment. The result of our Adaptive HDR (AHDR) approach is compared to those obtained using our reference HDR and two single exposure times.

The results in Table 4.2 are coherent with our previous comments on Fig. 4.10 results and give a quantitative evaluation of the depth error reduction that our adaptive approach brings, with respect to adapted single exposure times. The RMSEs obtained for the adaptive approach are very close to those obtained using the reference exhaustive HDR approach; even more, in terms of maximal RMSE, the former outperforms the latter. Fig. 4.12 illustrates how well the AHDR result (Fig. 4.12c) fits the GT (Fig. 4.12a) and contrasts it with the result obtained using the default exposure time (Fig. 4.12b).

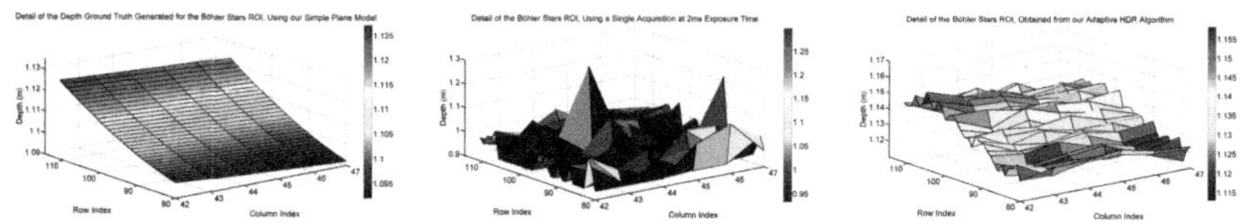

(a) Ground Truth (GT) (b) Single Exp.: 2 ms (c) AHDR

Figure 4.12.: Surface of the Böhler stars' panel for part of the pixels contained in the first ROI, which is totally planar and does not contain any star. The ground truth plane (a) is compared to the results obtained from a single acquisition (b) and from our AHDR approach (c). ©2015 IEEE.

Laboratory Experiment The complex scene in Fig. 4.9b corresponds to a large laboratory. Different objects, such as tables, chairs, boxes, etc. are located at medium and large distances from the camera, requiring a high depth range. Far areas, e. g., the ceiling or the front wall, are often at more than 13 m from the camera. The camera points orthogonally to one of the laboratory walls. A relatively long default exposure time of 10 ms is chosen in order to accurately sense the whole scene. In this experiment, the intrusive object is an apple, which is unexpectedly placed into the scene, at 1.30 m from the camera. The apple requires a lower exposure time of 1 ms to avoid saturation. In Fig. 4.13, the depth result given by our AHDR algorithm is presented and compared to those obtained using simple HDR and single acquisitions with exposure times adapted to the entire scene and to the intrusive object, respectively. The figure is to be complemented with Fig. A.8.

Fig. 4.13b shows no noise in far areas, thanks to the combination of a powerful illumination system and a long exposure time. On the other hand, it is easy to observe a large amount of noise for the apple pixels. Depth differences up to 80 cm between pixels of the apple are registered, while the apple size is not higher than 7 cm in depth direction. In Fig. 4.13c very low levels of noise are registered for the foreground object, the lowest in the image. The largest depth difference between apple pixels is around 4 cm, i. e., in perfect agreement with the real apple geometry. The depth map follows the apple shape (closer in the middle and farther in the borders). Note, on the

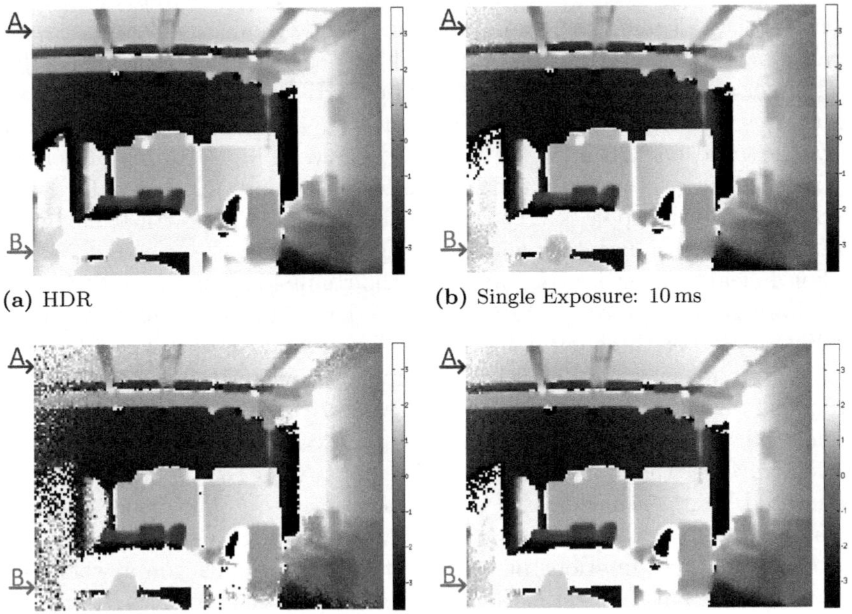

(a) HDR

(b) Single Exposure: 10 ms

(c) Single Exposure: 1 ms

(d) AHDR

Figure 4.13.: Depth maps obtained for the Exp. 2. The corresponding MultiCam color image is given in Fig. 4.9b. The depth result obtained from the intensive HDR algorithm is shown in (a). A single exposure time of 10 ms is used to capture the background in (b), and 1 ms to properly capture the intrusive apple in (c), respectively. Image (d) shows the result of our AHDR approach. ©2015 IEEE.

other hand, how intense is the noise in pixels belonging to non-perpendicular planes, such as ceiling and side walls, due to the insufficient illumination for such a low exposure time. The left part of the depth image is visually lost due to the noise. Compare now the result of our AHDR algorithm (Figure 4.13d) to the results for fixed exposure times in Fig. 4.13b and Fig. 4.13c. Note how the good background estimation observed in Fig. 4.13b is preserved, while accurately recovering the surface of the apple, as in Fig. 4.13c. The depth map on the apple is very close to that in Fig. 4.13c, with a maximum depth variation of also $4\,cm$ within the apple.

For a clearer pixelwise comparison of the results, and similarly to Exp. 1, two horizontal cuts are performed in the depth images, named A and B in Fig. 4.13. For these cuts, depth plots and standard deviation plots are generated and shown in Fig. 4.14. The upper graphs plot the depth and the standard deviation for the pixels of cut A, corresponding to a region which is far from the camera, for most pixels. The lower ones correspond to cut B, which traverses the apple in its central region. As in Exp. 1, deviations are computed by repeating the experiment 100 times.

In Fig. 4.14, it is easy to observe that the AHDR algorithm offers an accurate depth estimation there where the single acquisition at reference exposure time fails (apple pixels). Note that, out of the saturating mask—in this case limited to the apple—, the curves for 10 ms exposure time are hidden by the AHDR ones (in red). Observe the wrong depth estimation for the apple pixels in cut B, in comparison to the relatively moderate noise observed for the far pixels of cut A, at the reference exposure time (10 ms). Complementary, observe the very high noise levels for left pixels of cut A and the also high noise levels for background pixels in cut B when using 1 ms exposure time, in contrast to the accurate depth profile obtained for apple pixels, close to the AHDR result.

4.2.2.5. Overview of Hardware Implementations

As well as the Wyckoff film is a hardware solution for HDR analogical photography, there is room in digital imaging for hardware-based HDR methods. The existing approaches can be classified depending on the hardware level at which the HDR method is implemented. The lowest level would be directly at pixel level, by means of pixels with extended dynamic range, such as the wide dynamic range CMOS image sensor of [47], which is able to offer a dynamic range of 80 dB at 30 fps. For *in pixel* HDR approaches in multitap pixels, the reader is referred to Section 2.3.1. One simple hardware solution is to implement a basic HDR algorithm in an embedded system within the

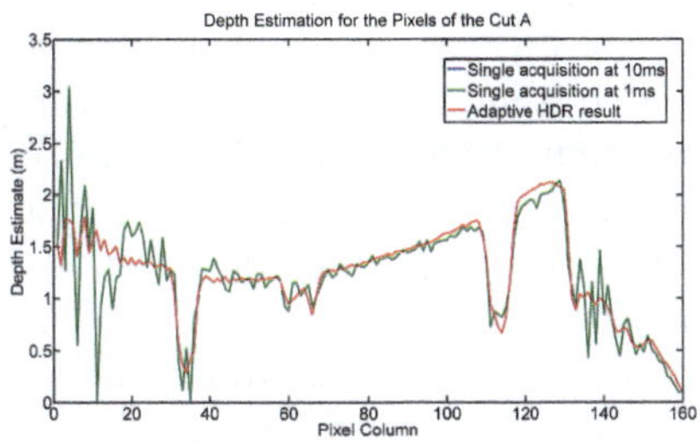

(a) Cut A: Depth Profile

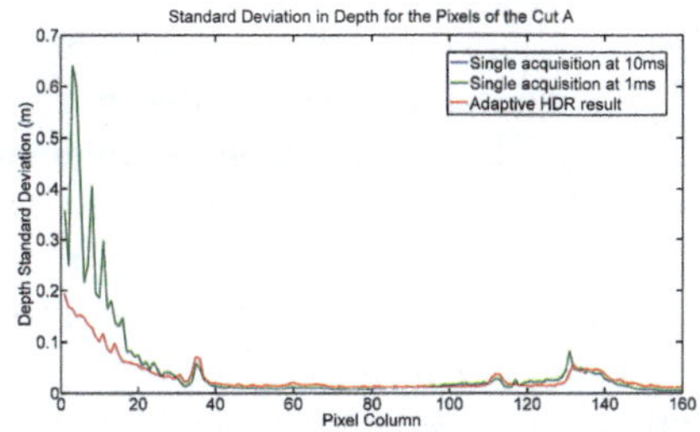

(b) Cut A: Depth Standard Deviation

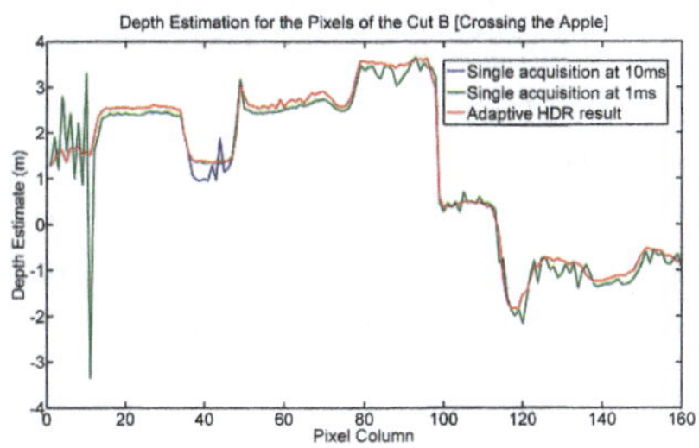

(c) Cut B: Depth Profile

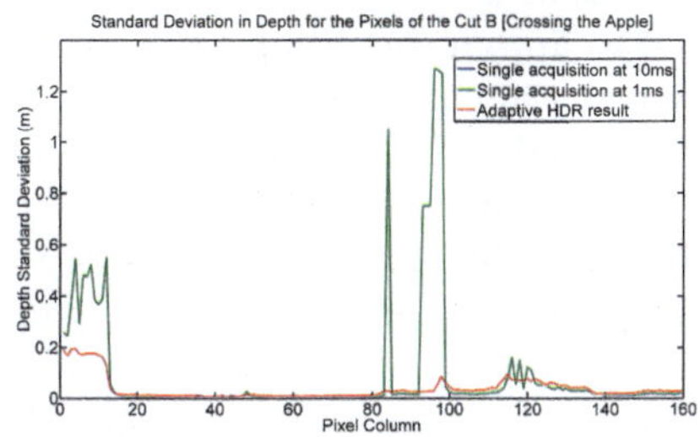

(d) Cut B: Depth Standard Deviation

Figure 4.14.: The plots at the left depict the depth profiles of cuts A (a) and B (c) in Fig. 4.13, while the plots (b) and (d), at the right, show the corresponding standard deviations. Three experimental cases are considered: depth estimation from single raw data acquisitions at 10 ms (blue) and at 1 ms (green) exposure time and using our AHDR algorithm (red). ©2015 IEEE.

camera, as suggested in [441], where images taken at different exposure times are stored in a memory unit, which is accessed by a CPU. If the pixel value at a reference exposure time reveals saturation, then it is replaced by the value obtained at lower exposure time. If underexposure is detected, then the pixel value is replaced by another obtained at higher exposure time. Note that this basic algorithm delivers a pseudo-HDR image, since its adapted pixel values are false, and therefore cannot be applied to our case of study. A hardware approach that accounts both for HDR imaging and tone mapping using an FPGA architecture is presented in [285]. Interestingly, the pixel response function is precomputed and stored into a look-up table in the FPGA, thus, overcoming the main difficulty of HDR imaging in a fast manner. The system delivers tone-mapped HDR images ready to be displayed in conventional monitors. The spatially-varying exposures method of [342] combines neighboring pixels acquired using different exposure times to get a HDR image from a single acquisition at the cost of image resolution. An optical mask with cells with different transparencies is enough to implement this method on an existing sensor. An implementation of this idea have been recently patented by the authors in [343]. In [340], the fixed pattern mask is replaced by a liquid crystal light modulator. Each cell of the modulator is set to a transmittance value such that pixels work in the highest sensitivity area of the response function. This way avoids resolution losses at the cost of some adaptation time and eventual blurring due to an additional optical system. Closed hardware solutions exist too, such as that patented in [16], where two different pixel charge times are considered and a composite image with extended dynamic range is delivered. Of special interest is the approach in [4], which combines for the first time HDR imaging and CS. Their setup, inspired in [156], requires only a single pixel, opening the possibility of composing HDR images from wide-spectrum or special wavelength sensing elements, that cannot usually be provided in high resolution imaging arrays. This is actually the case of ToF imaging systems, whose lateral resolution is low (cf. Section 2.4.2) and the dynamic range of the pixels limits the operative depth range of the camera. The approach seems promising in areas where HDR is required for wavelengths that cannot be captured by conventional cameras, such as biological imaging [3].

4.2.3. Shot Noise: Enabling Operation in Low-Light Conditions

The reduced limited dynamic range of the pixels can be artificially extended via HDR algorithms, avoiding SBI-related noise in the raw images. The FPN can be removed by per-pixel linear calibration. Nevertheless, shot noise due to operation in low-light conditions is related to the random nature of the photon arrival and cannot be easily rejected. Shot noise becomes relevant when the product of exposure time and irradiance at the pixel surface becomes too low and only few photogenerated carriers are stored at the end of the integration period. This is a common case when the illumination system is not powerful enough to project sufficient light over objects that are too far or exhibit too low reflectivity. Clearly, an appropriate HDR approach should ensure, by using different exposure times, that every pixel provides a value within the *safe* region of the response curve for some exposure time, i. e., higher than the noise floor (high enough to neglect photon shot noise) and lower than the saturation level. Nevertheless, if there exist strict restrictions on the exposure time, number of acquisitions and optical power of the illumination system, shot noise might be dominant in the raw images, increasing the uncertainty of the depth measurements (recall Eq. 2.29).

In this section we present the method for denoising PMD raw data included in [H8]. The core idea of the method is the similarity between raw images. The underlying signal should look similar, while the noise is different for each raw image. The generality of the method allows applying it to data gathered with any kind of multitap image sensor (e. g., the Xbox One sensor) or with combined multiaperture-multitap systems. An example of such hybrid hardware is, for instance, the two-tap multiaperture system presented in [329].

4.2.3.1. Bilateral Filtering for Multitap Systems

The denoising framework is based on the concept of bilateral filtering. A conventional bilateral filter working on a single image generates a filtered image, where the value of each pixel is a weighted combination of the values of the pixels contained in a certain neighborhood, where the weights depend on the distance with respect to the central pixel, both in spatial domain (closeness) and intensity domain (similarity), as shown in Eq. 4.14,

$$\begin{aligned} \hat{i}(\vec{x}) &= \frac{\sum\limits_{\vec{x}' \in \Omega_{\vec{x}}} i(\vec{x}')w(\vec{x}', \vec{x})}{\sum\limits_{\vec{x}' \in \Omega_{\vec{x}}} w(\vec{x}', \vec{x})}, \text{ where:} \\ w(\vec{x}', \vec{x}) &= \exp\left(-\frac{\|\vec{x}' - \vec{x}\|_2^2}{2\sigma_s^2} - \frac{\|i(\vec{x}') - i(\vec{x})\|_2^2}{2\sigma_i^2}\right) \end{aligned} \tag{4.14}$$

where $\vec{x}$ and $\vec{x}'$ are 2D coordinates of image pixels, $\Omega_{\vec{x}}$ is a neighborhood of pixel $\vec{x}$, $i(\vec{x})$ is the intensity value of the pixel indexed by $\vec{x}$ and $\hat{I} = \left[\hat{i}(\vec{x})\right]_{\vec{x} \in [1, n_{\text{rows}}] \times [1, n_{\text{cols}}]}$ denotes the filtered image. The variables σ_s and σ_i are the smoothing parameters in spatial and intensity domain, respectively. The filter converges to a Gaussian filter when $\sigma_i \to \infty$.

In the case of multitap systems, the raw images are expected to be different from each other, but, since the scene being sensed is the same in all them, they are also expected to be highly correlated. The filter has to exploit this fact, while preserving each one of the images separately. Multimodal bilateral filtering for ToF systems have been already studied in [281], where the filter operates with a vector of intensities per pixel, instead a single intensity value. In that work a MultiCam was used to acquire registered depth and color images simultaneously. The depth was treated as another channel, together with the color channels. Provided that the depth image has a much lower resolution than the color modality, the bilateral filter was intended to transfer the high resolution from the color image to an upscaled version of the depth image. This approach relies on the hypothesis that the depth modality is highly correlated with the color modality, which is a valid assumption for many natural and man-made scenes, but it does not always hold. A more meaningful alternative would be to filter before depth calculus, considering the raw images as different intensity channels. We adopt the multichannel approach proposed in [281] for bilateral filtering of the raw images of multitap systems. Since we focus on the low-light scenario, we do not suppose having an additional high-resolution color image to include in the joint filtering. Let $\vec{s}(\vec{x}) = [i_k(\vec{x})]_{k\in[1,N]} \in \mathbb{R}^N$ be the vector of raw intensities for the pixel indexed by $\vec{x}$, where N is the number of raw images. The bilateral filter is formulated in Eq. 4.15,

$$\begin{aligned} \hat{\vec{s}}(\vec{x}) &= \frac{\sum\limits_{\vec{x}'\in\Omega_{\vec{x}}} \vec{s}(\vec{x}')w(\vec{x}',\vec{x})}{\sum\limits_{\vec{x}'\in\Omega_{\vec{x}}} w(\vec{x}',\vec{x})}, \text{ where:} \\ w(\vec{x}',\vec{x}) &= \exp\left(-\frac{\|\vec{x}'-\vec{x}\|_2^2}{2\sigma_s^2} - \frac{(\vec{s}(\vec{x}')-\vec{s}(\vec{x}))^\top \Pi_{\vec{s}}^{-1}(\vec{s}(\vec{x})-\vec{s}(\vec{x}))}{2}\right) \end{aligned} \tag{4.15}$$

where $\Pi_{\vec{s}} \in \mathbb{R}^{N\times N}$ is the weighting matrix that accounts for the confidence of the intensity measurements. As in [281], we adopt a diagonal matrix, i. e., assume statistical independence between channels. In low-light imaging, where the variance is equal to the expected value, the diagonal elements of $\Pi_{\vec{s}}$ can be related to the average intensity level of the corresponding raw image.

4.2.3.2. Variance Stabilization

Shot noise, or quantum noise, appears due to the fact that, in low-light conditions, the process of photocarrier generation in the pixels follows a Poisson distribution. A discrete random variable following a Poisson distribution, $X \sim P(\lambda)$, is characterized by a single parameter, λ, and we have that $\mathrm{E}(X) = \mathrm{Var}(X) = \lambda$. In other words, the variance of the measurements is as large as the expected value. This poses a situation that does not fit the widely-adopted assumption of additive zero-mean Gaussian noise. If the variance depends on the signal, a filter with unique smoothing parameters might not be an appropriate solution, since the optimal value of σ_i should depend on the expected standard deviation. Consequently, the joint-bilateral filter might not perform equally well along all image areas. In dark areas, the smoothing might wash away the signal, while being insufficient in bright areas, where the noise exhibits higher variance.

Fortunately, a number of classical variance-stabilizing transformations exist, which aim to stabilize the variance of data following a binomial or Poisson distribution. Most of them are evolutions of the basic square root transformation, such as the Barlett transformation [30], the Anscombe transformation [13] and the Freeman-Tukey transformation [190]. For completeness, we collect the corresponding formulas for stabilizing the variance of Poisson variables in Fig. 4.15a. Their relative performance is evaluated using synthetic data following Poisson distributions $X \sim P(\lambda), \lambda \in (0, 10]$, with step $\Delta\lambda = 0.01$. For each case, the variances of the transformed data are computed, using a large population size of 10^5 data points. The results, in terms of the ratio between stabilized variance and limiting stabilized variance, are plotted in Fig. 4.15b. Recursive methods have also been proposed to achieve optimal variance stabilization [178], taking classic stabilizing techniques, e. g., Anscombe and Freeman-Tukey transformations, as starting point.

For λ not too low, the square root transformation offers both an acceptable performance and simplicity. For low values of λ, the performance of the square root transformation degrades and an additional term within the square root is required to achieve good variance stabilization. Of this kind are the Barlett and the Anscombe transformations, performing the latter slightly better than the former. Unfortunately, in both cases the variance stabilization degrades rapidly when approaching to unit λ. Note that a Poisson distribution with close-to-unit expected value is a worst-case scenario of our case of study, i. e., photon shot noise in extremely low-light imaging. In such cases, the Freeman-Tukey (F-T) transformation provides the best stabilization (from Fig. 4.15b, 95% at $\lambda = 1$, being the second best the Anscombe transformation, with only 70%), at the cost of a more complex expression that does not allow for a direct inversion. For this reason, we derive an inverse F-T transformation for the Poisson case, using as starting

Transformation	Formula
Square Root	$\sqrt{x}$
Barlett	$\sqrt{x + \frac{1}{2}}$
Anscombe	$\sqrt{x + \frac{3}{8}}$
Freeman-Tukey	$\sqrt{x} + \sqrt{x+1}$

(a) Variance Stabilization Formulas

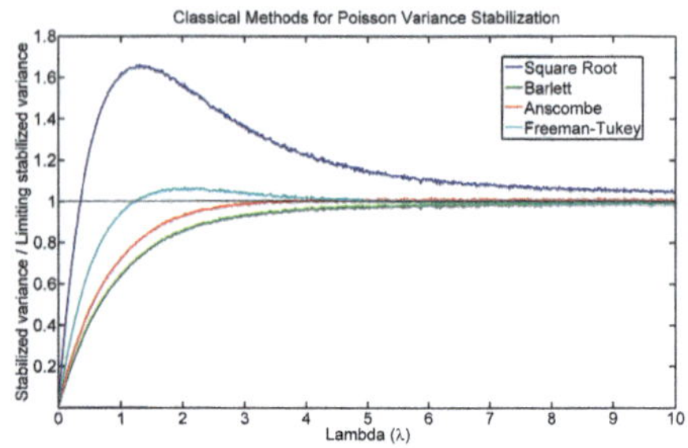

(b) Performance Comparison

Figure 4.15.: Classic square-root-based variance stabilization transformations for Poisson data. The performance of the different methods in stabilizing the variance of synthetic Poisson data is given in (b). The Freeman-Tukey transformation offers the best performance for data following Poisson distributions of very low λ ($\lambda \leq 3$). ©2016 IEEE.

point the original double arcsine transformation for the binomial case. The derivation is provided in Appendix A.9, yielding Eq. 4.16, where y denotes the transformed value, obtained from the direct F-T formula in Fig. 4.15a, and $\hat{x}(y)$ the estimate of the original x value given by the inverse transformation.

$$\hat{x}(y) = \left(\frac{y - y^{-1}}{2}\right)^2 = \sinh^2 \ln y \tag{4.16}$$

Numerical simulation confirmed that Eq. 4.16 is an exact inverse F-T transformation for Poisson data. By means of the direct and inverse F-T transformations, we are able to apply our bilateral filter-based denoising approach in a domain where the data exhibits an approximately constant variance and then project the result back to the original domain at no cost in terms of reprojection error.

4.2.3.3. Performance Evaluation with Real PMD Data

In this section we evaluate our method with real raw data from a PMD sensor, acquired with our MultiCam. Therefore, the set of raw images to process contains $N = 8$ elements. The filtering procedure is carried out both in the original intensity domain and the F-T transformed domain. In the latter case, the results are also evaluated in the original domain, after applying Eq. 4.16. The experimental setup exhibits depths ranging from 1.5 to 2.0 m, approximately. It contains objects of known geometry, such as two balls of different sizes on a table. The camera looks frontally to a plain wall. At the right side a panel with two Böhler stars [60] is placed, parallel to the wall and closer to the camera. The stars are used to confirm that the filtering does not lead to degradation of the angular resolution in the final depth image. Fig. 4.16 provides the output of the MultiCam when recording the scene in presence of some ambient light.

(a) Color Image **(b)** Depth Image

Figure 4.16.: Color (a) and depth (b) images of the experimental setup delivered by the MultiCam. Both images are acquired at exposures times that are adjusted to optimize the image quality, i. e., high enough to avoid strong photon shot noise. The depth scale is in meters. ©2016 IEEE.

In this experiment, for simplicity, but without loss of generality, we assign the same value, σ_i^2, to all the diagonal elements of $\Pi_{\vec{s}}$ in Eq. 4.15. This is an acceptable assumption in general, but note that its validity is signal-dependent. For example, in the case of PMD ToF, depths corresponding to phase shifts close to 0° and 180° might lead to very asymmetric charge distribution between the two integration channels. In such cases, an independent choice of $\sigma_{i_k}^2$, $k \in [1, N]$ for each of the N raw images is recommendable. This unique σ_i and the σ_s of Eq. 4.15 are jointly optimized within a certain feasible range, so that we make sure that we attain the best possible result, independently from a wrong parameter selection. The parameter optimization [H8] seeks minimizing the distance between the depth image obtained from the filtered raw data and a reference depth image (Fig. 4.16b), obtained using a long exposure time.

The datasets are acquired using abnormally low exposure times, in order to obtain data heavily corrupted by photon shot noise. Three different exposures are considered: 10, 20 and 50 μs. For 10 μs, the scene is hardly distinguishable in the raw images and the signal level close to the noise level of the sensor. The results of the experiments are presented in Fig. 4.17 in the shape of depth images. The first column shows the depth images obtained from the original raw data, while the second and third columns show the depth images obtained from jointly filtered data, in original and F-T domains, respectively. These results are to be compared to the reference depth image in Fig. 4.16b.

For all exposure times considered, the depth images obtained from filtered data exhibit much lower noise than those obtained from unfiltered raw data. Far from being oversmoothed by the filter, depth gradients are preserved or even enhanced (observe, for instance, the fields of the upper Böhler star). In the 10 μs case (first row), the round shape of the balls becomes visible. In the 20 μs case (second row), the table surface is clearly recovered and depth estimation becomes possible also for the left part of the image, under poorer illumination conditions. Finally, in the 50 μs case (third row), the visual quality of the depth images obtained from filtered data is as good or even better than that of Fig. 4.16b. No loss of angular resolution in the depth images was observed. In order to provide a quantitative evaluation of the improvement achieved by filtering prior to depth calculus, we compute the RMSE between the results in Fig. 4.17 and the reference depth map in Fig. 4.16b.

Exposure Time (μs)	RMSE (cm)		
	No Filter	Filtering Domain: Original Intensity	Filtering Domain: Freeman-Tukey
10	64.46	31.86	32.14
20	36.99	12.30	12.37
50	9.02	5.37	5.39

Table 4.3.: Root Mean Square Error (RMSE) of the depth images in Fig. 4.17 with respect the reference depth map (Fig. 4.16b) ©2016 IEEE.

The results are given in Table 4.3 and are coherent with the visual quality of the depth images in Fig. 4.17. The proposed method leads to an approximate depth error reduction of 50% for 10 μs exposure time, 77% for 20 μs and 40% for 50 μs. We observe that the depth error is independent

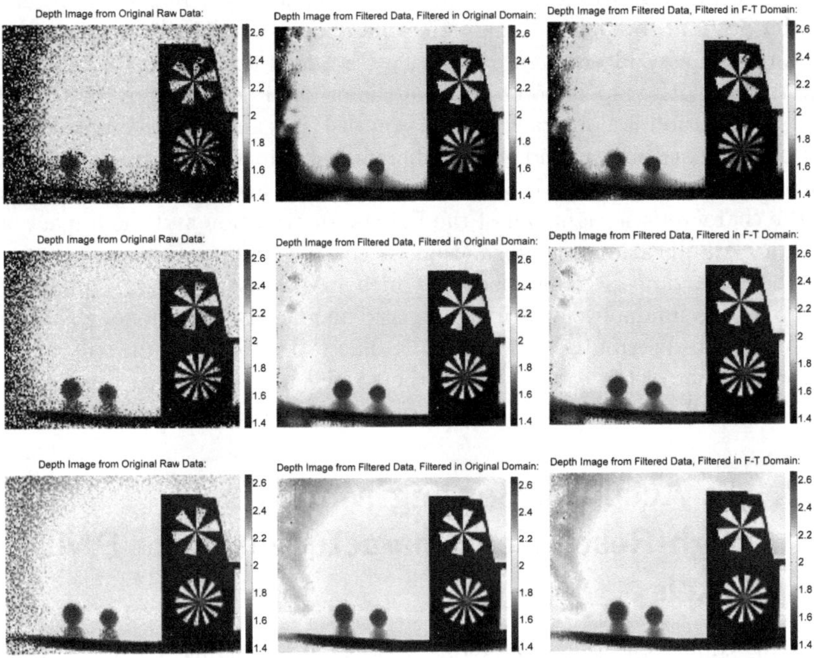

(a) From Raw Data **(b)** From Filtered Data **(c)** From F-T Filtered Data

Figure 4.17.: Results of the low-light image enhancement experiment. The images are the final depth images obtained using very short exposure times, namely, 10 μs (first row), 20 μs (second row) and 50 μs (third row). The depth images were computed from the original raw data (a), from raw data filtered in the original intensity domain (b) and from raw data filtered in the Freeman-Tukey (F-T) transformed domain (c). All scales are in meters. ©2016 IEEE.

from the domain where the filtering was carried out, being the filtering in the original intensity domain the one delivering slightly better results. One reason why filtering in a variance stabilized domain did not improve the results with respect to doing it in the original domain may be that the real distribution is not exactly a Poisson distribution, but a composition of a Poisson distribution and some other distributions (e. g., normal) whose variances are independent of the expected value. This would reduce the effect of the characteristic right tail of the Poisson distribution and the impact of a variance stabilization transformation. A complementary explanation would be that the joint-filtering schema, which makes use of all raw images to compute the unique weights in synergistic manner, is robust enough to cope with Poisson-distributed data directly. The latter explanation only applies if the product of the cardinality of the spatial neighborhood considered in the filter and the cardinality of the set of raw images is large enough.

4.3. An Accurate Sensing Model: High-Resolution Characterization of PMD Pixels

Existing simulation frameworks for PMD-like ToF imaging sensors [256, 277] seem to avoid including a realistic model for the pixel response in spatial domain and implicitly assume it to be constant, probably for the sake of generality. Nevertheless, the detailed model in [277] (posterior to all the work presented in this section) does consider binary spatial responses by means of a per-pixel mask in an oversampled spatial domain, which is intended to model the sensitive and blind areas of a pixel.

If CS is to be applied in spatial domain to achieve higher lateral resolution than the native lateral resolution of the sensor, knowledge on the pixel response function in spatial domain has to be integrated in the sensing model. While it is clear that the transition from a low resolution image to a high resolution image is, in general, an ill-posed problem, modeling the process of low-resolution image generation from a high-resolution (eventually infinite resolution) scene is a mandatory step before trying to invert the problem and recover the underlying high-resolution scene from few measurements. In the continuous case, the level at the pixel channel Γ of the pixel indexed by $\vec{x}$ at the end of the integration time for the PMD acquisition is given by Eq. 4.17, which is an extension of the integral in Eq. 2.11 to the spatial dimensions within the pixel. The subpixel spatial domain is accessed by the Cartesian vector $\vec{x}_{\text{sub}}$.

$$i^{\Gamma}(\vec{x}) = \int_{t_0}^{t_0+t_{\text{exp}}} \int_{\vec{x}_{\text{sub}} \in \Omega_{\vec{x}}^{\text{sub}}} p^{\Gamma}(\vec{x}_{\text{sub}}, t) q_{\Gamma}^{*}(t) r(t + \tau(\vec{x}_{\text{sub}})) d\vec{x}_{\text{sub}} dt \quad (4.17)$$

where $\Omega_{\vec{x}}^{\text{sub}}$ is the subset of $\mathbb{R}^2$ containing the subpixel spatial points ($\vec{x}_{\text{sub}}$) of the pixel indexed by $\vec{x}$. t_0 is an arbitrary initial time and $t_{\text{exp}} \gg \tau(\vec{x}_{\text{sub}}) \forall \vec{x}_{\text{sub}}$ is the exposure time. Regarding the three functions within the integrals, $q(t)$ and $r(t)$ are the reference signal and the reflected optical signal received by the pixel, respectively, implied in the original cross-correlation in Eq. 2.11, while $p^{\Gamma}(\vec{x}_{\text{sub}}, t)$ models the pixel response in spatio-temporal domain for the channel Γ. In a conventional camera, such response function is assumed to be a constant parameter, which does not vary in time or subpixel space. In other words, regardless of where or when a photon arrives to the pixel, it contributes in the same manner to the pixel level. Due to the structure of PMD pixels, it is expectable that $p^{\Gamma}(\vec{x}_{\text{sub}}, t)$ exhibits a remarkable dependency on $\vec{x}_{\text{sub}}$, different for each pixel channel. If the dependency of $p^{\Gamma}(\vec{x}_{\text{sub}}, t)$ on time is negligible, then we have that the term $p^{\Gamma}(\vec{x}_{\text{sub}}, t) \approx p^{\Gamma}(\vec{x}_{\text{sub}})$ can be extracted from the temporal integral in Eq. 4.17. Additionally, in a superresolution framework the aim is not recovering the image as a continuous function of infinite spatial bandwidth, but to increase the resolution in a certain factor, i. e., obtaining a certain number of subpixels per pixel. Consequently, the spatial integral can be substituted by a summation in a discrete formulation:

$$\begin{aligned} i^{\Gamma}(\vec{x}) &= \sum_{\vec{x}_{\text{sub}} \in \Omega_{\vec{x}}^{\text{sub}}} p_{\text{sub}}^{\Gamma}(\vec{x}_{\text{sub}}) \int_{t_0}^{t_0+t_{\text{exp}}} q_{\Gamma}^{*}(t) r(t + \tau(\vec{x}_{\text{sub}})) dt \\ &= \sum_{\vec{x}_{\text{sub}} \in \Omega_{\vec{x}}^{\text{sub}}} p_{\text{sub}}^{\Gamma}(\vec{x}_{\text{sub}}) i^{\Gamma}(\vec{x}_{\text{sub}}) \end{aligned} \quad (4.18)$$

where $p_{\text{sub}}^{\Gamma}(.)$ is the discrete equivalent of $p^{\Gamma}(.)$ and can be defined as the integral of $p^{\Gamma}(.)$ over the corresponding subpixel area. That is, $p_{\text{sub}}^{\Gamma}(\vec{x}_{\text{sub}})$ assigns a weight to the subpixel indexed by $\vec{x}_{\text{sub}}$, being $\vec{x}_{\text{sub}} \in [1, n_{\text{rows}}^{\text{sub}}] \times [1, n_{\text{cols}}^{\text{sub}}]$ now a discrete 2D subpixel index and $n_{\text{rows}}^{\text{sub}} \times n_{\text{cols}}^{\text{sub}}$ the size of the image patch to be recovered per pixel. In short terms, Eq. 4.18 shows that, in the discrete case and under the hypothesis of a time-independent pixel response, the readout of a pixel channel can be expressed as a weighted sum of the readouts of imaginary PMD subpixels, performing the cross-correlation between the signal received at the subpixel and the reference signal. Note that this is only valid if $p^{\Gamma}(.)$ does not depend on time. Otherwise, $p^{\Gamma}(.)$ could not be extracted from the temporal integral and this would be equivalent to having a different reference signal $q_{\Gamma}(t)$ per subpixel.

Our approach to obtain an accurate characterization of $p^{\Gamma}(\vec{x}_{\text{sub}}, t)$ is described in Section 4.3.1, together with results obtained from a real PMD 19k-S3 sensor [H1]. In Section 4.3.2, the discrete PMD pixel response functions obtained in Section 4.3.1 are used to model crosstalk effects and compensate them [H5].

4.3.1. Subpixel Spatial Response

In Section 2.4.2 the structure of the active areas of PMD pixels was briefly presented in order to give an estimate of the maximum lateral resolution of a PMD sensor. The schematic representation in Fig. 2.8b clearly shows that the sensitive area of PMD pixels is rectangular and not square as one could expect. Additionally, approximately half of the pixel is blind, concretely, the fill factor is 56%, according to the simple geometrical model in Fig. 2.8b. The colors in Fig. 2.8b denote rectangular areas centered at the A (red) and B (blue) readout wires. One would expect that the pixel response is stronger for channel A in red areas and for channel B in blue areas, but, in principle, if the PMD pixel operates ideally (constant response function), all the photogenerated carriers are stored either in the A or B channel, depending exclusively on the (binary) reference signal and not on the location of the photon arrival point. Prior to characterizing the response, in Section 4.3.1.1 we present images of a PMD array gathered with a microscope and use them to determine the real active areas of the PMD chip and provide an accurate value of the effective fill factor. The methodology is exposed in Section 4.3.1.2 and the results in Section 4.3.1.3. The same PMD chip, a 19k-S3 equal to the one in our MultiCam, is used in all experiments.

4.3.1.1. Initial Observation of a PMD Chip

Both in order to observe the structure of a PMD pixel array and to determine the spatial responses of the pixels, the daylight-blocking filter has to be removed from the chip. This is due to the fact that we do not have an appropriate NIR laser source in our lab. The filter removal is a delicate process that has to be carried out with care. Eventual direct contact with the pixels would destroy them. Even without direct contact, an electrostatic discharge would equally destroy the sensor, so measures against static electricity are to be observed. The filter is glued on top of the chip with epoxy resin. Removing it by mechanical methods like, e. g., milling, apart from generating vibrations, would surely cover the pixels with epoxy dust. The use of chemical agents to dilute the epoxy resin poses also a danger for the chip. The alternative we choose is to heat up the package close to its melting point and manually remove the filter applying pressure on the edges with a sharp object. The temperature should be between 80 and 100 °C. Our PMD chip is soldered in its respective frontend. Using higher temperatures may damage other electronic components.

In our case, the filter removal was not as smooth as desired due to the relatively low temperature, and microscopic particles of epoxy resin fell over the pixel array. Most of these dust particles were removed using compressed air, still few particles stayed there, in most cases out of the sensitive area of the pixels. Most PMD pixels remained clean of any particle. Immediately after the cleaning, the sensor was protected by a transparent glass of equivalent width to that of the filter. Images of the experiment are provided in Fig. 4.18. Dust particles are very rare, but they are clearly visible in fluorescence lifetime microscopy (FLIM) images. The FLIM image in Fig. 4.18c covers a $256 \times 256\,\mu$m area and was obtained using a 477 nm blue laser (Bluephoton® A350 with ultra-fast analogue modulation, from Omicron®) as illumination source. The image was generated using a lens of $NA = 0.4$ and $0.5\,\mu$m step size, yielding 512×512 pixels. Note the dust particle in the non-active area at the bottom-right of the image.

Apart from detecting intrusive particles, Fig. 4.18c shows the power of FLIM to determine the active areas of PMD pixels. To that end, a lens of $NA = 0.6$ is used to achieve better resolution in the FLIM images. For the given wavelength and $NA = 0.6$, Eq. 2.34 delivers a radius of the first Airy disc of 485 nm, being of 727 nm for the lens of $NA = 0.4$ used before. Fig. 4.19 provides two high-resolution FLIM images.

Despite Fig. 4.19a covers the same area as Fig. 4.18c, the structure of PMD pixels is better resolved in Fig. 4.19a. Fig. 4.19b covers a $64 \times 64\,\mu$m area with a step size of $0.25\,\mu$m. The image partially shows the active areas of four different pixels. The finger structure of the pixels is accurately recovered. The pink gaps between sensitive areas are the readout wires of the channels. Within each rectangular active area between wires, three *fingers* are visible. Those next to an A readout wire belong to an A gate, those next to a B readout wire belong to a B gate and those in between the previous are an intermediate area where the demodulation takes place, which we call mod. C areas. The two A gate areas at both sides of the A readout wire

(a) Daylight-Blocking Filter (b) PMD 19k-S3 without Filter (c) FLIM Image of the 19k-S3

Figure 4.18.: Images of the experiment for removing the daylight-blocking filter from the PMD chip of a 19k-S3 commercial frontend. The filter, after being removed, is depicted in (a). The filter removal took place with the chip soldered to the frontend. After it, a protective glass was glued to cover the sensor (b). Some micrometric particles fell on the sensitive area and are clearly observable in FLIM images, e. g., bottom-right of (c).

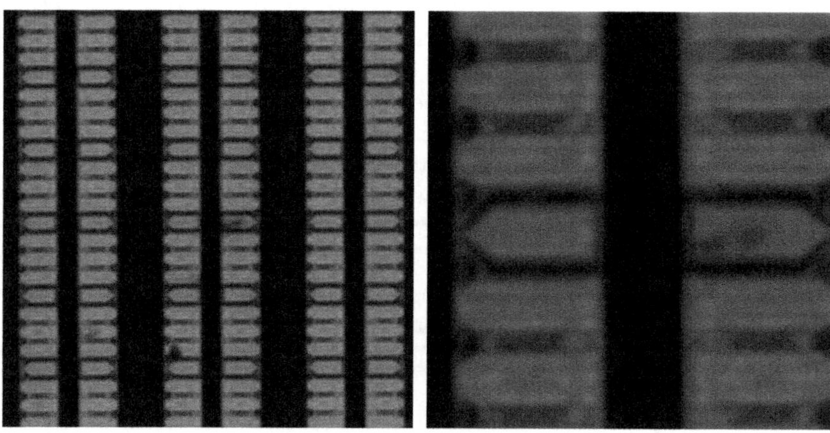

(a) FLIM: PMD 19k-S3 Array (b) FLIM: Detail

Figure 4.19.: Images of a PMD 19k-S3 obtained using fluorescence lifetime microscopy (FLIM). The daylight-blocking filter of the chip was substituted by a transparent glass. The numerical aperture of the lens was 0.6 and the illumination wavelength 477 nm.

plus the wire correspond to the red areas in Fig. 2.8b and equivalently for the B channel and the blue areas in Fig. 2.8b. It is clear that the black and pink areas in Fig. 4.19a and Fig. 4.19b are blind areas, i. e., only photons hitting the khaki areas contribute to the pixel measurements. In order to distinguish between active and blind areas, a binary mask is generated from Fig. 4.19a by appropriate thresholding. The result is shown in Fig. 4.20b, where the white areas are sensitive (silicon) and the black areas are blind (metal-shielded). A photographic image of the PMD array, obtained with a DSLR camera coupled to the microscope under diffuse illumination, is included for comparison in Fig. 4.20a. The vertical strips in Fig. 4.20a are metallic shields (blind regions in Fig. 4.20b).

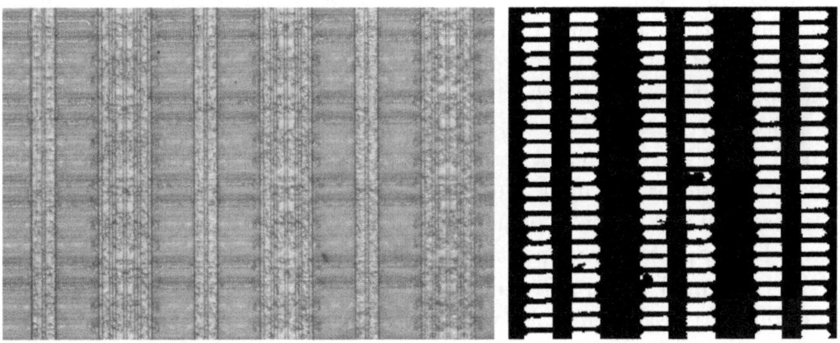

(a) Photographic Image **(b)** Active Areas

Figure 4.20.: Active areas of the pixels of a PMD 19k-S3 sensor (b),in white. This binary image was obtained by appropriate thresholding of the FLIM image in Fig. 4.19a. Note the large blind areas (in black) covering most of the sensor surface and leading to a poor fill factor of around 31%. The active areas cannot be so sharply segmented in a photographic image (a).

The black inclusions within the active areas in Fig. 4.20 are due to dust particles. The fill factor is computed from Fig. 4.20 as the ratio between white image pixels and total number of pixels, excluding those belonging to PMD pixels whose active areas are partially covered by some particle. We obtain a fill factor of $31.2531 \pm 0.0016\%$, being the ± 0.0016 the uncertainty due to the image resolution. It was observed that this uncertainty is negligible with respect to that originated by the threshold selection. Nevertheless, regardless from the threshold, the obtained fill factor was higher than 30%

and lower than 32%. This coincides with the 31% fill factor specified for the PMD 19k chips in [388]. The images in Fig. 4.19 and Fig. 4.20 clearly show the two type of PMD pixels already introduced in Fig. 2.8b. The two different pixels are grouped together two by two, forming symmetric sets of pixel columns. The geometry of the two types is the same, but mirrored with respect to each other. We refer the pixels with readout terminals at the left as *left* pixels and the pixels with readouts terminals at the right as *right* pixels. Obviously, different spatial responses will be obtained for each pixel type, in the ideal case, also mutually mirrored.

Using the high resolution color and FLIM images of the PMD chip, a simple geometrical model can be instantiated and used to predict the spatial responses of PMD pixels, e. g., supposing that photons hitting the gate area of one channel will contribute to the charge of that channel with higher probability than photons hitting the gate area of the other channel, for any given state of the reference signal. Note that this differs from the ideal behavior, according to which, that probability is the same regardless of where the photon arrives and is exclusively determined by the state of the reference signal. For illustration purposes, we have generated a simple theoretical pixel response for DC light, i. e., time-independent. This illustrative model supposes that the electrons generated in the A gate area contribute to the A channel with probability one, those generated in the B gate area contribute to the B channel with probability one, those generated in the mod. C areas contribute to both channels with equal probability and those eventually generated elsewhere do not contribute to any channel. The obtained 2D discrete responses are shown in Fig. 4.21 in a differential way (A-B), for both types of PMD pixels. The assignment of A and B areas is coherent with Fig. 2.8b and might not correspond to the real one.

The naïve responses in Fig. 4.21 show that the spatial dependence of the PMD pixel responses can be ignored up to a SR factor of 2 (Fig. 4.21a), since it is constant due to structural symmetry, but comes into play for higher SR factors. Fig. 4.21d shows the theoretical response at the maximum spatial resolution allowed by the microscopy images.

4.3.1.2. Methodology

The main goal of this section is to provide a method to confirm or reject our hypothesis about spatial dependency of the PMD pixel response and, in positive case, characterize it as accurately as possible. Additionally, the dependency of the spatial response with the phase delay is to be studied,

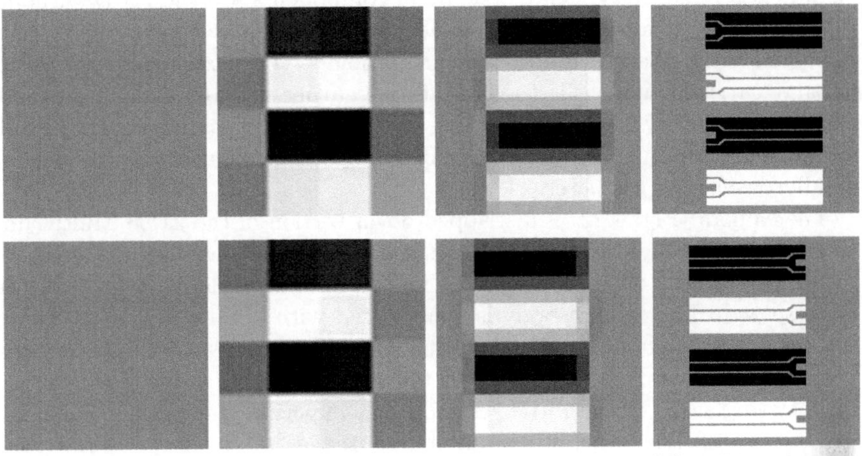

(a) SR Factor: 2 (b) SR Factor: 4 (c) SR Factor: 16 (d) Full Resolution

Figure 4.21.: Theoretical differential (A-B) PMD pixel responses in spatial domain for DC illumination. Photons arriving to the A gates are expected to contribute to the A response (maximum positive response, in white), while photons hitting the B gates are expected to contribute to the B response (maximum negative response, in black). Images in the first row correspond to *left* pixels and in the second row to *right* pixels. The superresolution (SR) factors considered are 2 (a), 4 (b) and 16 (c). Last column (d) shows the full-resolution theoretical response. ©2014 IEEE.

since it will unveil eventual divergences with respect to the expected behavior and point the limits of the PMD technology.

We suppose that the superposition principle holds in spatial domain for the A and B pixel responses, i. e., that the sum of the responses generated by several photons hitting the pixel in different positions is the same as the response generated by all those photons arriving simultaneously to the pixel surface at the same locations. With this in mind, we want to gather the values of A and B channels when illuminating the pixel with focused light at a certain point. If we can focus the light accurately enough and we repeat the experiment for an extensive grid of points that covers the entire pixel area, we will get, after normalization, an empirical discrete response function for the A and B pixel channels. We design an experimental setup that allows performing such experiment and keeping control on the illumination signal shape and phase delay. The main components are the PMD without daylight blocking filter, a function generator, a laser source and a microscope equipped with a stepper unit for the probe holder. This experiment is similar to that carried out in [11] to characterize earlier versions of PMD pixels. Nevertheless, they did not cover the entire pixel area or neighboring pixel areas. Additionally, we obtain twice better spatial resolution by means of a smaller laser spot diameter.

The camera (Fig. 4.22) is a stripped-down version of the ZESS MultiCam, featuring the PMD sensor from which the daylight blocking filter was removed in the experiment of Section 4.3.1.1. The original backend of the MultiCam is preserved. The camera body is constructed with aluminum pieces over a microscope probe holder. The mounting does not include any optical element and, instead, a wide aperture allows the microscope objective reaching the protective glass over the PMD surface. This experimental camera was used in Section 4.3.1.1 to acquire preliminary images of the PMD chip. The function generator is an Agilent® 81150A, which allows a maximum pulse frequency of 120 MHz and is used for signal conditioning and as accurate phase delay generator. The illumination control signal of the camera is used as external trigger and the output signal as input of the laser controller. The laser beam is conducted to a Nikon® TE300 inverted microscope, equipped with an ELWD 40x objective. The laser source, microscope and objective were already used in the experiment described in Section 4.3.1.1. The experimental setup is depicted in Fig. 4.23a. Observe the camera, mounted on the stepper unit of the microscope, pointing downwards. The function generator, the laser source and the computer used as control unit are not visible in the picture. The DSLR camera attached to the microscope allows taking pictures of the chip surface, e. g., Fig. 4.20a.

The radius of the first Airy disc for the lens of $NA = 0.6$ is 485 nm, i. e., when the laser is perfectly focused, the diameter of the spot is approximately 1 μm. We adopt this as the step size in our experiments. The grid of points where to perform measurements is always rectangular, with equal step size in both directions. Fig. 4.23b was acquired during a dummy-experiment run in order to provide a graphic illustration of the process. The image was gathered by the DSLR camera and the illumination combines an intense diffuse illumination with the focused laser spot. Even with the diffuse illumination, only a bright spot was visible in the original image and no pixel structure could be recognized. The image in Fig. 4.23b is the result of a gamma correction of the original, with $\gamma = 0.5$. The spot is focused at the border of one of the active areas of a pixel. The real experiments were carried out in the darkness, being the laser the only source

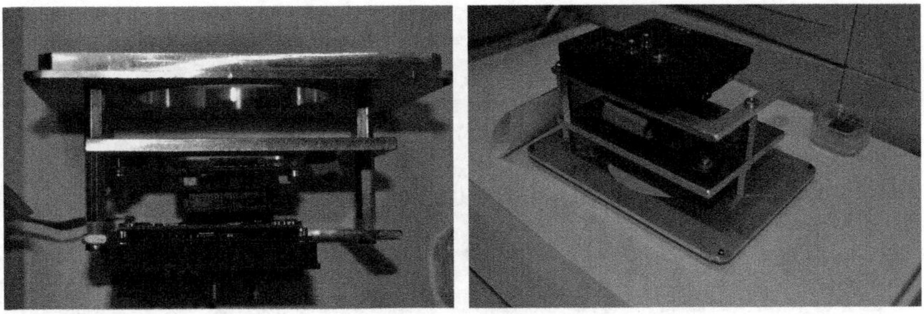

(a) Side View (b) Perspective View

Figure 4.22.: Prototype of PMD camera for use in pixel characterization experiments. The camera has no optical elements and exhibit an structure in three levels (a): the first level is the microscope probe holder, the second level contains the PMD frontend and the third level a MultiCam backend. The large aperture in the probe holder allows the microscope objectives reaching the second level. A wide rectangular hole in the aluminum plate that supports the frontend allows the objectives getting as close as desired to the protective glass, even for pixels at the borders of the array. The camera is to be mounted pointing downwards in the inverted microscope, as in (b).

of light, and keeping closed the optical paths corresponding to the DSLR camera and the binoculars. The function generator delivers a square wave at 20 MHz modulation frequency, triggered at each new period by the ICS of the MultiCam backend.

For each experiment, before starting the acquisition, the probe holder is displaced horizontally so that the laser spot is on the pixel located in the center of the grid. We call this pixel *pixel of interest*, whose outputs are recorded for the different positions of the laser spot. With the laser spot on the pixel, the microscope is focused by displacing the probe holder in vertical direction. Theoretically, a very good focus is possible thanks to the submicrometer resolution of the stepper. The program that runs in the computer and that controls all the hardware and collects the data during the experiments also includes a routine of autofocus, which tries to find the vertical displacement that maximizes the contrast between A and B channels of the pixel of interest. This works well when the spot is on the A or B gate areas, close to the horizontal borders of the region, but not when it

(a) Experimental Setup **(b)** Focused Light Spot on a PMD Pixel

Figure 4.23.: Full experimental setup used to acquire the spatial responses of PMD pixels. The camera prototype is mounted on the stepper unit of the inverted microscope (a). The PMD chip points downwards. The ICS from the MultiCam backend is used as trigger for the function generator that generates the signal to drive the laser. The camera, stepper unit and function generator are connected to a computer that acts as control unit. During the experiment, the laser is precisely focused (b) on points of a predefined grid covering the pixel surface and the outputs of the A and B channels at the end of the exposure time are recorded.

is centered in these areas, e. g., on a readout wire, on the mod. C area or in blind regions (recall these cover 69% of the pixel area). Consequently, the program includes a parameter that decides whether the focusing will be manual or automatic. The best results were obtained with manual focusing.

4.3.1.3. Results

The results presented in this section confirm the existence of a structured phase-dependent response function in spatial domain for the PMD pixels. Our best characterization has 1 μm spatial resolution in both directions, i. e., provided that the PMD pixel size is $45 \times 45\,\mu$m, it enables modeling a SR factor up to 45.

Wide Area with Saturation: 160×200 **Grid** A grid of 160×200 points is considered, leading to an area of $160 \times 200\,\mu$m, much larger than the pixel area itself. In this experiment, for the different positions of the laser spot, the outputs of the PMD pixel of interest and a frame of 10 pixels around it,

i. e., a 21 × 21 patch, are recorded, for comparison of the obtained responses. For each grid point, 38 cases are considered: DC illumination and modulated light with square waveform at 20 MHz with phase delays ranging from 0 to 360 °, both included, with a step of 10 °. For each one of the 38 cases, the raw images from the PMD sensor are averaged over 10 acquisitions to reduce noise of random nature. Eight raw images are produced per each case and position, corresponding to the A and B channels of the PMD pixels at each one of the four phases typically gathered by PMD sensors (0°, 90°, 180° and 270°). Therefore, 38 × 8 patches of size 21 × 21 are saved per each one of the 160 × 200 positions. If a 4-byte float is used to store each value, this experiment generates 15.98 Gb of data.

With the data, spatial and phase-spatial response functions can be obtained for all PMD pixels in the 21 × 21 patch. Here we present those obtained for the pixel of interest (in the center). In Fig. 4.24 graphical representations of the A, B and differential (A-B) responses for a phase delay of 0° are depicted by rows. We provide four images per case, corresponding to the four phases acquired by the PMD sensor.

As expected, for the A channel (first row of Fig. 4.24), the most intense response in spatial domain is obtained at 0° and the weakest response at 180°. Complementary, for the B channel (second row) we get maximum spatial sensitivity at 180° and minimum at 0°. The differential (A-B) response (last row) is of capital importance, since it contains the final correlation value for each phase. Observe the equivalence of the phases 90° and 270°. It is also clear that the sensitive areas contributing to the A response at 0° and to B at 180° are not the same, as one could expect, although they widely overlap. The differential responses are to be compared to the naïve theoretical response provided in Fig. 4.21d. Note that the theoretical responses were built supposing that the first gate of the pixel was of type A, while in the experiments it was found to be of type B. Consequently, black areas in Fig. 4.21d should correspond to red areas in the differential responses of Fig. 4.24 and white areas to blue.

Exploiting the information we have on the pixel responses in phase domain we can explore the behavior of the different key areas of the PMD pixels with respect to the phase shift of the incoming signal. We consider five different regions in the pixel: The A gate, the B gate, the readout wires of the A and B gates and the C neutral region between gates. Fig. 4.25 shows the response curves in phase domain for reference points taken in the center of these regions.

Note that, for each pixel channel (A or B), the response for points in the corresponding gate or gate wire is always over the response for points in the

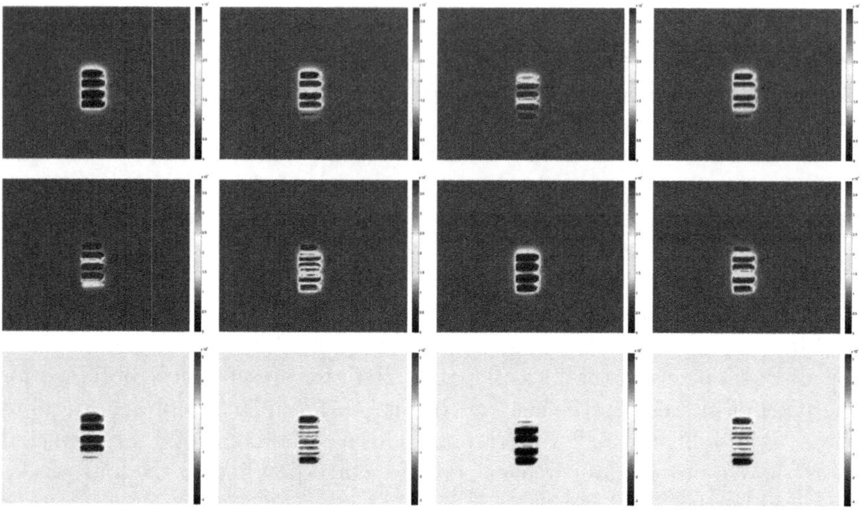

(a) Phase Shift: 0 ° (b) Phase Shift: 90 ° (c) Phase Shift: 180 ° (d) Phase Shift: 270 °

Figure 4.24.: Raw responses of a PMD pixel of type *right* in spatial domain for the four phases typically acquired by a PMD chip: 0° (a), 90° (b), 180° (c) and 270° (d). The images cover an area of 160 × 200 μm, with 1 μm step size. The first and second rows of images show the responses for the pixel channels A and B, respectively, while the third row provides the differential (A-B) response functions. The phase delay between ICS and received light signal is always zero. ©2014 IEEE.

corresponding areas of the other channel, i. e., the red and magenta lines (A gate and A gate wire, respectively) are always over the blue and cyan lines (B gate and B gate wire), respectively, in the plot in Fig. 4.25a, and inversely in the plot in Fig. 4.25b. This is of crucial importance, since it confirms the coupling between pixel structure and subpixel spatial response: the same light signal, with the same power, shape and phase shift produces a stronger A response if it falls within A gates than if it falls within B gates, and reciprocally for the B response. This does not mean any loss of linearity in the phase domain, in fact the curves look like a triangular function when operating far from saturation (observe the cyan and magenta curves in Fig. 4.25). In such cases, the subtraction of A and B responses also

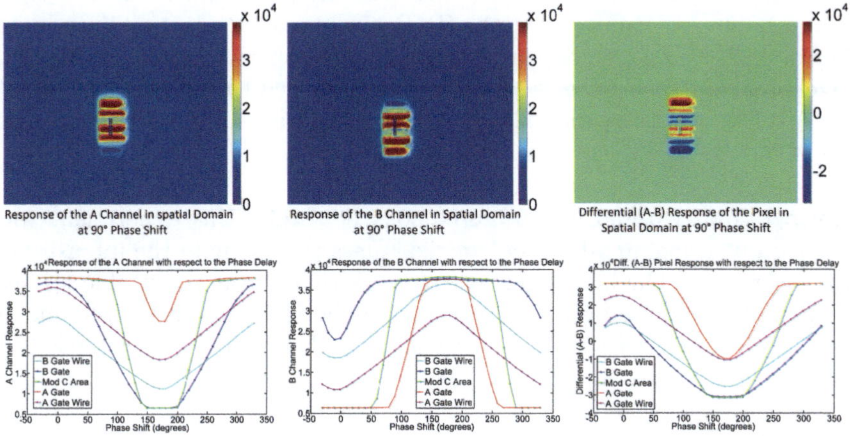

(a) A-Channel (b) B-Channel (c) Differential (A-B)

Figure 4.25.: Experimental PMD pixel responses obtained for A (a) and B (b) channels and differential (A-B) (c) responses in phase domain at relevant pixel points. The location of the points of interest is surrounded by a black square in the upper images, which are the corresponding responses in spatial domain at 90° phase shift (equivalent to DC response). The legend of the right plots corresponds to these points from top to bottom. ©2014 IEEE.

lead to a triangular differential response, i. e., a triangular cross-correlation. Taking into account that the light signal used in our experiments was square, a triangular cross-correlation function suggests that the reference signal is approximately square at 20 MHz modulation frequency.

Another effect that is observable in Fig. 4.25 is the SBI activation and its limit of operation, in those phase ranges where one channel is saturated, while the other stays at zero level. This is observable in the responses for the points on the wires (red and blue). For the A channel, the SBI critical area is around 0°, while for the B channel is around 180°. Operating symmetrically between channels, the SBI preserves the A-B difference (Fig. 4.25c) and, therefore, the correlation result, but it breaks the linearity of the A and B responses (Fig. 4.25a and Fig. 4.25b). Nevertheless, when the SBI is active and one of the pixel channels reaches zero level, the correlation (A-B) delivers wrong values, since no more charge can be removed from that channel. This is the case for the non-triangular areas in the plot of Fig. 4.25c.

Provided that the curves in Fig. 4.25 are close to triangular when no saturation occurs, we can say that the PMD pixel response function introduced in Eq. 4.17, $p^{\Gamma}(\vec{x}_{\text{sub}}, t)$, depends strongly on the spatial location within the pixel area $\vec{x}_{\text{sub}}$, being the response in time governed by the reference signal, $q_{\Gamma}(t)$. This confirms the validity of the superposition approximation in Eq. 4.18.

Narrow Area without Saturation: 50×50 **Grid** This experiment was carried out using an additional smoked glass to attenuate the intensity of the laser spot. The laser light was so weak that it was not visible for the human eye, but the optical power, concentrated in a tiny $1\,\mu$m-diameter spot, was enough to get high pixel readouts when pointing over the active areas, while avoiding saturation. The grid was reduced to 50×50 points, in order to speed up the experiment and reduce the volume of data to gather.

Also for time reasons, the phase step size was set to 20° and phase shifts from 0 to 360°, both included, were considered. The modulation frequency was kept at 20 MHz. As in the previous case, ten acquisitions were averaged per grid point.

Since the spatial responses look visually similar to those presented in Fig. 4.24, they are omitted for brevity. The new dataset, free of SBI activation, opens the possibility of generating an aggregate characterization of the pixel in phase space. The response functions can be integrated in spatial domain over the entire pixel area. This means computing an average response function for each pixel channel, exclusively in phase domain. The resulting curves, after offset compensation and amplitude normalization, are plotted in Fig. 4.26a. Note the clean triangular shape of the average responses, free of any sign of saturation (cf. Fig. 4.25).

We can go even further and use our discrete aggregate differential pixel response function to infer the real reference function the PMD pixels correlate the light input with. A method to determine this function is given in [304]. In that work, a (near-)Dirac impulse function is used as illumination signal at different phase shifts, sampling in this way the PMD reference signal. Note that this way, most of the integration time there is no incoming signal and the situation might be different from having a periodic signal with 50% duty cycle, in terms of carriers generation and diffusion. We use a square signal for driving the laser and, therefore, our average differential response is the result of the cross-correlation between that signal and the reference signal, which we must undo if we want to recover the latter. Signal recovery from correlations is a well-known topic in image recovery [400, 401], but, in our case, the knowledge of all values of the discrete correlation function

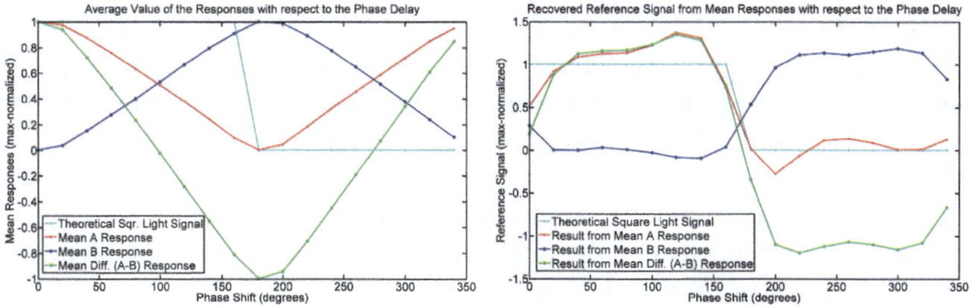

(a) Phase Responses

(b) Recovered Reference Signals

Figure 4.26.: Average experimental PMD pixel responses obtained for A and B channels and differential (A-B) responses in phase domain (a). Note the similarity between the average responses of the A channel (red) and the B channel (blue), which appear displaced 180° with respect to each other. The shape of the average differential response is almost a perfect triangular signal and gives the results of the cross-correlation between the light signal (theoretical, in cyan) and the reference signal, for each phase shift value. The empirical PMD reference signal is recovered from the average differential response and depicted in green in (b). ©2014 IEEE.

and the original light signal allows for a much simpler approach. Since both original signals and correlation result are periodic and discrete with equal phase step, it is possible to formulate the problem as the linear system in Eq. 4.19,

$$\begin{gathered}\vec{c} = \boldsymbol{R}\vec{q},\ \vec{q} = \left[q(\theta_k)\right]_{k\in[1,n]},\ \vec{c} = \left[c(\theta_k)\right]_{k\in[1,n]} \\ c(\theta_k) = \frac{1}{n}\sum_{i=1}^{n} r(\theta_i)q(\theta_i+\theta_k)\end{gathered} \tag{4.19}$$

where $\boldsymbol{R} \in \mathbb{R}^{n\times n}$ is a circular matrix, which first row is the discrete illumination signal $\vec{r} = \left[r(\theta_i)\right]_{i\in[1,n]}^{\top}$, $\vec{q}$ is the reference signal we want to recover and $\vec{c}$ is the vector of correlations (given by A-B), also sampled at n positions. $\theta_i = \frac{2\pi i}{n}$ and $\theta_k = \frac{2\pi k}{n}$ are just the discrete phases for each one of the n sampling points. Note that, by means of the circulant matrix $\boldsymbol{R}$, we consider displacements of the light signal, instead of shifting the reference signal, as in the conventional PMD operation. The system in Eq. 4.19 is solved numerically using the biconjugate gradients method [28] for all three average pixel responses in Fig. 4.26a. The corresponding recovered reference signals are plotted in Fig. 4.26b using the same line colors. The recovered curves are close to square signals, as the triangular shape of the aggregate responses suggested. The green curve in Fig. 4.26b is, therefore, the empirical reference function used by the PMD pixel. The relatively large phase shift step and the existence of irregularities in the triangular shape of the aggregate responses, together with the instability of deconvolution, visibly affect the quality of the recovered reference signals. Still, the trapezoidal shape is coherent with previous results presented in [304].

Provided that this dataset is free of saturation, the A and B spatial responses obtained at 0° or 180° phase shift between received light signal and reference signal at pixel level can be used to obtain a subpixel scale characterization of the demodulation contrast (Eq. 2.23). As one could expect, the demodulation contrast function in spatial domain looks similar to the differential (A-B) responses, i. e., it exhibits the same finger structure. We use the data obtained for 0° phase shift to compute the demodulation contrast and obtain similar results to those in [11], with average values around 50% within the sensitive area of the pixel. Differently from [11], we observe much greater oscillations, thanks to our superior spatial resolution.

Inter-Pixel Response: 200×200 **Grid** A natural question that arises from the previous experiments is how much area is to be considered to generate the discrete spatial responses of the PMD pixels. In principle, a $200 \times 200\,\mu$m region is larger than necessary, provided that the pixel size is only $45 \times 45\,\mu$m. Indeed, the spatial responses presented in Fig. 4.24 seem to have, after offset removal, a value close to zero for all grid locations outside the sensitive areas of the pixel. Here we present the results of an experiment destined to check the existence of unexpected mutual influences between pixels. In other words, we study the effect that photons arriving to the sensitive areas of pixels around the pixel of interest have, not just on the level of the channels of the pixel they hit, but also on the channels of the pixel of interest (in the center). To this end, a 200×200 grid is defined, centered at the narrow non-sensitive area between two contiguous pixels, in order to have two pixels of interest, one of type *left* and another of type *right*. The step size is kept at $1\,\mu$m. Differently from previous experiments, no phase-dependency was considered. The laser source was driven with a constant signal and with a square signal with a phase shift such that the displacement between light signal at pixel surface and reference signal was 90°. As expected, the same results were obtained in both cases and only the DC results are presented here. The obtained differential (A-B) spatial responses for both pixels considered are shown in Fig. 4.27, using two different scales to make the inter-pixel responses visible.

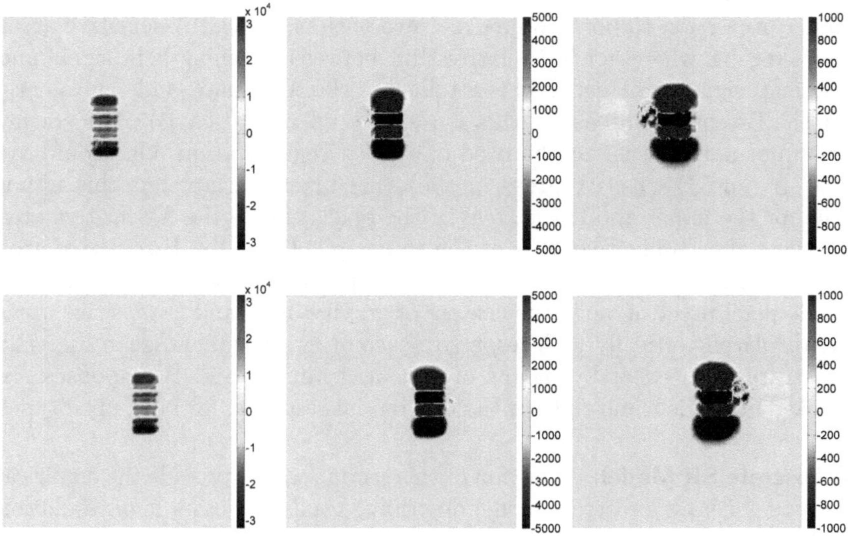

Figure 4.27.: Differential (A-B) spatial responses obtained for a pixel of type *left* (first row) and a pixel of type *right* (second row) using DC illumination. By columns, different scales have been used to make visible the inter-pixel response, i. e., how photons arriving to active areas of neighboring pixels affect the outputs of the pixel of interest. Note that only the scale used in the left plots cover the full response range.

Fig. 4.27 clearly shows that certain coupling between neighboring pixels exists and that this coupling happens in a structured way. For instance, the *left* pixel is coupled with the neighboring *right* pixel at its left, while the *right* pixel is coupled with the neighboring *left* pixel at its right. Also, it seems that no coupling exists between pixels of different rows. In other words, our spatial responses allow obtaining a characterization of crosstalk effects in PMD pixel arrays at subpixel scale. This finding is the motivation for our extensive crosstalk detection and characterization experiments, exposed in Section 4.3.2. In the right plots of Fig. 4.27 we observe that the crosstalk occurs mainly between A channels. Photons reaching the A-sensitive areas of the *coupled* pixel (e. g., the A gate) produce a stronger effect on the (A) response of the pixel of interest than photons reaching the B-sensitive areas (e. g., the B gate).

Sensu stricto there are no disjoint A- and B-sensitive areas, since, for every point in the sensitive areas, photons arriving to it will contribute in some proportion to both channels. Nevertheless, we call A-sensitive areas those regions where, for 90° relative shift between incoming light signal and reference signal, a higher level is attained in the A channel, and equivalently for the B-sensitive areas. In this sense, the differential (A-B) plots convey this information without the need of a strict segmentation. One could say that A- and B-sensitive areas of the same pixel overlap, but this is not true for the upper and lower areas of the pixels, where the A-sensitive area overlaps the B-sensitive area of the upper pixel and the B-sensitive area overlaps the A-sensitive area of the lower pixel. This means that, from a single-pixel point of view, the centers of gravity of A- and B-sensitive areas are displaced vertically with respect to each other. The montage in Fig. 4.28 illustrates this effect by means of a pentagram. A and B responses are stacked at the left and right of the differential response, for both pixel types.

A Discrete SR Model: The aim of this section was to provide the empirical data to generate a discrete model describing the pass from a high resolution scene to a low resolution image. Once the time-invariance of the response functions have been checked (i. e., approximately triangular shape of the cross-correlation function for all grid points in absence of saturation), the response functions obtained for DC illumination can be used to generate the discrete functions $p_{\text{sub}}^{\Gamma}(\vec{x}_{\text{sub}})$ introduced in Eq. 4.18. For simplicity, we keep the notation $p_{\text{sub}}^{\Gamma}(\vec{x}_{\text{sub}})$, but, since there are two pixel types with different geometry and different responses, four different basic response functions are to be generated for a PMD pixel. Then, the set $\Gamma \in \{A, B\}$ needs to be extended to $\Gamma \in \{A_{\text{left}}, A_{\text{right}}, B_{\text{left}}, B_{\text{right}}\}$. We generate the four discrete response patches, for any desired SR factor lower than 45, from the DC result of the 200×200 grid experiment. In practice, the full 200×200 is not always considered, since most values are close to zero after offset removal and normalization.

In Eq. 4.18 it was supposed that the output of the pixel channels was only determined by photons arriving to that pixel and not affected by photons arriving to neighboring pixels, in which case $\vec{x}_{\text{sub}} \in \left[1, n_{\text{rows}}^{\text{sub}}\right] \times \left[1, n_{\text{cols}}^{\text{sub}}\right]$, being $\left[1, n_{\text{rows}}^{\text{sub}}\right] \times \left[1, n_{\text{cols}}^{\text{sub}}\right]$ the size of the image patch to be recovered per pixel in a SR framework. Provided that mutual influences between pixels exist and it might be desirable to take them into account implicitly in the

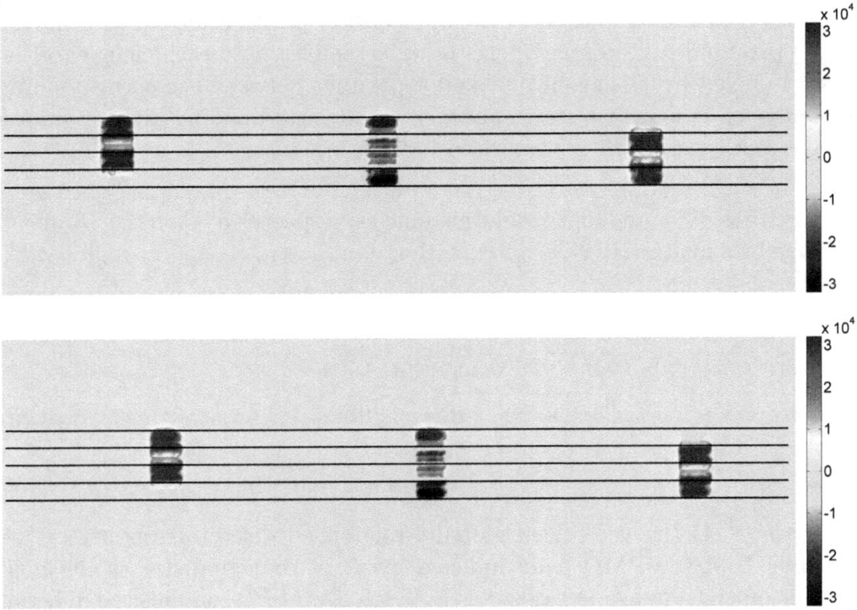

Figure 4.28.: Montage of spatial responses to illustrate the vertical displacement between A- and B-sensitive areas, for *left* and *right* pixels (first and second row, respectively). A and B responses are stacked at the left and right of the differential response (middle). All responses were obtained using DC illumination and 90° relative phase shift between incoming light signal and reference signal. The pentagram is intended to help the reader to establish correspondences between A, B and differential (A-B) responses.

discrete SR model, we change the coordinates center of the discrete responses $p_{\text{sub}}^{\Gamma}(\vec{x}_{\text{sub}})$ from local to global and extend the domain accordingly to the full high resolution image to recover, i. e., $n_{\text{rows}}^{\text{sub}} = k_{\text{SR}} n_{\text{rows}}$, $n_{\text{cols}}^{\text{sub}} = k_{\text{SR}} n_{\text{cols}}$, where k_{SR} denotes the desired SR factor. The globally-referenced responses are clearly different for each PMD pixel, i. e., they are different for each $\vec{x}$, and are, therefore, denoted by $p_{\text{sub}}^{\vec{x},\Gamma}(\vec{x}_{\text{sub}})$. This also allows us coming back to the original use of Γ, exclusively denoting the pixel channel, i. e., $\Gamma \in \{A, B\}$, since a single pixel, indexed by $\vec{x}$, is either of type *left* or *right*, being this completely determined by the architecture of the PMD array.

One can think of the globally-referenced discrete responses, $p_{\text{sub}}^{\vec{x},\Gamma}(\vec{x}_{\text{sub}})$, as an array of zeros, where the locally-referenced (and low-size) response

$p^{\Gamma}_{\text{sub}}(\vec{x}_{\text{sub}})$ has been located in the appropriate position, obtained from the PMD pixel index vector $\vec{x}$. Then, the value delivered by each channel of the pixel indexed by $\vec{x}$ is given by the dot product between the corresponding globally-referenced discrete response and the corresponding high resolution image, both conveniently vectorized in the same manner. If we denote with $\vec{i}^{\Gamma}_{\text{LR}} = \left[i^{\Gamma}_{\text{LR}}(\vec{x})\right]_{\vec{x}\in[1,n_{\text{rows}}]\times[1,n_{\text{cols}}]}$ and $\vec{i}^{\Gamma}_{\text{HR}} = \left[i^{\Gamma}_{\text{HR}}(\vec{x}_{\text{sub}})\right]_{\vec{x}_{\text{sub}}\in[1,n^{\text{sub}}_{\text{rows}}]\times\left[1,n^{\text{sub}}_{\text{cols}}\right]}$ the vectorized low and high resolution images, respectively, then the SR model is given by a matrix $\boldsymbol{P}^{\Gamma} \in \mathbb{R}^{n\times n_{\text{sub}}}$, with $n = n_{\text{rows}}\times n_{\text{cols}}$, $n_{\text{sub}} = n^{\text{sub}}_{\text{rows}}\times n^{\text{sub}}_{\text{cols}}$, as:

$$
\begin{aligned}
i^{\Gamma}_{\text{LR}}(\vec{x}) &= \langle \vec{p}^{\,\vec{x},\Gamma}_{\text{sub}}, \vec{i}^{\Gamma}_{\text{HR}} \rangle \\
\vec{i}^{\Gamma}_{\text{LR}} &= \boldsymbol{P}^{\Gamma}\vec{i}^{\Gamma}_{\text{HR}} \text{ , being:} \\
\boldsymbol{P}^{\Gamma} &= \left[\vec{p}^{\,\vec{x},\Gamma}_{\text{sub}}\right]^{\top}_{\vec{x}\in[1,n_{\text{rows}}]\times[1,n_{\text{cols}}]}
\end{aligned}
\tag{4.20}
$$

where $\vec{p}^{\,\vec{x},\Gamma}_{\text{sub}}$ is the vectorized globally-referenced discrete response of the channel Γ of the PMD pixel indexed by $\vec{x}$, of the same size as the high resolution (HR) image to recover, $\vec{i}^{\Gamma}_{\text{HR}}$. Each row of $\boldsymbol{P}^{\Gamma}$ contains the different $\vec{p}^{\,\vec{x},\Gamma}_{\text{sub}}$ for each one of the pixels of the low resolution (LR) image, $\vec{i}^{\Gamma}_{\text{LR}}$, indexed by $\vec{x}$. The vectorization schema used to construct $\boldsymbol{P}^{\Gamma}$ has to be the same used to vectorize $\vec{i}^{\Gamma}_{\text{LR}}$. Each discrete response is zero-offset and normalized.

Generating the model matrices $\boldsymbol{P}^{\Gamma}$ explicitly requires large amounts of memory, even for relatively low k_{SR}. For instance, the size each $\boldsymbol{P}^{\Gamma}$ for a PMD 19k-S3 array and $k_{\text{SR}} = 3$ is $19,200 \times 172,800$, i. e., more than 12 Gb if 4-bits floats are used. Consequently, if the matrix needs to be explicitly generated, a sparse matrix representation should be used. In our case, even when using the characterization of the full $200 \times 200\,\mu$m region obtained from the experiment, only 3 Mb would suffice to store that matrix, since the $200 \times 200\,\mu$m area would be characterized by a patch of only 13×13 pixels.

In order to show the appearance of the final discrete responses avoiding excessively large images, we consider a patch of 3×4 PMD pixels and generate the twelve discrete functions $\vec{p}^{\,\vec{x},\Gamma}_{\text{sub}}$ for each of the two pixel channels. We use a SR factor $k_{\text{SR}} = 9$. In Fig. 4.29 we provide the 2D representation, i. e., non-vectorized, of the obtained $\vec{p}^{\,\vec{x},\Gamma}_{\text{sub}}$ for $\vec{x} \in \{(2,2),(2,3)\}$, $\Gamma \in \{A, B\}$.

4.3.2. Crosstalk Characterization

The complex two-channel structure of PMD pixels, together with the existence of *in pixel* circuitry (e. g., the SBI system), induces to think that

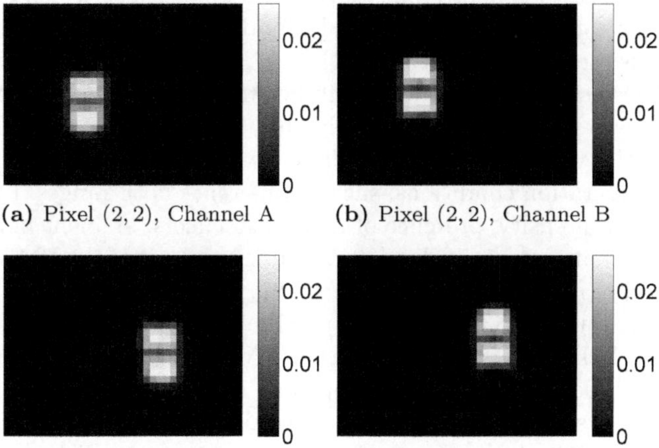

(a) Pixel (2, 2), Channel A (b) Pixel (2, 2), Channel B

(c) Pixel (2, 3), Channel A (d) Pixel (2, 3), Channel B

Figure 4.29.: Normalized discrete PMD pixel responses obtained for both channels (A and B) of both pixel types using a SR factor of 9. The responses are intended to model the pass from a high resolution to a low resolution image domain. For visibility reasons, the original size of the PMD images is considered to be only 3×4 pixels for the figure generation. The pixel (2, 2) (a),(b) is of type *right* and the pixel (2, 3) (c),(d) is of type *left*.

mutual influences between pixel channels of the same or neighboring pixels might occur. The readout wiring has to be as compact as possible, in order to keep an acceptable fill factor. This is another reason to expect systematic crosstalk effects between PMD pixels. The experiments presented in Section 4.3.1.3 showed the first evidences of a systematic crosstalk pattern between neighboring PMD pixels. A full and accurate characterization of crosstalk between PMD pixels for both pixel channels is of capital importance, not just for crosstalk compensation, but to generate a trustworthy model of the PMD sensing process. This characterization should not be limited to crosstalk with direct neighbors of the pixel of interest, but has to be extended to a wider neighborhood, as wide as necessary to cover all observable crosstalk effects. In this section, an experiment similar to those presented in Section 4.3.1.3 is carried out, with lower step size and covering a wider area. The methods and results are presented in Section 4.3.3. In Section 4.3.4 the subpixel scale crosstalk characterization obtained in Section 4.3.3 is used to generate pixel level crosstalk kernels, which can be used to remove crosstalk effects from real PMD data.

There exist several publicly available works providing a thorough analysis of the PMD technology, but crosstalk between PMD pixels seems to have been left away from them, probably due to the technically challenging experimental setup required for a reliable evaluation. A study on the effects of different operation conditions, such as different depth ranges, illumination waveform and intensity, presence of multipath effects, ambient light, etc., on the PMD systems is contained in [304]. A brief analysis of the PMD operation is also found in [372]. Additionally, PMD modeling and simulation have been the object of study in [368], but the pixel model does not include crosstalk. The subpixel characterization of earlier PMD pixels in [11], conceptually close to ours, does not consider inter-pixel influences. Basic concepts and evaluation of systematic errors and statistical uncertainties of a PMD system are given in [383]. A deeper statistical analysis of the ToF measurement process is given in [335], where realistic noise models for amplitude, phase and intensity allow predicting errors in a PMD camera. Still, crosstalk is not considered as an error source in any of these works. Very complete models [398], where even the SBI is taken into account, neglect the effect of pixel crosstalk. Therefore, to the best of our knowledge, we are the first providing a reliable characterization of PMD pixel crosstalk.

4.3.3. PMD Crosstalk Characterization at Subpixel Level

The method we propose to detect and characterize the crosstalk between PMD pixels is that already presented in Section 4.3.1.3. Also the experimental setup is the same and will not be described again here. A difference with respect to the experiments presented in Section 4.3.1.3 is that we come back to the original orientation of PMD pixels supposed in the simple model of Fig. 2.8b (B gate up, A gate down) and used to generate the naïve responses in Fig. 4.21. In the experiment presented in this section the PMD array is oriented in the right way to match this hypothesis. In order to acquire data that reveals crosstalk effects, we gather the values of A and B channels of a *pixel of interest* when illuminating with focused DC light at different points within a large area around the pixel. Ideally, when illuminating the sensitive area of the pixel of interest, a non-zero response will be observed in A or B channels, but no response should be observed when illuminating out of the sensitive area of the pixel, unless crosstalk between pixels exists. In presence of crosstalk phenomena, non-zero responses might be obtained when the light spot is on the sensitive area of other—probably neighboring—pixels.

In our experiments we ensure that the laser spot is focused, with a diameter close to 1 μm. The step size of the grids is set to 2 μm in both directions. Several grid sizes were used, in order to look for eventual long-range crosstalk, but here we only present the results obtained for a grid of 250×250 positions, i. e., $500 \times 500\,\mu$m. Provided that the pixel size is $45 \times 45\,\mu$m, we study the crosstalk within a 10×10 pixels area, which was found to be large enough to cover all crosstalk phenomena.

A first step, which was also necessary in the experiments presented in Section 4.3.1.3 but omitted there for brevity, is to find the exact location of the pixel of interest in the obtained raw responses. Of course, the starting and end position of the grid are known, but its reference system is relative, and its position with respect to an arbitrary point of the pixel of interest, e. g., the geometrical center of the pixel, is unknown. All we know is that this pixel will appear approximately in the center of the responses if the grid was designed to this end. In order to find the exact location of the pixel of interest we make use of our knowledge of the pixel geometry to generate naïve high resolution spatial responses for both pixel types (see Fig. 4.21d). Then the theoretical responses are downsampled to match the resolution of our empirical differential (A-B) spatial responses. For a step size of 2 μm, the size of a PMD pixel is only 22.5×22.5 subpixels. The exact locations of the pixel of interest are determined by finding the maximum of the 2D cross-correlation between the downsampled theoretical responses and the empirical responses. The theoretical responses match very well the empirical ones, specially in the differential (A-B) case and the obtained cross-correlation maximum is observed to be unique and at the expected location. Two pixels of interest are taken into consideration: one of *left* type and one of *right* type, since they happen to exhibit different crosstalk effects. Provided that the size of the PMD pixels is known, once the location of the pixels of interest is determined, the locations of neighboring pixels can be calculated and a theoretical geometric model matching the empirical response can be instantiated. A representation of such model obtained in the experiment is provided in Fig. 4.30. Instead simply marking the locations of the pixel centers, the theoretical differential responses are placed there where they should appear, for all pixels in the same row as the pixels of interest. Therefore, this plot can be interpreted as the differential spatial response function that would be obtained for any of the pixels of interest supposing full crosstalk between pixels of that row, i. e., all A sensitive areas contribute to all A pixel channels and all B areas to all B pixel channels.

Fig. 4.30 is intended to be a *roadmap* for the reader to understand the crosstalk results we obtain and determine which channels of which pixels

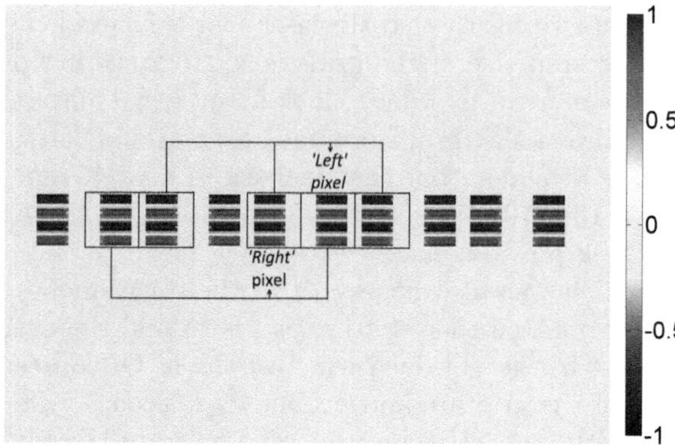

Figure 4.30.: Location of the sensitive areas of the pixels implied in the crosstalk experiment. The locations are determined by cross-correlation with the theoretical differential responses. The plot shows the real locations of the different sensitive areas, represented in a differential way (A-B).

have an influence in the responses of the pixels of interest. The black squares delimit the areas of the pixels exhibiting crosstalk with one or both pixels of interest. The raw crosstalk results obtained for the pixel of *left* type for both A and B channels are summarized in Fig. 4.31, while the results for the *right* pixel are given in Fig. 4.32. In the figures, additional plots with reduced range and a different colormap are used in order to make crosstalk visible, similarly to Fig. 4.27. All images show a partial $360 \times 500\,\mu$m area, since no crosstalk between rows was observed. The location of neighboring pixels producing crosstalk is indicated for each pixel type with black squares, horizontally centered at the pixel center, as in Fig. 4.30. Differently from Fig. 4.30, in Fig. 4.31 and Fig. 4.32 the squares are vertically centered at the center of the sensitive area of the corresponding channel (different for A and B).

Fig. 4.31b shows that the A channel of the PMD *left* pixels is affected by crosstalk with the A sensitive areas of three neighboring pixels: the two immediate neighboring pixels and the third pixel at the left (of type *right*) in the current orientation (B gate up, A gate down). Fig. 4.31d also confirms crosstalk with the third pixel at the left and the immediate neighbor at the right for the B channel, but shows almost negligible crosstalk with the left neighbor.

Fig. 4.32 clearly shows that *right* pixels are affected by crosstalk with the

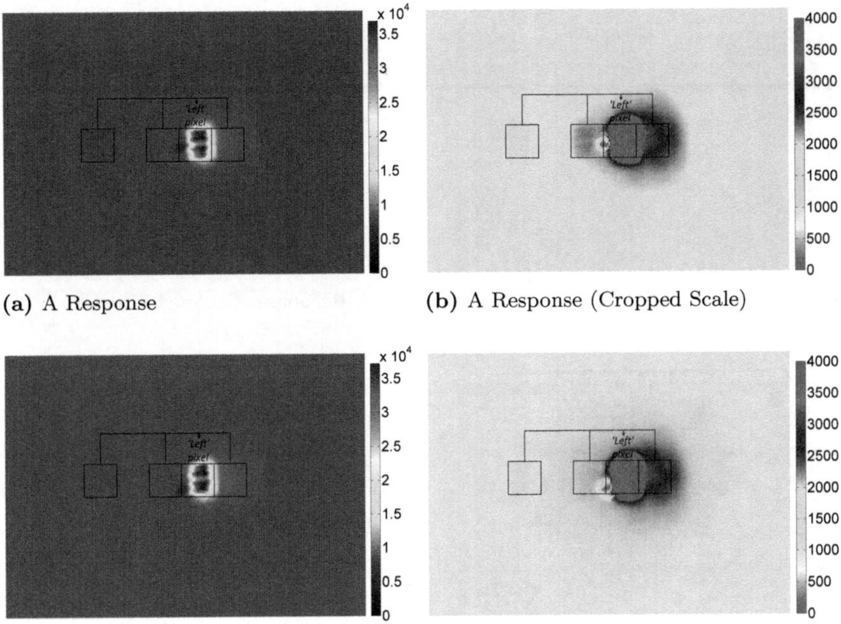

(a) A Response

(b) A Response (Cropped Scale)

(c) B Response

(d) B Response (Cropped Scale)

Figure 4.31.: Crosstalk for the *left* pixels. Each image pixel corresponds to a point on the chip surface. The color encodes the value obtained at the A or B channel of the pixel of interest of type *left* when the laser spot illuminates that point. Image pixels belonging to the sensitive areas of the pixel of interest depict its DC response function, while image pixels belonging to sensitive areas of other PMD pixels depict the subpixel crosstalk functions.

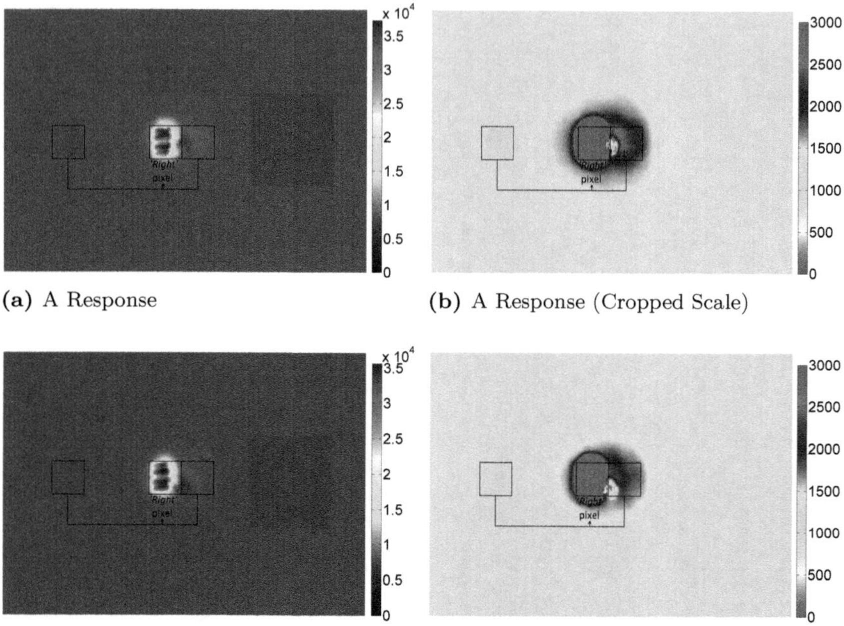

(a) A Response (b) A Response (Cropped Scale)

(c) B Response (d) B Response (Cropped Scale)

Figure 4.32.: Crosstalk for the *right* pixels. Each image pixel corresponds to a point on the chip surface. The color encodes the value obtained at the A or B channel of the pixel of interest of type *right* when the laser spot illuminates that point. Image pixels belonging to the sensitive areas of the pixel of interest depict its DC response function, while image pixels belonging to sensitive areas of other PMD pixels depict the subpixel crosstalk functions.

immediate neighboring pixel at the right (type *left*) and with the third pixel at the left (type *right*). This holds for both pixel channels, as can be derived from the similarity between Fig. 4.32b and Fig. 4.32d. Still, it is interesting to observe that the B-crosstalk with the immediate neighbor is visibly lower than the A-crosstalk.

It is remarkable that there is no inter-channel crosstalk. Crosstalk disturbance in one of the channels of a PMD pixel is produced by photons falling in the sensitive area of the same channel of specific pixels in its neighborhood. Taking profit of our knowledge of the location of the sensitive areas of each of the channels, we can integrate the cross-response over each sensitive area of each channel in order to obtain accurate average crosstalk values per pixel channel. We compute such mean cross-responses and compare them to the mean response over the sensitive area of the corresponding channel in the considered pixel itself. This way, we obtain a percentage of each mean crosstalk with respect to the mean response of each pixel channel. The results obtained for both pixel types are given in Table 4.4. Average crosstalk values ranging from −0.36% to 6.92% were obtained for *left* pixels and from −0.40% to 4.38% for *left* pixels.

	Crosstalk by Pixel Pairs (%)				
Channel	Left 1	Left 2	Left 3	Right 1	Right 2
A	-0.28	1.78	6.92	-0.40	4.38
B	-0.36	0.89	6.44	-0.23	2.92

Table 4.4.: Aggregate results of the crosstalk evaluation. Each column represents the combination of the reference pixel (*left* or *right*) and each one of the neighboring pixels producing crosstalk, numbered from left to right. First row gives the average crosstalk values for channel A and second row for channel B.

4.3.4. Crosstalk Kernels and Crosstalk Compensation

The subpixel-scale characterization of crosstalk effects between PMD pixels can be used to generate a simple but reliable crosstalk model at pixel level. That model would characterize the pass from a hypothetical crosstalk-free original image to the image delivered by the PMD chip, affected by crosstalk. As in the case of the SR model, modeling the direct transformation does not imply making possible to revert it, but constitutes a necessary foundation stone. The feasibility of an inversion depends both on the direct process itself and the characteristics of the data. For a better understanding consider that in the SR model the transfer function has low-pass filtering characteristics (a very real assumption) and does not exhibit any response for spatial frequencies higher than a certain cutoff frequency. Then, regardless of the SR factor for which we generate the discrete SR model, the maximum resolution we can recover is bounded by that cutoff frequency. Obviously, if the data we acquire lacks such high spatial frequencies, such limitations become irrelevant. Similarly, crosstalk is expected to attenuate the high spatial frequencies, since it couples neighboring pixels. Consequently, the effects of crosstalk are expected to become visible in areas with high-frequency texture, especially where strong gradients exist. Conversely, only in areas where some texture exists in the image affected by crosstalk there is hope for crosstalk removal. If low-contrast texture faded away due to strong crosstalk, it cannot be recovered by using an inverse crosstalk model.

Since the crosstalk only occurs with few neighboring pixels, low-size 2D kernels arise as a natural way to efficiently model the crosstalk effects. Four different kernels are required: for A and B channels in *left* and *right* pixels. As pointed out before, there is no crosstalk between rows, so only one row per kernel is needed. The number and relative location of non-zero kernel columns depend on the pixel type. The longest-range crosstalk is registered with pixels located three pixels away from the one considered, so an universal kernel size of 1×7 is enough to model all observed crosstalk effects. The aggregate crosstalk values in Table 4.4 are used to fill the non-zero elements of the kernels. The central pixel of each kernel is instantiated with 100% value and all kernels are l_1-normalized. A bar plot for easy visualization of the four kernels is provided in Fig. 4.33. The height of the bars is the logarithm of the absolute value of the crosstalk percentage. Using logarithmic scale makes visible all the non-zero elements of each kernel. Values for the longest-range crosstalk (third pixel left) are lower than 1% in absolute value and yield a negative logarithm.

These low-size and sparse normalized kernels allow formulating the crosstalk phenomena as a convolution. PMD pixels in the 19k-S3 chip are oriented inversely as in our response functions (as pointed out in Section 4.3.1.3, the real orientation is A channel up, B channel down), but the mirroring effect of the convolution over the kernels allows us using them directly. For clarity, we stick to the general definition of discrete convolution and overcome the difficulty of two pixel types with a general spatially-dependent kernel, as summarized in Eq. 4.21,

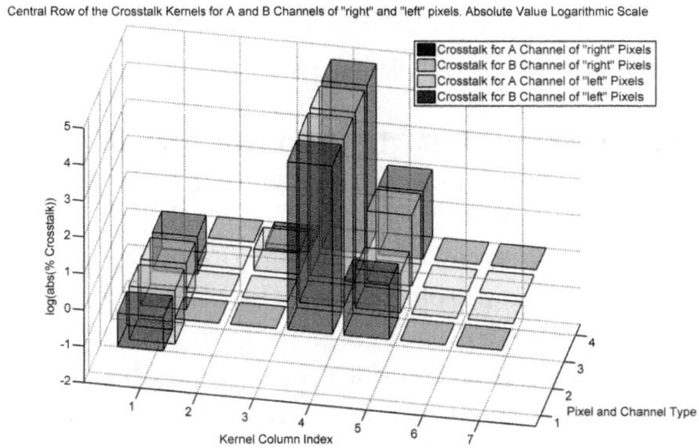

Figure 4.33.: Bar diagram representing the four PMD crosstalk kernels in logarithmic scale. The four kernels model the crosstalk for the A and B channels of *left* and *right* pixels (see plot legend). The height of each bar is the logarithm of the absolute value of the crosstalk percentages.

$$
\begin{aligned}
i^{\Gamma}(x,y) &= \sum_{\tau=1}^{n_{\text{cols}}} i^{\Gamma}_{\text{clean}}(x,\tau)k^{\Gamma}(y-\tau) \\
k^{\Gamma}(t) &= \begin{cases} k^{\Gamma}_{\text{left}}(t), & \text{if } \exists a \in \mathbb{N} \mid y = 2a \\ k^{\Gamma}_{\text{right}}(t), & \text{if } \exists a \in \mathbb{N} \mid y = 2a+1 \end{cases}
\end{aligned}
\tag{4.21}
$$

where $k^{\Gamma}_{\text{left}}, k^{\Gamma}_{\text{right}}$ are the crosstalk kernels for the channel $\Gamma \in \{A, B\}$ of the *left* or *right* pixel type, conveniently padded with zeros to reach n_{cols} length. The kernel changes according to whether the pixel column, y, is even or odd, following the alternating distribution of pixel types observed in Fig. 4.20. $i^{\Gamma}_{\text{clean}}(x,y)$ denotes the value of the pixel indexed by x, y of the hypothetical image that the chip would deliver in absence of crosstalk and $i^{\Gamma}(x,y)$ the value of the same pixel of the image we actually obtain, distorted by crosstalk, both images of size $n_{\text{rows}} \times n_{\text{cols}} = n$. Consequently, $x \in [1, n_{\text{rows}}]$ and $y \in [1, n_{\text{cols}}]$.

The challenge is to recover $i^{\Gamma}_{\text{clean}}(x,y)$ from $i^{\Gamma}(x,y)$, $\forall [x,y] \in [1, n_{\text{rows}}] \times [1, n_{\text{cols}}]$. We develop Eq. 4.21 as a system of n linear equations with n unknowns. In the case of border pixels, the image is padded with zeros, which is realistic, since there are no further pixels to have crosstalk with.

The resulting $n \times n$ matrix is invertible and the system can be solved either directly or using least squares or iterative methods.

Crosstalk Compensation Through Deconvolution in Real PMD Data Due to the nature of the kernels (a central maximum and a few non-zero elements close to it), the effect they produce is similar to a very soft low-pass filter, slightly smoothing areas of great contrast. The proposed crosstalk deconvolution aims to accurately recover the original high contrast of the scene in those areas. In ToF imaging, high contrast in the raw data is often related to sharp changes in depth, e. g., object borders. In such cases, crosstalk between neighboring pixels in the raw images leads to a worse depth estimation around sharp edges. In order to test our method in a challenging setup, we make use of Böhler stars [60]. The edges of the stars are regions of high contrast, both in intensity and depth domains. The ToF camera we use in our experiments is, as in previous experiments, the ZESS MultiCam, a multimodal sensor that provides fully-registrable depth and color images. Recall that the PMD sensor of the camera is a 19k-S3, the same for which the pixel responses and crosstalk effects have been characterized. In this experiment, a lens of 50 mm focal length is used to obtain a narrow field of view, covering the whole image with a Böhler star, even at several meters distance. The camera is placed at 3.5 m from a wall and the Böhler star panel at 2.6 m from the camera.

After applying the intensity calibration presented in Section 4.2.1, the eight raw images delivered by the PMD sensor (recall, two channels per image, four sequential acquisitions) are deconvolved separately with the appropriate kernels. Fig. 4.34 shows the raw image corresponding to the channel B at 0° phase shift, before (Fig. 4.34a) and after (Fig. 4.34b) applying our crosstalk-removal method. There is no visible difference between them and a difference image (Fig. 4.34c) is needed to observe the effect of the deconvolution. The effect is absolutely negligible for most pixels, where the difference is within the noise level of the sensor. As expected, non-negligible effects appear on the vertical edges of the stars, where a high contrast is registered due to a range gap of 0.9 m.

In order to evaluate the effect of the crosstalk compensation on the depth result, a depth image is computed from the raw data before and after crosstalk compensation, using the four phases algorithm, as usual. The two resulting depth images are shown in Fig. 4.35, together with a depth difference image. For most pixels, negligible variations in the millimeter or submillimeter range are registered, but, for some pixels lying on edges, variations up to 25 cm are found (observe Fig. 4.35c). Note that the maximum differences in depth are

(a) Raw Image Before Compensation (b) Raw Image After Compensation (c) Difference Image

Figure 4.34.: Effect of our crosstalk deconvolution on a raw image of the B channel at 0° phase shift. From left to right, original image, affected by crosstalk, recovered image, after crosstalk removal, and difference image, obtained subtracting the former from the latter.

obtained for gradients in horizontal direction and no variations are observed for gradients in vertical direction. This is coherent with the fact that there is no crosstalk between rows, only between columns.

(a) Depth Image Before Compensation (b) Depth Image After Compensation (c) Difference Depth Image

Figure 4.35.: Effect of our crosstalk removal method on the final PMD depth images. From left to right, depth image obtained from the original raw data, affected by crosstalk, depth image obtained from crosstalk-compensated data, and difference depth image, obtained subtracting the former from the latter. The 79th column is highlighted with a green rectangle around it, since it is an area of high depth contrast in horizontal direction, where crosstalk has non-negligible effects.

The existence of noticeable depth variations in the edges does not confirm that the crosstalk deconvolution leads to a better depth estimation. Provided that we know the real distances from the camera to the Böhler star panel and to the wall behind it, a comparison to ground truth (GT) is feasible. We select the pixels belonging to the 79th column of the image for the evaluation, highlighted with a green rectangle in Fig. 4.35. This column corresponds to the central vertical edge of the star. The first half of the edge belongs to a star hole, i. e., to the wall, and the second half to the panel. The depth estimation for this column is plotted in Fig. 4.36, where the results with and without crosstalk compensation are compared to GT.

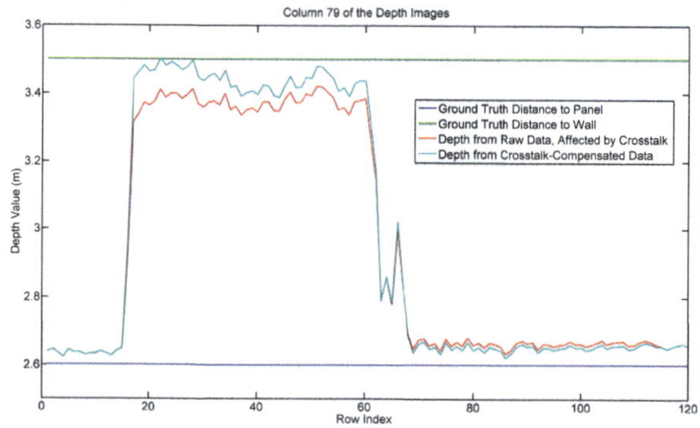

Figure 4.36.: Depth estimation obtained for the 79th column (green rectangle in Fig. 4.35), using raw data with and without crosstalk compensation. The depth obtained using original raw data, affected by crosstalk, is plotted in red, while the result achieved using our approach for crosstalk compensation is plotted in cyan. The ground truth distances to the Böhler star panel and to the wall behind it are depicted as blue and green lines, respectively.

The removal of crosstalk clearly pushes the depth estimate towards its real value. Depth RMSEs with respect to GT distances are computed separately for column pixels belonging to the panel and to the wall and the results are summarized in Table 4.5. Last column shows the percentage of depth RMSE reduction provided by our method. For the pixels belonging to the wall, 48.50% depth error reduction is achieved.

	Depth RMSE (cm)		−ΔRMSE (%)
Area	Non-Compensated	Compensated	Difference
Panel	5.70	4.94	13.07
Wall	12.88	6.64	48.50

Table 4.5.: Depth RMSE for the 79th column, obtained using non-compensated and compensated data. First row shows the result for pixels belonging to the Böhler star panel surface, while the results in the second row are for pixels belonging to the wall behind it. Last column shows the relative RMSE reduction.

4.4. Sparse Recovery in Spatial Domain

In this section we cover the application of CS exploiting the sparsity (or compressibility) of images in spatial domain. Spatial domain refers to the image projected on the sensor chip. Since the sensing area is two-dimensional, the spatial domain is, in principle, also 2D. Clearly, the pixels are responsible for the conversion from a continuous domain, whose resolution in both dimensions is only bounded by the PSF of the optical system (Eq. 2.34), to a discrete domain, where the total number of discrete points and, in consequence, the spatial resolution, is determined by the number of pixels in the sensor array.

In the case of PMD-based ToF imaging, the relatively low number of pixels, which is a direct consequence of the large pixel size, leads to poor lateral resolution in the depth images (Section 2.4.2). Modeling the transformation from a high resolution spatial domain to the low resolution pixel domain was the focus of Section 4.3.1.

The application of CS to conventional imaging started in 2D spatial domain, and was later extended to a third dimension (time-frequency, phase shift or depth, depending on the formulation). Probably the first milestone in the field of *compressive imaging* was the advent of the single pixel camera [156]. The original setup, replicated later in many related works, uses a DMD as SLM, in combination with a single photodetector. The image is first focused on the DMD, which superimposes a 2D sensing pattern. A lens focuses the light reflected by the DMD on the photodetector. Consequently, the measurements are the result of the scalar product between the (typically stochastic) sensing patterns and the image. The image can be accurately recovered from few measurements through l_1-minimization, provided that natural images are sparse in, e. g., wavelet domain. The maximum image resolution is limited by the native resolution of the DMD and, in real applications, by the allowable total exposure time, which bounds the maximum number of measurements.

The single pixel camera becomes attractive in cases of special pixels, e. g., those sensing non-conventional radiations, for which silicon-based pixels are blind. Often it is not possible to build large arrays of such detectors and the images are of poor resolution. In the following we introduce some recent works that apply the idea of [156] to ToF imaging. In general, a question to be considered is whether it is worth at all to adopt the single-pixel architecture when applying CS to ToF imaging. This is a fundamental choice between two opposite trends of applying CS: on one side, achieving the same quality with a simpler measurement technology or less measurements and, on the other side, achieving better quality preserving the same measurement technology or similar number of measurements. While the mainstream research tends to the former focus, we are more tempted by the latter, since our ultimate goal is to obtain depth images of better quality, e. g., in terms of depth accuracy and lateral resolution, while fully exploiting the capabilities of the given technology.

Related Work A compressive depth acquisition camera framework is proposed in [262, 263]. The authors use a single photodetector and perform measurements of the scene response to spatiotemporally-modulated illumination. The spatial modulation requires a 2D LCoS SLM, while the temporal modulation is achieved by means of a femtosecond laser. Additionally, a 20 GHz oscilloscope is needed to sample the scene response in time domain. This design is presented as an alternative to LIDAR or ToF cameras, but only depth maps of size 64×64 are recovered and no values on acquisition and recovery times are provided. Furthermore, hard restrictions on the scene geometry are imposed, such as piecewise planarity and negligible azimuth and altitude angle variations within an object, which make possible a sparse parametrization of the scene response in time domain. The restriction on angular variation, crucial for the model, poses a limitation on the opening angle of the camera or, complementary, on the ratio between size of an object and distance to the camera. In their experiments only textureless objects are considered and a generalization to textured objects would require a 2D image of the scene.

A more feasible implementation of a single pixel ToF camera is presented in [160], where a pure sinusoidal wave is used as illumination signal and a DMD allows using random binary patterns as measurement kernels in spatial domain. Simultaneous recovery of amplitude and phase of the reflected signal is achieved thanks to an appropriate complex representation of the sinusoidal light signal. The recovery method is a modified version of the model-based algorithm in [25] that includes a TV-regularization term in the signal estimation step.

In [115] and [230] a single photon counting detector is used as single pixel. The system and methods presented in [115] derive from those in [263], where the photodetector has been substituted by a single photon counting detector and the LCoS SLM by a reflective DMD of higher resolution. The depth recovery framework also comes from [263], inheriting its intrinsic limitations, e. g., restriction to fronto-parallel facets. Additionally, their approach is evaluated using a toy situation with only two depths. The system in [230] also uses pulsed laser illumination and a DMD to apply pseudorandom (PR) binary patterns as sensing kernels. The depth map recovery implies solving two l_1-minimizations, where the sparse vectors are supposed to share the same support. Surprisingly, the sparse vectors are recovered sequentially and not jointly in an MMV framework, as their joint-sparsity suggests. This requires blindly forcing the support of the second sparse signal to be the same as of the first. A hard thresholding on the sparse coefficients is performed after l_1-minimization, followed by a least squares estimation of the overdetermined system that results from considering only the dimensions with non-zero coefficients after thresholding. The use of one single detector comes at the price of longer acquisition times, e. g., seconds for toy situations (three planar objects) or minutes for natural scenes. Pulsed-laser approaches require either a photon counter or dense sampling of a photodetector output in time domain.

It looks more advisable to take profit of the state-of-the-art ToF imaging arrays, such as the PMD sensor or the Xbox One sensor, instead of creating an unnecessary bottleneck in the shape of a single detector. This seems to be the direction of the most recent approaches, which draw from light transport analysis and transient imaging, and aim to achieve robust separation of the individual light transport components (e. g., direct reflections, inter-reflections, caustics, etc.). This is a crucial point that was missing in previous work using a single photodetector [263, 115, 230], enforcing unrealistic requirements regarding the scene, such as restriction to planar facets with few possible depth values. These novel light-transport-inspired approaches [359, 209, 338] require simultaneous modulation of the projected light in temporal and spatial domain. The temporal modulation requires a laser light source, while the spatial modulation is carried out by an SLM, typically a DMD. The corresponding temporal demodulation is carried out using low-cost ToF sensors, such as the PMD 19k-S3 [359] or the Xbox One [338]. A complex formulation allows transferring the concept of light transport matrix from conventional imaging to ToF imaging in multifrequency approaches [359, 209]. The light separation between direct reflection and other components allows estimating the depth exclusively from the direct reflection, achieving immunity to diffuse MPI. For details on these methods, the reader is referred to the multipath part of Section 2.4.1.

In the following, we show that it is also possible to apply CS sensing to PMD sensors in spatial domain without requiring external SLMs or projectors. In Section 4.4.1 we remain faithful to the initial concept of compressive imaging and show that it is possible to recover PMD raw images from few PR measurements. This allows leveraging the volume of data to be transmitted, as in conventional compressive imaging. A more ambitious goal is to integrate the discrete spatial responses of PMD pixels obtained in Section 4.3.1 in the sensing model and try to cross from a low resolution *PMD pixel domain* to a higher resolution *scene domain*. In other words, use CS as a superresolution framework, taking profit of a realistic sensing model. Such a framework is presented in Section 4.4.2, together with encouraging results. In Section 4.4.3 we show that redundancy and joint sparsity of the raw images gathered by any phase-shift-based ToF imaging system can be exploited to reduce both the noise and the data flow, leading to more accurate depth estimation, even from reduced data. A *greedy* algorithm, based in ORMP (Section 3.4.2.4), is specifically developed to handle the joint sparsity and the structured sparsity in multiscale dictionaries, such as wavelet dictionaries, while *a priori* information on the redundancy of the raw data is integrated as a rank constraint.

4.4.1. Sparse Recovery from Few Measurements

In this section we consider the case of sparse recovery of PMD raw images from few measurements and study the effects of the representation matrix and the number of measurements on the quality of the final depth images. The m measurements are generated as different PR combinations of the pixel values, $m < n$, $n = n_{\text{rows}} \times n_{\text{cols}}$, being $n_{\text{rows}} \times n_{\text{cols}}$ the resolution of the PMD chip. In principle, it does not make sense to perform an *a posteriori* compressed sensing of the raw images for subsequent sparse recovery, since they have been already generated and all the pixel values are known. The process emulates a single pixel camera with an SLM to superimpose the PR codes on the image, where the virtual SLM has the resolution of the PMD chip. Provided that, in the case of exact recovery, one ends up with the initial set of raw images, the only advantage of such framework would be to reduce the data flow between the sensor and the processing unit where the depth images are computed. Nevertheless, if we think of the whole PMD chip (or an area of it containing several pixels) as a single pixel, where each of the actual pixels corresponds to pixels of an SLM placed before the chip, this evaluation constitutes a way to determine the eventual applicability of the concept of single pixel imaging to each of the pixels of a low resolution PMD array by means of an SLM of higher resolution.

There are some fundamental differences, which cause that the results obtained with this framework do not directly apply to the case of using a high resolution external SLM. Consider, for instance, the large blind areas in the PMD pixels reported in Section 4.3.1.1 and confirmed later by means of the spatial responses of the pixels. While the pixel-level CS would not be directly affected by this fact, using an SLM of higher resolution would imply that a large fraction of the SLM pixels would project the light onto non-sensitive areas of the PMD pixels. As a consequence, regardless of the PR code used as sensing kernel, the effect of those SLM pixels is the same as if they were always set to zero, i. e., they do not contribute to the measurements. Provided that the fill factor of the 19k-S3 was found to be around 31%, approximately 69% of the number of SLM pixels per PMD pixel would be affected by this issue.

4.4.1.1. An Appropriate Dictionary in Spatial Domain

Signals that are sparse in temporal domain can be represented using a dictionary of spikes, i. e., the identity matrix as representation matrix, while periodic signals admit a compressible representation, if not sparse, in frequency dictionaries, such as the DCT, Walsh or Fourier basis. Finding an optimal representation for images is a more challenging task. One could argue that most natural images are bandlimited and, therefore, can be sparsely represented in a frequency dictionary. Indeed, most natural images exhibit 2D Fourier decompositions whose coefficients decay with the frequency, being at least compressible in this domain. Nevertheless, image details are supported by high spatial frequencies, while being localized in a small region. That is, a good representation should combine frequency with localization. An example of such representation is the Windowed Fourier Transform (WFT), also known as Short-Time Fourier Transform (STFT), which is just a simple extension of the Fourier transform that includes a weighting function in the integral to restrict the temporal (or spatial) domain to a certain neighborhood of the point where it is applied. Building upon this concept, the wavelet transform was proposed. A function is called a wavelet if it can be used to define a Hilbert basis, i. e., a complete orthonormal basis, by means of dyadic translations and dilations. Two-dimensional wavelet bases are known to offer an efficient representation for natural images. In fact, the JPEG 2000 compression standard adopted wavelets to attain a sparser representation than that achievable with the DCT transform, implemented in the original JPEG standard.

From now on, we restrict our analysis to wavelet bases. Higher compressibility may be attained by means of dictionary training, at the cost of loosing generality of the representation. For a review on the possibilities for creating an appropriate dictionary, we refer to Section 3.3. The wavelet bases are named according to the wavelet functions used to create them. In practice, most wavelet bases offer similar performance in terms of approximation error for a given sparsity requirement when representing natural images. Natural images are, in most cases, compressible and not exactly sparse in wavelet dictionaries. A common way of quantifying how compressible a signal is in a given dictionary or, equivalently, comparing the performance of different dictionaries for a given signal or signal family is the so-called *best k-term approximation error* (Eq. 3.3). The best k-term approximation error is the minimum error that can be attained when a compressible signal $\vec{x} \in \mathbb{R}^n$ is approximated by only $k \leq n$ coefficients. If the dictionary is an orthonormal basis, thresholding yields the best k-term approximation of a compressible signal.

We consider the orthonormal bases derived from the following wavelets: Haar, Meyer and Daubechies-4, 6 and 8. For comparison, we also include the Fourier basis, which lacks the localization capability of the wavelet bases. Fig. 4.37 is intended to provide a comparison between the performance of the Fourier basis (Fig. 4.37a) and that of the different wavelet bases, for raw images of a PMD 19k-S3 sensor. There are eight raw images, corresponding to two channels per pixel, in four phases operation. The scene being sensed is the office environment already used for testing FPN removal by linear calibration of raw data (see Fig. 4.5d). Additionally, the plots in Fig. 4.37 show how compressible, or even sparse, the raw data generated by a PMD sensor is in spatial domain, if an appropriate representation is adopted. All images were calibrated using the method in Section 4.2.1 before the analysis.

In Fig. 4.37a, only the argument of the complex coefficients is plotted. The coefficient corresponding to zero frequency (DC) exhibits an outstandingly large value, but the rest of the coefficients show a relatively slow decay. Clearly, most of the signal energy is contained in few coefficients, but a great part of it is distributed all over the Fourier domain. While exact sparsity is not achieved, the Fourier representation yields an extremely compressible

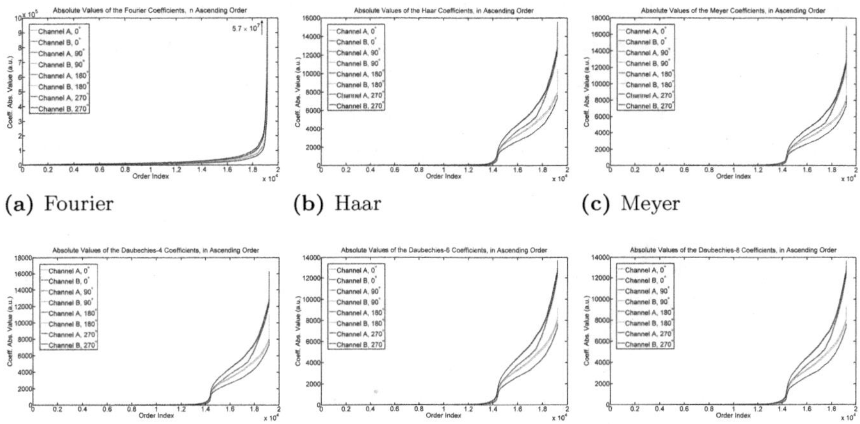

(a) Fourier **(b)** Haar **(c)** Meyer

(d) Daubechies-4 **(e)** Daubechies-6 **(f)** Daubechies-8

Figure 4.37.: PMD raw images represented using different orthonormal 2D bases. The plots show the absolute value of the corresponding coefficients, plotted in ascending order. The steepness of the decay is an indicator of compressibility. Curves with many zero elements correspond to exactly sparse representations. Compare the performance of the 2D Fourier basis (a) to that of wavelet bases.

signal. In contrast, the representations in wavelet bases are much more efficient, since they allow concentrating almost all signal power in a small number of coefficients. An ideal representation should concentrate all the area below the curves in a short coefficients range, at the right end of the plots. The choice of the wavelet used to generate the wavelet bases seems to have a negligible effect in the distribution of absolute value of the coefficients. Also, the three types of Daubechies wavelets considered, with 4, 6 and 8 vanishing moments, yield similar plots.

For further comparison, we evaluate the best k-term approximation error obtained for all PMD raw images when using each one of the dictionaries. The results are plots showing how the error decays with increasing k, depicted in Fig. 4.38. We compute the error as the l_2-norm of the difference between the original raw image and its best k-term approximation. Since all the dictionaries we consider are orthonormal basis, the best k-term approximation is obtained by simple thresholding, i. e., selecting the highest k coefficients in that representation. We denote by $\vec{i} \in \mathbb{R}^n$ the original image and by $\hat{\vec{i}} \in \mathbb{R}^n$ the approximated image, both conveniently vectorized. If we denote the dictionary by $\mathbf{\Psi}$, we have that $\vec{i} = \mathbf{\Psi}\vec{x}$ and $\hat{\vec{i}} = \mathbf{\Psi}\hat{\vec{x}}$, being $\vec{x}$ and $\hat{\vec{x}}$ the corresponding sparse or compressible representations of the images. The dictionary $\mathbf{\Psi}$ is constructed by columns using the same vectorization scheme as for $\vec{i}$ and $\hat{\vec{i}}$. Then the best k-term approximation is given by:

$$\begin{aligned} \sigma_k(\vec{x})_2 &= \min_{(\mathbf{\Psi}^{-1}\hat{\vec{i}})\in\Sigma_k} \|\vec{i} - \hat{\vec{i}}\|_2 = \min_{\hat{\vec{x}}\in\Sigma_k} \|\vec{x} - \hat{\vec{x}}\|_2 \\ &= \|\vec{x} - \hat{\vec{x}}\|_2 \mid \|\hat{\vec{x}}\|_0 = k,\ \hat{x}_i = x_i > x_j\ \forall i \in \Omega_{\hat{\vec{x}}},\ j \in \overline{\Omega}_{\hat{\vec{x}}} \end{aligned} \tag{4.22}$$

where Σ_k denotes the set of all k-sparse vectors of $\mathbb{R}^n$. The right hand side of the first line follows from the fact that $\mathbf{\Psi}$ is orthonormal and, consequently, defines an isometry. The second line is also due to orthonormality. The set $\Omega_{\hat{\vec{x}}}$ is the support of $\hat{\vec{x}}$ and $\|\hat{\vec{x}}\|_0 = |\operatorname{supp}(\hat{\vec{x}})|$.

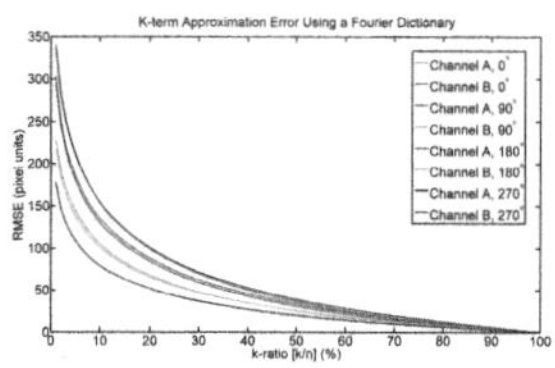

(a) Fourier

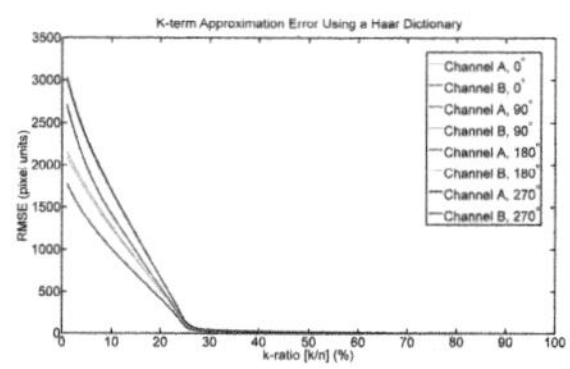

(b) Haar

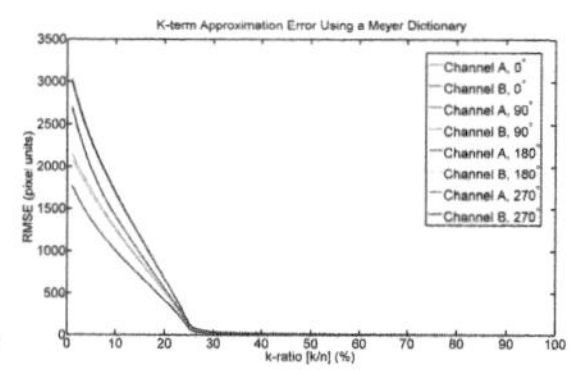

(c) Meyer

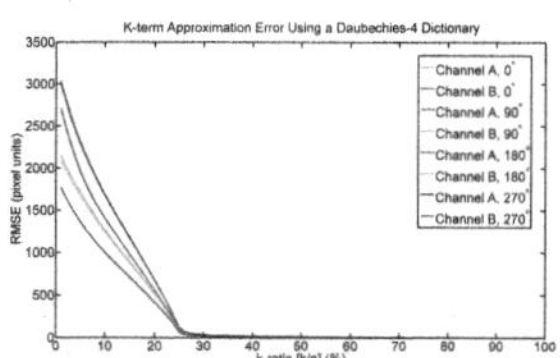

(d) Daubechies-4

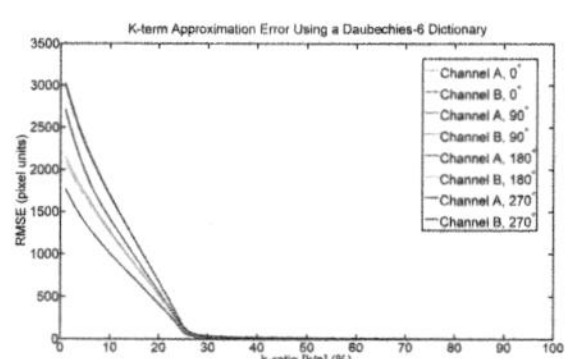

(e) Daubechies-6

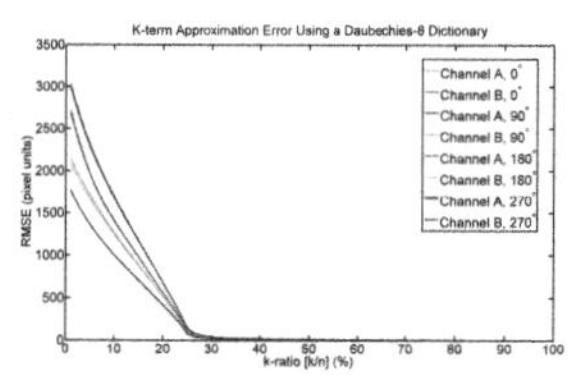

(f) Daubechies-8

Figure 4.38.: k-term approximation errors obtained for real PMD raw images using different orthonormal 2D bases. The raw images and dictionaries are the same as in Fig. 4.37. The abscissa is the ratio between k and the signal dimensionality $n = 19200$, in percentage. Note the fast decay obtained using wavelet bases, almost linear, and compare it to the slow decay obtained for the Fourier basis (a), close to an inverse function of k.

As one could expect from Fig. 4.37, the best k-term approximation error decays in a similar way for all wavelet bases. The curves obtained for the Fourier basis (Fig. 4.38a) exhibit a great negative slope for very low values of k (i. e., the highest coefficients retain a great part of the information), which tends to zero as k increases (that is, a non-negligible amount of information is spread over many low coefficients). The fact that we obtain equivalent results for all Daubechies bases reveals that a wavelet with 4 vanishing moments is sufficient to represent the structure in the images and considering wavelets with more vanishing moments does not reduce the approximation error. From Fig. 4.38, it is clear that, unless the signal is to be approximated with an abnormally low number of coefficients, in which case the Fourier representation yields the best approximation error, wavelet bases are the best choice to achieve approximately sparse representations of PMD raw images. Regarding the sparsity, the k-term approximation error starts to be negligible for k/n between 30% and 40%. Depending on the admissible reconstruction error, one could provide more optimistic sparsities, but a clear lower bound is found at approximately 25%, i. e., $s_{\min} \approx 0.25n$.

From now on, at the light of the results presented in Fig. 4.37 and Fig. 4.38, we select the Daubechies-4 orthonormal basis as dictionary or representation matrix in spatial domain for PMD raw data and we denote it as DB4 for brevity. The plots in Fig. 4.38 quantify the approximation error and give an upper bound for the reconstruction error that can be achieved via sparse recovery for a given sparsity. Nevertheless, the plots provide no graphic information on how the image degrades with decreasing k. For completeness, Fig. 4.39 shows the images obtained by thresholding the coefficients at different ratios k/n, using the Fourier basis and the DB4 basis.

Note how the Fourier-thresholded images (first row) degrade globally with decreasing k, due to the complete lack of spatial resolution of the Fourier basis. On the contrary, the DB4-thresholded images (second row) degrade locally with decreasing k, thanks to the tradeof between non-zero spatial resolution and reduced frequency resolution of wavelets. Fig. 4.39a corresponds to a case where k is too low even for the DB4 representation. While the thresholding in Fourier domain leads to an uniform degradation of the image, thresholding in DB4 domain preserves an acceptable image quality in well-illuminated areas, while the image is completely lost in dark areas. For $k \geq s_{\min} \approx 0.25n$, the quality of the DB4-thresholded images is visibly superior to that of the Fourier-thresholded ones and actually the k-term approximation can be considered exact for the former.

(a) $k = 0.2n$ (b) $k = 0.4n$ (c) $k = 0.6n$ (d) $k = 0.8n$ (e) $k = n$

Figure 4.39.: Images obtained from the best k-term approximation of a real PMD raw image using the Fourier (first row) and DB4 bases (second row).

4.4.1.2. Influence of the Number of Measurements in the Sparse Signal Recovery

Consider the classical CS formulation, where the measurement process follows a linear model (Eq. 3.8) and the sparse recovery is relaxed to the optimization given by Eq. 3.10. At this point we consider each of the raw images as an independent signal, which has to be recovered separately. Consequently, we simplify the notation leaving away that related to pixel channel and phase and denote by $\vec{i} \in \mathbb{R}^n$ a generic vectorized raw image to recover, which admits a sparse (or compressible) representation $\vec{x} \in \mathbb{R}^n$ in the dictionary $\mathbf{\Psi} \in \mathbb{R}^{n \times n}$, which is chosen to be the DB4 basis, i. e., $\vec{i} = \mathbf{\Psi}\vec{x}$.

The measurements are obtained from $\vec{i}$ as projections to a m-dimensional space using a sensing matrix $\mathbf{\Phi}$, i. e., $\vec{y} = \mathbf{\Phi}\vec{i}$. The sensing matrix is chosen to be binary, with elements of value ± 1. If the matrix elements are generated randomly, with equal probability for both occurrences, this binary matrix performs similarly to a Gaussian matrix for n high enough (compare, for instance, their inter-column coherence in Fig. 3.1). Instead of a purely random generation, we derive $\mathbf{\Phi}$ from a Hadamard matrix of size $n \times n$, carrying out random permutations of its columns and randomly selecting m rows out of the n possibilities (Section 3.2.3). The sensing kernels are the rows of $\mathbf{\Phi}$. The column scrambling avoids any possible coherence between the kernels (dense and random) and the dictionary elements (localized and structured). The measurement matrix $\boldsymbol{A}$ of Eq. 3.8 is $\boldsymbol{A} = \mathbf{\Phi}\mathbf{\Psi}$. An advantage of obtaining $\mathbf{\Phi}$ from a Hadamard matrix is that the different measurement kernels are orthogonal between each other, that is, the information gathered by each measurement kernel does not overlap that gathered by the others.

The quality of the recovered images depends on the number of measurements and the sparsity. The sparser the signal to recover is, the less measurements are needed to achieve equal reconstruction error, and vice versa. Upper bounds on the sparsity s have been proposed that depend directly on m and inversely on $\log(n/m)$ (e. g., Eq. 3.31 for Gaussian and binary measurement matrices). In other words, the required number of measurements to achieve exact reconstruction of exactly sparse signals in the noiseless case, or a certain error bound in the noisy case, depends directly on the signal sparsity and in a weaker way on the signal dimensionality. This is a fundamental concept of CS theory: the required number of measurements mostly depends on the information contained in the signal, and not on the dimensionality. In practice, this translates into the fact that, for a given signal, a threshold on m exists, such that higher values of m do not bring a significant reduction of the reconstruction error, while lower values lead to wrong recovery.

Provided that most PMD raw data from real scenes analyzed yield sparse representations similar to those depicted in Fig. 4.37, similar levels of approximate sparsity are to be expected for most scenes. Of course, unnaturally complex scenes, e. g., containing many artificial objects with high frequency texture or non-smooth depth profiles, may worsen the sparsity of the raw images. We evaluate the evolution of the reconstruction error with m for the raw data analyzed in Fig. 4.37 and Fig. 4.38, which was obtained from an office environment and contains several small objects on a table. Fig. 4.40a plots the reconstruction errors for all raw images against the ratio m/n, in percentage. The sparse recovery was carried out via l_1-minimization using the well-known l_1-magic library [89], which provides an implementation of the primal-dual interior point method for solving the convex problem in Eq. 3.10. At the light of the plots in Fig. 4.40a for the range $5\% \leq m/n \leq 55\%$, one could think that different raw images exhibit different compressibilities and, therefore, different RMSEs are obtained for a given m. In reality, differences in compressibility are negligible, which translates in a sharp decay of the RMSE between $55\% < m/n < 60\%$, common for all raw images. If different images exhibit different compressibilities, the location of this point would be different for each one. From Fig. 4.40a it seems that successful recovery is attained for $m/n > 55\% \pm 5\%$.

Despite the results in Fig. 4.40a are encouraging, they do not assure, by themselves, that a good depth image can be obtained from recovered raw data. In order to check this, depth images are computed from the sets of recovered raw images for different values of m. The depth RMSE is then calculated with respect to the depth image obtained from the original set of raw images. The results are plotted in Fig. 4.40b. The depth RMSE behaves as a monotonically-decreasing function of m, with a slope approaching to zero for $m/n > 55\% \pm 5\%$. The minimum depth error is approximately 2 cm, which is not too high, provided the typical depth accuracy of PMD-based sensors (cf. Fig. 2.3b). Note that this error is RMSE with respect to a reference, not the real error with respect to GT. The correct way of interpreting Fig. 4.40b is that, as soon as the RMSE is lower than the sensor accuracy, i. e., $m/n > 55\% \pm 5\%$, the depth image can be considered exact. In practice, the fact that the depth RMSE is not exactly zero in that range means a denoising effect that improves the depth estimation.

Another important question is to determine how far the *successful recovery* results are from the optimal results. To that end, we compute the ratio between the RMSEs of the recovered images and their corresponding best k-term approximation error, for $k/n \in \{0.2, 0.3, 0.4\}$. The resulting plots are provided in the second row of Fig. 4.40. When this ratio reaches the value of one (black horizontal line in the plots), the recovered image is as good as its best k-term approximation. Supposing that the 20% most important DB4 coefficients retain all the information in the raw images is an optimistic assumption and, consequently, the RMSE ratio in Fig. 4.40d reaches values far below one for m high enough. An approximate sparsity of 30% seems to be a more realistic assumption and the plots in Fig. 4.40e tend to one for m high enough. Adopting the best 40%-term approximation error as reference is a pessimistic scenario and, in fact, unit values cannot be achieved in the plots of Fig. 4.40f (minimum value of approximately 1.4).

Fig. 4.40c shows the average time required to recover each raw image by l_1-minimization, as a function of m/n. One would expect some close-to-exponential behavior, which seems to be the case up to $m/n = 65\%$. For higher m the runtime does not increase significantly. This may have several causes. The simulations were carried out in a multi-core machine, with a non-deterministic OS and with other programs and OS processes eventually running in parallel. One cannot predict issues such as how the OS is going to manage the large amounts of RAM required by the processes, which has an important effect in the runtime. The only conclusion to be drawn from Fig. 4.40c is that, due to the prohibitive increase in runtime with m, this has to be as low as possible, as far as it allows successful signal recovery, i. e., $m/n \geq 60\% \pm 5\%$ in this case.

Fig. 4.41 shows one of the recovered raw images for three different ratios $m/n \in \{0.4, 0.6, 0.8\}$ (first row) and the corresponding depth images from

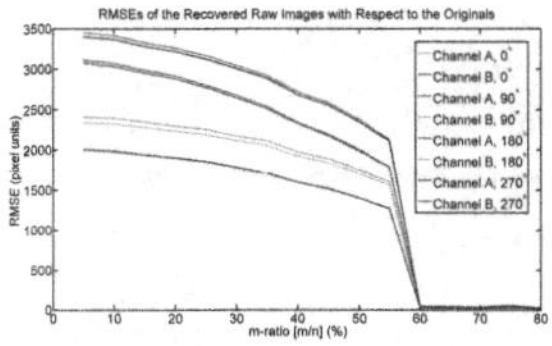

(a) RMSE

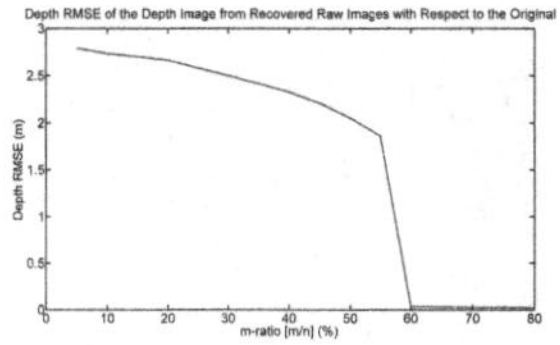

(b) Depth RMSE

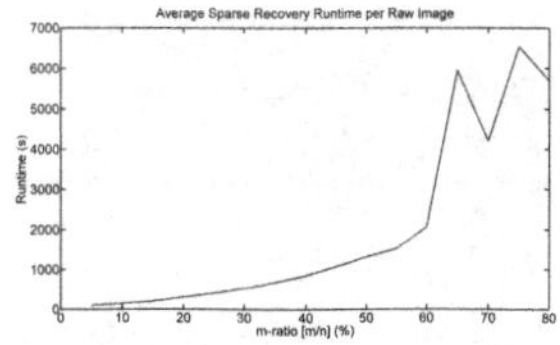

(c) Runtime

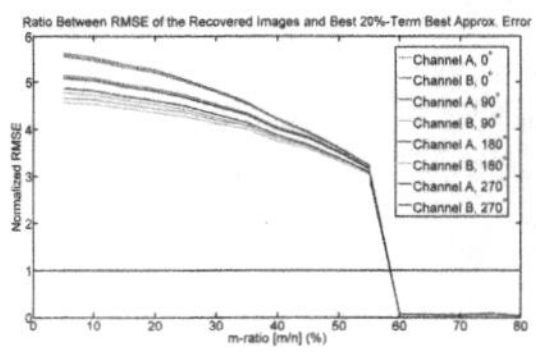

(d) Normalized RMSE $k = 0.2n$-term

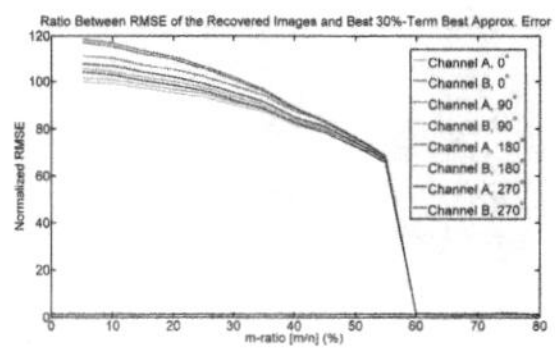

(e) Normalized RMSE $k = 0.3n$-term

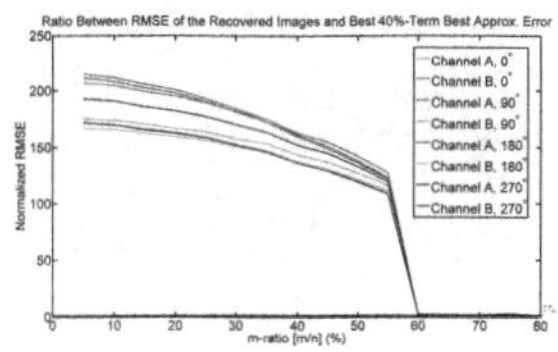

(f) Normalized RMSE $k = 0.4n$-term

Figure 4.40.: (a): RMSE of the sparse-recovered raw images with respect to the original ones for different ratios between number of measurements m and signal dimensionality $n = 19200$. (b): RMSE of the depth image obtained from the recovered raw images with respect to the depth image obtained from the original raw data. (c): average time needed to recover each raw image through l_1-minimization, against the ratio m/n. In order to provide an objective measurement of how good or bad the errors in (a) are, we normalize them by the best k-term approximation error for three values of k/n: 20% (d), 30% (e) and 40% (f), which can be understood as optimistic, realistic and pessimistic scenarios (cf. Fig. 4.38), respectively.

recovered data (second row). Clearly, $m = 0.4n$ measurements are not enough to achieve successful recovery (a). Both in the raw and depth images it is hard to distinguish any structure, apart from a dominant low depth at the left half and a dominant high depth at the right half. For the other two cases, the recovery can be considered successful and the scene is clearly recognizable. In fact there is not much difference between the images of Fig. 4.41b and Fig. 4.41c, apart from a light noise reduction.

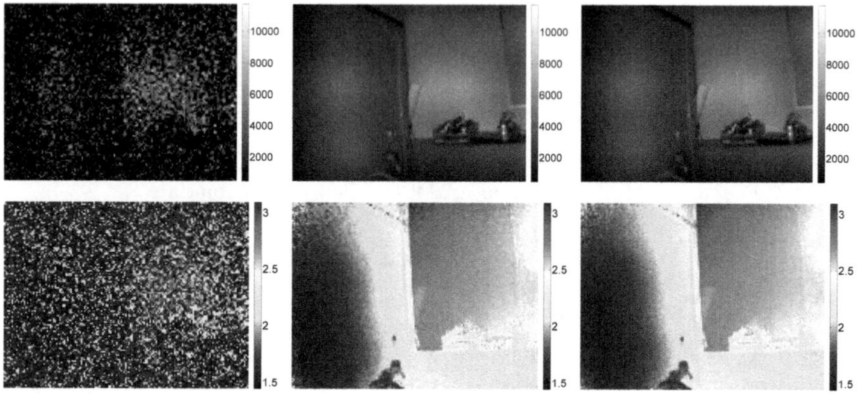

(a) $m = 0.4n$ (b) $m = 0.6n$ (c) $m = 0.8n$

Figure 4.41.: The images in the first row show one of the PMD raw images, recovered by l_1-minimization using different numbers of measurements (m). The corresponding depth images, obtained from recovered raw data, are given in the second row. The ratios m/n considered are 40% (a), 60% (b) and 80% (c).

4.4.2. Superresolution Using the Subpixel Spatial Responses

The results obtained in Section 4.4.1 showed that PMD raw data exhibit high compressibility and can be represented by approximately sparse signals in an appropriate basis, e. g., wavelet bases. As a consequence, the raw images can be recovered from few measurements via l_1-minimization in a CS framework. That is, the same information can be transferred using a lower volume of data. The next and much more interesting question that arises from the results is whether more information can be recovered from the same data, using a similar CS framework. Adopting this focus, one could think of the measurements of each pixel of the PMD chip as *compressed* measurements of a higher resolution signal and hope to recover that signal from the low resolution PMD images exploiting the underlying sparsity in an appropriate domain. This is of great interest, since the resolution of PMD sensors is very low (e. g., 120×160 for the 19k series, 200×200 for the 41k) when compared to color sensors.

While superresolution is an appealing idea, the success of the sparse recoveries in Section 4.4.1 does not guarantee that higher resolution images can be obtained from the original low resolution ones. The first step in this direction is to model the process that transforms a continuous signal in spatial domain—the image of the scene, as projected onto the sensor surface—in a set of pixel measurements. Modeling the spatial responses of PMD pixels was the motivation of Section 4.3.1, where discrete response functions with spatial resolution up to 1 μm are provided. Provided that the PMD pixel size of the 19k-S3 sensor, from which the responses were obtained, is $45 \times 45\,\mu$m, this allows for superresolution (SR) factors up to 45. Custom discrete SR models can be generated using different SR factors. The SR model transfers a high resolution vectorized image $\vec{i}^{\Gamma}_{\mathrm{HR}} \in \mathbb{R}^{n_{\mathrm{sub}}}$, to a low resolution image $\vec{i}^{\Gamma}_{\mathrm{LR}} \in \mathbb{R}^{n}$, where $n_{\mathrm{sub}} = n^{\mathrm{sub}}_{\mathrm{rows}} \times n^{\mathrm{sub}}_{\mathrm{cols}}$ is the dimensionality of the high resolution image domain. The ratio between dimensionalities is: $n_{\mathrm{sub}} = k^2_{\mathrm{SR}} n$, where k_{SR} denotes the SR factor. The transformation is linear and, therefore, modeled with a matrix $\boldsymbol{P}^{\Gamma} \in \mathbb{R}^{n \times n_{\mathrm{sub}}}$, where $\Gamma \in \{A, B\}$ denotes the PMD pixel channel. Each row of the matrix models how photons arriving to each of the fictitious subpixels affect to the level of a specific PMD pixel (see Section 4.3.1.3 for details on how $\boldsymbol{P}^{\Gamma}$ is constructed). This formulation naturally accounts for close-range crosstalk effects, which are inherently modeled in the discrete response functions used to generate the rows of $\boldsymbol{P}^{\Gamma}$. The neighborhood for which crosstalk is included in the model depends exclusively on the area for which the discrete pixel responses were computed.

The matrices $\boldsymbol{P}^{\Gamma}$, modeling the PMD pixel response for both pixel channels, are the key element of our SR formulation, but not the only one. Any imaging device has an optical system (a lens in the simplest case), which has, in turn, an associated point spread function (PSF), which acts over the irradiance function, convolving it in spatial domain and spreading any punctual signal in a certain way over the sensing array. Most optical systems, specially simple lens systems such as those in most color and ToF cameras, yield a PSF that follows an Airy pattern (Eq. 2.33), whose radius of the first Airy dark ring can be calculated from the numerical aperture of the system (Eq. 2.34). In practice, the pattern can be well-approximated by a Gaussian function. This brings the need for an optical model that complements the pixel model. Provided that the optical transformation is just a convolution, it can be modeled as a matrix product, where the model matrix, denoted by $\boldsymbol{O} \in \mathbb{R}^{n_{\mathrm{sub}} \times n_{\mathrm{sub}}}$, is a square matrix whose rows are vectorized versions of the conveniently-shifted 2D discrete PSF. $\boldsymbol{O}$ acts over the high-resolution signal $\vec{i}^{\Gamma}_{\mathrm{HR}} \in \mathbb{R}^{n_{\mathrm{sub}}}$ and does not modify its dimensionality, despite it has a certain low-pass filtering effect. In practice, even for high k_{SR}, the discrete

PSFs are close to a Dirac delta function and, consequently, $\boldsymbol{O}$ is close to the identity matrix. For instance, the 8.5 mm lens we typically use with our MultiCam has an F-number $f\# = 1.3$, which yields a diameter of the first Airy dark ring of 1.35 μm. Provided that PMD pixels measure 45 μm in both directions and that a PMD pixel widths is equal to k_{SR} subpixel widths, we have that $d_{\text{Airy}} = 3 \times 10^{-2} k_{\text{SR}}$ subpixels. In other words, using that lens, $\boldsymbol{O} \approx I_{n_{\text{sub}}}$, $\forall k_{\text{SR}} \leq 33$.

At this point, if we simply construct the sensing matrix as $\boldsymbol{\Phi} = \boldsymbol{PO}$, strong correlations will appear between the sensing kernels (rows of $\boldsymbol{\Phi}$) and the elements of any wavelet basis, since both are localized. As in Section 4.4.1.2, a random projector is needed to keep the coherence between the sensing kernels and the dictionary elements low. Let's denote this randomization or scrambling operator by $\boldsymbol{S} \in \mathbb{R}^{n \times n}$. Further dimensionality reduction is always possible by randomly selecting m rows of $\boldsymbol{S}$, yielding a matrix $\boldsymbol{S}_m \in \mathbb{R}^{m \times n}$, as in Section 4.5.1.1. Then the vectors of measurements would be obtained from the PMD raw images as $\vec{y}^{\Gamma} = \boldsymbol{S}_m^{\Gamma} \vec{i}_{\text{LR}}^{\Gamma}$. In principle it is not necessary to have an independent scrambling operator per channel and one can set $\boldsymbol{S}_m^{\Gamma} = \boldsymbol{S}_m$ $\Gamma \in \{A, B\}$. Then the sensing matrix is given by:

$$\boldsymbol{\Phi}^{\Gamma} = \boldsymbol{S}_m \boldsymbol{P}^{\Gamma} \boldsymbol{O} \tag{4.23}$$

If $\boldsymbol{S}_m$ in Eq. 4.23 is square, i. e., if $m = n$, the number of measurements is equal to the number of PMD pixels. The final measurement matrix is $\boldsymbol{A}^{\Gamma} = \boldsymbol{\Phi}^{\Gamma} \boldsymbol{\Psi}$, where $\boldsymbol{\Psi} \in \mathbb{R}^{n_{\text{sub}} \times n_{\text{sub}}}$ denotes here a high resolution dictionary, e. g., the DB4 orthonormal basis. This way we have a linear measurement model as in Eq. 3.8, which implicitly models the transition from high to low resolution domains. Consequently, recovering the sparse signals $\vec{x}_{\text{HR}}^{\Gamma} \in \mathbb{R}^{n_{\text{sub}}}$ means obtaining the superresolved images: $\vec{i}_{\text{HR}}^{\Gamma} = \boldsymbol{\Psi} \vec{x}_{\text{HR}}^{\Gamma}$. From Eq. 3.10, $\vec{x}_{\text{HR}}^{\Gamma}$ can be obtained solving a constrained l_1-minimization.

Unfortunately, due to the extensive blind areas present in the PMD chip (see Fig. 4.20b), depending on where the photons reach the pixel surface, they might produce no effect on the level of any pixel channel. This means that no sensing is carried out in these areas and it is not possible to determine values of a superresolved image for subpixels located in these areas. One could argue that interpolation techniques may be used to obtain an upscaled image, but upscaling differs from SR in that the resolution limit is not improved. Scene objects that are smaller than a pixel when projected on the pixel surface cannot be resolved by interpolation, while they should be resolved in a SR framework if the SR factor is large enough. From a CS perspective, the extensive blind areas yield a matrix $\boldsymbol{A}^{\Gamma}$ that does not satisfy the s-RIP (Eq. 3.24). This is easy to show: recall that the blind areas are distributed by columns, covering the entire sensor height and their total area is approximately 69% the total area of the pixel array. Consequently, without further assumptions, it is clear that approximately 69% of the high resolution dictionary elements are totally incoherent with all measurement kernels (dot product equal to zero), i. e., they are *invisible* for them. This is because these basis elements are spatially-localized in blind areas of the sensor. Provided that the sparsity of PMD raw images was shown to be around 30% of the signal dimensionality and supposing that this also holds for the superresolved images, non-zero s-sparse vectors can be easily constructed that violate the s-RIP, simply selecting coefficients corresponding to wavelets located in blind areas. Such vectors will be projected to the zero vector by the sensing matrix—i. e., lie in its null space—and the RIP does not hold.

There are two trivial ways of overcoming this issue without modifying the design of the PMD array: either displacing the camera with respect to the scene or displacing an object of interest with respect to the camera and gathering more than one image per raw image, in such way that there are no areas of the scene that are projected to blind areas of the sensor in all the displaced images. This seems to contradict the CS paradigm, since we are forced to gather more samples and, depending on the number of relative displacements we consider, one could reach the point of acquiring the same number of samples as the dimensionality of the HR signal to recover, n_{sub}, in which case the entire CS framework would be just an unnecessary artifact, since there is no real compression and one could trivially recover the HR signal from the measurements.

For the sake of simplicity and without loss of generality, we propose an experimental setup in which the scene is a sample that can be accurately positioned within the FOV of the camera. The main component of the experimental setup is the inverted microscope equipped with a stepper unit, already used in the experiments described in Section 4.3.1. The displacements are only in the image plane, i. e., they produce image shifts with subpixel resolution, but not scale changes. With such a setup, the easiest way to obtain a superresolved image is to define a regular grid of displacements, where the displacement step in both directions is the pixel pitch divided by the desired SR factor. The SR image is immediately obtained as composition of the individual displaced acquisitions.

A simple way of bringing CS into the game would be to substitute the regular grid of size $k_{\text{SR}} \times k_{\text{SR}}$ by an irregular grid of lower size where the locations of the grid points are selected at random within the pixel size. Due to practical considerations, we decide to acquire images for a regular grid of size $k_{\mathrm{SR}} \times k_{\mathrm{SR}}$ and perform the compression afterwards by means of the scrambling operator $\boldsymbol{S}_m$. Note that the effect of the sample displacements is to be modeled as shifts of the spatial response function of the pixels. Consequently we end up with $k_{\mathrm{SR}} \times k_{\mathrm{SR}}$ response functions per pixel, which are, like usually, displaced versions of a unique response function per channel and pixel type. This means that we synthetically increase the resolution of the PMD chip and $\boldsymbol{P}^{\Gamma}$ is no longer of size $n \times n_{\mathrm{sub}}$, but $n_{\mathrm{sub}} \times n_{\mathrm{sub}}$. If one has access to the full signal after projection by $\boldsymbol{P}^{\Gamma}$, the problem is solved and no method for recovering the original image is required. Nevertheless, we only have the measurements after further projection by the matrix $\boldsymbol{S}_m$, i. e., we only have $m < n_{\mathrm{sub}}$ measurements. The binary matrix $\boldsymbol{S}_m$ simply selects m different random combinations of PMD pixel values, regardless of whether the pixels are *real* (acquisition with $(0,0)$ displacement) or generated *synthetically* by non-zero 2D relative displacements of the sample with respect to the image plane.

Note that, if the grid of displacements is dense enough, performing the compression after sensing by means of $\boldsymbol{S}_m$ is equivalent to sensing in a compressed manner by only considering a reduced number of random displacements ($< k_{\mathrm{SR}} \times k_{\mathrm{SR}}$). An important consideration is that the sensing process has to ensure sufficient bandwidth in spatial domain, that is, enough to sense the highest spatial frequency we want to recover in the superresolved image. For this reason, the minimum size of the displacements grid is $k_{\mathrm{SR}} \times k_{\mathrm{SR}}$, which is the one we adopt in our experiments. Acquiring according to a regular grid and performing the compressed sensing later yields much shorter acquisition times and allows considering custom values of m without the need of acquiring new datasets.

Experimetal Results. An Application Case: Fluorescence Lifetime Microscopy The experimental setup with a microscope and a stepper unit to produce displacements in the sample being observed already suggests FLIM as a natural application where to test our CS-based SR methodology without hardware modifications. A brief introduction to the concept of FLIM is given in Appendix A.10, together with an explanation on why ToF cameras can be used as a low-cost alternative to gated cameras. One of the main drawbacks of using a ToF camera for FLIM is undoubtedly its low resolution in comparison to commercial gated CCD cameras. Consequently, our CS-based SR framework can be valuable in this area and pave the way for the consolidation of ToF as an appropriate technology for FLIM.

In our FLIM setup the sample does not show a challenging depth profile. In contrast, it is a cellulose layer, painted with two different fluorescent markers. Each color exhibits a different fluorescence decay time (see Appendix A.10), which translates into different phase shifts in the reflected modulated light and, as a result, different depths in the final depth image. Also the cellulose structure, which is hardly visible in an intensity image becomes distinguishable in the phase (or depth) image.

The sample is mounted on the stepper unit of the inverted microscope. See Fig. 4.23a for a view of the microscope with a DSLR camera in the camera port. In this case, we use a PMD-based camera instead. The camera is an assembled MultiCam featuring the PMD chip without daylight-blocking filter that is being tested in Fig. 4.23a. The MultiCam is attached to the camera port by means of a C-mount adapter. Without the filter, the PMD sensor can capture the visible wavelengths emitted by the fluorescent materials.

In that setup, 2D displacements of the stepper unit correspond to complementary displacements of the sensor, but not of the same magnitude, being necessary to rescale the desired displacements by the magnification of the optical system. We determine the effective magnification factor empirically with the following experiment. Using a sample with high contrast, four PMD acquisitions are carried out, for four grid points describing a square of $450\,\mu$m side. Provided that the PMD pixel pitch is $45\,\mu$m, the pixel displacements between PMD raw images acquired at the four corners are $10\times$ the magnification. The displacements are computed by finding the maximum of the 2D-cross-correlation between images. Provided that in all cases relative displacements of 80 pixels were observed, the system has an empirical magnification of $8 \pm 0.1\,\times$.

Several datasets were acquired with different samples and different displacement grids, but in this section we only present the results obtained for a cellulose sample using a 11×11 grid with $0.625\,\mu$m step size, i. e., $5\,\mu$m on the PMD chip. With *compressed* measurements of these data we can recover HR images with maximum $k_{\mathrm{SR}} = 9$. In practice, we only use 9×9 PMD raw acquisitions from the 11×11 available, since the full size of the 11×11 displacement grid corresponds to slightly more than one pixel size and is, therefore, redundant.

The final result of our CS-SR-FLIM experiments are not depth images, as in conventional ToF imaging, but phase images, which are obtained from the superresolved raw images using Eq. 2.6 with $N = 4$ phases. The lifetime image can then be calculated from the phase shift image using Eq. A.12. Obviously, the distance between the sample and the measurement device produces that the measured phase shift is a superposition of depth-related and fluorescence-related shifts [44] and lifetime retrieval requires either calibration or a more intrincated method (see Appendix A.10). Provided that in our microscopy setup the depth-related shift produces just a constant offset in the phase shift image that does not degrade its quality, we consider the problem solved as soon as the phase image is obtained.

First, in order to provide an idea of the best possible reconstruction attainable via the proposed framework, we carry out a first reconstruction with unit compression ratio, that is, with a number of pseudorandom measurements that is equal to the dimensionality of the signals we aim to recover. This yields a determined system, for which the least squares solution is the optimal solution. Consequently, the solution can be obtained in a closed form via the Moore-Penrose pseudoinverse of the measurement matrix $\boldsymbol{A}^{\Gamma}$, which is different for each pixel channel Γ. The phase image computed from superresolved raw data obtained this way is given in Fig. 4.42c. This image is to be taken as reference, in the sense that further results obtained for the underdetermined case, i. e., for compression ratios $m/n < 1$, should be compared to it. The pseudorandom projector $\boldsymbol{S}_m$ is chosen to be a Hadamard-derived matrix with ± 1 elements, the SR factor $k_{\mathrm{SR}} = 9$ and the dictionary is the orthonormal DB4 basis. Fig. 4.42c is to be compared, in turn, to Fig. 4.42b, which is the low resolution version of the same ROI. The ROI is a 30×40 pixels area centered in the middle of the original image, as highlighted in Fig. 4.42a, which shows the original phase image obtained from a single PMD acquisition by conventional means, without any processing.

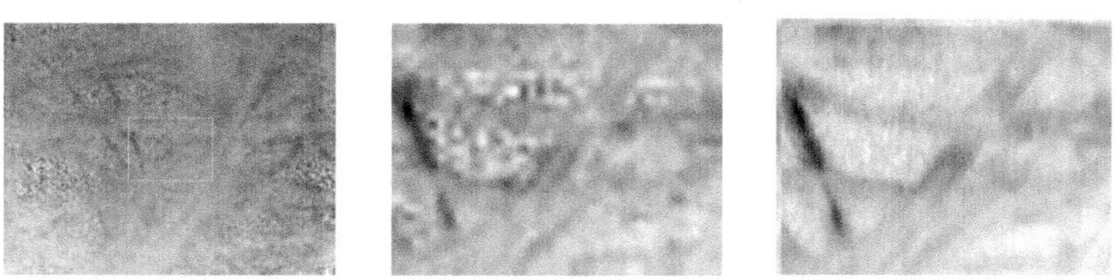

(a) Original Resol. **(b)** ROI Detail **(c)** SR ROI

Figure 4.42.: Comparison between the original phase image, obtained from a conventional PMD acquisition, and the least squares solution obtained from our CS-based superresolution framework, for a SR factor $k_{\mathrm{SR}} = 9$. The superresolved image (c) covers only the ROI highlighted in (a), which is reproduced in (b) to ease the comparison. The color scales have been adapted to each case and, consequently, differ from each other.

Fig. 4.42 clearly shows the gain in resolution, from a 30×40 pixels ROI to 270×360 superresolved pixels, which makes distinguishable the finest cellulose fibers, which cannot be resolved in Fig. 4.42a. For a compression ratio of one, we can reconstruct an image with 19.2×10^3 times more pixels than using a single time-resolved detector, just due to the multichannel acquisition in a PMD array. Additionally, if the images are compressible in some basis or frame—as natural images are—, higher compression ratios at sensing can be used, increasing the ratio between number of pixels of the HR recovered image and number of acquisitions.

Despite the use of the pseudorandom projector $\boldsymbol{S}_m$, the results obtained using an explicit DB4 wavelet dictionary as sparsity basis are affected by severe artifacts, due to the high correlation between the PMD spatial pixel responses and some high-frequency dictionary atoms. In order to solve this issue, and show the general validity of the proposed framework, we carry out the recovery process seeking sparsity in a more general gradient domain, that is, we adopt a TV-minimization, instead of l_1 minimization of the vector of coefficients in an explicit dictionary, as recovery method. This way any coupling between the sensing method and the sparse recovery procedure vanishes. Indeed, the superresolved raw images obtained via TV-minimization are generally free of the pattern-like artifacts that were observed when using an explicit DB4 dictionary. Despite in both cases the recovery can be considered successful, the reconstruction quality is superior in the TV-minimization case. Additionally, the presence of slightly different patterns in the recovered raw data yields to non-negligible errors in the final phase image. For this reason and for brevity, we omit the results obtained via l_1 minimization in DB4 domain and only provide here the final phase images obtained from the superresolved raw images recovered via TV-minimization. The superresolved phase images obtained for the 30×40 ROI highlighted in Fig. 4.42a using a SR factor $k_{\mathrm{SR}} = 9$ are given in Fig. 4.43 for different compression ratios m/n, where m is, as usual, the total number of compressed measurements and n is the number of pixels of the superresolved images.

Note that, when recovering the superresolved raw images by TV-minimization, one can effectively reduce the data flow, while preserving the quality of the reconstructed image. In fact, for compression ratios higher than $m/n = 2/3$, the quality of the final phase image decreases, and some vertical pattern eventually appears (see Fig. 4.43e), which is related to the distribution of active and blind areas in the PMD chip. As commented before, modeling the spatial responses of the different pixel channels for the two pixel types is a necessary condition to attain accurate signal recovery, but does not ensure exact recovery by itself. Fig. 4.43e is an example of how the TV-minimization converged to a solution that, despite being coherent

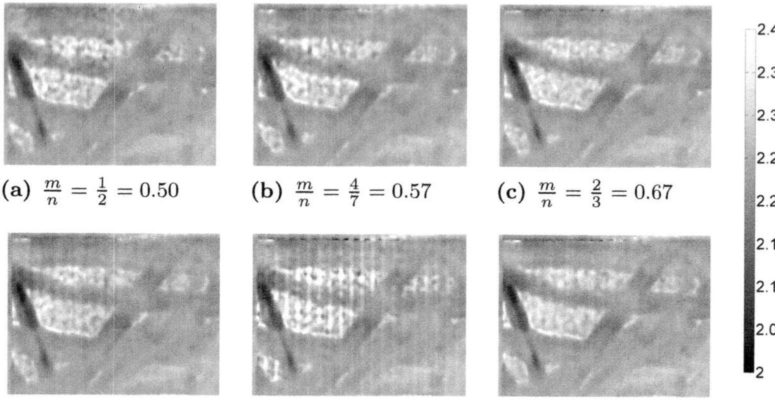

(a) $\frac{m}{n} = \frac{1}{2} = 0.50$ (b) $\frac{m}{n} = \frac{4}{7} = 0.57$ (c) $\frac{m}{n} = \frac{2}{3} = 0.67$

(d) $\frac{m}{n} = \frac{3}{4} = 0.75$ (e) $\frac{m}{n} = \frac{4}{5} = 0.8$ (f) $\frac{m}{n} = 1$

Figure 4.43.: Phase images obtained from the superresolved raw data obtained applying the proposed CS-based SR framework on measurements from a cellulose sample. Only the 30×40 (in original pixel units) region of interest highlighted in Fig. 4.42a was considered. The pseudorandom projector $\boldsymbol{S}_m$ is a Walsh-derived binary matrix with elements ± 1. The signal recovery was carried out via TV-minimization using the TVAL3 library.

with the measurements, is obviously not optimal. Nevertheless, and very interestingly, for lower number of measurements (first row of Fig. 4.43) our framework yields phase images whose quality is comparable to the least squares solution in Fig. 4.42c. The same number of cellulose fibers can be distinguished and their borders are as sharp, or even sharper than those of Fig. 4.42c. See, for instance, Fig. 4.43a, which was obtained from half of the measurements needed to obtain Fig. 4.42c.

Clearly, one could process the 16 regions of size 30×40 in which the original domain (Fig. 4.42a) can be divided in parallel and compose the final phase image from the superresolved ones afterwards. This way one could obtain a FLIM image of size 1080×1440 pixels from approximately only 40 (for $m/n = \frac{1}{2}$) PMD acquisitions of our adapted camera, acquired at different 2D displacements of the stepper unit. Note that we are simultaneously taking profit of the multichannel nature of our sensor (the PMD array) and the compressibility of the data being sensed to push the number of acquisitions to the minimum. In a case of study such as FLIM, in which the bottleneck lies on the large number of sequential acquisitions (and corresponding 'slow' mechanical displacements of the stepper unit) required, the proposed method shows great potential.

4.4.3. SRA-ORMP: A Method for Data Flow and Noise Through Joint-Recovery of PMD Raw Images

In Section 4.4.2 we have shown that CS can be used to increase the resolution of the phase (or depth) image provided by a ToF imaging system. The sparsity or compressibility of the raw data determines how many measurements are required to recover a superresolved image of a given size. Unfortunately, off-the-shelf ToF cameras do not allow randomly positioning their pixels to obtain appropriate *compressed* measurements. Consequently, instead of aiming to recover an image of higher resolution from a number of measurements that is lower than the dimensionality of the superresolved image, in this section we focus on exploiting the structured sparsity of the PMD raw images and the redundancies between them to reduce both the data flow between the sensor and the processing unit and the noise contained in the images. This idea has already been introduced in [H7], where a greedy algorithm is presented to that end. The algorithm is transcribed and commented in Section 4.4.3.3 and its performance on real PMD data is evaluated in Section 4.4.3.4. Intuitively, one can imagine that there is a tradeoff between transmitting redundant information, which can be used to reduce the noise and transmitting less data at the cost of lower noise-reduction capabilities.

4.4.3.1. Introduction

Phase-shift-based ToF cameras make use of multi-tap pixels to correlate the illumination signal reflected by the scene with some reference signals. Most commercial devices feature two channels or taps per pixel. Additionally, to obtain the scene-related phase shift, at least three acquisitions are required, as in classical interferometry (see Section 2.2.1 for details). PMD cameras acquire four equally-spaced phases $\theta \in \{0°, 90°, 180°, 270°\}$ and obtain the depth from Eq. 2.19. From the data flow point of view, this means generating eight (four phases with two channels per pixel) raw images, which have to be read out and transmitted to a processing unit to compute the depth image out of them. This set of raw images conveys redundant information, but typically nothing is done at this respect, probably for the sake of simplicity. Another example is the new Kinect sensor, which makes use of three phases ($\theta \in \{0°, 120°, 240°\}$) at three different modulation frequencies. Provided that a DC IR image is also transmitted, ten raw images are used by this sensor to generate the depth image, even more than in the PMD case.

Due to the similarity between raw images, processing them in a joint manner arises as a natural idea to reduce the noise or efficiently compress the information they convey. Joint bilateral filtering [435, 281] of the raw images would profit from the similarity between modalities (phases) and provide robust noise reduction. Of this flavor is the method presented in Section 4.2.3. Here we propose the use of CS to reduce the volume of data to be transmitted. This way we can directly profit from the sparsity of the raw images and their similarity, while taking into account eventual redundancies in the data collection. Inter-image similarity translates into overlapping sparse supports, while redundancy affects the rank of the set of raw images. In other words, we propose a qualitative jump from a *visual perception*-driven approach, e. g., image filtering, to an *information*-driven approach, where knowledge on the sensing process and the nature of our data is used to reduce the data, preserve the information and discard the noise in a robust way.

4.4.3.2. Rank and Structured Sparsity

From now on we focus on the PMD case to provide concrete derivations, but our method applies to any phase-shift-based ToF sensor. For simplicity, let the illumination signal be sinusoidally modulated and the same signal used as reference in the PMD pixels. Then (see Section 2.2.2), the measurements provided by the PMD pixels are just samples of Eq. 2.12. Let $i_\theta^\Gamma(\vec{x})$ be the value delivered by the channel $\Gamma \in \{A, B\}$ of the PMD pixel indexed by $\vec{x} \in [1, n_{\text{rows}}] \times [1, n_{\text{cols}}]$ for the phase $\theta \in \{0°, 90°, 180°, 270°\}$. Provided that the DC offset is supposed to remain constant during all acquisitions, the trivial cosine property $\cos(\phi) = -\cos(\phi + 180°)$ can be used to derive the following dependencies between (calibrated) raw images:

$$\begin{aligned}
\cos(\phi) &= -\cos(\phi + 180°) \Rightarrow \vec{i}_0^{\text{A}} &= \vec{i}_{180}^{\text{B}} \\
\cos(\phi + 90°) &= -\cos(\phi + 270°) \Rightarrow \vec{i}_{90}^{\text{A}} &= \vec{i}_{270}^{\text{B}} \\
\cos(\phi + 180°) &= -\cos(\phi) \qquad\quad\; \Rightarrow \vec{i}_{180}^{\text{A}} &= \vec{i}_0^{\text{B}} \\
\cos(\phi + 270°) &= -\cos(\phi + 90°) \;\; \Rightarrow \vec{i}_{270}^{\text{A}} &= \vec{i}_{90}^{\text{B}}
\end{aligned} \tag{4.24}$$

where $\vec{i}_\theta^\Gamma \in \mathbb{R}^n$ denotes the vectorized raw image for channel Γ and phase θ, being $n = n_{\text{rows}} \times n_{\text{cols}}$. Additionally, if the scene is static and the DC offset due to background illumination remains constant, the sum of the A and B channels (the DC IR image) is constant, i. e., independent from θ. This fact translates into an additional constraint, which can be formulated as in Eq. 4.25, where both right hand side equations are equivalent due to Eq. 4.24.

$$\vec{i}_0^{\mathrm{A}} + \vec{i}_0^{\mathrm{B}} \approx \vec{i}_{90}^{\mathrm{A}} + \vec{i}_{90}^{\mathrm{B}} \approx \vec{i}_{180}^{\mathrm{A}} + \vec{i}_{180}^{\mathrm{B}} \approx \vec{i}_{270}^{\mathrm{A}} + \vec{i}_{270}^{\mathrm{B}} \Rightarrow \begin{cases} \vec{i}_0^{\mathrm{A}} + \vec{i}_{90}^{\mathrm{A}} \approx \vec{i}_{180}^{\mathrm{A}} + \vec{i}_{270}^{\mathrm{A}} \\ \vec{i}_0^{\mathrm{B}} + \vec{i}_{90}^{\mathrm{B}} \approx \vec{i}_{180}^{\mathrm{B}} + \vec{i}_{270}^{\mathrm{B}} \end{cases} \tag{4.25}$$

Let's define now a matrix $\boldsymbol{I} \in \mathbb{R}^{n \times N}$, constructed with the N vectorized raw images by columns. From Eq. 4.24 and Eq. 4.25, we get five linear dependencies between raw images, which reduce the maximum rank of the matrix I from eight to three in the PMD case ($N = 8$). Let $\boldsymbol{\Psi}$ be an orthonormal basis that transfers the dense signals $\vec{i}_\theta^{\mathrm{T}}$ to a domain where they admit a sparse representation. Then, $\boldsymbol{I} = \boldsymbol{\Psi}\boldsymbol{X}$, being $\boldsymbol{X} \in \mathbb{R}^{n \times N}$ a sparse matrix fulfilling Eq. 4.26. This inequality becomes a strict equality unless a plain depth image equivalent to 90° phase shift is being sensed, in which case, $\mathrm{rank}(\boldsymbol{X}) = \mathrm{rank}(\boldsymbol{I}) = 1$.

$$\mathrm{rank}(\boldsymbol{X}) = \mathrm{rank}(\boldsymbol{I}) \leq 3 \tag{4.26}$$

Note that the upper bound on the rank given by Eq. 4.26 can be computed beforehand for any sensor because it is *system*-dependent and not *scene*-dependent.

The sparsity of typical PMD raw images in appropriate dictionaries is obviously scene-dependent and was studied in Section 4.4.1.1. Their high compressibility and almost exact sparsity is to be observed in Fig. 4.37. One can affirm that all raw images exhibit equivalent sparsity, but this does not imply, in principle, that they also share the same sparse support.

In this section we reuse the test scene developed for the experiment in Section 4.5.1.4 (see Fig. 4.52a). Our methods work in a per-depth-image basis and, therefore, only a single acquisition of raw data is required. The dictionary $\boldsymbol{\Psi}$ is chosen to be a 2D-DB4 basis, but other wavelet-based bases provide equivalent results. Fig. 4.44a depicts the four raw images of the scene delivered by the A channel of the PMD sensor. The corresponding B channel images are equivalent to those in Fig. 4.44a with a phase displacement of 180° due to Eq. 4.24 and are omitted. Fig. 4.44b plots the absolute values of the DB4 coefficients of the images in Fig. 4.44a, in ascending order. If the DB4 coefficients of the raw images of both pixel channels are plotted together, a complete overlap of the curves is observed, as in Fig. 4.37.

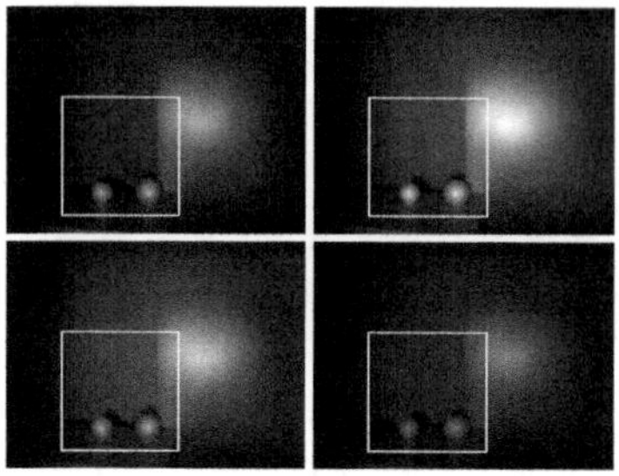

(a) Raw Images (A Channel)

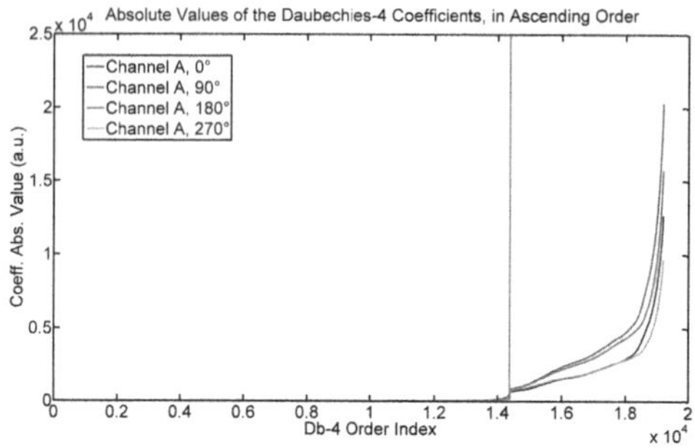

(b) Absolute Values of DB4 Coeffs.

Figure 4.44.: PMD raw images of the test scene obtained from the A channel at phases $\theta \in \{0°, 90°, 180°, 270°\}$ (a) and their corresponding DB4 coefficients (b), in absolute value and plotted in ascending order. The raw images in (a) are ordered by θ, from left to right and from top to bottom. The area within the white rectangle is the most challenging and is used for evaluation. The magenta line in (b) separates negligible and non-negligible coefficients, giving a visual estimate of the sparsity, provided that the total number of coefficients is $n = 120 \times 160 = 1.92 \times 10^4$. ©2015 IEEE.

Note that the simplicity of the structure in the test scene translates into a steeper decay of the DB4 coefficients in Fig. 4.44b (higher compressibility), with respect to the plots in Fig. 4.37, corresponding to a more cluttered scene. The approximate sparsity, in turn, does not change much with respect to that observed in Fig. 4.37 and we observe that $\left|\text{supp}\left(\vec{i}_i\right)\right| \leq s \approx \frac{1}{4}n \; \forall i \in [1, N]$ (magenta line in Fig. 4.44b).

In order to check that the raw images share the same sparse support, we approximate their support independently by considering only those dictionary elements whose corresponding coefficients are higher than a threshold in absolute value and then checking whether the selected support is the same for all raw images. In all cases considered, the support overlap is almost complete, which is coherent with the fact that the underlying scene is the same. This is a stronger condition than equal sparsity and implies that $\boldsymbol{X}$ is k joint sparse, according to the definition of joint sparsity given in Section 3.4.2.8, where $k \approx s$.

Wavelets are subject to position and scale parameters. In other words, wavelet bases are multiscale bases. This produces that the non-zero coefficients of wavelet decompositions of natural signals typically follow a tree structure, where deeper levels of the tree correspond to finer scales. This leads to a structured sparsity, where zero wavelet coefficients cannot have non-zero *children*—finer scale at the same position—. This sparse tree representation can be used as *a priori* information within the recovery algorithm [273] to improve the reconstruction quality.

Fig. 4.45 provides a multiscale DB4 wavelet decomposition of three of the images in Fig. 4.44a, corresponding to $\theta \in \{0°, 90°, 180°\}$. The number of wavelet scales is set to three, which gives four levels, taking into account the approximation coefficient image (A), of equal size to the first scale detail coefficient images. There are three detail coefficient images per scale, depending on the main direction of the wavelet: horizontal (H), vertical (V) and diagonal (D). In order to provide a graphical cue of the joint sparsity of the signals, the DB4 decompositions of the three selected raw images are encoded as a single RGB image, where the three channels are assigned in order of increasing θ. Consequently, grayscale images would mean equal values in the three raw images, while pure red, green or blue pixels would mean that some DB4 coefficients are non-zero for one raw image and zero for the others, violating this way the joint sparsity assumption. None of the pixels of the images in Fig. 4.45 shows this behavior, despite differences between the channels do lead to more pink or green tones in some areas. Note that sparsity is structured through scales, i. e., black areas in a coarse scale do not lead to non-black pixels in finer scales. If this happens and parents with zero value lead to non-zero children, as it seems to occur in the central area of the detail coefficient images of the third scale, it is due to the presence of noise. Indeed, that central area corresponds to a plane in the scene, without any structure. This fact already points out the denoising potential of the structured sparsity when properly taken into account in a sparse recovery framework.

In order to show that the theoretical constraint on the rank given by Eq. 4.26 applies also in practice, we check how well it suits our dataset of PMD raw images. Fig. 4.46 is intended to illustrate the results of an SVD of the set of eight vectorized raw images and subsequent projection of the data to the space spanned by the first three eigenvectors. Fig. 4.46b shows the extremely fast decay of the corresponding eigenvalues. The ratio between the each eigenvalue and the first one is plotted in dB. Most of the signal power is retained by the first three eigenvalues (mostly the first), while the last ones are of similar importance, much lower than that of the first. Note that there is a gap of approximately 10 dB between the 3rd and 4th

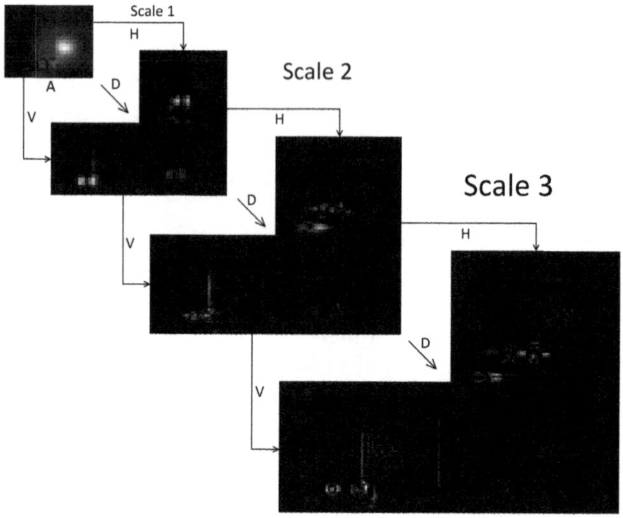

Figure 4.45.: Multiscale DB4 decomposition of three of the raw images in Fig. 4.44a, corresponding to $\theta \in \{0^\circ, 90^\circ, 180^\circ\}$. Three scales are used and the results are shown from coarse to fine from top-left to bottom-right. Each channel of the RGB images corresponds to each of the raw images. The values of the non-zero coefficients in finer scales are much lower than those of coarser scales. For visibility, all images have been maximum-normalized. The original maximums of the detail coefficient images with respect to that of the approximation coefficient image are 1.49×10^{-1}, 4.71×10^{-2} and 1.36×10^{-2}, from coarser to finer scales.

eigenvalues, while the last 5 eigenvalues lie all in a range of 10 dB. Fig. 4.46a shows the four raw images of the A channel (cf. Fig. 4.44a) after projection onto the space spanned by the first three eigenvectors. The differences with respect to the original images are negligible and not visible to the naked eye. Quantitatively, the average power loss due to rank-restriction was observed to be -10.8 ± 7.5 dB, or, equivalently, an average signal power gain of -0.65 ± 0.48 dB. Note that, when working with signal powers, the prefactor for computing gains and losses in dB is 10 and not 20. Additionally, the pixel values of the difference images were observed to be within the noise level of the PMD sensor (RMSE of 23.05 ± 5.46, in post-calibration PMD pixel units). In short terms, constraining the rank to its theoretical maximum does not seem to produce information loss in real PMD raw data.

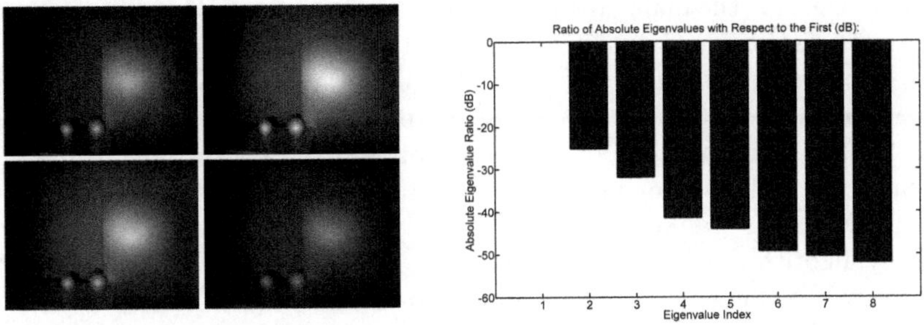

(a) Rank-restricted Raw Images (A Channel) **(b)** Eigenvalues with Respect to the First (dB)

Figure 4.46.: Effect of constraining the rank of PMD raw data to its theoretical limit (three). The rank-restricted raw images in (a) (channel A) are to be compared with the originals in Fig. 4.44a. The perfect reconstruction is due to the high concentration of the signals along the first three eigenvalues, mostly the first one, as shown in (b).

4.4.3.3. Structure and Rank Aware Order Recursive Matching Pursuit

In Section 4.4.3.2 the cornerstones of our approach have been enunciated, namely, rank-awareness, joint sparsity of the raw images and structured sparsity when using a wavelet basis. Our goal is to provide a framework for compressed sensing and sparse recovery of the raw data that takes profit of the former considerations in order to keep or improve the signal quality, with as few measurements as possible.

In order to perform measurements on our set of signals, $\boldsymbol{I}$, we use a sensing matrix, $\boldsymbol{\Phi} \in \mathbb{R}^{m \times n}$, which derives from an $n \times n$ Hadamard matrix with randomly permuted columns, where m rows are selected at random. Here we consider a case where $n = 2^r$, $r \in \mathbb{N}$, which is a sufficient—but not necessary—condition for the existence of a Hadamard matrix. Our matrix of measurements is, therefore $\boldsymbol{Y} = \boldsymbol{\Phi I}$, i. e., each vector of measurements is $\vec{y}_i = \boldsymbol{\Phi}\vec{i}_i$, $i \in [1, N]$. Consequently the problem of recovering the joint sparse matrix $\boldsymbol{X}$ from $m \ll n$ measurements per vector can be formally written as the MMV sparse recovery in Eq. 3.128, where $\boldsymbol{A} = \boldsymbol{\Phi\Psi}$ and $\mathrm{supp}(\boldsymbol{X})$ denotes joint support.

Some MMV extensions of greedy algorithms are presented in Section 3.4.2.8 that can efficiently solve Eq. 3.128. Our approach builds upon the RA-ORMP algorithm, which exploits the fact that, in general, $\mathrm{rank}(\boldsymbol{X}) > 1$ in the MMV problem, which is the main advantage with respect to SMV. From now on, we talk about dictionary atoms, instead of basis vectors, since our method also applies to redundant representations. The adaptation of the RA-ORMP to our case of study is two-fold: On one side, we explicitly force the orthonormal basis for the residual to be of the rank imposed by our system—three, in the case of a PMD sensor in four-phases operation, as given by Eq. 4.26—, since $\mathrm{rank}(\boldsymbol{R}^{(0)}) = \mathrm{rank}(\boldsymbol{Y}) \leq \mathrm{rank}(\boldsymbol{X}) = 3$. On the other side, we force the recovery algorithm to follow the sparse-tree structure induced by the wavelet dictionary. To this end, the algorithm operates in l sequential states, where l is the number of levels in the tree, i. e., the number of scales in the wavelet dictionary. When the algorithm reaches the wth scale, the set of *candidate* atoms to be added to the joint support of the sparse signals is given by the atoms that belong to the scale w and whose parents have been included in the support at the scale stage $w-1$.

Our 2D DB4 dictionary was chosen to be an orthonormal basis and, therefore, shows only two scales (corresponding to approximation and detail coefficients), but overcomplete wavelet dictionaries with more scales can be used instead. An increased number of scales may deliver a more efficient representation, in terms of sparsity of the solution, but at the cost of a higher runtime. The pseudo-code for our Structure and Rank Aware ORMP (SRA-ORMP) is given in Algorithm 19.

The vectors $\vec{a}_i$ and $\vec{u}_i^{(\cdot)}$ denote the ith column of the matrices $\boldsymbol{A}$ and $\boldsymbol{U}^{(\cdot)}$, respectively. Note the integration of *a priori* system-related information (maximum allowable rank, $r_{\max}$) in line 9, after the SVD of the residual matrix, $\boldsymbol{R}^{(k-1)}$. A major difference with respect to conventional RA-ORMP is the existence of a candidate set, Π^k, containing the indices of those atoms to be considered at iteration k. For the first scale ($w=1$), Π^k contains the indices of all dictionary atoms belonging to that scale, except from those already indexed in the temporal support, Ω^k, but for higher scales only children from previous scale atoms that were indexed in the support are allowed to be included in Π^k. When the mean (μ)-normalized value of the standard deviation from the mean (σ_μ) of $\left\{\left\|\vec{a}_{i^j}\boldsymbol{U}^{(k-1)}\right\|_2\right\}_{j=k-\Delta k}^{j=k}$ is lower than a noise-related threshold, $\varepsilon_{\text{noise}}$, all remaining atoms for the current scale in Π^k are discarded and the algorithm automatically jumps to the next scale. This is just a robust alternative to the derivative of $\|\vec{a}_{i^k}\boldsymbol{U}^{(k-1)}\|_2$ in k domain, since its smoothness was found to be a very reliable indicator of a correct atom selection process. The shape of the μ-normalized standard deviation curve is quite independent from Δk, allowing for very low values, e. g., $\Delta k = 3$. The iterative process ends when certain tolerance on the

Algorithm 19 Structure and Rank Aware Order Recursive Matching Pursuit (SRA-ORMP)

Initialize: $\boldsymbol{R}^{(0)} = \boldsymbol{Y}, X^{(0)} = 0, \Omega^0 = \emptyset, \Pi^0 = \emptyset, \vec{\tilde{a}}_i = \vec{a}_i, \forall i \in [1, n], k = 0$

for $w = 1; w := w + 1$ **to** $w = l$ **do**
  if $\Omega^k \neq \emptyset$ **then**
    $\Pi^k = \left\{ i | \, \text{parent}(i) \in \Omega^k, \text{scale}(\vec{\psi}_i) = w \right\}$
  else
    $\Pi^k = \left\{ i | \, \text{scale}(\vec{\psi}_i) = w \right\}$
  end if
  while $\left(\left\|\boldsymbol{R}^{(k)}\right\|_2 > \varepsilon_{\text{tol}}\right)$ **and** $\left(\left|\Omega^k\right| < s_{max}\right)$ **and** $\left(\Pi^k \neq \emptyset\right)$ **do**
    $k := k + 1$
    $\left[\boldsymbol{U}^{(k-1)}, \boldsymbol{\Sigma}^{(k-1)}\right] = \text{SVD}\left(R^{(k-1)}\right), \vec{u}_i^{(k-1)} = \vec{0}, \forall i > r_{\max}$
    $i^k = \underset{i \in \Pi^{k-1}}{\arg\max} \left\| \vec{\tilde{a}}_i^\top \boldsymbol{U}^{(k-1)} \right\|_2$
    if $\dfrac{\sigma_\mu\left(\left\{\left\|\vec{\tilde{a}}_{i^j} \boldsymbol{U}^{(k-1)}\right\|_2\right\}_{j=k-\Delta k}^{j=k}\right)}{\mu\left(\left\{\left\|\vec{\tilde{a}}_{i^j} \boldsymbol{U}^{(k-1)}\right\|_2\right\}_{j=k-\Delta k}^{j=k}\right)} > \varepsilon_{\text{noise}}$ **then**
      Update support: $\Omega^k = \Omega^{k-1} \cup i^k$
      Update candidate set: $\Pi^k = \Pi^{k-1} \setminus i^k$
      $\boldsymbol{X}_{\Omega^k}^{(k)} = \boldsymbol{A}_{\Omega^k}^\dagger \boldsymbol{Y}$
      Calculate orthogonal projector: $\boldsymbol{P}_{\Omega^k}^\perp := \left(\boldsymbol{I}_m - \boldsymbol{A}_{\Omega^k} \boldsymbol{A}_{\Omega^k}^\dagger\right)$
      Update residual: $\boldsymbol{R}^{(k)} = \boldsymbol{P}_{\Omega^k}^\perp \boldsymbol{Y} = \boldsymbol{Y} - \boldsymbol{A}_{\Omega^k} \boldsymbol{X}_{\Omega^k}^{(k)}$
      Project measurement matrix: $\boldsymbol{A}' = \boldsymbol{P}_{\Omega^k}^\perp \boldsymbol{A}$
      Normalize: $\vec{\tilde{a}}_i = \dfrac{\vec{a}_i'}{\left\|\vec{a}_i'\right\|_2}, \forall i \in \Pi^k$
    else
      $\Pi^k = \emptyset$
    end if
  end while
end for

residual norm, ε_{tol}, is met, the cardinality of the temporal support reaches the maximum sparsity, $s_{\max}$, or Π^k is empty.

4.4.3.4. Experimental Evaluation

In this section we provide an evaluation of the performance of our SRA-ORMP recovery framework on the dataset introduced in Section 4.4.3.2 (see Fig. 4.52a for a color image of the scene) and compare it to that of a conventional SMV l_1-minimization with equality constraints. The chosen baseline method for exact l_1-minimization is that provided by the l_1-magic [89] software package, which uses the primal-dual interior point method. For the evaluation, we consider only a challenging 64×64 region within the images and, consequently, $n = 4096$. The location of this region is marked with a white rectangle in the raw images of Fig. 4.44a. Four experimental cases are considered, for different number of measurements per raw image, $m \in \{\frac{1}{2}n, \frac{2}{3}n, \frac{3}{4}n, n\}$.

For each value of m considered, the raw images recovered using our method and the baseline approach are used to compute the respective depth images. The depth images obtained from recovered data are shown in Fig. 4.47. Note how the depth images computed from SRA-ORMP-recovered raw data (second row) are visibly better than their SMV l_1-minimization counterparts (first row). The superiority of the proposed method becomes more noticeable for higher values of m, for which conventional l_1-minimization starts to recover the noise present in the original images, while the SRA-ORMP robustly rejects it.

A quantitative evaluation of the quality of the recovery provided by each method can be obtained by simply computing the l_2 distance between recovered and original raw images. Similarly, the distance can be measured in depth domain, between the depth image computed from recovered raw data and that computed from original raw data. The results of such calculus are given in the first two sections of Table 4.6 (by rows). Distances in raw image domain are given in calibrated PMD pixel units (p.u.). From these results one can conclude that our method performs as well as the baseline, while allowing considerable noise suppression for m high enough (Fig. 4.47). Despite this is *per se* an achievement, a quantitative evaluation of the error reduction due to the denoising capabilities of our method is still missing.

To that end, we consider a plain region belonging to the background plane within the 64×64 ROI and fit a plane model to the original depth image. This way we count with instantiated GT that can be used to obtain depth errors. The third section of Table 4.6 provides the depth RMSEs of that

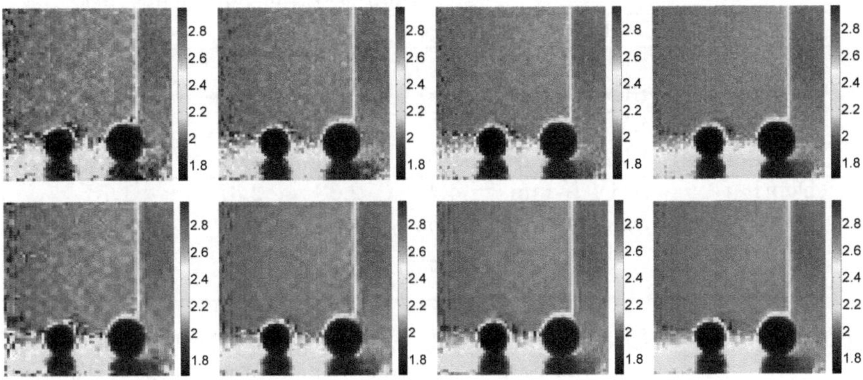

(a) $m = \frac{1}{2}n$ (b) $m = \frac{2}{3}n$ (c) $m = \frac{3}{4}n$ (d) $m = n$

Figure 4.47.: Depth images obtained from PMD raw data recovered from few pseudorandom measurements using conventional SMV l_1-minimization per raw image (first row) and the proposed SRA-ORMP algorithm (second row, ©2015 IEEE). The results are ordered in increasing number of measurements: $m \in \{\frac{1}{2}n, \frac{2}{3}n, \frac{3}{4}n, n\}$, being $n = 64 \times 64 = 4096$ the image size. Scales are in meters.

plain region with respect to the instantiated GT for all experimental cases considered. The last row of the table shows the depth RMSE differences of the SRA-ORMP with respect to the SMV l_1-minimization.

For $m = n$ (last column of Table 4.6), the SLE to solve is no longer underdetermined and, consequently, the baseline approach achieves exact recovery of all the original raw images and the final depth image coincides with that computed from the original raw data. Obviously, this means recovering both the signal and the noise present in the original data. In contrast, the effect of increasing m on the SRA-ORMP algorithm is better robustness to noise and, therefore, the best denoising results are attained for $m = n$.

The last row of the Table 4.6 quantifies the outstanding noise reduction effect visible in Fig. 4.47. The proposed method performs better than the baseline in terms of depth RMSE regardless of m. Additionally, provided that the depth RMSE for $m = n$ is lower than that of the original depth image (0.943 cm vs. 1.73 cm, since the baseline achieves exact recovery for $m = n$), one could adjust m to reduce the data flow while achieving a

		Number of Measurements (m)			
	Method	$\frac{1}{2}n$	$\frac{2}{3}n$	$\frac{3}{4}n$	n
Raw l_2 dist. (p.u.)	SMV l_1-min.	72.3	36.6	27.0	1.82×10^{-4}
	SRA-ORMP	78.2	35.7	33.3	20.4
Depth l_2 dist. (cm)	SMV l_1-min.	7.59	4.44	3.60	1.63×10^{-7}
	SRA-ORMP	8.55	3.92	3.66	2.21
Depth RMSE (cm)	SMV l_1-min.	4.43	2.68	2.37	1.73
	SRA-ORMP	3.65	1.63	1.54	0.943
Δ Depth RMSE (%)	—	−17.7	−39.3	−35.1	−45.5

Table 4.6.: The first section provides the l_2 distance between the recovered and the original raw data. The second section provides the l_2 distance between the depth image computed from recovered raw data and that computed from original raw data. The last section provides the depth RMSE of the depth images computed from recovered raw data with respect to instantiated GT. The results are given for different numbers of measurements (m), using sequential SMV l_1-minimizations and our MMV SRA-ORMP algorithm as recovery methods.

depth error equal to that one would obtain from the original PMD raw data without any compression or processing. For instance, with only $m = \frac{2}{3}n$ measurements, the SRA-ORMP algorithm still achieves slightly better depth RMSE (1.63 cm) than that of the original depth image.

4.5. Sparse Recovery in Time-Frequency Domain

In this section the show that the sparsity of the emitted light signal in frequency domain can be exploited to recover more than one frequency from few measurements in a CS framework. In Section 4.5.1 we show that a commercial PMD 19k-S3 frontend, without any hardware modification, can be used to that end, with the only need of an external delay generator. This way, the quality of the illumination waveform, which is a critical factor in any ToF system, can be densely evaluated over the entire FOV of the camera without the need of external equipment such as photodiodes and oscilloscopes, typically used to acquire the illumination signals. In Section 4.5.2 we provide a more generic and feasible framework for sparse recovery of the received light signal in frequency domain, which would only require a modification of the program in the CPLD of the PMD frontend to be implemented in practice. Encouraging simulation results motivated the development of an experimental prototype of a CS-based PMD camera, presented in Chapter 5.

4.5.1. Simultaneous Multichannel Waveform Recovery of Illumination Signals

As pointed out in Section 2.2.2, when either the emitted light signal or the reference signal used at pixel level are sinusoidal, then the harmonic content of the non-sinusoidal one has no effect on the measurements, since the corresponding terms vanish in the cross-correlation function. Nevertheless, if both signals are *close to* sinusoidal but both of them have a non-negligible harmonic content, then additional non-vanishing terms appear in the cross-correlation function (Eq. 2.15) and the so-called *wiggling* effect (see Section 2.4.1) spoils the depth estimation. The four phases algorithm and, in general, any phase retrieval algorithm building upon the hypothesis that the measurement process is a cross-correlation between pure sinusoidal signals will fail, to a greater or lesser extent, since they assume a wrong sensing model.

Therefore, the illumination system is one of the most critical elements of ToF imaging devices. Most commercial devices operate in the short range and are designed for indoor use. Such devices use simple illumination setups featuring one or few LEDs or laser diodes (e. g., three in the Xbox One sensor), grouped together and demanding low power. Recent developments towards medium and long range ToF imaging [283], oriented to outdoor operation, have brought the need for powerful illumination setups, constituted by many emitters, which might be grouped in distributed modules. This makes even more difficult to give a reliable evaluation of the quality of the illumination waveform, since it introduces variability in 3D space, i. e., the illumination waveform might strongly depend on where it is being measured. Combinations of many waveforms coming from different emitters with different location and orientation might lead to a wide range of waveform variations within the illuminated volume. The effect on the depth estimation is a non-linear error that depends not only on the depth, but on the 3D position of the point being measured. In most applications, emitters are IR LEDs, whose rising and falling characteristics might differ between each other, leading to non-negligible waveform differences when operating at high frequencies. Additionally, signals coming from different emitters might not always be in phase (see Appendix A.6).

Full knowledge of the illumination system allows creating a model and simulating the light intensity distribution [368]. Similarly, the waveform distribution in 3D might also be simulated if the waveform of each emitter and the system geometry is known. Nevertheless, in illumination systems with many emitters, where the waveform variability is due to unmodeled differences between LEDs and external factors (such as temperature, humidity, etc.), an empirical evaluation becomes necessary.

A common way of checking the quality of the illumination signal is to acquire it by means of a fast photodetector and an oscilloscope. Using a photodetector to sense the waveform at different 3D locations is time-consuming and requires a mechanical setup with at least three DOFs for accurately placing the detector at the desired locations. Additionally, we are much more interested in determining the waveform *as seen by the pixels* of the camera, that is, the signal being correlated at the pixel. The similarity of these signals with respect to a sinusoidal determines the correctness of the depth measurements. In this section we present the method in [H6], which covers the lack of a methodology for simultaneously gathering the light signal at different spatial locations. The framework recovers the waveform of the light received by each pixel of a PMD array from the pixel measurements, taking profit of the sparsity of AMCW illumination signals in frequency domain.

4.5.1.1. Methodology

A typical illumination waveform, as measured at one of the LEDs of our wide-area illumination system is plotted in blue (in the color version) in Fig. 4.2. Clearly, the signal is not perfectly sinusoidal. A DCT decomposition (in red in the color version of Fig. 4.2) delivers non-negligible coefficients for the lowest-frequency harmonics. Looking at the DCT plot, one could say that the signal is approximately sparse in frequency domain, since the 10% highest DCT coefficients (at the left of the vertical red line) retain approximately all signal energy. Nevertheless, if the considered frequency domain (feasible bandwidth) is reduced, then the sparsity of the signal might be of the same order of magnitude as the dimensionality. Even in that case, one can profit from the fact that the signal is highly compressible, since the k-term approximation error, as defined in Eq. 3.3, decays extremely rapidly in k. This is due to the low-pass filtering effect of the LEDs, whose frequency transfer function is monotonically decreasing (see Fig. 4.1). As a result, the DCT coefficients decay with frequency.

As usually, let's denote by $\mathbf{\Phi} \in \mathbb{R}^{m\times n}$ the sensing matrix $(m < n)$, which models how the measurements are obtained, and $\mathbf{\Psi} \in \mathbb{R}^{m\times n}$ the sparsity basis. Since we aim to recover periodic signals, $\mathbf{\Psi}$ is a frequency basis, e. g., Fourier or DCT basis. Then, the vector of measurements $\vec{y}$ can be formulated as in Eq. 3.8, which is the linear model required in a CS framework, where $\boldsymbol{A} \in \mathbb{R}^{m\times n}$ is the measurement matrix: $\boldsymbol{A} = \mathbf{\Phi}\mathbf{\Psi}$. In CS theory, $\boldsymbol{A}$ is expected to satisfy the so-called Restricted Isometry Property (RIP) (Section 3.1). Additionally, the rows of the sensing matrix, $\mathbf{\Phi}$, which model how the measurements are performed, should be incoherent with the columns of the dictionary or representation matrix, $\mathbf{\Psi}$, which spans the space where the signal admits a sparse representation. We adopt the mutual coherence, as defined in Eq. 3.39. It is clear that the higher the incoherence, the less measurements needed. If the previous conditions are satisfied, the sparse signal $\vec{x}$ can almost always be recovered by solving the convex program given by Eq. 3.10.

In a PMD chip the pixel measurements are the result of the cross-correlation between the incoming light signal and a reference signal, which is a shifted version of the ICS. In a four phases framework, the phase shifts are $\theta \in \{0°, 90°, 180°, 270°\}$. The effective reference signal at pixel level is known to be close to square (see corresponding results in Section 4.3.1.3), being the incoming signal integrated in one pixel channel or another (A or B), depending on the level of the reference signal (high or low). Considering differential measurements (A-B) we get rid of the constant level due to background illumination. This increases the sparsity of our signal, by eliminating the highest coefficient in the sparse domain, which tends to be always the one of zero frequency. Note that the sensing kernel we are implicitly using to get this measurements is a square wave between $\pm c$, being c some constant related to the pixel sensitivity and ADC gain.

The four acquisitions gathered by the PMD chip are not enough to recover a high resolution waveform, especially since only two of them are usable in a CS recovery framework. This is due to the fact that any two consecutive measurement kernels are fully correlated with the other two, and is a direct consequence of using the differential (A-B) measurements. That is, if we denote by $\vec{\phi_i}$ the differential measurement kernel for a phase shift $\theta_i = \frac{2\pi(i-1)}{p}$, $i \in [1, p]$, where $p = 4$ is the number of phases, then we have that $|\langle\vec{\phi_i}, \vec{\phi_j}\rangle| = 1,\ \forall i, j \in [1, p] \mid |i - j| = \frac{p}{2}$.

We propose gathering more measurements at different phase shifts, in order to have enough information for recovery. Then we formulate the problem as recovering a discrete periodic waveform $\vec{r} \in \mathbb{R}^n$ from $m < n$ equally distributed correlation measurements, knowing that the signal admits a sparse representation $\vec{r} = \boldsymbol{\Psi}\vec{x}$. If the phase shifts are $\theta_i = \frac{2\pi(i-1)}{m}$, $i \in [1, m]$, i. e., equally distributed between 0 and 2π radians, then we run into the same problem of full correlation between measurement kernels displaced π radians between each other: $|\langle\vec{\phi_i}, \vec{\phi_j}\rangle| = 1, \ \forall i, j \in [1, m] \mid |i - j| = \frac{m}{2}$. That is, half of the measurement kernels are fully correlated with the other half. Consequently, similarly to the four phases case, where only $\frac{p}{2} = 2$ consecutive measurements were usable, we can gather, at maximum, $m = \frac{n}{2}$ measurements, with phase shifts $\theta_i \in [0, \pi)$. Since we consider regular phase shifts, $\theta_i = \frac{\pi(i-1)}{m}$, $i \in [1, m], m \leq \frac{n}{2}$.

Custom phase shifts of the PMD reference signal are not achievable without modifying the program of the CPLD of the PMD frontend or modifying the frontend itself. Therefore, for simplicity, we displace the ICS before using it to drive the LEDs, which is equivalent to displacing the reference signal at pixel level in opposite direction. The normalized sensing kernels are, therefore, n-dimensional vectors representing a $\pm\frac{1}{n}$ square signal and displaced one with respect to the previous one position left. We aim to recover the light signal received by each pixel, $\vec{r} \in \mathbb{R}^n$, from, at maximum, $m = \frac{n}{2}$ measurements. This means that the simplest sensing matrix, $\boldsymbol{\Phi}$, is a circulant matrix composed by the sensing kernels by rows.

The dictionary $\boldsymbol{\Psi}$ was chosen to be the DCT basis, which was found to be a simple and efficient alternative to the Fourier basis. The CS requirement of incoherence between $\boldsymbol{\Phi}$ and $\boldsymbol{\Psi}$ is not met, since the square signals of the measurement kernels are highly correlated with the first non-zero frequency of the DCT dictionary. In order to solve the coherence issue, we introduce a scrambling operator, $\boldsymbol{S}$, which is chosen to be a Hadamard matrix with randomly permuted columns and normalized by rows. This way we achieve a pseudorandom (PR) binary matrix with orthonormal rows. For simplicity and without loss of generality, we consider signal sizes $n = 2^k$, $k \in \mathbb{N}$ and, therefore, the maximum required size of the Hadamard matrix is $n_{\text{Had}} = 2^{k'}$ $k' = k - 1$. As pointed out in Section 3.2.3, this is a sufficient but not necessary condition for the existence of a Hadamard matrix of size n_{Had}. For the design of the scrambling operator for arbitrary signal size, we recall again the optimum binary signature set design procedure proposed in [253]. Note that, despite mutual incoherence is highly desired, it is not strictly necessary that the rows of the scrambling operator are orthogonal and purely random operators may perform equally well.

The number of measurements is now adjusted by randomly selecting m rows of $\boldsymbol{S} \in \mathbb{R}^{n_{\text{Had}} \times n_{\text{Had}}}$, while the original size of the Hadamard matrix, equal to the number of columns of the non-scrambled $\mathbf{\Phi}$, is kept to the maximum, i. e., $n_{\text{Had}} = \frac{n}{2}$. We denote the m-row version of the scrambling operator by $\boldsymbol{S}_m \in \mathbb{R}^{m \times \frac{n}{2}}$, which is applied to the sensing matrix $\mathbf{\Phi} \in \mathbb{R}^{\frac{n}{2} \times n}$. The resulting sensing matrix is denoted by $\mathbf{\Phi}_m \in \mathbb{R}^{m \times n}$ and obtained as $\mathbf{\Phi}_m = \boldsymbol{S}_m \mathbf{\Phi}$. Therefore, the final measurement matrix is given by:

$$\boldsymbol{A} = \mathbf{\Phi}_m \mathbf{\Psi} = (\boldsymbol{S}_m \mathbf{\Phi}) \mathbf{\Psi} \tag{4.27}$$

Obviously, the measurements have to be equally projected through the scrambling operator: $\vec{y} = \mathbf{\Phi}_m \vec{c}$, where $\vec{c} \in \mathbb{R}^{\frac{n}{2}}$ denotes the vector of cross-correlation samples, i. e., the differential measurements.

4.5.1.2. Signal Recovery Results

As in previous experiments, we make use of the MultiCam sensor, equipped with our medium range NIR illumination system (Appendix A.6), to gather the raw data. In order to perform custom phase shifts on the ICS, we make use of an external function generator, acting as signal conditioner and pure delay generator. The ICS, which is a square signal generated by the CPLD of the PMD frontend of the MultiCam, is used as external trigger of the function generator. The delayed output signal, after amplification, triggers simultaneously all the illumination modules. A single program was developed to control the function generator, acquire the PMD raw data from the MultiCam, perform the illumination signal recovery and plot the results. The user can select one or several pixels of interest and the corresponding recovered illumination signals will be simultaneously plotted on the screen during the execution, allowing comparison between the illumination waveform received at different regions of the sensor, corresponding to different spatial locations. A block diagram of the experimental setup is given in Fig. 4.48.

The results we present here were obtained placing a plain white wall in front of the camera. The high reflectivity of the wall, together with the high power of the illumination system, allow using very short exposure times, e. g., 0.1 ms for distances between 1 and 2 m, in absence of background illumination. This theoretically opens the possibility for taking all necessary measurements in a very short time and run in real time. Unfortunately, there exists a considerable time overhead after the acquisition of each raw image, due to analog-to-digital conversion and data transmission, that is independent from the exposure time and constitutes the main bottleneck of the approach. The modulation frequency is set to 20 MHz, producing a periodic waveform of 50 ns period. We consider three different sizes of the discrete waveform to recover: $n \in \{32, 64, 128\}$. For each case we

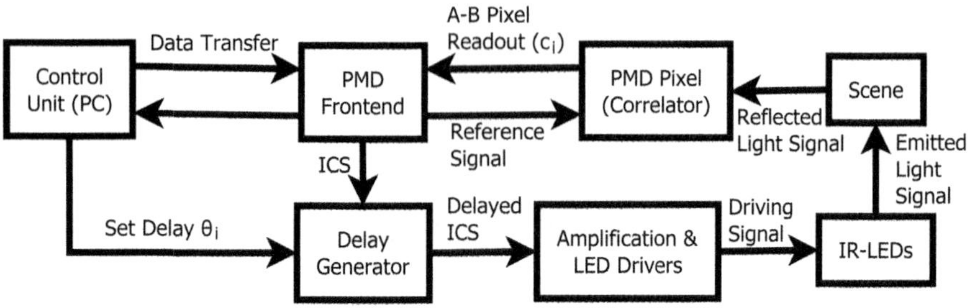

Figure 4.48.: Block diagram of the experimental setup. A correlation measurement (c_i) is gathered for each phase delay considered (θ_i).

carry out the recovery process via l_1-minimization using different number of measurements, $m = \frac{n}{k}$, $k \in \{2, 3, 4\}$. We use the primal-dual interior point method to solve the convex optimization in Eq. 3.10. The illumination waveform is then directly obtained from the recovered vector of coefficients: $\vec{r} = \Psi\vec{x}$. The waveforms obtained for the central pixel of the PMD chip are given in Fig. 4.49. In each plot, the waveforms have been maximum-normalized and centered with respect to each other, using the maximization of the cross-correlation as centering criterion.

The ICS is almost a perfect square signal, while the real light signal, as acquired by a fast photodetector located next to the illumination system, exhibits the characteristic asymmetric rising and falling times of the LEDs. The recovered illumination waveform is approximately the same, regardless of m. In Fig. 4.49a and Fig. 4.49b all recovered signals exhibit negligible variations between each other. In the case of 128 time steps (Fig. 4.49c), the high frequency components of the noise that affects the real signals are included in the DCT dictionary and, therefore, this noise appears in the recovered waveforms. The results using 50% and 33% measurements are still correct, but the waveform obtained from 25% measurements shows already non-negligible differences with respect to the others. Note that the magenta line (GT) is the light waveform as acquired in a position close to the camera and is only for reference and cannot be used for error calculus, since it might differ from the signal we want to recover.

4.5.1.3. Waveform Quality Evaluation

In this section we show that the method for recovering the waveform of the light signal per pixel of a PMD array can be exploited to use any

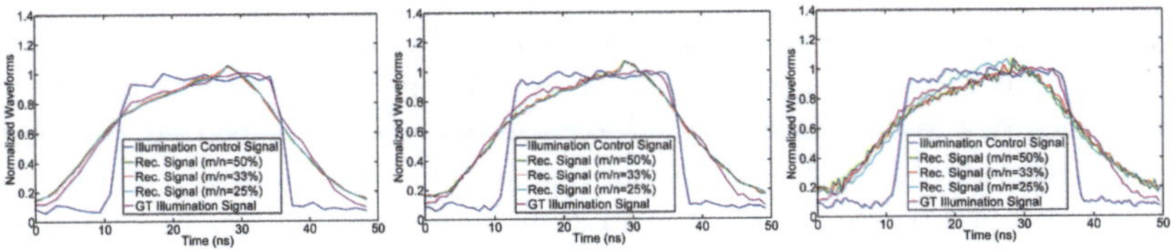

(a) $n = 32$ **(b)** $n = 64$ **(c)** $n = 128$

Figure 4.49.: Recovered waveforms of the real AMCW illumination signal (20 MHz frequency), as received by the central pixel of the PMD chip. Discrete signal sizes of 32 (a), 64 (b) and 128 (c) were used. In each plot, the blue curve is the illumination control signal (ICS) at the input of the illumination system. The curve in magenta is a ground truth (GT) illumination signal acquired with a fast photodetector, located next to the camera. Curves in green, red and cyan are the results of our CS-recovery approach, using a number of measurements that is 50%, 33% and 25% of the signal size, respectively. All signals have been maximum-normalized and centered. ©2015 IEEE.

PMD-based ToF system as its own *illumination tester* [H2]. That is, a complete evaluation of the waveform quality can be obtained per pixel, without any additional signal acquisition equipment. Since the sinusoidality of the illumination signal is a *sine qua non* condition for using the four phases algorithm, the content of high order harmonics is an appealing indicator of the signal distortion in our context. The Total Harmonic Distortion (THD) is a well-known parameter in electrical engineering that quantifies the ratio between the signal power due to the harmonic components and the power of the fundamental frequency. This parameter is often used to evaluate the degradation of the waveform in AC power networks supplying power electronic devices switching at high frequencies. Such high-frequency switching, necessary for, e. g., PWM speed control of motors, generates short-duration voltage peaks in the network that increase the THD of the signal and might even trigger upstream protection devices. For completeness, we provide a general formula for the THD in Eq. 4.28,

$$\mathrm{THD} = \frac{\sqrt{\sum_{i=2}^{n} A_i^2}}{A_1} \tag{4.28}$$

where A_i, $i \in [1, n]$ is the RMS amplitude of the ith-harmonic, being A_1 the amplitude of the fundamental frequency. As stated in Section 2.4.1, the wiggling effect due to harmonic content is difficult to compensate beforehand and typically a lookup table is used for *a posteriori* compensation. Nevertheless, an illumination system with many emitters brings many free parameters into play, e. g., phase shift between signals emitted by different emitters, emitter positions and orientations, which can be adjusted to minimize the harmonic distortion and, therefore, the depth error due to wiggling effect. Being able to reconstruct the light signal received by each pixel allows online calculation of the THD per pixel and performing adjustments in the illumination hardware and control signal in order to minimize it or, at least, to get a distribution of THD as uniform as possible in pixel domain. Our experiments revealed that the THDs computed from CS-recovered waveforms are stable and constitute a good method to evaluate the quality of the illumination. The results of a THD evaluation of the results presented in Fig. 4.49 are given in Table 4.7.

	THD(%)					
Size(n)	Th. Square	ICS	Rec. 50%	Rec. 33%	Rec. 25%	GT
32	47.93	39.43	18.16	17.84	19.58	24.72
64	48.24	38.75	18.39	17.66	18.73	25.64
128	48.32	38.10	19.35	18.23	9.44	27.35

Table 4.7.: Total Harmonic Distortion (THD) of the recovered illumination waveforms compared to those of a theoretical square signal, the real square illumination control signal (ICS) and the ground truth (GT) light signal. ©2014 IEEE.

The second column of Table 4.7 shows the THD of a discrete theoretical square signal, which is known to be 48.34% in the continuous case, in order to expose the approximation error due to discretization. Note the positive effect of the waveform smoothing produced by the LED response, which brings the illumination waveform closer to a sinusoidal. It is also observable that the GT light signal exhibits higher THD than the recovered signals. This is due to the fact that the light signal arriving to the central pixel is a mix of the signals emitted by almost all the LEDs of the illumination system. Different LEDs exhibit slightly different rising and falling characteristics, therefore, such signal superposition leads to a waveform which is smoother at the borders (closer to a sinusoidal), as it is visible in Fig. 4.49.

Full-array Evaluation Until now, results have been presented for the central pixel of a PMD array, i. e., the PMD chip has been used as a monochannel oscilloscope to sense the waveform of the light signal and compute the THD online. We take a step forward and, preserving the simple setting described above, with the illumination system in front of a white plain wall, we recover the illumination waveform for *all* the pixels of the PMD sensor. This provides an *image of waveforms* that cannot be visualized directly but that characterizes the performance of the illumination system in spatial domain. This has no additional cost in terms of acquisition time, since every PMD acquisition delivers full images, containing the values of A and B channels of all pixels at the end of the integration.

We evaluate the light intensity distribution calculating the RMS value of all recovered waveforms in time domain. This way we obtain an image where each pixel value is equal to the RMS value of the light signal that is receiving. If the signals are correctly recovered, the RMS image should not differ from the illumination intensity image in more than a linear scaling. Both images are provided in Fig. 4.50, where Fig. 4.50a shows the real illumination intensity, computed as the sum of A and B pixel channels, and Fig. 4.50b is the RMS image, computed from recovered waveforms. Indeed, the images do not differ in more than a constant scaling factor. Only very few pixels show an appreciable difference, revealing a very good performance of the CS recovery framework.

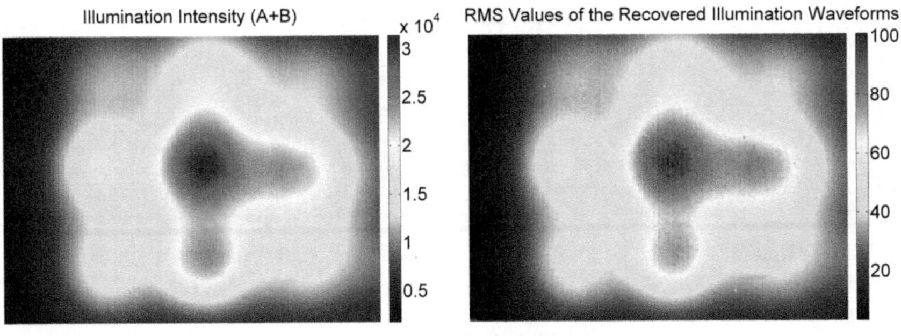

(a) Intensity (A+B)

(b) RMS Image

Figure 4.50.: Root Mean Square (RMS) value of the recovered illumination waveforms (b) in comparison to the real light intensity (a), computed as the sum of both pixel channels (A+B), in pixel units. The proportionality between the RMS values and the real intensities is an indicator of successful recovery. ©2014 IEEE.

The THD provides a characterization of how far the waveform is from a sinusoidal. A THD image will reveal in which areas the wiggling effect is more significant. Nevertheless, the THD does not characterize irregularities in the phase of the signals, which might appear if the illumination modules do not emit synchronously (Appendix A.6). Consequently, the phase of the recovered waveforms is also a parameter to be evaluated. We compute the phase shift as the phase value for which the cross-correlation between the recovered signal and a reference signal exhibits the absolute maximum. The reference signal is chosen to be a square signal (ideal ICS) with falling edge at the middle of the period. The phase shift with respect to this arbitrary signal is irrelevant for a single pixel, but becomes relevant when computed for all pixels, since it allows easy identification of the phase shifts introduced by the different illumination modules.

The THD image is shown in Fig. 4.51a and the phase shift image in Fig. 4.51b. The parameters were calculated from recovered waveforms of size $n = 128$. This yields a relatively poor phase resolution of 3°, which explains the lack of smoothness in the images of Fig. 4.51. Higher signal dimensionality leads to smoother images, at the cost of longer acquisition and recovery times.

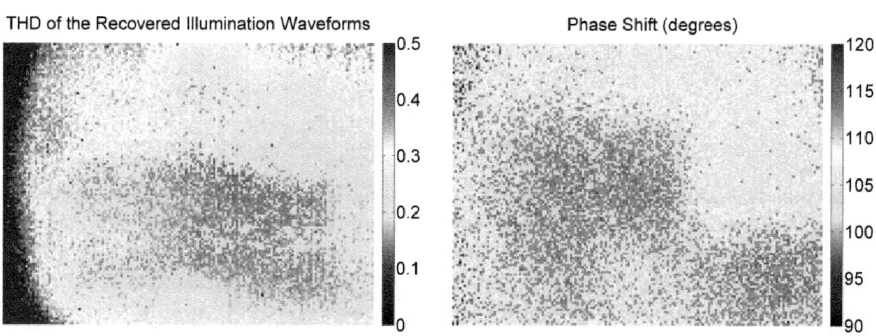

(a) THD Image (b) Phase Shift Image

Figure 4.51.: Total Harmonic Distortion (THD) of the recovered illumination waveforms (a) and their phase shift with respect to an arbitrary square signal (b). The lower the THD, the closer the illumination signal is to a sinusoidal and the lower will be the error in depth due to the presence of harmonics. Illumination modules adding phase delay to the illumination signal cause a depth offset in the areas of the scene they illuminate. ©2014 IEEE.

Fig. 4.51a reveals a good quality of the illumination signal shape, with typical THDs ranging from 0.15 to 0.25. The lowest values are found in the central and right part of the image, which corresponds to the place where the resulting illumination signal is a combination of signals coming from almost all emitters. Areas close to the borders exhibit higher THD due to the fact that the signal is coming from few or a single emitter and the smoothing effect of the mixing of slightly different signals is not present. The left border is an area where the poor illumination intensity (see Fig. 4.50) produces very noisy recovered waveforms.

Fig. 4.51b can be interpreted as a graphical representation of a depth offset that will be superimposed to the depth profile of any scene being sensed. Note that a phase offset difference of 5°, at 20 MHz modulation frequency, is equivalent to a depth difference of more than 10 cm. From Fig. 4.51b one can deduce that the central-left modules, together with the bottom-right ones, are significantly delayed with respect to the rest and, especially, to the central-right modules. Left and right refer to the images and corresponds to the left and right of the illumination system when observed from behind, not frontally.

4.5.1.4. Compensation of Depth Distortions

A natural idea is to use the phase shift map (Fig. 4.51b) to compensate the effect of phase inhomogeneity of the illumination system in real PMD depth images. If the phase image is correct, this should decrease the depth error. Nevertheless, the result in Fig. 4.51b cannot be used directly for compensation, since the phase shift image also depends on the scene being sensed during the experiment (plain wall) and contains an additional constant offset. Therefore, the phase shift image in Fig. 4.51b, $\Delta\Theta_{\text{total}}$, is the superposition of three fundamental phase shift images, as established in Eq. 4.29:

$$\Delta\Theta_{\text{total}} = \Delta\Theta_{\text{offset}} + \Delta\Theta_{\text{wall},d} + \Delta\Theta_{\text{illum}} \tag{4.29}$$

where $\Delta\Theta_{\text{offset}}$ is a constant image and models the constant delay added by the common electronics and cables from the camera to the illumination modules, assuming ideal behavior thereof. $\Delta\Theta_{\text{wall},d}$ is the theoretical phase shift map obtained when sensing a wall, orthogonal to the focal line of the camera, at a distance d from it. $\Delta\Theta_{\text{wall},d}$ can be computed from the FOV and d, but the real d is unknown, since it is tied to the constant phase offset, also unknown. $\Delta\Theta_{\text{illum}}$ is the real phase shift pattern that the illumination system is superimposing, which we want to obtain. In order to solve the problem, and without loss of generality, we need to assign an arbitrary zero delay at some point of $\Delta\Theta_{\text{illum}}$, i. e., a point where the illumination system has no influence on the depth estimation, exhibiting all the other areas a positive or negative illumination-related phase delay with respect to it. We choose this point to be the center of the image, indexed by $\vec{x_c}$, where the quality of the illumination signal is highest (see Fig. 4.51a). In that case, Eq. 4.29 yields:

$$\begin{aligned} \Delta\Theta_{\text{total}}(\vec{x_c}) &= \Delta\Theta_{\text{offset}}(\vec{x_c}) + \Delta\Theta_{\text{wall},d}(\vec{x_c}) \\ &= \delta\theta_{\text{offset}} + \delta\theta_d \end{aligned} \tag{4.30}$$

where $\delta\theta_{\text{offset}}$ is the scalar offset value and $\delta\theta_d$ the phase shift that corresponds to the distance d. Eq. 4.30 ties the offset and the distance camera-wall together and, together with the camera model, which is needed to generate the depth map of the wall for a given distance (recall Eq. 4.12), is used to solve Eq. 4.29 in an iterative fashion. In reality, for distances camera-wall not too short and not too wide FOV, $\Delta\Theta_{\text{wall},d}$ is close to a constant image and the pure illumination-related phase shift image, $\Delta\Theta_{\text{illum}}$ looks similar to the original $\Delta\Theta_{\text{total}}$. Nevertheless, if $\Delta\Theta_{\text{illum}}$ is to be used for *a posteriori* compensation of depth images, the depth map induced by the wall setup should not be omitted in Eq. 4.29.

As an additional check of the quality of the obtained illumination-related phase shift image, $\Delta\Theta_{\text{illum}}$, we use it to compensate a real PMD depth image, acquired using the same illumination system. To this end, a simple scene is set up, with a vertical panel and a table placed in front of a wall and two balls of different sizes on the table. The camera used for this experiment is also the MultiCam, which provides a color image (Fig. 4.52a) and a depth image (Fig. 4.52b) of the scene. The phase shift pattern produced by the illumination, $\Delta\Theta_{\text{illum}}$, is subtracted from the original depth image to obtain the compensated depth image, shown in Fig. 4.52c. The original and illumination-compensated depth images look similar to each other, but differences can be observed in the surface of the panel at the right (area delimited by the red square). The depth maps in Fig. 4.52b and Fig. 4.52c are projected into the point clouds in Fig. 4.52d and Fig. 4.52e, respectively, using the known camera parameters. Only the area within the cyan rectangles is shown in the point clouds. The slightly noisier depth image obtained after compensation (Fig. 4.52c) is due to the originally noisy phase shift image obtained from the recovered waveforms (Fig. 4.51), which is a direct consequence of the large discretization step. This effect is also observable in Fig. 4.52e.

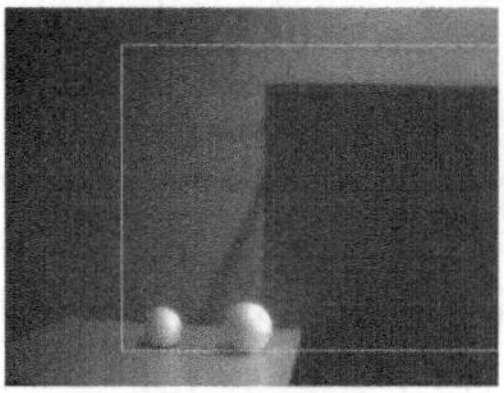

(a) Color Image

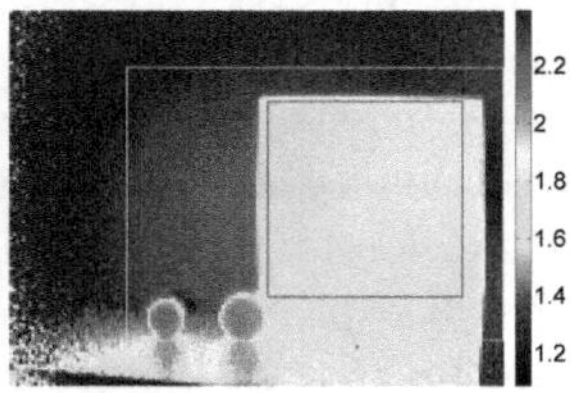

(b) Original Depth Image

(c) Compensated Depth Image

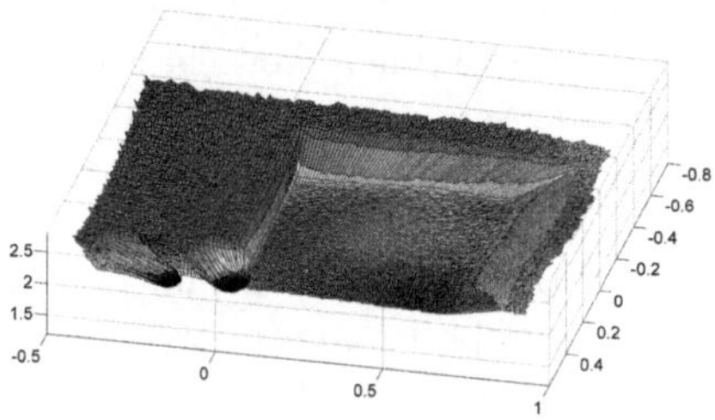

(d) Original Point Cloud

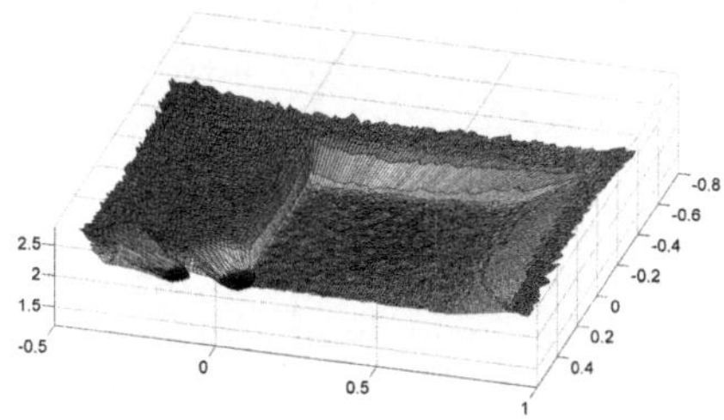

(e) Compensated Point Cloud

Figure 4.52.: Color (a) and depth images of the scene, before (b) and after (c) per-pixel compensation of the depth offset introduced by the illumination system, and corresponding point clouds. Large errors in the surface of the panel placed at the right (protuberance in cyan in (d)) disappear after compensation (uniform color in (e)). All scales are in meters. ©2014 IEEE (only (a), (d) and (e)).

In order to quantify the improvement in terms of depth error, we consider the area delimited by the red rectangle in the color image and compute the standard deviation of the z coordinate for both point clouds. Provided that the area corresponds to a plain panel, parallel to the image plane, the standard deviation should be close to zero. We obtain a value of 3.6 cm for the non-compensated point cloud and 2.0 cm for the compensated one, i. e., a depth error reduction of 44%.

4.5.2. Compressed Sensing Framework for Waveform Reconstruction and Phase Retrieval

The good quality of the AMCW illumination waveform of a PMD-based ToF camera in a CS fashion presented in Section 4.5.1.2 encouraged the development of a more general framework for waveform reconstruction using CS. The main drawback of the simple approach introduced in Section 4.5.1.1 is that the measurements are not actually gathered in a random manner, incoherent with the sparsity dictionary. Incoherence is achieved *a posteriori*, considering as measurements different pseudorandom (PR) combinations of the original measurements. Despite this improves the conditioning of the problem, making the sensing matrix look random and allowing a CS recovery, it poses some issues. Since the effective measurement kernels are binary weighted (± 1 weight values) combinations of the real measurement kernels, the frequency spectrum of the effective kernels is bounded by that of the real ones. Provided that the real measurement kernels are just shifted square signals, all of them share the same frequency spectrum. The frequency spectrum of a square wave is known to follow an inverse function, being the even harmonics zero. In other words, the amplitude of the even harmonics is inversely proportional to its frequency. This imposes a severe limitation, since one cannot hope to recover frequencies that cannot be actually captured by the measurement kernels, regardless of the sparse recovery method. Additionally, using shifted square waves as real measurement kernels lead to eventually high coherence between them, i. e., redundancy of the measurements, which translates into the need of a relatively large number of *real* measurements.

Such limitations cannot be solved without having full control on the shape of the reference signal, which requires a hardware modification of the PMD frontend. Using custom waveforms instead of shifted versions of the ICS as reference signals in a PMD frontend requires reprogramming the CPLD of the frontend and is contemplated in the CS-PMD prototype camera presented in Chapter 5 (see Section 5.2.2 for implementation details). Since the output signal of the CPLD used as reference signal by the PMD chip to control the two-channel integration process is a binary signal, the best we can do is to use PR binary codes as sensing kernels. In this section we describe this framework in detail and provide simulation results that show that recovering all relevant frequencies of a periodic but non-sinusoidal AMCW illumination signal (i. e., fully recovering the signal, implicitly including its phase) from few PR measurements is feasible, thanks to the compressibility of such signal in frequency domain [H3].

4.5.2.1. PMD Revisited: Turning Disadvantages into Opportunities

As pointed out in Section 4.5.1.1, commercial PMD cameras run in four phases operation, i. e., four acquisitions using four shifted versions of the ICS as reference signals are carried out. From a CS perspective, the measurements are equivalent to apply different sensing functions (or kernels, in the discrete case) to the signal to recover. Since both functions are periodic with equal fundamental frequency, restricting the analysis to a single period suffices. A set of measurements obtained over several signal cycles only differs in a scaling factor from another taken over a single period. Considering differential (A-B) measurements and provided that the four phases are equally spaced ($\theta \in \{0°, 90°, 180°, 270°\}$), the set of four sensing kernels is redundant, since two of them are equivalent to the other two. If one of these pairs is eliminated, the remaining two kernels are orthogonal between each other. Consequently, the maximum number of frequencies one can recover from the measurements is one, since there are two free parameters per frequency component: amplitude and phase shift, provided that the offset is removed using differential measurements. This is coherent with the basic assumption of the four phases algorithm (Eq. 2.19), which assumes that the illumination signal is purely sinusoidal. In the common case that the illumination signal is not exactly sinusoidal, the only chance to get a correct phase shift estimation from these four measurements is that the effective reference signals are sinusoidal, despite being originated as square, due to low-pass filtering effects of the circuitry. Unfortunately this also does not hold. Consequently, the four phases estimation frontally collides with the following facts regarding the PMD technology:

1. The reference signal used for correlation at pixel level is not sinusoidal, neither a perfect square signal. The shape was shown to be slightly trapezoidal in Section 4.3.1.3, closer to square than to sinusoidal. Equivalent results were obtained in [304] using a different methodology. The non-zero rising and falling times of the trapezoidal waveform are a coherent explanation for the relatively low values of PMD demodulation contrast presented in [330].

2. LED-based illumination systems impose a limitation on the modulation frequency. Additionally, the typically asymmetric rising and falling times observed when using a square driving signal invalidate the assumption of a quasi-sinusoidal illumination.

3. MPI is present in most real environments, due to multiple reflections on the floor and neighboring walls, or simply to the presence of translucent objects.

Note that they are presented as facts of the technology, which are to be handled appropriately, but not as intrinsic disadvantages. The first two invalidate Eq. 2.19 and require considering the measurements as samples of the cross-correlation between two non-sinusoidal signals (Eq. 2.15). Additionally, regardless of the sinusoidality of the illumination waveform, the third fact will make Eq. 2.1 deliver a wrong result. Even more, it is not possible to natively deal with multipath effects with only these four (effectively two) measurements, because a single frequency does not suffice to solve multipath ambiguities. From a CS point of view, the previous facts can be reinterpreted as systematic advantages as follows:

1. The trapezoidal shape of the reference signals is an indicator of the relatively high bandwidth of the electronics, which allows for using sensing waveforms with wider frequency spectra and capturing more information per measurement.

2. The periodic AMCW illumination signals are extremely sparse in frequency domain and can be recovered from few measurements. The low-pass filtering effect of the LEDs helps in this direction.

3. If the measurements contain enough information to accurately recover the illumination waveform in a sparse domain, the phase shift with respect to a reference can be determined and multiple paths might be detected and separated.

Note the fundamental change in the sensing paradigm: in a four phases framework, targeted information is gathered in few measurements, that are redundant. Unfortunately, the effect of unmodeled systematic irregularities cannot be eliminated by redundancy, as noise of random nature is reduced by averaging. Our proposal is to maximize the gathered information by capturing non-redundant information, while keeping a low number of measurements. This way the multiple frequencies contained in the illumination signals can be recovered and used in our own interest, e. g., to tackle the MPI, instead of being just a source of systematic irregularities.

4.5.2.2. Methodology

Conceptually, our method combines CS theory with phase-shift-based ToF imaging. Most works in the literature trying to bring the advantages of a CS problem formulation to the area of ToF imaging are based on the concept of the single pixel camera, introduced in [156]. Consequently, they require an external spatial modulator and gather measurements of random combinations of the signals reaching the different pixels of the modulator. Since these approaches operate in spatial domain, they are briefly introduced in Section 4.4. Our approach is closer to [248] than to single pixel approaches. They adopt the PMD 19k-S2 sensor, similar to the 19k-S3 we use. Differently to them, our goal is not to recover time profile movies that monitor the light in motion, but to improve the depth estimation in ToF cameras. They use m-sequencies as coding strategy, both for the illumination and the reference signal, achieving an autocorrelation function that approaches a Dirac delta function as the length of the code grows. We do not code the illumination signal, in order to soften the requirements on the illumination system and allow for low-cost LED solutions. The methods in [45] and [189] use sinusoidal waves of different frequencies as illumination signals and a sparse backscattering vector is recovered for each pixel. Both approaches focus on recovering multiple paths under sparsity and compressibility conditions, respectively. The method in [189] deals with diffuse multipath and was demonstrated with an Xbox One sensor [471], with only three modulation frequencies, while the method in [45] requires 77 frequencies to solve a multipath of known sparsity of 3, using a PMD 19k-S2 sensor. Differently, we propose to take profit of the harmonic content of the illumination signal in conventional ToF imaging cameras, rather than using laser illumination to emit sequentially pure sinusoidal signals of different frequencies.

General Formulation The measurements delivered by a PMD pixel are just samples of the cross-correlation function between the reflected light signal, as received by the pixel, $r(t)$, and a certain reference signal $q(t)$. Let both signals be periodic with equal period $T = f_{\text{mod}}^{-1}$, being f_{mod} the fundamental frequency of the ICS and, therefore, also of the emitted illumination signal $i(t)$, from which $r(t)$ is just a reflection (recall Eq. 2.9). Then, the cross-correlation in time domain (Eq. 2.11) can be rewritten in phase domain as:

$$c_{q,r}(\theta) = \frac{k_{\text{exp}}}{2\pi} \int_0^{2\pi} q^*(\phi) r(\phi + \theta) d\phi \Leftrightarrow (q \otimes r)(\phi), \ \forall \phi \in [0, 2\pi] \qquad (4.31)$$

where k_{exp} is just a constant that depends on the exposure time t_{exp}, i. e., $k_{\text{exp}} = t_{\text{exp}} T^{-1}$. As in most real applications, let's suppose that the illumination system is composed by LEDs, driven synchronously with a square signal of period T. Each emitter superimposes a smoothing effect on the signal, given by the function $g_{\text{illu}}(\phi)$, which vanishes for increasing $|\phi|$. Then the emitted light signal is given by $i(\phi) = (ICS * g_{\text{illu}})(\phi)$. If the rising and falling times of the LEDs are similar, g_{illu} can be approximated by a Gaussian function. The scene geometry shifts the signal phase according to the depth, so its effect in the received signal can be modeled as a displaced Dirac delta function, under the hypothesis of a single bounce (Eq. 2.8). If the signal received at the pixel is the superposition result of multiple signals following different paths (non-negligible MPI effects), the scene response $e(t)$ is a finite sum of Dirac delta functions, which in phase domain yields:

$$e(\phi) = \sum_{k=1}^{P} a_k \delta(\phi - \phi_k) \tag{4.32}$$

where ϕ_k is the phase shift that the illumination signal undergoes when following the path $k \in [1, P]$ and a_k is the attenuation factor due to non-unit reflectance of the surface point that produced that reflection. We restrict our analysis to a finite number of paths P. Then, the periodic signal $r(\phi)$ received by a PMD pixel can be written as:

$$\begin{aligned} r(\phi) &= (i * e)(\phi) = (ICS * g_{\text{illu}} * e)(\phi) \\ ICS(\phi) &= \begin{cases} v_{\text{max}}^{\text{ICS}} & \text{if } 0 \leq \phi < \pi \\ v_{\text{min}}^{\text{ICS}} & \text{if } \pi \leq \phi < 2\pi \end{cases} \end{aligned} \tag{4.33}$$

where $v_{\text{min}}^{\text{ICS}}$ and $v_{\text{max}}^{\text{ICS}}$ are signal levels corresponding to a logic '0' and '1' in the ICS, respectively. The pixel measurements are the result of the cross-correlation between the received light signal $r(\phi)$ and a reference signal $q(\phi)$,which controls the two-tap integration process. Despite the PMD chip handles binary signals, we have shown that the effective reference signal at pixel level shows a trapezoidal shape, rather than square, with quite symmetric rising and falling characteristics, whose width is close to $\frac{\pi}{3}$ in phase, at $f_{\text{mod}} = 20\,\text{MHz}$. Therefore, we can write the effective reference function $q(\phi)$ in terms of the theoretical one $q_{\text{th}}(\phi)$ as:

$$\begin{aligned} q(\phi) &= (q_{\text{th}} * g_{\text{pix}})(\phi) \\ g_{\text{pix}}(\phi) &= \text{rect}\left(\frac{\phi}{\frac{\pi}{3}}\right) = \begin{cases} 0 \text{ if } \phi < -\frac{\pi}{6} \vee \phi \geq \frac{\pi}{6} \\ 1 \text{ if } -\frac{\pi}{6} \leq \phi < \frac{\pi}{6} \end{cases} \end{aligned} \tag{4.34}$$

where $q_{\text{th}}(\phi)$ denotes the theoretical square reference signal and $\text{rect}\,(x)$ is the rectangular function, one-valued for $-\frac{1}{2} \leq x < \frac{1}{2}$ and zero elsewhere.

Consequently, $g_{\text{pix}}(\phi)$ is just a rectangular function of $\frac{\pi}{3}$ width. Eq. 4.34 is valid for conventional PMD operation at f_{mod} close to 20 MHz. The theoretical reference $q_{\text{th}}(\phi)$ is a displaced version of the ICS, so, for each one of the phases θ to be sampled (four in the classical four phases framework), we have: $q^{\theta}_{\text{th}}(\phi) = (ICS * \delta^{\theta})(\phi)$, where $\delta^{\theta}(\phi)$ is a Dirac delta function centered at $\phi = \theta$.

Sensing in a Compressed Manner with PR Codes The novelty of our approach is to use PR binary codes as reference signals. The goal of such codes is to maximize the gathered information, by means of a relatively wide frequency spectrum, and minimize measurement redundancy, by assuring certain orthogonality between codes. The length of the codes is given by the ratio $m_{\text{max}} = \frac{2f_{\text{CLK}}}{f_{\text{mod}}}$, being f_{CLK} the frequency of the clocking signal used to trigger the transitions between one code value and the next. Consequently, the total length of the code is equivalent to $T = f_{\text{mod}}^{-1}$ in time domain or 2π in phase domain.

Provided that each acquisition takes $t_{\text{exp}} = k_{\text{exp}}T,\ k_{\text{exp}} \gg 1$, the code is repeated k_{exp} times during the exposure time. In this case, the real reference signal is given by Eq. 4.35, where $i_k \in [1, m_{\text{max}}]$ is the code index, indexed in turn by the measurement index k, and $j \in [1, m_{\text{max}}]$ the element index within the code. The number of measurements is $m \leq m_{\text{max}}$. Consequently, $k \in [1, m]$ and the mapping $k \rightarrow i_k$ only ensures that m codes are picked randomly between the m_{max} possibilities. The rectangular function $g_{\text{pix}}(\phi)$ is linearly rescaled according to m_{max}, so that it does not impose an excessive limit in frequency, being the new width $\frac{1}{m_{\text{max}}}\frac{\pi}{3}$.

$$
\begin{aligned}
q^{k}(\phi) &= (q^{k}_{\text{th}} * g_{\text{pix}})(\phi) \\
q^{k}_{\text{th}}(\phi) &= q_{i_k,j} \Leftrightarrow (j-1) \leq \frac{\phi}{2\pi} m_{\text{max}} < j
\end{aligned}
\tag{4.35}
$$

We propose to derive the codes from a Hadamard matrix of size $m_{\text{max}} \times m_{\text{max}}$, as the scrambling operator $\boldsymbol{S}$ introduced in Section 4.5.1.1, i. e., randomly permuting its columns. Similarly, and for the sake of simplicity, we only consider sizes $m_{\text{max}} = 2^{k},\ k \in \mathbb{N}$, but other sizes are also feasible [253]. The resulting matrix is denoted by $\boldsymbol{Q}_{\text{th}}$, and contains the PR codes by rows, i. e., $q_{i,j}$ (Eq. 4.35) is the element i, j of $\boldsymbol{Q}_{\text{th}}$. From a CS perspective, our sensing functions are $q^{k}(\phi)$ and each measurement, y_k is obtained as the scalar product between that sensing function and the signal $r(\phi)$, in phase domain, which still can be seen as a sample of their cross-correlation function at the origin.

Until this point, all signals were considered continuous functions of ϕ, including the binary PR codes, in order to emphasize that the measurement process carried out in the PMD pixels is continuous. Since in our CS framework the signals are always considered as vectors of finite dimensionality, the continuous measurement model needs to be discretized. The number of discrete steps per signal period, $n \gg m$, imposes a bound on the maximum achievable phase resolution and, therefore, has to be chosen according to the waveform reconstruction quality we want to achieve. Supposing that the discretization step is chosen high enough to not to have an influence on the reconstruction quality, then the limit on the reconstruction accuracy is given by the frequency spectra of the sensing functions. In other words, we cannot reconstruct what we do not measure, that is, if the signal to recover, $r(\phi)$, contains frequencies that are not present in the measurement functions, exact recovery is not feasible. Conversely, as a rule of thumb, the discretization step should be chosen to be at least twice as small as the period of the highest frequency contained in the sensing functions.

Fig. 4.53 is intended to illustrate the frequency spectrum covered by our PR binary sensing functions. Plots (b), (c) and (d) depict the absolute values of the DCT coefficients obtained for all possible Hadamard-derived sensing functions, $q^k(\phi)$, $k \in [1, m_{\max}]$, for $m_{\max} = 16$, which was found to be large enough to reach state of the art phase resolution[1] in our simulations. The envelope of the spectra is plotted in green (in the color version) and resembles the profile of the frequency spectrum of a pulse function. This is due to the fact that the theoretical codes are not more than the superposition of displaced pulse functions. For completeness, in Fig. 4.53a we provide the DCT spectrum of a pulse function of width equal to that of each code element in time domain. All functions analyzed in Fig. 4.53 were realistic, i. e., they were generated considering rising and falling edges of finite slope ($\frac{\pi}{3}$ width each), including the pulse in Fig. 4.53a. Plots (b), (c) and (d) of Fig. 4.53 are the result of three independent experiments, in which three different random patterns were used to scramble the columns of the original Hadamard matrix. Clearly, the set of sensing functions cover the same regions of the frequency spectrum in a similar way in all three experiments. The aggregate sensing spectrum densely covers the low frequencies until the $2m_{\max} = 32^{nd}$ harmonic (vertical red line), where the amplitude of the coefficients approaches to zero in the three cases.

[1]In private communications between the author and PMD, phase resolution values of approximately 0.5° in optimal conditions were reported. Recently, a value of 0.58° was published for the novel PMD-based distance sensor in [235]

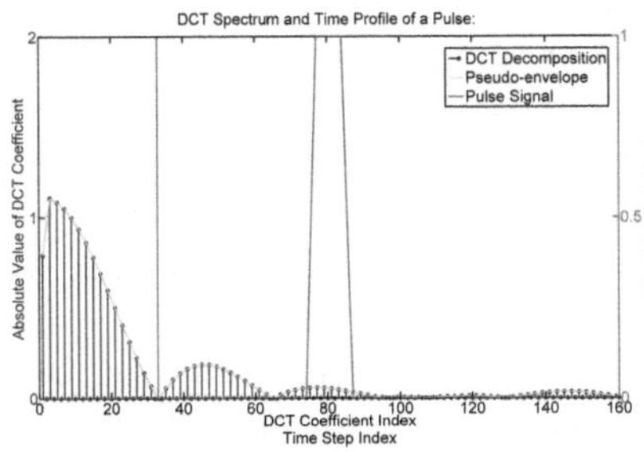

(a) Pulse DCT Spectrum

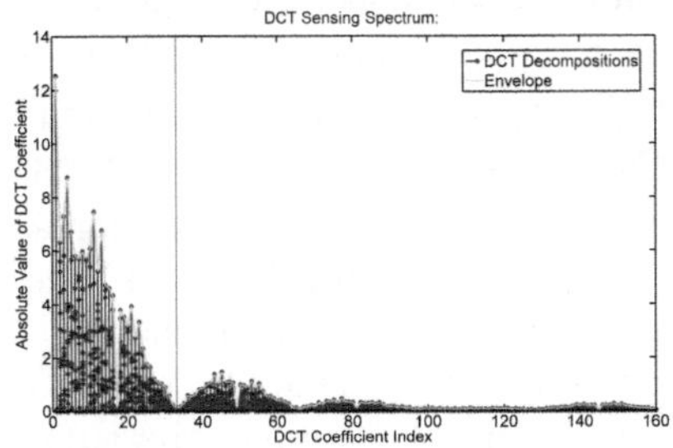

(b) PR Codes DCT Spectra (Exp. 1)

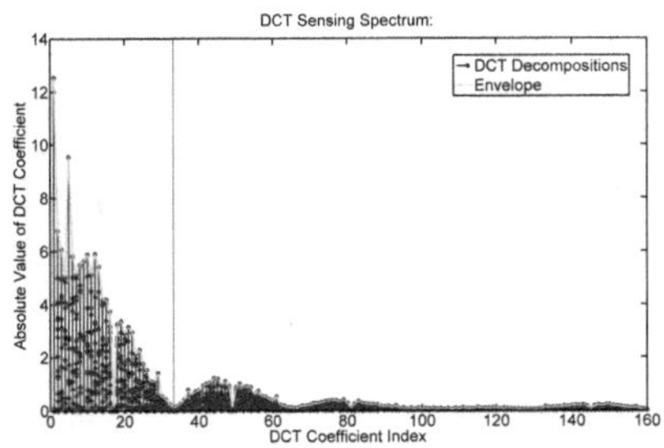

(c) PR Codes DCT Spectra (Exp. 2)

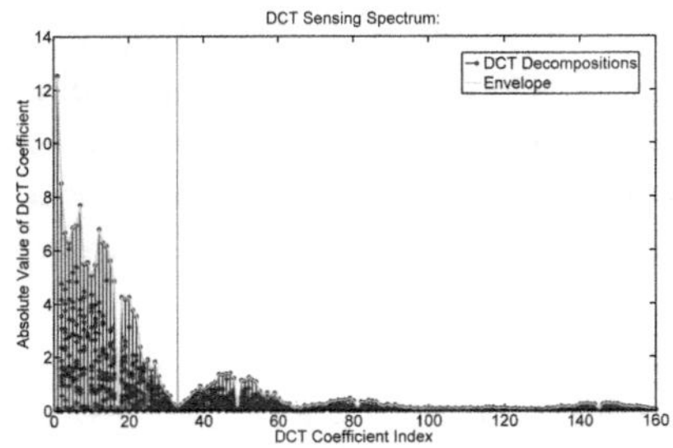

(d) PR Codes DCT Spectra (Exp. 3)

Figure 4.53.: Absolute values of the DCT coefficients of our measurement functions, obtained in three independent code generations, (b), (c) and (d). Realistic rising and falling times equivalent to $\frac{\pi}{3}$ phase were imposed to PR codes of length $m_{\max} = 16$. The basic signal underlying the codes is a pulse function, whose DCT decomposition is given in (a) for the pulse represented in the middle of the plot. Note the similarity between the envelope of the code spectra (in green in the color version) and the envelope of the pulse spectrum. In all cases the resolution was 160 steps per ICS period. In each plot a vertical red line is located at the $2m_{\max} = 32^{nd}$ harmonic (DCT index 33) in order to point the first zero-crossing. Do not confuse with the centered pulse in (a)

Compressive Sensing Formulation The sensing matrix of our CS model, $\boldsymbol{Q} \in \mathbb{R}^{m \times n}$, is, therefore, defined by rows as $Q_{i,j} = q^i(\phi_j)$, being each row $\vec{q^i} = \left[q^i(\phi_j)\right]_{j \in [1,n]}$ an n-dimensional discrete version of the reference signal in Eq. 4.35, for a custom number of discretization steps per period, n. The columns of the dictionary or representation matrix define a basis—or, more generally, a tight frame—where the signal to recover, $r(\phi)$, admits a sparse representation. We exploit the periodicity of the AMCW illumination signals and adopt a frequency-based dictionary, which is chosen to be the DCT basis for simplicity and coherence with Section 4.5.1.1, but other choices such as Fourier or Walsh basis are equally valid. In the discrete case, the aim is to recover an n-dimensional vector, $\vec{r} = [r(\phi_k)]_{k \in [1,n]}$, in its s-sparse representation, $\vec{x} \in \mathbb{R}^n \mid \|\vec{x}\|_0 := |\operatorname{supp}(\vec{x})| \leq s$, such that $\vec{r} = \boldsymbol{\Psi}\vec{x}$, being $\boldsymbol{\Psi} \in \mathbb{R}^{n \times n}$ the DCT basis by columns, in order of increasing frequencies. Then, we can obtain the measurement matrix of our CS model as $\boldsymbol{A} = \boldsymbol{Q}\boldsymbol{\Psi}$ and formulate the estimation of $\vec{x}$ as the convex program given by Eq. 3.10, where $\vec{y} \in \mathbb{R}^m$ is the vector of measurements.

Sparse Recovery As in Section 4.5.1.1, the primal-dual interior point method can be used to solve Eq. 3.10, but a variety of greedy algorithms have been proposed that achieve similar performance while reducing the runtime. With a future real implementation in mind, we consider the use of suboptimal greedy algorithms in order to keep an acceptable frame rate. Specifically, we adopt OMP (Section 3.4.2.2) for selecting the sparse support, which offers a good compromise between simplicity and recovery accuracy. Once $\vec{x}$ is determined, $\vec{r}$ can be immediately obtained from it: $\vec{r} = \boldsymbol{\Psi}\vec{x}$. The recovered signal, both in sparse or dense representation, contains, in principle, all the information related to the scene structure sensed by the pixel that can be obtained from a multifrequency AMCW ToF system, encoded as phase delays of the different frequencies. Note that we have uncoupled the acquisition from the phase shift estimation and, at this point, we count with full freedom regarding the methodology for phase retrieval. In the simplest case, cross-correlation with the original illumination waveform at some arbitrary zero phase can be used to retrieve the phase shift, admitting a fast FPGA implementation. A general simplification of this method would be to correlate against a delta function, i. e., finding the position of the maximum, which is unique within a period. If $s > 1$, multiple paths can be retrieved, formulating the multipath estimation as a $2P$-dimensional maximization problem, being P the number of paths (recall that there are two free parameters per path).

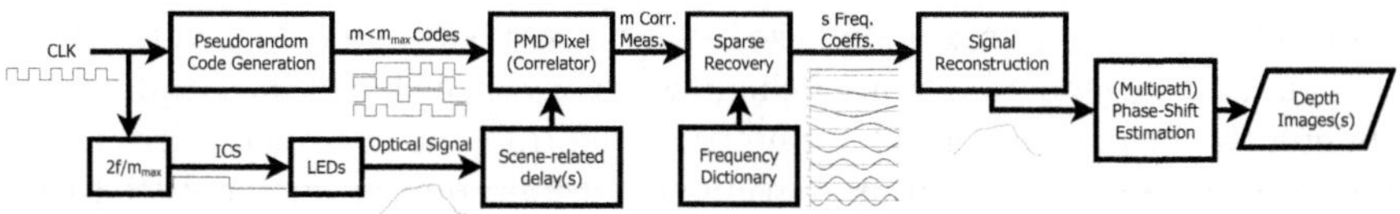

Figure 4.54.: Block diagram of the method exposed in this section. PR binary codes with $m_{\max}$ elements are generated from the high frequency clock signal CLK. Code transitions are triggered by rising and falling edges of CLK. The ICS is generated from CLK by frequency division. Note the factor 2 in the ICS frequency, due to bidirectional triggering. This allows for higher modulation frequencies. The illumination signal reaches the surface of the PMD pixel after being reflected by the scene and is cross-correlated against the binary codes. During the exposure time, the code is continuously repeated. The number of codes, $m < m_{\max}$, determines the number of PMD acquisitions and, therefore, the number of *compressed* measurements per pixel. A sparse representation of the signal is then recovered from the measurements in frequency domain. A single or multiple scene-induced phase shifts can be then retrieved from the reconstructed signal.

4.5.2.3. Simulation Results

In this section we evaluate the performance of our method in computer simulations that realistically recreate operation with a CS-adapted PMD camera. For a description of a real prototype featuring the necessary hardware modification, we refer to Section 5.2.2. The simulation framework provides measurements of reconstruction accuracy and runtime and was designed to evaluate feasibility of a hardware implementation. In short terms, the simulation can be divided in four sequential steps: generation of the ground truth (GT) data, sensing the GT data in a compressed way and sparse recovery as described in Section 4.5.2.2, (single- or multiple-path) phase retrieval and evaluation of the results.

Ground Truth Generation The GT is a 3D matrix containing the light waveforms received by each pixel, where the first two dimensions are the size in pixels of the PMD raw images and the third is n, the number of discrete points per period of the ICS. We adopt the PMD 19k-S3 chip, with a resolution of 120×160 pixels. Two different values of n are considered: 720

and 1440, yielding theoretical phase resolutions of 0.5° and 0.25°, respectively. The dataset is generated from a single waveform, which is scaled and shifted differently for each pixel. The *mother* waveform is not synthetic, but a real acquisition of a AMCW illumination signal, obtained from the NIR LEDs of our wide-area illumination system, driven with a square ICS of 20 MHz frequency. For the acquisition we use a fast photodetector together with a 2.5 Gs/s oscilloscope. The equipment is the same used for the characterization of systematic asynchronism between LEDs, reported in Appendix A.6. The result, as captured by the oscilloscope, is plotted in blue (in the color version) in Fig. 4.2. The waveform, shown only for one period (50 ns), contains high frequency noise that is not filtered and preserved after resizing to $n \in \{720, 1440\}$. In other words, the considered step sizes are small enough to preserve all information after discretization. The plot also contains the DCT decomposition (in red in the color version), which unveils the extreme sparsity of the signal in frequency domain. The original signal can be accurately reconstructed from as few as the highest 10% DCT coefficients (in green in the color version). Note that the sets of PR binary codes of length 16 analyzed in Fig. 4.53 would be appropriate to recover the waveform in Fig. 4.2, since their frequency spectra are wider than that of the signal to sense, which is a necessary condition for exact recovery.

The per-pixel scalings and phase shifts that are applied to the light signal to build the cubic dataset are not random. A real PMD acquisition of a reference scene is used as a guide, i. e., the IR intensity (A+B) image is used to scale the amplitude of the waveforms, while the phase image provides the phase shifts. The PMD data was acquired using our MultiCam, which features a PMD 19k-S3. The reference scene, as observed by the color sensor of the MultiCam, is depicted in Fig. 4.52a, while Fig. 4.52b provides the corresponding depth image. For brevity, we omit the phase image, which only differs from the depth image in a scaling factor. The modulation frequency of the illumination was also set to 20 MHz.

In our experiments we consider three different frequency ratios, that determine the size of the PR codes: $m_{\text{max}} = \frac{2f_{\text{CLK}}}{f_{\text{mod}}} \in \{8, 16, 32\}$. The higher this parameter is, the more measurements $m \leq m_{\text{max}}$ can be gathered and the wider the frequency spectrum of the sensing kernels is. Nevertheless, for a given f_{mod}, it is bounded by the maximum f_{CLK} of the CPLD, responsible for the clocking signal generation in the PMD frontend. For this reason, we choose $m = m_{\text{max}}$ for our experiments, in order to maximize the gathered information for relatively low values of m_{max}, provided that the frequency spectrum of the sensing kernels is wide enough (Fig. 4.53). For each PMD pixel, the sensing matrix $\boldsymbol{Q} \in \mathbb{R}^{m \times n}$ is used to obtain the vector of measurements, $\vec{y}$, from the GT data. Recall that the representation matrix $\boldsymbol{\Psi}$ is the DCT basis by columns and the full measurement matrix is $\boldsymbol{A} = \boldsymbol{Q}\boldsymbol{\Psi}$.

Recovery Considerations The fact that the frequency spectrum of our set of PR codes covers that of the signal we aim to recover means that we have the ability of sensing what we want to recover, i. e., that we are not blind to any component of the signal. Nevertheless, this is not a sufficient condition for exact recovery and a deeper analysis becomes necessary.

Our Hadamard-based PR sensing matrix is expected to be highly incoherent with any periodic basis, such as the DCT basis. Consequently, the entries of the measurement matrix $\boldsymbol{A}$ are expected to follow a distribution close to a zero-mean Gaussian. As a result, the coherence of $\boldsymbol{A}$ (Eq. 3.42) should be low [150, 439]. Consider the following limit-case scenario, in terms of coherence: $m = m_{\max} = 32$ measurements are performed, being the signal size per period $n \gg m$. For simplicity, and without loss of generality, consider a partial version of the dictionary, $\boldsymbol{\Psi}_{\Omega_s}$, where Ω_s is a set of indexes corresponding to the s lowest frequencies of $\boldsymbol{\Psi}$. If we adopt $s = s_{\max} = 32$, we obtain a square matrix $\boldsymbol{A}_{\Omega_s} \in \mathbb{R}^{m \times s}, m = s$. A histogram of the entries of $\boldsymbol{A}_{\Omega_s}$ is given in Fig. 4.55a, which shows clear resemblance to a zero-mean Gaussian distribution. Nevertheless, a close look to our measurement matrices reveals low variance along the first row and the first column, together with a relatively high value for the first element. This is due to the fact that the first measurement kernel (constant) and the first dictionary element are highly correlated between each other (due to its zero frequency), while equally uncorrelated with the rest. The first measurement kernel comes from the first row of a Hadamard matrix, which is a constant vector of ones. This row is always selected in the subset of m rows, because it generates the only measurement kernel gathering information on the DC offset of the signal, since all other rows of the Hadamard matrix exhibit zero mean. A direct consequence is the rise of the coherence of $\boldsymbol{A}_{\Omega_s}$, as depicted in Fig. 4.55b, where the absolute values of the coefficients of the Gram matrix are plotted in ascendant order. The Gram matrix of any set of vectors $\boldsymbol{A} = [\vec{a}_i]_{i \in [1,n]}$, stacked by columns in the matrix $\boldsymbol{A}$, is given by Eq. 3.44. In our case, $\boldsymbol{G}_{\Omega_s} = \boldsymbol{A}_{\Omega_s}^{\top} \boldsymbol{A}_{\Omega_s}$, after columnwise normalization of $\boldsymbol{A}_{\Omega_s}$. Obviously, values corresponding to diagonal positions are excluded from the plot. The coherence of $\boldsymbol{A}_{\Omega_s}$ is given by the maximum of this graph, which is as high as 0.245. Recall that the minimum coherence of $\boldsymbol{A}_{\Omega_s}$ is given by the Welch bound [461], which is zero for a square measurement matrix. Note that this issue can be easily addressed, since the first measurement kernel (DC) is totally incoherent with all the dictionary elements of frequency greater than zero, i. e., the value of the first coefficient of $\vec{x}$ is uniquely given by the first measurement. Therefore, we can refocus the coherence evaluation considering $s - 1 = 31$ dictionary elements. In such case, the plot in Fig. 4.55b would end at the red horizontal line and the coherence would be reduced to 0.031, i. e., in an 87%.

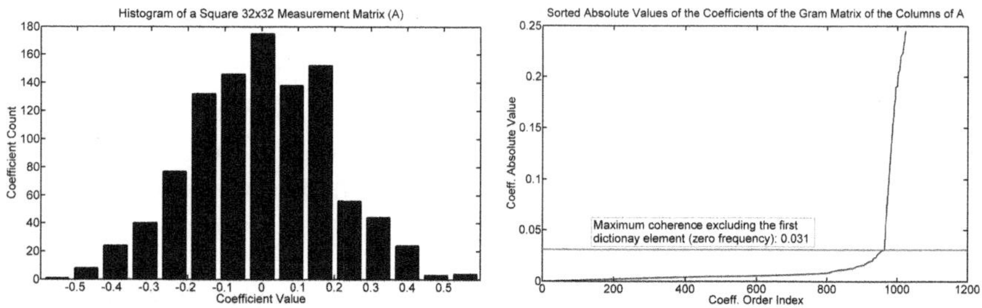

(a) Histogram of $\boldsymbol{A}$ **(b)** Gram Coeffs. of $\boldsymbol{A}$ cols.

Figure 4.55.: (a): histogram of the 32×32 measurement matrix ($\boldsymbol{A}_{\Omega_s}$) coefficients, after columnwise normalization. The number of bins is 15. Note the zero-mean and the close-to-Gaussian shape. (b): absolute values of the off-diagonal coefficients of the Gram matrix generated from the normalized columns of $\boldsymbol{A}_{\Omega_s}$, sorted in increasing order. The highest values are due to the high correlation between the first measurement kernel (DC) and the first dictionary element. If they are omitted, the coherence of $\boldsymbol{A}_{\Omega_s}$ boils down to 0.031 (red line). ©2015 IEEE.

If we denote by $\mu(\boldsymbol{A}_{\Omega_s}) = 0.031$ the coherence of $\boldsymbol{A}_{\Omega_s}$ neglecting the first sensing kernel and the first dictionary element, we have that $\mu(\boldsymbol{A}_{\Omega'_s}) \approx \mu(\boldsymbol{A}_{\Omega_s})$, $\forall s' > s$, since the frequency spectrum of the sensing kernels decays with frequency and additional columns of $\boldsymbol{A}$, originated by higher frequency elements of $\boldsymbol{\Psi}$, will tend to the zero vector and do not affect the coherence of the matrix. Consequently, we can conclude that $\mu(\boldsymbol{A}) = 0.031$. From the coherence and Eq. 3.54, we can derive an upper bound on the sparsity of $\vec{x}$. We obtain $s < 17$. For signals of $s > 17$ the uniqueness of the solution to Eq. 3.10 is not guaranteed.

Despite $\vec{r}$ can be well-approximated by the fundamental frequency plus 16 harmonics, we still can profit from an additional constraint on $\vec{x}$ that has not been taken into account yet, related to its *structured sparsity* [25]. If the signal is s-sparse, $\vec{x}$ should have only $s \ll n$ non-zero coefficients.

Additionally, the location of the non-zero coefficients is not completely free, but influenced by the empirical fact that the absolute values of the coefficients follow a power law decay. This is due to the square shape of the ICS and the low-pass filtering effect of the light emitters, typically LEDs (Eq. 4.33). Consequently, $|x_{k+1}| < |x_k|$, $1 \leq k \leq n$, being x_k the kth-element of $\vec{x}$ and the elements of the dictionary $\mathbf{\Psi}$ in order of increasing frequency by columns. This fact can be used to generate an *oracle-given* support for $\vec{x}$, provided that s is known beforehand. If the support is known, the only condition for the uniqueness of the solution to Eq. 3.10, is that $s \leq m$, being the problem solvable by least squares (LS). For comparison, in our simulations we consider both recovery with conventional OMP, stopping when $\text{supp}(\vec{x}) = s$, and recovery with known support, in which case, OMP converges to LS.

Performance Evaluation (Single Path) The GT waveforms, generated by shifting and rescaling a realistic reference waveform, emulate the reflected signal due to a single direct reflection in the scene. We do not consider multiple paths at this point, in order to provide an evaluation of the best performance in absence of depth ambiguities. The phase shift—or, equivalently, depth—is calculated by finding the maximum of the discrete normalized cross-correlation between the recovered signal, $\vec{r}$ and an illumination signal at zero phase. As introduced above, two different experiments are considered, one using OMP with maximum sparsity ($s = s_{\text{max}} = 32$) as stopping criterion and another using LS and an *oracle-given* support (s lowest-frequency DCT components) to recover the signal, denoted in the following as Exp. 1 and Exp. 2, respectively.

The performance of our method is evaluated in terms of phase RMSE, using the GT natural depth image as reference. The results are provided in Table 4.7a. The total time needed for signal recovery for all pixels is collected in Table 4.7b, in order to provide a speed comparison between recovery alternatives. The equivalence between Exp. 1 and Exp. 2 in terms of phase RMSE is an indicator of the validity of the support oracle for the natural signal considered. In terms of runtime, *a priori* knowledge of the sparse support yields an average runtime reduction of 11%, for the same sparsity. Consequently, in the following we only consider oracle-given supports.

In order to have a reference to compare the effects of the trapezoidal shape of the reference signal on the recovery, we consider an additional experiment (Exp. 3), where perfectly square reference signals are considered, i. e., with theoretical infinite slope ($g_{\text{pix}}(\phi) = \delta(\phi)$). Better phase RMSEs are achieved (Table 4.7a), while the runtimes coincide with those of Exp. 2 and are omitted in Table 4.7b.

The runtimes obtained for Exp. 1, 2 and 3 are so high that the final depth images would be delivered at one frame every $2-4$ minutes, precluding a real-time implementation. This is due to fact that our method works in a per-pixel basis, i. e., the signal recovery process is performed sequentially for each one of the 19200 pixels of the PMD 19k-S3 chip. Fortunately, there exists a simple way to boost it: provided that measurement matrix, $\boldsymbol{A}$, is the same regardless of the PMD pixel, we can recast the sequential SMV recoveries into a single MMV recovery (see Section 3.4.2.8). A fourth experiment (Exp. 4) is carried out to characterize the runtime improvement achieved using a known-support MMV recovery framework. The results show an average runtime reduction of 99.86% with respect to Exp. 1—i. e., three orders of magnitude—, while achieving exactly the same phase RMSEs. Runtimes in Table 4.7b are only for relative comparison, since they may differ depending on the compiler, operative system, machine and work load of the machine at the time the simulation runs. Nevertheless, the low runtimes achieved in Exp. 4 and the high parallelization potential of the MMV implementation suggest that a real-time GPU implementation is feasible.

What remains to be improved, at the light of the results for Exp. 1-4 in Table 4.7a is the phase RMSE for low m. Note the large gap between the phase RMSE obtained for Exp. 3 and for the rest of the experiments in the case of $m = 8$. This is due to the restricted frequency spectrum of the sensing kernels, unable to sense some of the highest-frequency harmonics present in the signal. For clarity, imagine a sensing spectrum such as those presented in Fig. 4.53, where the horizontal axis has been scaled by a factor $\frac{1}{2}$ (since $m_{\max} = m = 8 = \frac{1}{2}16$). The vertical red line would appear at the 16th harmonic and no further frequencies would be properly acquired. Including dictionary elements for frequencies that are poorly or not sensed by the sensing kernels means leaving more degrees of freedom than those strictly necessary, since there is no hope to recover what was not sensed. In order to avoid unnecessarily high errors due to wrong recovery of high-frequency harmonics that cannot be properly measured, we carry out a fifth experiment (Exp. 5), where the maximum sparsity of the signal to recover is bounded by the number of measurements, i. e., we impose $s = m$. This forces $\boldsymbol{A}$ to be square and the recovery problem to be determined. The last column of Table 4.7a and Table 4.7b for Exp. 5 was taken from Exp. 4, since they are equivalent. The results show a dramatic reduction of the phase RMSE for the case $m = 8$ (74%) and $m = 16$ (between 65% and 71%).

The MMV recovery framework with sparsity adaptation used in Exp. 5

Exp.	n	m: 8	16	32
1	720	2.85	1.05	0.144
	1440	2.84	1.01	0.115
2	720	2.85	1.05	0.144
	1440	2.84	1.01	0.115
3	720	1.66	0.630	0.144
	1440	1.66	0.579	0.119
4	720	2.85	1.05	0.144
	1440	2.84	1.01	0.115
5	720	0.762	0.375	0.144
	1440	0.745	0.297	0.115

(a) Phase RMSE (°)

Exp.	n	m: 8	16	32
1	720	121	191	262
	1440	198	239	270
2	720	113	139	262
	1440	144	172	337
3	720	—	—	—
	1440	—	—	—
4	720	0.138	0.148	0.141
	1440	0.326	0.676	0.386
5	720	0.131	0.179	0.141
	1440	0.405	0.397	0.386

(b) Runtime (s)

Table 4.7.: Phase RMSE (a) of the recovered signals with respect to GT and aggregate recovery runtimes (b), for different discrete signal sizes (n) and number of measurements (m), and for different experimental cases (Exp. #). Exp. 1 uses OMP with $s = s_{\max} = 32$. Exp. 2 replicates Exp. 1 with known support corresponding to the s lowest frequencies. Exp. 3 uses perfectly square reference signals. Exp. 4 jointly recovers the signal for all pixels using an MMV framework. Exp. 5 adapts the sparsity to the number of measurements: $s = m$. The aggregate runtimes are for all $120 \times 160 = 19200$ PMD pixels.

offers the best results, both in terms of RMSE and runtime. RMSEs lower than 1° are achieved in all cases considered, being as low as 0.12° for $m = 32$ and $n = 1440$, much lower than the 0.5° phase resolution limit characteristic of PMD-based ToF cameras. Fig. 4.56 shows the discrete signals $\vec{r}$ recovered in Exp. 4 (a) and Exp. 5 (b), for the pixel exhibiting lowest phase shift. These waveforms are to be compared to the GT waveform, plotted in blue (in the color version) in Fig. 4.2.

Note the high errors for $m = 8$ in Fig. 4.56a, produced by wrong estimation of the highest-frequency DCT coefficients, and compare it to the corresponding waveform in Fig. 4.56b, where the contribution of those high frequency components is neglected in order to keep a balance between sensed and recovered information. Consequently, waveforms corresponding to lower m in Fig. 4.56b are less faithful to the real signal than those for higher m.

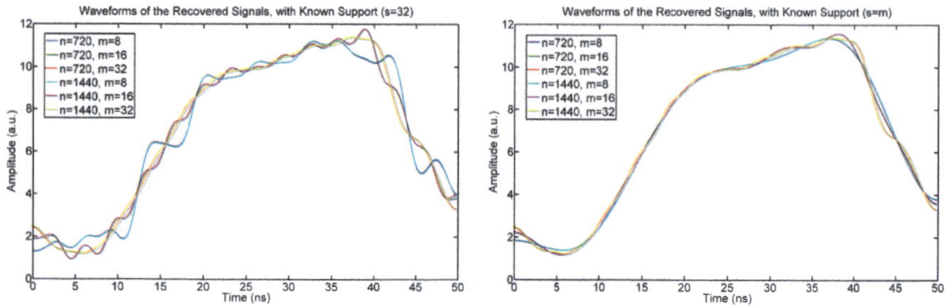

(a) Known Support $s = s_{\max} = 32$ **(b)** Known Support $s = m \leq s_{\max} = 32$

Figure 4.56.: Recovered waveforms for the pixel with lowest phase shift in Fig. 4.52b (52°) using an MMV framework with known support of fixed sparsity $s = s_{\max} = 32$ (a) and known support of adaptive size $s = m$ (b). Note that the waveforms for discrete signal size $n = 720$ are behind those for $n = 1440$, showing that n has a negligible effect on the quality of the recovery, being the number of measurements, m, the crucial parameter. Compare the waveforms to the GT in Fig. 4.2 (blue). ©2015 IEEE.

That is, a compromise is to be established between frame rate, i. e., number of measurements m, since the exposure time grows with m, and reconstruction quality. The final selection of the frequency ratio $m_{\max} = \frac{2f_{\text{CLK}}}{f_{\text{mod}}}$ and m depends on the application.

An Application Case: Multiple Path Retrieval It is out of the scope of the thesis to provide an efficient MPI compensation method. Nevertheless, our CS framework naturally allows for an additional step of multipath estimation from the recovered waveforms. The full knowledge of the waveforms ensures that multipath recovery is feasible, as far as the recovered waveform is an unambiguous combination of rescaled and displaced versions of the emitted illumination waveform. In the continuous case, the unambiguity requirement is met by an illumination signal of enough bandwidth. In principle one needs, at least, as many non-zero frequencies as number of paths to estimate. From Eq. 4.32, there are two unknowns per each of the P paths to estimate: a phase shift ϕ_k and an attenuation factor a_k, $k \in [1, P]$. Provided that the frequency spectrum of the illumination signal is wide enough ($s \gg P$), the paths can be retrieved if Eq. 4.36 holds.

$$|\phi_i - \phi_j| \geq \varepsilon_\phi,\ a_i, a_j > 0,\ \forall i, j \in [1, P], i \neq j \tag{4.36}$$

where the threshold ε_ϕ depends on the method used to solve the multipath estimation and, in a discrete implementation, is lower-bounded by the dimensionality of the signal, n: $\varepsilon_\phi \geq \frac{2\pi}{n}$. For n large enough, ε_ϕ depends exclusively on the quality of the recovered signal and, therefore, the phase RMSE of the recovered single-path signals (Table 4.7a) can be used as a lower bound. Using the scene response from Eq. 4.32, we cast the multiple path estimation into the minimization problem in Eq. 4.37

$$\begin{aligned} [\vec{\phi}, \vec{a}] &= \underset{\substack{\phi_k \in [0,2\pi) \\ \alpha_k \in (0,1]}}{\arg\min} \sqrt{\frac{1}{2\pi}\int_0^{2\pi} (r(\phi) - r'_{\vec{\phi},\vec{a}}(\phi))^2 d\phi} \\ r'_{\vec{\phi},\vec{a}}(\phi) &= i(\phi) * e_{\vec{\phi},\vec{a}}(\phi) \end{aligned} \tag{4.37}$$

where $\vec{a} = [a_k]_{k\in[1,P]}$, $\vec{\phi} = [\phi_k]_{k\in[1,P]}$ are the vectors of attenuation factors and phase shifts to recover, $r(\phi)$ is the real waveform received at the pixel surface and $r'_{\vec{\phi},\vec{a}}(\phi)$ is an estimate of $r(\phi)$ generated by multipath composition. Since we do not have access to the continuous function $r(\phi)$, but to the discrete recovered signal $\vec{r}$, we solve a discrete version of the minimization in Eq. 4.37, given by

$$[\vec{\phi}, \vec{a}] = \underset{\substack{\phi_k \in [0,2\pi) \\ \alpha_k \in (0,1]}}{\arg\min} \left\| \vec{r} - \vec{r}'_{\vec{\phi},\vec{a}} \right\|_2 \tag{4.38}$$

We adopted the simplex method [275] to solve Eq. 4.38, but custom minimization techniques can be used instead.

The multipath GT is generated using the PMD phase and intensity data obtained for the scene in Fig. 4.52a to delay and scale the original illumination signal, as before. The peculiarity is that in this case we also model a fictitious transparent panel, with a (diffuse) reflective coating toward the camera. The panel is located at 1 m from the camera, which is equivalent to an induced phase shift of $\phi_{\text{panel}} = 0.84$ rad at $f_{\text{mod}} = 20$ MHz. The panel is square of a size such that at 1 m distance covers 26×26 pixels. The reflectivity and transmitivity are set to $\rho_{\text{panel}} = 0.5$, $\tau_{\text{panel}} = 1 - \rho_{panel} = 0.5$, respectively. That is, the panel transmits half of the light, diffusely reflects the other half and does not absorb any light. For pixels belonging to the panel, there are, at least, $P = 2$ paths: one due to the panel reflection, plus another due to the transmitted light, which is reflected by the scene behind the panel. For simplicity, we assume that the light reflected by the scene is transfered through the panel back to the camera without attenuation. This way we avoid modeling a third path due to a second reflection on the back side of the panel. Consequently, for pixels within the area covered by the panel, the initial single-path GT waveforms are further rescaled according to τ_{panel} and summed to the waveform coming from the panel reflection, which is, in turn, obtained from the empirical waveform (Fig. 4.2), after the phase shift ϕ_{panel} and the scaling given by ρ_{panel}.

The recovery framework of Exp. 5 is used to reconstruct the composed signal from $m = s = 32$ measurements, with $n \in \{720, 1440\}$. Most minimization methods, including simplex, require an initial guess. We use the single-path phase shift estimation (Fig. 4.57a) as initial phase shift estimation for all paths and the RMS value of the recovered waveform to compute the initial guess on the intensity: $a_k^{(0)} = \frac{\text{rms}(\vec{r})}{P}, \forall k \in [1, P]$. Note that not all pixels are affected by MPI, while the presented approach estimates P paths for every pixel. In order to detect the single-path cases, we implement a simple threshold on the standard deviation of the vector of multipath phases, $\vec{\phi}$. If the threshold is not achieved, the estimated multipath parameters are discarded and the initial monopath estimation is adopted. In general, for $P \geq 2$ a threshold on some distance between elements of $\vec{\phi}$ can be used to detect how many paths are active for each pixel, and combine those that are repeated.

The results of the multipath recovery for $P = 2$ are shown in Fig. 4.57. Images in the first row were obtained using $n = 720$ and those in the second using $n = 1440$. Note that the monopath estimation is automatically adopted for those pixels exhibiting very low standard deviation of the vector $\vec{\phi}$ (pixels outside the panel area in Fig. 4.57c). In visual terms, the path separation seems to be perfect for both experimental cases. Fig. 4.57b is to be compared with Fig. 4.52b (note the different scale units). Regarding Fig. 4.57c, if the path retrieval is exact, the panel area should appear in the darkest blue of the scale, corresponding to $\phi_{\text{panel}} = 0.84\,\text{rad}$.

Provided that we have GT for both the panel and the scene behind, phase RMSEs can be computed for each experimental case. Note that, for the first path (Fig. 4.57b), the error is computed over the entire image, while for the second path (Fig. 4.57b) it is computed only for the area where this second path exists (the panel). The performance of our multipath estimation in terms of phase RMSE and waveform recovery runtime is given in Table 4.8.

Note the outstanding ratio between system complexity and quality of

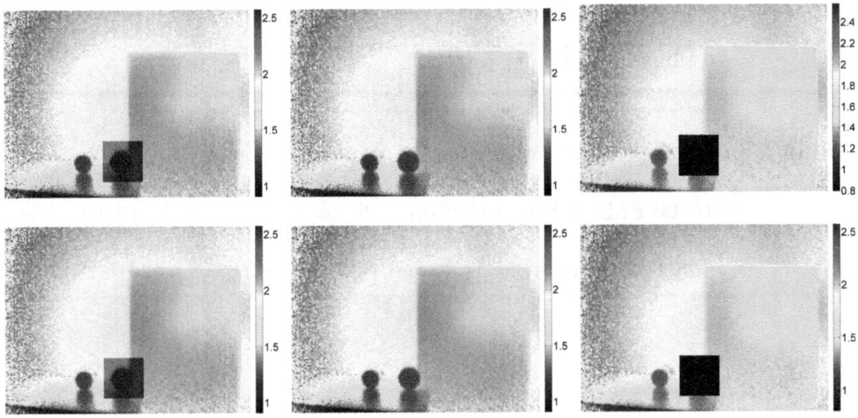

(a) Single Path **(b)** Multipath: Path 1 **(c)** Multipath: Path 2

Figure 4.57.: Phase images recovered in the multipath simulation, in radians. The first row shows the results obtained for a discrete signal size $n = 720$ and the second for $n = 1440$. In both cases $m = s = 32$ measurements were used. The phase images obtained from the recovered waveforms under single-path assumption (a) exhibit wrong phases in the area of the semitransparent panel, whose values are between those corresponding to the panel and those corresponding to the scene behind it. Our simple framework for multipath retrieval is able to accurately separate the path corresponding to the scene behind the panel (b) from the path corresponding to the panel (c), reconstructing its plain surface. ©2015 IEEE.

the results. No additional frequencies are required. The harmonic content of the non-sinusoidal illumination signal is used instead to achieve the capability of recovering multiple paths per pixel. Compare, for instance, to the multifrequency method in [45], where sequential acquisitions using 77 different frequencies are required to recover $P = 3$ paths.

n	720		1440	
Path ID	SCN	PNL	SCN	PNL
Phase RMSE (°)	0.252	1.11	0.175	0.834
Runtime (s)	0.232		0.497	

Table 4.8.: Recovery performance for the multipath case with $P = 2$ paths. Phase RMSE and aggregate recovery runtimes for all $120 \times 160 = 19200$ PMD pixels, for discrete signal sizes $n \in \{720, 1440\}$. The paths corresponding to the background scene and the foreground panel are denoted by SCN and PNL, respectively. ©2015 IEEE.

5. CS-PMD: A Compressive Sensing ToF Camera based on the Photonic Mixer Device

In this chapter we present a prototype of a CS-oriented ToF camera based on the PMD. The main goal of the prototype, which has been fully developed at the ZESS, is to serve as a platform where to implement the concepts of CS and sparse recovery in spatial and frequency domains presented in Section 4.4 and Section 4.5, respectively.

This chapter is organized as follows: Section 5.1 provides a general description of the system. Section 5.2 describes the hardware novelties with respect to a conventional PMD camera, namely the optical system allowing modulation in spatial domain, and the customized PMD module. Finally, Section 5.3 describes the methodology used for simultaneous sparse recovery, in both spatial and frequency domain, from the compressed raw data delivered by the camera and provides some implementation details.

5.1. General System Description

The CS-oriented PMD camera we introduce in this section is a prototype developed at ZESS with the intention of serving as test platform for the methods presented in Section 4.4 and Section 4.5, which operate either in the time-frequency domain or in the spatial-spatial frequency domain, respectively. These methods aimed to apply CS to conventional PMD-based ToF systems in order to obtain better results while keeping constant the number of measurements or, conversely, achieve the same quality with a reduced number of measurements. Exploiting prior knowledge on the system and the signals we deal with within the sparse recovery stage eventually allows both reducing the number of measurements while still being able to improve the result with respect to the non-CS result. This was, for instance, the case presented in Section 4.4.3, where the low rank of the set of PMD raw images, together with their structured joint sparsity was synergically exploited in a novel joint recovery framework. In that case, the improvement was measured in terms of depth accuracy and the framework did not contemplate superresolution, i. e., the size of the final depth image was determined by the size of the PMD array, which was also the maximum number of non-redundant measurements. If the improvement of the final depth image is also pursued in terms of lateral resolution, some additional hardware is required to perform custom modifications on the spatial responses of PMD pixels, that is, to control whether a photon generates or not a carrier to be integrated by the pixel (be the integration in the A or B channel). A similar situation occurs in time-frequency domain. The PMD pixels are supposed to perform the cross-correlation between two sinusoidal signals of equal frequency. This imposes a severe limitation in the sensing process and, in an ideal case, PMD measurements can be interpreted as Fourier measurements of the reflected illumination signal, as it arrives to the pixel surface. Unfortunately, neither the illumination signal nor the reference signal at the PMD pixels are exactly sinusoidal. A non-negligible harmonic content precludes the correct performance of PMD ToF systems. CS frameworks such as that presented in Section 4.5.2 aim to recover the whole harmonic content of the signal from few measurements. Recovering high order harmonics can help to achieve better depth resolution, due to its higher frequency, and to resolve more than one path per pixel, i. e., cope with finite reflective MPI.

Spatial Domain Regarding the *compressive* sensing in spatial domain, there are at least two simple ways of modifying the response function of the PMD pixels at subpixel level, so that superresolution from compressed measurements becomes possible. One way is to displace the PMD array with custom 2D displacements in the image plane, as performed in the application case presented in Section 4.4.2. The resolution of the stepper unit (or equivalent device used to produce the displacements) has to be lower than the pixel size, i. e., has to provide subpixel resolution. The maximum lateral resolution that can be recovered is then bounded by the resolution of the stepper unit and no longer by the number of pixels in the PMD array. In terms of frequency, this simply means that spatial frequencies that are out of the bandwidth of the measurement device cannot be recovered from the compressed measurements. A high resolution image can be then recovered from a number of measurements that is lower than its number of pixels. This way, the modified PMD responses in spatial domain are just shifted versions of the originals. The subpixel shifting eventually allows recovering higher spatial frequencies than those given by the original PMD responses. As in any other CS framework, the sensing has to be performed incoherently with respect to the sparsity dictionary. This can be achieved, for instance, by means of randomly chosen subpixel 2D displacements or, alternatively, by means of an artificial sensing matrix that generates few random combinations of measurements obtained from a fine displacement grid, as in Section 4.4.2. Unfortunately, a camera with mobile parts that should allow for micrometer displacements may be of limited applicability in real-world scenarios, where vibrations and exposure to dust may limit the effective resolution of the positioning system.

The other way to modify the subpixel spatial response of PMD pixels is including additional hardware to modulate the light reaching the sensor in spatial domain, with higher lateral resolution than that of the PMD array for equal FOV. In other words, to include an SLM that is able to superimpose custom patterns on the images of the scene, right before they are projected onto the PMD array. For an overview of the different technologies that can be used for light modulation the reader is referred to the paragraph of Section 2.2.2 dedicated to EOM. Most commercial SLMs are based on some of the EOM technologies presented there. Prominent examples are mecano-optic devices, such as DLPs, or electro-optic devices, such as LCoS arrays. While the former can only operate in reflective mode, the latter can operate both in transmissive or reflective mode. Provided that ToF imaging uses active illumination, the SLM can be used to spatially-modulate the light being projected onto the scene, as done in [262] for CS ToF imaging. While spatially-modulating the emitted light instead of the received one leads to a great simplification of the optical system of the camera, it moves the complexity towards the illumination system. Wide-area and long-range illumination systems are often composed by several emitters with different positions and orientations (consider, for instance, our medium-range illumination system, analyzed in Appendix A.6) and providing unique spatial modulation of the resulting light field may not be feasible. If a transmissive SLM is placed at the camera itself, three main possibilities exist, as exposed in [340]: placing the SLM adjacent to the image detector, in front of the imaging lens or within the imaging optics.

In the first case, the SLM is placed as close as possible to the sensor, yielding a very compact setup. Normally, SLM pixels are bigger than sensor pixels, but this is not the case when using a PMD sensor. Nevertheless, the typical ratios between PMD and SLM pixel sizes are not large enough to enable high SR factors. One should take profit of the much larger number of pixels of the SLMs to project a large number of SLM pixels per PMD pixel, which is impossible by means of an adjacent setup. Additionally, when using optics with very low DOF, non-zero distances between the SLM and the PMD sensor may lead to significant image degradation. If the SLM is placed in front of the lens, image distortion can be avoided, but depending on the aperture of the lens, the attenuation of an specific SLM pixel affects to a different number of PMD pixels. In other words, the image of the SLM is diffusely projected on the PMD chip and, therefore, SLM patterns are not directly superimposed on the image formed at the PMD array, as in the adjacent setup. The last setup is the most complex but allows perfect projection of the SLM pixels on the PMD pixels. The image is first focused on the SLM and then a second lens or optical system is used to refocus it on the PMD array. This setup also allows a perfect adjustment of the SLM array to the the PMD array and, therefore, the only parameter that bounds the maximum attainable SR factor is the ratio between the resolutions of both.

Unfortunately, transmissive SLMs exhibit much lower fill factors than their reflective counterparts. In ToF imaging, where the power budget is often the main criterion to design the system, loosing a large fraction of the scene-reflected light due to the poor fill factor of the SLM is out of the question. For this reason, reflective SLMs are to be considered as an alternative, even at the cost of a bulkier optical system. Two of the most widely-used reflective SLM technologies are DLPs and reflective LC-based SLMs. Both of them provide notoriously higher fill factors than, e. g., transmissive LC-based SLMs, higher than 90%. Regardless of the technology, there are at least two possibilities to integrate a reflective SLM within the optical system of the camera so that a custom HR pattern can be superimposed to the scene image before it is (re)projected onto the PMD chip, as depicted in Fig. 5.1. Both setups in Fig. 5.1 are equipped with a pair of polarizers, surrounded by a dashed rectangle. The dashed line means that these elements are only required if the SLM is based on the change of the light polarization (an LC-based SLM, for instance) and can be left out if the SLM modulates the light intensity directly (e. g., mechanically, by means of DMDs). In the former case, the first polarizer is used to ensure that the light reaching the SLM has the desired polarization, while the second polarizer is intended to allow (or, conversely, block) the transmission of light whose polarization angle has been rotated, while blocking (or, conversely, allowing) the transmission of light whose polarization remained the same after reflection at the SLM.

The simplest option is to generate the image first on the SLM surface by means of a lens system with large focal length (L1-L2) and form then the image of the SLM surface on the active area of the PMD chip by means of a second lens system (L3-L4) that provides a large focal length with reduced system length (a telephoto lens system). This allows for the

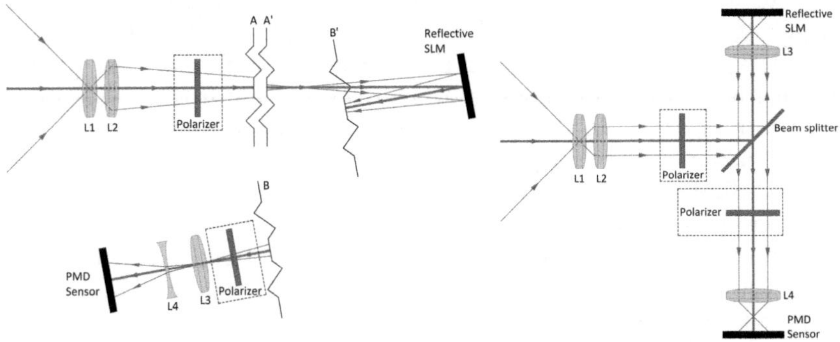

(a) Single-reflection Setup **(b)** Split Setup

Figure 5.1.: Two possible setups for the integration of a reflective SLM in the CS-PMD camera. The simplest option is to generate the image first on the SLM and later reproject it on the PMD sensor by means of a second lens system (a). This requires a long distance (low angle between optical paths) in order to avoid image distortion due to width shrinking. A-A' and B-B' are schematic cuts for a clearer and more compact representation. Using a beam splitter (b) solves the alignment issue, at the cost of loosing 75% of the optical power.

very low opening angle required for imaging the small SLM surface at a relatively large distance. The main drawback of this setup (Fig. 5.1a) is the impossibility of aligning the incident and reflected optical paths. Provided that the lenses have a non-negligible size, the best one can do is to increase the distance between the lens groups and the SLM, so that the angle between optical paths becomes as small as possible. In general, if the SLM is aligned with the sensor, as in Fig. 5.1a, the image of the scene is shrunk in width before the pattern is superimposed. That is, the pattern itself does not get distorted by the reprojection on the PMD array, but it is applied over a distorted image. Conversely, if the SLM is placed orthogonal to the incident optical path, the SLM pattern is applied over an undistorted scene image, but width shrinkage of the result will occur when reprojecting on the PMD sensor. The only advantage of the setup in Fig. 5.1a is the reduced loss of optical power, an important factor in ToF imaging. The alignment issue can be overcome by means of a beam splitter, which allows for a more compact setup, as shown in Fig. 5.1b. The lenses L1 and L2 are just an open representation of the lens system of the camera, which should generate parallel rays that are directed to the beam splitter. The beam splitter produces a power loss of 50% in the first reflection towards the SLM and another 50% in the transmission of the reflected light from the SLM to the PMD sensor. That is, this setup transmits, at maximum, only 25% of the optical power to the PMD sensor. In a more realistic scenario, e. g., binary (on/off) SLM codes with equal probability of occurrence of both values plus slight losses due to the polarizers, the split setup could easily loose more than 90% of the optical power collected by the first lens. The main advantage of the setup is the possibility of a perfect alignment between SLM and PMD chip and the absence of image distortions.

Time Domain As commented above, the time-frequency sensing of the light signal received by a PMD pixel can be considered to happen in frequency domain, since the measurements are believed to be samples of the cross-correlation between the received signal and a sinusoidal reference signal. If one wants to perform measurements according to custom reference signals, e. g., some sort of temporal code, the PMD frontend has to be modified in order to generate such signals as references and synchronize the ICS accordingly. In the commercial 19k [374] and 41k PMD frontends, both the reference signal for the PMD pixels and the ICS for the illumination system are generated out of the PMD chip, in an external CPLD. The newest generation of PMD chips [385], manufactured by Infineon®, seems to have integrated the generation of all necessary signals and the ADCs within the chip. This way the external circuitry has only to account for communications with the upper level, reading out the data and computing the depth image. While this allows for higher modulation frequencies (up to 100 MHz), once the chip is manufactured, no *a posteriori* custom modifications of the circuitry for reference signal generation are possible. Consequently, we select a PMD 19k-S3 chip from the previous generation, requiring an externally-generated reference signal, for our CS-PMD prototype.

Provided that the input for the reference signal is binary, the best we can do is to use custom binary codes as reference signals. One can argue that analog reference signals can be emulated by means of a PWM, but this is nothing but an specific case of a custom code, where the instantaneous duty cycle is determined by the desired amplitude of the signal. We suppose that a low-cost illumination system with reduced bandwidth, e. g., LED-based, is used. Consequently, the modulation frequencies are restricted to tens of MHz and the emitted signal is expected to have a considerable harmonic content when the illumination system is driven with a square signal, as in conventional ToF cameras. The maximum length of the binary code is, therefore, given by the ratio between the period of the modulation signal and the minimum achievable pulse width of the reference signal. Similarly to the spatial sensing, the maximum frequency that can be recovered from the measurements depends on the bandwidth of the sensing system. In theory, a sensing system with infinite temporal bandwidth, i. e., allowing binary codes with pulse widths arbitrarily close to zero (Dirac impulses), would allow recovering any signal that is sparse or compressible in frequency domain from a set of measurements that is much lower than the dimensionality of the signal. The goal is, therefore, to provide a system that generates a reference signal according to selectable custom binary codes, keeping the minimum pulse width as low as possible. Since the PMD pixels integrate the result of the scalar product between received light signal and reference over many cycles of the modulation signal, the reference signal has to repeat the custom code in each new period. Consequently, a perfect synchronization between the reference signal and the ICS is a critical system requirement. Implementation details are given in Section 5.2.2.

System Overview The CS-based PMD camera prototype, named CS-PMD camera in the following for brevity, aims to sense in a compressed manner both in 2D spatial domain and in temporal domain. From the compressed raw data a dense three-dimensional array of higher dimensionality can be recovered provided that the data admits a sparse representation in an appropriate dictionary (per domain). From the dense raw data a depth image can be computed, where the lateral resolution is given by the spatial dimensionality of the recovered raw data and is, therefore, no longer bounded by the number of pixels of the PMD array. Fig. 5.2 provides a block diagram of the CS-PMD camera, from the low-level signal generation to the depth image. Note the similarity between Fig. 5.2 and Fig. 4.54. The CS-PMD camera is intended to be an inherently-superior system to that presented in Fig. 4.54 and, in fact, for $m_{\text{spatial}} = 1$, i. e., $k_{\text{SR}} = 1$, the system in Fig. 5.2 performs equivalently to the CS framework in time-frequency domain presented in Fig. 4.54. Fig. 5.3 provides a timing diagram of the principal signals involved in the system in Fig. 5.2.

The process in Fig. 5.2 is controlled by the clocking signal CLK. This signal refers to the internal CPLD CLK signal and, therefore, its frequency can be arbitrarily set. In order to provide the maximum time-frequency sensing bandwidth, the CLK frequency f_{CLK} should be the highest frequency for which the CPLD shows a stable behavior. The CLK signal triggers the transitions between the elements of the custom code. As in the framework

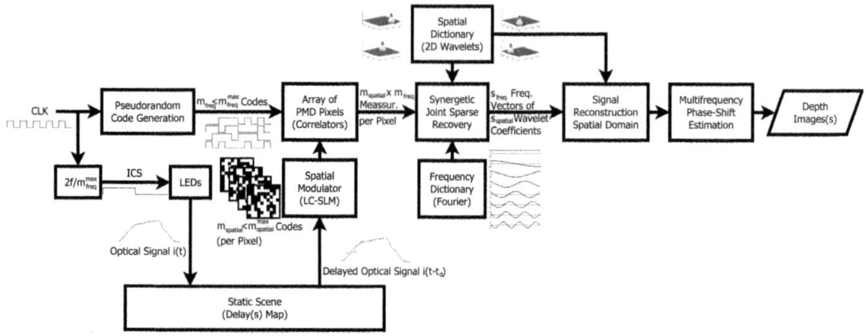

Figure 5.2.: Block diagram showing the principle of operation of our CS-PMD camera prototype. The camera senses in a compressed way in three dimensions simultaneously: two spatial dimensions and the temporal dimension. The required raw data is then recovered from the compressed measurements exploiting the sparsity of the raw images in wavelet domain and of the illumination signal in frequency domain. The system gets rid of the four phases algorithm and estimates the depth from multiple frequency (e. g., Fourier) measurements.

presented in Section 4.5.2.2, code transitions are supposed to be triggered both in the rising and falling edges of CLK, in an attempt to maximize the maximum length of the binary codes for a given modulation frequency. Clearly, the number of measurements m_{freq} is bounded by the maximum number of orthogonal codes, which obviously coincides with its dimensionality. Therefore we have that $m_{\text{freq}}^{\max} = \frac{2f_{\text{CLK}}}{f_{\text{mod}}}$, where the factor 2 is due to triggering the code transitions in both rising and falling edges. Consequently, a simple frequency divider by a factor of $\frac{m_{\text{freq}}^{\max}}{2}$ suffices to obtain the square ICS from CLK. Any set of m_{freq} codes with $m_{\text{freq}} > m_{\text{freq}}^{\max}$ is redundant. If the custom codes are designed to emulate displaced versions of the ICS, the system operates in the conventional interferometric mode. Additionally, if $m_{\text{freq}} \geq 3$ codes are used, corresponding to m_{freq} equally spaced phase intervals, the phase can be computed from the measurements using a simple closed form formula (Eq. 2.6). In the specific case of $m_{\text{freq}} = 4$, the system operates in a four phases mode, as a conventional PMD camera, and the phase can be obtained from the four phases algorithm (recall Eq. 2.18). In other words, the four phases operation is just a particular (suboptimal) operation mode of the system.

The ICS is the input of the LED drivers. The LEDs produce non-negligible distortions in the signal due to a decaying frequency response (cf. Fig. 4.1), which is, in turn, motivated by non-zero and typically asymmetric rising and falling times. The resulting optical signal $i(t)$ is projected onto the scene. We consider that each scene point produces a single or a finite number of reflections (see Eq. 4.32), so that the received reflection is given by the superposition of displaced versions of the incident optical signal $i(t)$. Considering a finite number of reflections per scene point accounts for reflective MPI and thin translucent objects. If the scene objects are opaque and in absence of MPI, the reflected signal generated by each scene point arrives to the camera as a delayed version of the emitted signal, $i(t - \tau_{\mathrm{d}})$, where τ_{d} is the delay due to the distance between the camera and the corresponding scene point.

Instead of generating the image directly on the PMD surface, it is formed first on the surface of an SLM, which is responsible for superimposing 2D custom codes to the image in spatial domain. This is intended to open the door for superresolution, provided that the size of the SLM is greater than that of the PMD array. If the ratio of array sizes is $k_{\mathrm{SR}} \in \mathbb{N}$ (per dimension) and both SLM and PMD pixels are square, then the maximum size of the 2D pattern to be superimposed per PMD pixel is $k_{\mathrm{SR}} \times k_{\mathrm{SR}}$. As before, the maximum number of orthogonal codes is bounded by the code size, i. e., $m_{\mathrm{spatial}} \leq m_{\mathrm{spatial}}^{\max} = k_{\mathrm{SR}}^2$. Independently from the sensing in time-frequency domain, the sensing in spatial domain also requires an acquisition per code. Consequently, the CS-PMD system requires a number of acquisitions equal to the number of per-pixel measurements $m = m_{\mathrm{spatial}} \times m_{\mathrm{freq}}$, being $m_{\mathrm{spatial}} \geq 1$ and $m_{\mathrm{freq}} \geq 3$. In practice, m_{spatial} has to be large enough to ensure successful recovery of the superresolved images and depends on the native SR factor k_{SR} of the system. If the required m_{spatial} is so high that leads to prohibitive acquisition times, one can apply pixel binning to the SLM pixels to reduce its effective resolution and, therefore, k_{SR}. The effective per-pixel SR after applying a factor k_{bin} binning to the 2D SLM patterns is $k'_{\mathrm{SR}} = \frac{k_{\mathrm{SR}}}{k_{\mathrm{bin}}}$, being k_{bin} a submultiple of k_{SR}.

If k_{SR} is high enough, the sparse recovery framework can work in a per-pixel basis, that is, recovering an image per pixel and frequency contained in the support of the light signals in the frequency dictionary. The images themselves are recovered exploiting the sparsity of natural images in, e. g., wavelet domain. This brings together the concept of the *single pixel camera* [156] and the CS framework from Section 4.5.2.2. Differently from *single pixel* approaches, we do not renounce the benefits of having more than one pixel, but use this concept in a per-pixel basis to increase the lateral resolution of

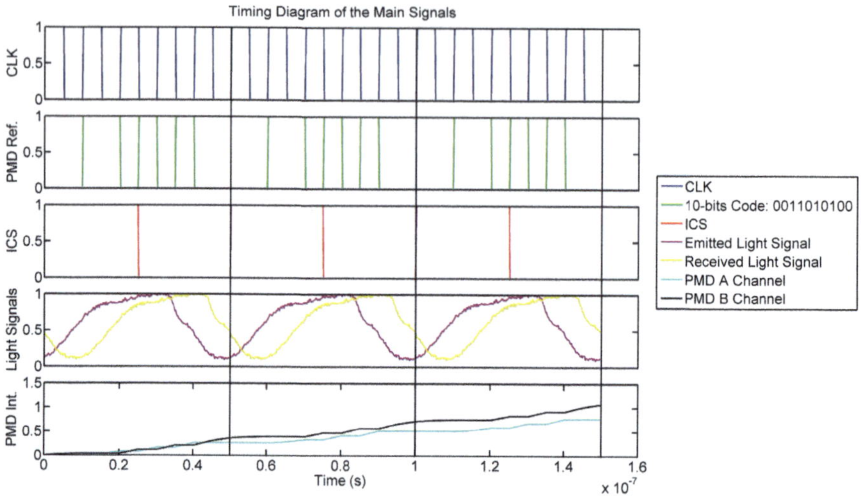

Figure 5.3.: Simulated timing diagram showing the main signals of our CS-PMD camera prototype during three periods of the ICS. The ratio between CLK and ICS signals is five. Consequently, the length of the codes used as PMD reference signals is ten, provided that transitions are triggered in both rising and falling edges of CLK. The emitted light signal is the output of a commercial NIR-LED when driven with the ICS. The received light signal exhibits a delay with respect to the former, related to the depth being sensed. The PMD pixel integrates the received signal in the A channel when the logic level of the reference signal is high and in the B channel when it is low. Signals in the first four plots are maximum-normalized.

the sensor. For details on the recovery process and how measurements in both domains can be combined in a synergistic manner to yield better signal recovery, the reader is referred to Section 5.3. Once the images corresponding to the different frequencies contained in the light signal are recovered, a single depth image is computed in a multifrequency framework.

5.2. Hardware

The hardware used to implement the CS-PMD system presented in Section 5.1 can be divided in the optical system and the customized PMD frontend. The optical system (Section 5.2.1) includes, apart from the necessary lenses, the SLM that allows using custom 2D patterns as sensing kernels in spatial domain, while the PMD module (Section 5.2.2) has been modified to also allow the use of binary sensing kernels in time domain. For further implementation details omitted in this section and real images of our prototype, the reader is referred to Appendix A.11.

5.2.1. Optical System

The main innovation in the optical system of the CS-PMD camera is the inclusion of an SLM to allow for 2D spatial modulation. As introduced in Section 5.1, SLMs can be classified in transmissive or reflective, depending on whether their pixels preserve the direction of the light (and its sense) or not. Some reflective SLMs (e. g., DLPs) can offer large reflectivities. In the case of LCoS SLMs, the lower fill factor of transmissive devices also translates in power loss. For these reasons, the SLM of our choice for the CS-PMD camera is a reflective LCoS SLM. The model chosen for our prototype is an LC-R 1080, from Holoeye®, which offers an array of 1200×1920 pixels with 90% fill factor. Provided that the PMD array features 120×160 pixels, the SLM is reconfigured to have a size of 1200×1600, i. e., a width of 160 SLM pixels is left inactive at both sides of the array. This way we achieve the same aspect ratio as the PMD chip and a 1:1 reprojection of the SLM image on the PMD active area becomes possible. In case of perfect adjustment, an SLM pattern of size 10×10 is projected on the surface of each PMD pixel. This means that using this hardware one can attain a maximum SR factor of $k_{\mathrm{SR}} = 10$. Obviously, custom $k_{\mathrm{SR}} \in \{1, 2, 5, 10\}$ are feasible without any hardware modification.

As pointed out in Section 5.1, at least two simple possibilities exist for integrating a reflective SLM in the optical path of a PMD camera (see Fig. 5.1). The simplest one avoids the use of splitting elements in an attempt of minimizing the power losses (Fig. 5.1a), while the other makes use of a beam splitter to offer a relatively compact setup, at the cost of large power losses (Fig. 5.1b). The first option requires long optical paths for the incident and reflected light and is, therefore, not appropriate for a final product. Nevertheless, since our prototype is not intended to be a commercial product, but a proof of concept of our CS-based ToF imaging techniques, we adopt the first option. This decision is mainly motivated by the massive loss of optical power that the setup in Fig. 5.1b would produce in the typically-weak ToF reflected signal.

The LCoS SLM does not modulate directly the amplitude of the incident light, but changes its polarization plane. When an SLM pixel is active, the polarization plane of the linearly-polarized light experiences a rotation between 0 (black pixel) and 90° (white pixel). Consequently, the light reflected by an active SLM pixel is cross-polarized with respect to the incident light and, therefore, a pair of crossed polarizers suffice to turn the *polarization rotation* into *amplitude modulation.* In the general setups of Fig. 5.1 the first polarizer is placed after the first lens group, before the SLM, while the second is placed after the SLM reflection, before the second image formation in the PMD chip. In practice, the exact location of the polarizers has effects in terms of image distortion and must be carefully selected. Provided that in a ToF system the light we are interested to receive is that being emitted, the initial polarizer can be placed at the illumination system. This alternative reduces the emission of useless optical power to the scene, which would be anyway blocked by the first polarizer at the camera. Consequently, this setup increases the optical power for which the illumination system is still eye-safe. Additionally, emitting polarized light provides some robustness to MPI, since direct reflections (like those arising from frontoparallel surfaces) do not induce polarization changes, while reflections with non-normal incidence (typical from multiple reflections) lead to large changes in polarization. This also helps in case of diffuse interreflections, arising from diffusive scene objects, since only the retroreflective component keeps the original polarization and the second polarizer will filter out the partial rotations. In fact, cross-polarization is a common approach to separate the diffuse and specular components of an image [207, 36].

The position of the second polarizer largely affects the quality of the PMD images. For instance, placing the polarizer right before the second lens group was observed to significantly reduce the sharpness of the images. In a setup whose principal mission is gaining lateral resolution obtaining burred images due to an incorrect optical setup is unacceptable. The second lens group is a telephoto lens. In our prototype we choose a Nikon® Nikkor® ED 300mm 1:2.8 objective, which features a built-in port for *drop-in* polarizers and filters. Placing the second polarizer in this location is the optimal choice, since the lens is designed accounting for a glass (or equivalent material) layer in that position, and leads to images that are almost as sharp as without the polarizer. A detailed block diagram of the optical setup of the prototype is given in Fig. 5.4. The elements belonging to the illumination system, external to the camera, are depicted within the red area, while the elements belonging to the camera itself are depicted within the blue area. For clarity, the incident and reflected optical paths are separated by means of a fictitious *ghost* SLM. In practice there is only one SLM, being both paths almost coaxial. The misalignment between both paths is given by the arctangent of the ratio between half the distance between the centers of the imaging lens and the telephoto lens and the distance between the lenses and the SLM (L in the diagram). Provided that the former distance cannot be adjusted, L is chosen to be large, with respect to the other dimensions of the system. Further details are given in Appendix A.11.

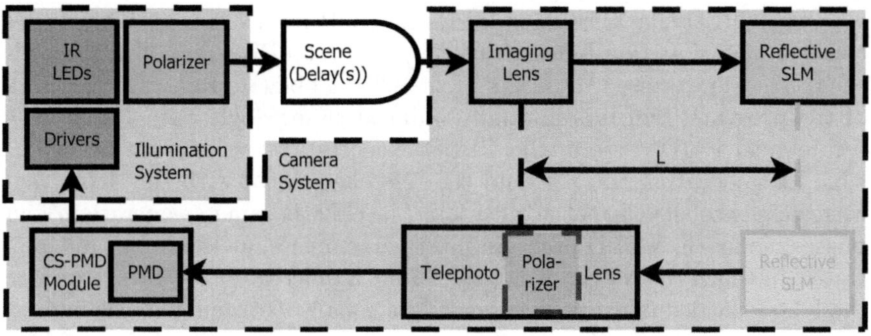

Figure 5.4.: Detailed block diagram showing the complete optical system of our CS-PMD camera prototype, including the illumination system. The core of the camera is the PMD module, adapted for CS operation. The module generates the ICS to drive the LEDs. The illumination system is equipped with a polarizer, so that the light projected onto the scene is linearly-polarized. The image of the scene is first formed on the surface of the SLM and then reprojected on the PMD sensor by means of a telephoto lens with built-in polarizer. The polarizers are in crossed configuration and translate the polarization change induced by the LCoS SLM into actual intensity modulation.

Superimposing custom patterns per PMD pixel also provides a mean of artificially increasing the dynamic range of the PMD pixels. For instance, in a binary pattern the maximum level can be adjusted per pixel according to the readouts of the corresponding PMD pixel, so that saturation is avoided. This way black and white binary patterns would become *black and gray*, where the gray level, corresponding to maximum transmittance can be adjusted dynamically according to the readouts of the pixel channels. This idea is similar to that in [340], being the main difference that our system aims to increase the lateral resolution of a poor resolution sensor and the HDR capability arises as a natural side effect, while the target application in [340] is HDR RGB video, where the resolution of the image sensor might be equal or higher than that of the SLM.

5.2.2. The CS-PMD Module

In the 19k-S3 frontend, both the ICS and the PMD reference signal are generated outside the PMD chip by a CPLD. A CPLD is a programmable logic device whose complexity ranges between that of PALs and FPGAs. In our case, the CPLD is a XC2C256 CoolRunner®-II from Xilinx®. This device consists of 16 function blocks featuring a 40×56 PLA and 16 macrocells each. The very limited resources of the CPLD obligate to a careful design of the program, which is normally reduced to few logic operations. In a commercial PMD camera, the CPLD executes a program that is responsible for generating the ICS and the reference signal with the delay that corresponds to the current acquisition. Since PMD cameras operate in four phases mode, the CPLD program must contemplate, at least, four different states, in which the reference signal exhibits a different delay with respect to the ICS. Provided that the phase shifts are equally distributed in the interval $[0°, 360°)$, the phase shifts for the four phases are $\theta \in \{0°, 90°, 180°, 270°\}$, applied in ascending order.

One can think of different efficient hardware implementations of a state machine that generates the right output signals using logic components. In any case, a selection signal that specifies which of the four phases is being acquired is required. A schematic showing a simple implementation is given in Fig. 5.5, where PHASE_SEL is the 4-bit signal responsible for phase selection, MOD_LED is the ICS and MOD_PMD is the so-called PMD reference signal. The system clock signal (CLK) triggers the transitions of a 2-bit counter, whose output signal is, in turn, used as control signal of a four channel multiplexer. The input channels of the multiplexer are fed with PHASE_SEL. Note that the signals PHASE_SEL are conveniently designed to emulate successive 90° phase shifts. The ICS (MOD_LED in Fig. 5.5) is taken from the MSB of the 2-bits counter output signal and, therefore, does not undergo any delay. Both output signals, MOD_LED and MOD_PMD, are reduced in frequency by a factor of four with respect to the frequency of CLK. This poses an intrinsic limitation on the maximum modulation frequency. Suppose, for instance, that the PMD pixels are able to operate at frequencies $f_{\text{mod}} > 100\,\text{MHz}$. With the CPLD configured for four phases operation, this would require a clock frequency $f_{\text{CLK}} > 400\,\text{MHz}$, which is infeasible with this CPLD. In other words, the periphery of the PMD chip in the frontend constrains its capabilities.

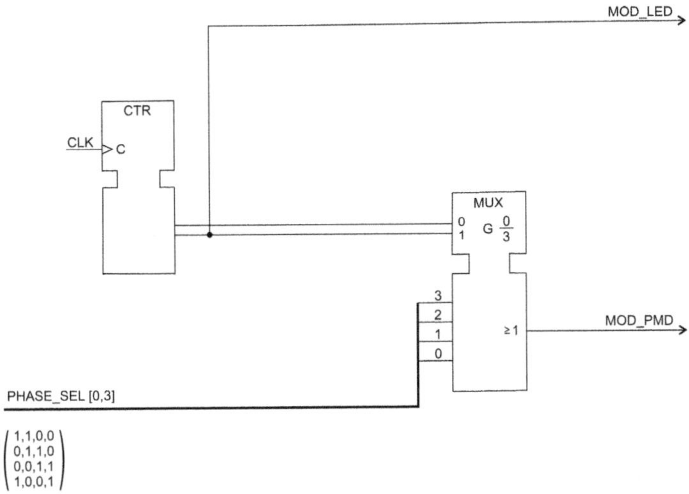

Figure 5.5.: Minimal schematic of the logic implemented in the CPLD of the PMD frontend for four phases operation. The schematic is just provided as reference and may not coincide with the real logic implemented in commercial PMD cameras. Note the simplicity of the setup, with only a 2-bit counter controlling a four channel multiplexer, which imposes a severe reduction of the maximum frequency of the output signals by a factor of four with respect to that of the CLK signal.

Our goal is to provide an alternative circuit design that allows generating a reference signal according to some custom binary codes for each period of the ICS, so that it allows implementing, for instance, the CS recovery framework presented in Section 4.5.2. Ideally, the code length, i. e., the number of code elements, should be adjustable. This allows for different compromises between maximum number of measurements $m_{\text{freq}}^{\max}$ and modulation frequency of the ICS, f_{mod}. This is so due to the fact that the maximum number of non-redundant codes is equal to $m_{\text{freq}}^{\max} = \frac{2f_{\text{CLK}}}{f_{\text{mod}}}$, as introduced in Section 5.1. The factor 2 only applies if the transitions between code elements are triggered both in the rising and falling edges of CLK, so that the system can operate twice as fast as without this feature.

Fig. 5.6 presents the schematic of the circuit implemented in the CPLD of our CS-PMD prototype. The output signals are named as in Fig. 5.5 and there are two input signals: one providing the code length (CS_LENGTH) and another for the code itself (CS_CODE). In our prototype, the maximum

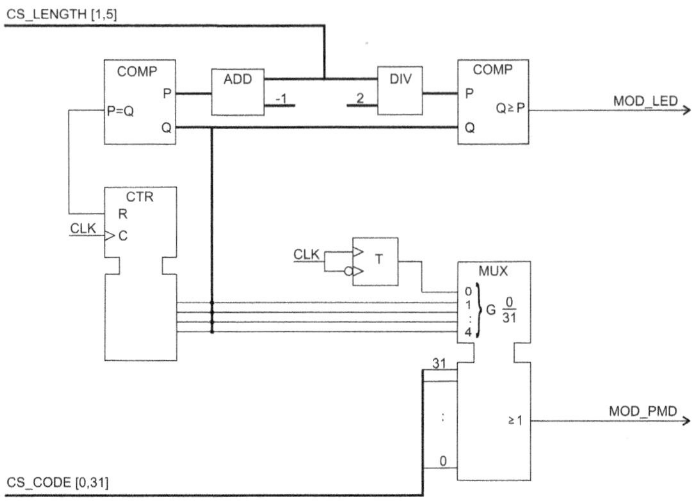

Figure 5.6.: Schematic of the logic implemented in the CPLD of the CS-PMD prototype. Similarly to the system in Fig. 5.5, the CS-PMD logic also makes use of a counter to control a multiplexer, which sequentially outputs the elements of our custom binary code as PMD reference signal. A comparator is used to control the reset of the counter when the last element of the code is reached. This allows custom code lengths. The edge detector at the LSB line of the multiplexer control signal allows triggering code transitions both in the rising and falling edges of CLK.

code length is $m_{\text{freq}}^{\max} = 32$ elements and, therefore, CS_CODE is a 32-bit signal. Consequently, CS_LENGTH requires six bits, in order to represent numbers from 2 to 32 (recall $32 = 2^5$). Provided that the codes are always of even length, the LSB of CS_LENGTH does not carry any information and is not used in the logic for efficiency. The core of the design in Fig. 5.6 is, similarly to Fig. 5.5, the pair counter-multiplexer, responsible for the transitions of the code elements. In this case, the counter features a 4-bit output signal, which is plugged to the four MSBs of the multiplexer control signal. The LSB of this signal is given by the output of an edge detector, which operates on CLK. This way the multiplexer triggers the code transitions in both the rising and falling edges of CLK, as desired. The multiplexer selects, one after another the different elements of the code given by CS_CODE. When the last code element is reached, a comparator is used to reset the counter, so that the first element of the code will be selected again at the next edge of CLK. The signal MOD_LED is a square signal whose period is equal to the code length in time domain and is generated by means of a comparator. The division by two at its first input just symbolizes a shift of the binary signal CS_LENGTH one bit to the left, so that only its four MSBs are used for comparison to the output of the counter.

Note that, for equal f_{CLK}, the system in Fig. 5.6 is able to generate an ICS of equal f_{mod} than that generated by the system in Fig. 5.5, while correlating against 8-bit custom binary codes. The CS-PMD intrinsically allows for higher modulation frequencies than its four phases counterpart. In fact, the four phases operation can be contemplated as a special operation mode of the system in Fig. 5.6, using 4-bit codes equal to the four PHASE_SEL signals depicted in Fig. 5.5. Note that the CS-PMD system operating in four phases mode duplicates the modulation frequency for the same f_{CLK}, with respect to the original design, thanks to the dual-edge triggering of the code transitions.

In order to provide an empirical value of the maximum modulation frequency of the ICS that can be attained for a desired code length, preliminary experiments are carried out to test the quality of the signals generated by the CS-PMD CPLD. Fig. 5.7 shows the results obtained for two high clock frequencies $f_{\mathrm{CLK}} \in \{80, 128\}$ MHz for code lengths $m_{\mathrm{freq}}^{\mathrm{max}} \in \{4, 8, 16, 32\}$. Lower f_{CLK} yields cleaner signals and, therefore, further results are omitted for brevity. The signals were acquired with an oscilloscope of 350 MHz bandwidth and are, therefore, bandlimited. The codes of different lengths were obtained by selecting the corresponding LSBs of a single code of length 32 (see figure caption).

Obviously, for $f_{\mathrm{CLK}} = 128$ MHz (Fig. 5.7c), the quality of the signals is worse than for $f_{\mathrm{CLK}} = 80$ MHz (Fig. 5.7b). The rising and falling times practically cover (together) the total time corresponding to a code element in the PMD reference signal. For $f_{\mathrm{CLK}} = 80$ MHz the situation is not so critical and the hypothesis of rising and falling times of $\frac{\pi}{3}$ width in phase domain, introduced in Section 4.5.2.2, shows to be perfectly valid in practice. Note that rising and falling times of $\frac{\pi}{3}$ in phase of CLK mean one third of the total time corresponding to a code element in the PMD reference signal, which is coherent with Fig. 5.7b. Testing the operation of the CS-PMD module for higher f_{CLK} reveals that the generation of the coded reference signal becomes unstable for $f_{\mathrm{CLK}} > 136$ MHz, which corresponds to a minimum width of the code elements of 3.7 ns in time domain.

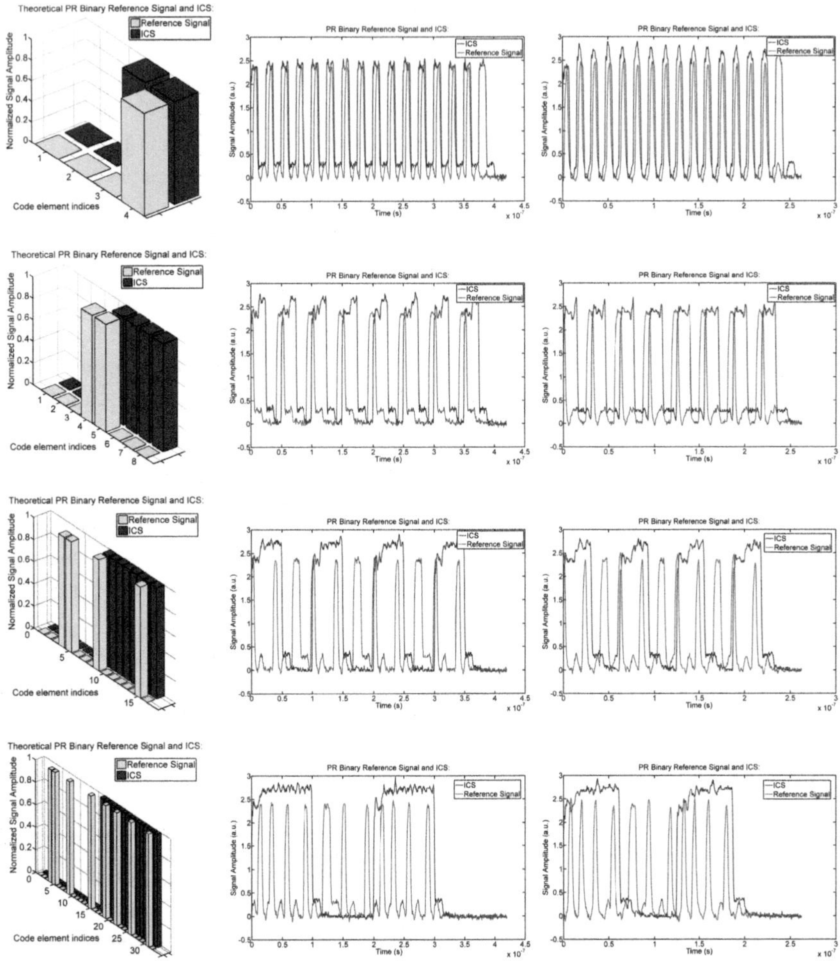

(a) Theoretical Signals (b) Real Signals $f_{\text{CLK}} = 80\,\text{MHz}$ (c) Real Signals $f_{\text{CLK}} = 128\,\text{MHz}$

Figure 5.7.: Theoretical and empirical signals generated by the CPLD of our CS-PMD prototype for different code lengths and clock frequencies. The considered code lengths are $m_{\text{freq}}^{\max} \in \{4, 8, 16, 32\}$, increasing by rows. A single code of length 32 (0001 1000 1000 0010 0010 0100 0100 0010) was used for all experiments and only CS_LENGTH was modified.

5.3. Software: 3D Sparse Recovery from Few Measurements

In this section we provide details on the software side of our CS-PMD prototype, which is responsible for performing the sparse recovery of the required raw data from the compressed measurements gathered by our custom hardware and computing the final depth image out of the recovered data. In order to ease the understanding of this section, a block diagram of the entire data processing pipeline is given in Fig. 5.8, from the raw data gathered by the hardware to the depth image.

5.3.1. Synergistic 3D Sparse Signal Recovery

As introduced in Section 5.1, the compressed sensing is carried out in three dimensions simultaneously: two spatial dimensions and the temporal dimension. The total number of acquisitions is given by the product of the number of measurements in spatial and time-frequency domains: $m_{\text{spatial}} \times m_{\text{freq}}$. Each measurement is gathered with a different combination of a spatial code and a time code. For a PMD sensor size of $n_{\text{rows}} \times n_{\text{cols}}$ pixels, each acquisition means an image of that size, that is, the full collection of raw data is of size $(n_{\text{rows}} \times n_{\text{cols}}) \times (m_{\text{spatial}} \times m_{\text{freq}})$. The m_{spatial} spatial codes are applied per pixel, that is, the same 2D code is superimposed on the surface of every PMD pixel. This is possible thanks to the integer ratio of SLM and PMD resolutions and a careful alignment. Consequently, the recovery process can be formulated per pixel and repeated for each of the $n_{\text{rows}} \times n_{\text{cols}}$ pixels of the PMD array. This allows for massive parallelization, since once the compressed data is acquired, the recovery process can run independently for each pixel.

From now on, equations are given for a generic PMD pixel and its 2D index is omitted for the sake of simplicity. Let $\boldsymbol{Y}_{\text{spatial}} \in \mathbb{R}^{m_{\text{spatial}} \times m_{\text{freq}}}$ be the matrix of measurements. The final challenge is to recover a sparse matrix $\boldsymbol{X} \in \mathbb{R}^{n_{\text{spatial}}^{\text{atoms}} \times n_{\text{freq}}^{\text{atoms}}}$ that explains the measurements jointly exploiting the sparsity of natural images in, e. g., some wavelet domain and of the periodic illumination signals in frequency domain. Here $n_{\text{spatial}}^{\text{atoms}}$ denotes the number of atoms of the spatial dictionary and $n_{\text{freq}}^{\text{atoms}}$ that of the frequency dictionary. If the dictionary is a basis for the signal space, the number of dictionary atoms (basis elements) coincides with the dimensionality of the signal to recover, i. e., the dictionary is a square matrix. In our case, the DB4 orthonormal wavelet basis is adopted as sparsity dictionary for the

Data Processing Chain in the CS-PMD System:

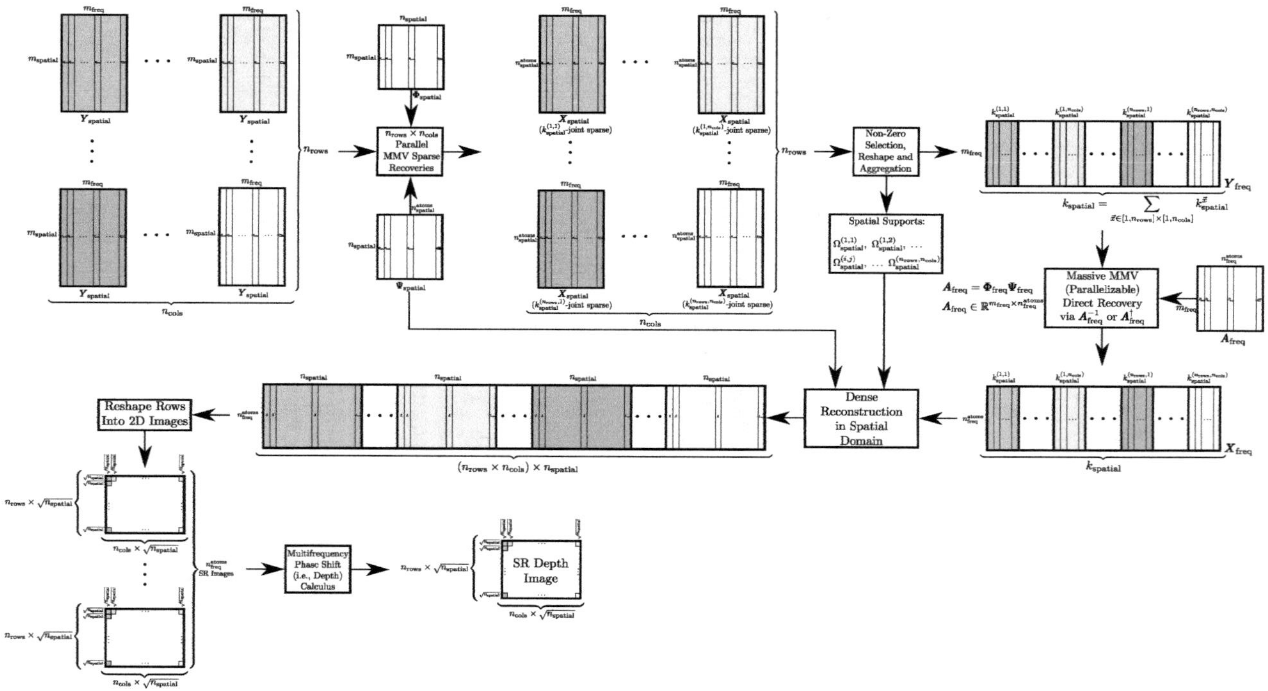

Figure 5.8.: Block diagram of the data processing chain carried out by the software of the CS-PMD system. The initial set of raw data (top-left) is composed by $n_{\text{rows}} \times n_{\text{cols}}$ matrices of size $m_{\text{spatial}} \times m_{\text{freq}}$, one per PMD pixel in the array. The final goal is to obtain a single superresolved depth image (bottom) from the gathered data. The pipeline can be conceptually divided in an initial 3D recovery stage, where sparsity in the 2D spatial domain and in the 1D (temporal) frequency domain is exploited in a synergistic manner (first two rows of the diagram), and a final step of depth estimation from multiple frequency measurements (last row). Different colors denote different PMD pixels.

2D spatial domain and, therefore, $n_{\text{spatial}}^{\text{atoms}} = n_{\text{spatial}} = k_{\text{SR}} \times k_{\text{SR}}$, where $k_{\text{SR}} \leq 10$ is the superresolution factor, bounded by the ratio of SLM and PMD resolutions. In contrast, the dimensionality of the discrete frequency atoms should be very high, in order to avoid imposing a limit on the phase resolution of the system due to severe discretization. Provided that the signal power of the illumination signal is known to be concentrated in a relatively narrow band of low frequencies, we have that $n_{\text{freq}}^{\text{atoms}} \ll n_{\text{freq}}$. In fact, in the ideal case, $n_{\text{freq}} \to \infty$, which corresponds to a continuous model, without discretization.

In general, the optimal way to recover a 3D signal that is sparse in its three dimensions is using some 3D basis. As well as the 2D DB4 wavelets can be generated from the 1D version, the 2D DB4 dictionary can be extended to a 3D spatio-temporal dictionary, where the third dimension is generated by multiplying the 2D DB4 basis by the 1D frequency dictionary, which can be, for instance, a partial DCT or Fourier basis. Nevertheless, note that our case differs from the general one, because we count with *a priori* information related to the sparsity in frequency domain. As shown in Fig. 4.2 and exploited later in the methods presented in Section 4.5, for an approximate sparsity s_{freq}, the support of the signal in frequency domain is given by the s_{freq} lowest-frequency atoms of the dictionary. This was observed to hold for all gathered illumination signals from our system and is a direct effect of the periodic ICS and the low-pass filtering effect of the LEDs. Consequently, provided that we have a bound on the maximum sparsity of the signal, $s_{\text{freq}}^{\text{max}} | s_{\text{freq}} \leq s_{\text{freq}}^{\text{max}}$, we are in the case of *oracle-given* support and as few as $m_{\text{freq}} \geq s_{\text{freq}}^{\text{max}}$ measurements are sufficient to recover the sparse signal. In practice, we set $m_{\text{freq}} = n_{\text{freq}}^{\text{atoms}} \geq s_{\text{freq}}^{\text{max}}$ in order to take the maximum profit from the measurements in terms of recovered frequencies. In real operation both m_{freq} and m_{spatial} are bounded by the maximum allowable acquisition time, which is application-dependent.

The previous means that the sparse recovery in frequency domain does not open the door for an eventual dimensionality reduction by considering only the non-zero components in the sparse domain, since both the measurement vector and the recovered vector are, in principle, dense and of equal dimensions, due to the fact that the frequency dictionary only contains elements belonging to the sparse signal support and $m_{\text{freq}} = n_{\text{freq}}^{\text{atoms}}$. Consequently, there is no advantage in recovering the sparse data jointly in a 3D domain, while this would translate into a dramatic rise of the memory requirements and computational cost, at the time that it would require an unnecessarily complex recovery algorithm. On the contrary, the recovery in the 2D spatial domain does bring the possibility of a dimensionality reduction, provided that, in case of unknown sparse support, the number of measurements to perform is necessarily greater than the expected signal sparsity, i. e., $m_{\text{spatial}} > s_{\text{spatial}}$. Therefore, the signals must be recovered first in the 2D spatial domain, in order to leverage the computational cost of the subsequent recovery in frequency domain.

The recovery in spatial domain has to be performed for all time measurements, gathered using different binary codes as PMD reference signals, that is, for every column of $\boldsymbol{Y}_{\text{spatial}}$, as defined above. The result of each recovery is a $k_{\text{SR}} \times k_{\text{SR}}$ image in its sparse DB4 representation. In Section 4.4.3.2, it was shown that PMD raw images acquired according to the four phases algorithm are joint sparse in wavelet domain. This fact was exploited in a joint sparse recovery framework to attain robust sparse support estimation. Similarly, the per-pixel images we want to recover for the different compressed acquisitions in temporal domain also share the same sparse support in any wavelet domain and, in consequence, the matrix to recover $\boldsymbol{X}_{\text{spatial}}$ is k_{spatial} joint sparse, according to the definition of joint sparsity given in Section 3.4.2.8, where $k_{\text{spatial}} \approx s_{\text{spatial}}$. If the support overlap is exact for all columns of $\boldsymbol{X}_{\text{spatial}}$, the latter approximate equality becomes strict. Then, the recovery in spatial domain can be formulated in a MMV manner (Eq. 3.128), as:

$$\begin{aligned} \hat{\boldsymbol{X}}_{\text{spatial}} = & \underset{\boldsymbol{X}_{\text{spatial}} \in \mathbb{R}^{n^{\text{atoms}}_{\text{spatial}} \times m_{\text{freq}}}}{\arg\min} \left|\operatorname{supp}\left(\boldsymbol{X}_{\text{spatial}}\right)\right| \\ & \text{subject to: } \boldsymbol{A}_{\text{spatial}} \boldsymbol{X}_{\text{spatial}} = \boldsymbol{Y}_{\text{spatial}} \\ & \text{being: } \boldsymbol{A}_{\text{spatial}} = \boldsymbol{\Phi}_{\text{spatial}} \boldsymbol{\Psi}_{\text{spatial}} \end{aligned} \tag{5.1}$$

where $\boldsymbol{\Phi}_{\text{spatial}}$ is the spatial sensing matrix, whose rows are the vectorized 2D codes in spatial domain, and $\boldsymbol{\Psi}_{\text{spatial}}$ is the 2D spatial dictionary, in this case the 2D DB4 orthonormal basis, whose vectorized elements are stacked by columns. The vectorization schema followed to vectorize the spatial codes and the 2D DB4 basis elements can be arbitrarily chosen, as far as it is the same in both cases. In our implementation, two methods are provided to solve the MMV problem, which are MMV extensions of the MP, OMP and ORMP algorithms (see Section 3.4.2.8). By simply editing a configuration file, the user can select the most appropriate method depending on the runtime requirements. For $\boldsymbol{A}_{\text{spatial}}$ with small RIP constant, ORMP might not perform significantly better than OMP, while being considerably slower.

The per-pixel recovery in spatial domain yields a k_{spatial} joint sparse matrix $\boldsymbol{X}_{\text{spatial}} \in \mathbb{R}^{n^{\text{atoms}}_{\text{spatial}} \times m_{\text{freq}}}$ per PMD pixel. In general, k_{spatial} and the spatial sparse support will be different from one PMD pixel to another. Let $\Omega^{\vec{x}}_{\text{spatial}}$ the set containing the indices of the joint support elements of the matrix $\boldsymbol{X}_{\text{spatial}}$ recovered for the PMD pixel indexed by $\vec{x}$. For clarity, if we denote by $\boldsymbol{X}^{\vec{x}}_{\text{spatial}}$ such matrix for the PMD pixel $\vec{x}$, then $\Omega^{\vec{x}}_{\text{spatial}} = \{i | \vec{X}^{\vec{x}}_{\text{spatial}_i} \neq \vec{0}\}$, being $\vec{X}^{\vec{x}}_{\text{spatial}_i}$ the ith row of $\boldsymbol{X}^{\vec{x}}_{\text{spatial}}$ and then we have that the joint sparsity of $\boldsymbol{X}^{\vec{x}}_{\text{spatial}}$ is $k^{\vec{x}}_{\text{spatial}} = |\Omega^{\vec{x}}_{\text{spatial}}|$. Consequently, in this sparse representation, the HR images obtained by composition of the per-pixel recovered images can be represented with only $k_{\text{spatial}} = \sum_{\vec{x} \in [1, n_{\text{rows}}] \times [1, n_{\text{cols}}]} k^{\vec{x}}_{\text{spatial}}$ coefficients (note the abuse of notation), typically much fewer than the actual dimensions of the HR images, given by $(k_{\text{SR}} n_{\text{rows}}) \times (k_{\text{SR}} n_{\text{cols}})$.

The second recovery step aims to recover the signals in frequency domain. The matrix of measurements is now constructed from the matrices $\boldsymbol{X}^{\vec{x}}_{\text{spatial}}$, $\vec{x} \in [1, n_{\text{rows}}] \times [1, n_{\text{cols}}]$ previously recovered. For clarity, the construction of the new matrix of measurements, denoted by $\boldsymbol{Y}_{\text{freq}} \in \mathbb{R}^{m_{\text{freq}} \times k_{\text{spatial}}}$, from the previous results is provided in Eq. 5.2.

$$
\begin{aligned}
\boldsymbol{Y}_{\text{freq}} &= \left[\boldsymbol{X}^{\vec{x}}_{\Omega_{\text{spatial}}}{}^{\top} \right]_{\vec{x} \in [1, n_{\text{rows}}] \times [1, n_{\text{cols}}]} \\
&= \left[\boldsymbol{X}^{1,1}_{\Omega_{\text{spatial}}}{}^{\top} \quad \boldsymbol{X}^{1,2}_{\Omega_{\text{spatial}}}{}^{\top} \quad \dots \quad \boldsymbol{X}^{i,j}_{\Omega_{\text{spatial}}}{}^{\top} \quad \dots \quad \boldsymbol{X}^{n_{\text{rows}}, n_{\text{cols}}}_{\Omega_{\text{spatial}}}{}^{\top} \right]
\end{aligned}
\tag{5.2}
$$

where $\boldsymbol{X}^{\vec{x}}_{\Omega_{\text{spatial}}}$ denotes the submatrix obtained from $\boldsymbol{X}^{\vec{x}}_{\text{spatial}}$ by selecting the rows indexed by $\Omega^{\vec{x}}_{\text{spatial}}$. Let $\boldsymbol{X}_{\text{freq}} \in \mathbb{R}^{n^{\text{atoms}}_{\text{freq}} \times k_{\text{spatial}}}$ be the matrix of frequency coefficient vectors to recover. Provided that $\boldsymbol{\Psi}_{\text{freq}}$ was designed using *a priori* knowledge on the sparse support, one can solve the recovery in a closed form, since the dictionary is already restricted to the support of the signals. Consequently, the MMV recovery differs from that in spatial domain and can be formulated as:

$$
\begin{aligned}
\hat{\boldsymbol{X}}_{\text{freq}} &= \boldsymbol{A}^{\dagger}_{\text{freq}} \boldsymbol{Y}_{\text{freq}} \\
\text{being: } \boldsymbol{A}_{\text{freq}} &= \boldsymbol{\Phi}_{\text{freq}} \boldsymbol{\Psi}_{\text{freq}}
\end{aligned}
\tag{5.3}
$$

where $\boldsymbol{\Phi}_{\text{freq}}$ is the sensing matrix in time domain, whose rows are the codes used as PMD reference signals, and $\boldsymbol{\Psi}_{\text{freq}}$ is the partial frequency dictionary by columns, restricted to the $n^{\text{atoms}}_{\text{freq}} \ll n_{\text{freq}}$ lowest-frequency components. $\boldsymbol{A}^{\dagger}_{\text{freq}}$ denotes the Moore-Penrose pseudoinverse of $\boldsymbol{A}_{\text{freq}}$, which provides the optimal solution in terms of least squares error. Since we set $m_{\text{freq}} = n_{\text{freq}}^{\text{atoms}}$, the matrix $\boldsymbol{A}_{\text{freq}}$ turns out to be square and singular (in which case, $\boldsymbol{A}_{\text{freq}}^{\dagger} = \boldsymbol{A}_{\text{freq}}^{-1}$). Consequently, the problem in Eq. 5.3 can be efficiently solved by inversion of a small matrix and a single matrix product.

Once $\boldsymbol{X}_{\text{freq}}$ is determined, one could obtain a large 3D block of data, where two dimensions are the superresolved spatial domain and the other is the high resolution time domain. Nevertheless, we are not interested in recovering the light signals in time domain for each superresolved pixel, but only a single depth value. For this reason, only $\boldsymbol{\Psi}_{\text{spatial}}$ is used to transform the data from the sparse 2D DB4 representation into dense superresolved images. Therefore, the final result of the sparse signal recovery is a set of $n_{\text{freq}}^{\text{atoms}}$ HR images, from which a single depth image is to be calculated. The size of these SR images is $\left(n_{\text{rows}}\sqrt{n_{\text{spatial}}}\right) \times \left(n_{\text{cols}}\sqrt{n_{\text{spatial}}}\right)$ and they are constructed by grouping together the $n_{\text{rows}} \times n_{\text{cols}}$ patches of size $\sqrt{n_{\text{spatial}}} \times \sqrt{n_{\text{spatial}}}$ that are obtained per PMD pixel after dense 2D reconstruction from the spatial dictionary coefficients.

Implementation Details For speed, greedy algorithms are used to estimate the sparse support of the signals in the 2D DB4 basis when recovering in spatial domain. In the last version of the software, the ORMP algorithm was not included as an option, for unnecesarily slowing down the recovery process. Instead, the program counts with two options to solve Eq. 5.1: an MP algorithm and an OMP algorithm. The functions implementing both algorithms are coded using functions from the GPU-accelerated linear algebra and solver library ViennaCL [393, 392]. For details on these algorithms, the reader is referred to Section 3.4.2. Probably the most critical point in a greedy search is to determine when to stop the algorithm. We adopt a threshold on the decreasing rate of the residual norm as stopping criterion. That is, if $\boldsymbol{R}^{(k)}$ denotes the residual matrix at iteration k and ε_{Δ} denotes the desired threshold, the algorithm stops if:

$$\frac{\left\|\boldsymbol{R}^{(k-1)}\right\|_2 - \left\|\boldsymbol{R}^{(k)}\right\|_2}{\left\|\boldsymbol{R}^{(k-1)}\right\|_2} < \varepsilon_{\Delta} \tag{5.4}$$

Once the algorithm stops, the result delivered is $\boldsymbol{X}_{\Omega_k}^{(k)}$ if the left hand side of the inequality in Eq. 5.4 is positive or $\boldsymbol{X}_{\Omega_{k-1}}^{(k-1)}$ otherwise. This way one avoids an eventual degradation of the result in the last iteration of the algorithm, because a negative immediate decreasing rate of the residual norm means that the residual norm obtained in the last iteration is greater the previous.

Another relevant implementation detail is how the measurement matrices $\boldsymbol{A}_{\text{spatial}}$ and $\boldsymbol{A}_{\text{freq}}$ are constructed. Clearly, provided that the spatial recovery is performed independently for each PMD pixel, for low k_{SR}, both $\boldsymbol{\Phi}_{\text{spatial}}$ and $\boldsymbol{\Psi}_{\text{spatial}}$ could be generated online at the start of the program and kept stored in RAM during execution. On the contrary, the dimensionality of the time-frequency domain n_{freq} is set to be very large, in order to avoid degrading the depth estimation by an intermediate limitation of the phase resolution of the rows of $\boldsymbol{\Phi}_{\text{freq}}$ and the columns of $\boldsymbol{\Psi}_{\text{freq}}$. Provided that $\boldsymbol{A}_{\text{freq}}$ is given by the product of $\boldsymbol{\Phi}_{\text{freq}}$ and $\boldsymbol{\Psi}_{\text{freq}}$, the large n_{freq} does not affect the size of $\boldsymbol{A}_{\text{freq}}$, but it does increase the time required to generate $\boldsymbol{\Psi}_{\text{freq}}$. For this reason, both dictionaries are generated offline and stored in files. The codes to generate the sensing matrices are stored in their binary form in light CSV files (size $< 1\,\text{Kb}$). Both the codes and the dictionaries are loaded at the program start. At that point, the sensing matrices are generated from their respective codes. Then, the measurement matrices, required to solve Eq. 5.1 and Eq. 5.3, are computed from their respective sensing matrices and dictionaries. Additionally, $\boldsymbol{A}_{\text{freq}}^{-1}$ (or $\boldsymbol{A}_{\text{freq}}^{\dagger}$), which is to be calculated to solve Eq. 5.3, can also be computed at this point and, together with $\boldsymbol{A}_{\text{spatial}}$, kept in memory during execution, provided that all these matrices remain constant. Note that, despite both dictionaries are needed for constructing the measurement matrices, only $\boldsymbol{\Psi}_{\text{spatial}}$ is required for reconstruction in spatial domain, while $\boldsymbol{\Psi}_{\text{freq}}$ is not needed because the depth can be computed from the coefficients in frequency domain, without the need for a costly step of reconstruction in the HR time domain.

For simplicity, both the 2D codes used for generating $\boldsymbol{\Phi}_{\text{spatial}}$ and the 1D codes for $\boldsymbol{\Phi}_{\text{freq}}$ are binary codes derived from a Hadamard matrix. The rows (or columns) of a Hadamard matrix are orthogonal to each other. Consequently, using them as sensing kernels (after reshaping them into 2D codes in the spatial case) ensures that the information gathered in each new measurement was not already contained at all in the previous. Note, nevertheless, that the elements of a Hadamard matrix take values ± 1, while the SLM does not have the capability of *subtracting* light: the light is either transmitted or blocked, but cannot be negated. For this reason, the 2D spatial codes, while derived from a Hadamard matrix, are scaled and offset to the range $\{0, 1\}$. One can argue that orthogonality is lost and redundancy between measurements exist. While this is true, note that only the DC component of the signal (constant offset) is being redundantly sensed in every measurement. Apart from this, the rest of information gathered by different measurements is still non-redundant. PMD pixels feature two integration channels and, consequently, when the reference signal is a binary signal repeating the same custom code for each period of the ICS, the A channels integrate the instantaneous product of that signal and the incoming light signal, while the B channels do the same, but using the negate of the reference signal. As a consequence, using the A-B values as measurements is equivalent to having a system able to deal with binary ± 1 codes. Our system operates this way and, therefore, the 1D binary codes to apply in time domain are randomly selected rows of a columnwise-scrambled Hadamard matrix, with ± 1 values. The rows of $\mathbf{\Phi}_{\text{freq}}$ are not just upscaled versions of these vectors up to n_{freq} dimensions. Valuable *a priori* knowledge of the CS-PMD hardware can be easily integrated. For instance, we apply rising and falling edges using the empirical values obtained when testing the CS-PMD modified CPLD in Section 5.2, which were observed to be $t_{\text{r/f}} \approx 2\,\text{ns}$ (see Fig. 5.7). Provided that the rising and falling behavior is linear, the final trapezoidal codes can be efficiently obtained from the original upscaled (n_{freq} discrete length) square codes by means of a convolution with a discrete square kernel of width $n_{\text{r/f}}$ given by:

$$n_{\text{r/f}} = \frac{2 f_{\text{CLK}} n_{\text{freq}}}{m_{\text{freq}}^{\text{max}}} t_{\text{r/f}} \tag{5.5}$$

where $m_{\text{freq}}^{\text{max}}$ is the number of code elements, i. e., the cardinality of the largest set of non-redundant codes, that is, the maximum number of non-redundant measurements. Note that the non-zero rising and falling times of the PMD reference signal are one of the reasons why $\mathbf{\Phi}_{\text{freq}}$ and, consequently, $\boldsymbol{A}_{\text{freq}}$ cannot be generated offline and loaded at program start, since $m_{\text{freq}}^{\text{max}}$ and n_{freq} are some of the many parameters that can be adjusted by the user through a configuration file and are, therefore, not known beforehand.

Regarding to the dictionaries, $\mathbf{\Psi}_{\text{spatial}} \in \mathbb{R}^{n_{\text{spatial}} \times n_{\text{spatial}}^{\text{atoms}}}$ is, in principle the DB4 orthonormal wavelet basis. If no dictionary atoms are discarded for some reason—e. g., integration of some *a priori* information—, we have that $n_{\text{spatial}} = n_{\text{spatial}}^{\text{atoms}}$ and $\mathbf{\Psi}_{\text{spatial}}$ is square. The frequency dictionary $\mathbf{\Psi}_{\text{freq}} \in \mathbb{R}^{n_{\text{freq}} \times n_{\text{freq}}^{\text{atoms}}}$ is a frequency dictionary made out of sines and cosines of frequencies that are multiples of the fundamental one. The first dictionary element is a constant function and the rest are alternating sines and cosines in order of increasing frequency. The fundamental frequency corresponds to the ICS fundamental frequency f_{mod} and, therefore, the maximum frequency contained in the dictionary, $f_{\mathbf{\Psi}_{\text{freq}}}^{\text{max}}$, and the maximum frequency for which both sine and cosine atoms are contained in the dictionary, $f_{\boldsymbol{\Psi}_{\text{freq}}}^{\text{max pair}}$, are given by the expressions in Eq. 5.6.

$$
\begin{aligned}
f_{\boldsymbol{\Psi}_{\text{freq}}}^{\text{max}} &= \left\lceil \frac{n_{\text{freq}}^{\text{atoms}} - 1}{2} \right\rceil f_{\text{mod}} \\
f_{\boldsymbol{\Psi}_{\text{freq}}}^{\text{max pair}} &= \left\lfloor \frac{n_{\text{freq}}^{\text{atoms}} - 1}{2} \right\rfloor f_{\text{mod}}
\end{aligned}
\tag{5.6}
$$

For notation simplicity, the number of different non-zero frequencies with a complete pair of atoms in the dictionary is denoted in the following as $n_{\text{freq}}^{\text{pairs}} = \left\lfloor \frac{n_{\text{freq}}^{\text{atoms}}-1}{2} \right\rfloor$. Note that $\boldsymbol{\Psi}_{\text{freq}}$ can be constructed only knowing n_{freq} and $n_{\text{freq}}^{\text{atoms}}$, completely independently from f_{mod}, which only brings the mapping to time domain and determines the unambiguous range in depth. In practice, a single frequency dictionary of size $n_{\text{freq}} \times n_{\text{freq}}^{\text{max atoms}}$ is generated offline and stored in a CSV file, where $n_{\text{freq}}^{\text{max atoms}} \ll n_{\text{freq}}$ is just an upper bound on the number of dictionary atoms that the system supports. The value of $n_{\text{freq}}^{\text{max atoms}}$ is chosen to be very conservative, i. e., high enough to obtain a dictionary able to retain the whole signal power of our illumination signals (see Fig. 4.2 and related discussion in Section 4.1.2). When the program starts, only the first $n_{\text{freq}}^{\text{atoms}} < n_{\text{freq}}^{\text{max atoms}}$ atoms of the dictionary are loaded, so that we obtain a $\boldsymbol{\Psi}_{\text{freq}}$ of the size specified by the user in the configuration file. One can easily see that representing periodic real signals (like the illumination signals we deal with) using such dictionary immediately yields the real and complex values of the first Fourier coefficients up to frequency $f_{\boldsymbol{\Psi}_{\text{freq}}}^{\text{max pair}}$.

5.3.2. Depth Calculus

The last step of the CS-PMD software pipeline is to compute the final depth image from the set of $n_{\text{freq}}^{\text{atoms}}$ superresolved dense images obtained by reconstruction in 2D spatial domain from the sparse coefficient vectors. By construction of $\boldsymbol{\Psi}_{\text{freq}}$, each pair of frequency coefficients, $x_{2i_{\text{freq}}}^{\vec{x}_{\text{SR}}}, x_{2i+1_{\text{freq}}}^{\vec{x}_{\text{SR}}}$, $i \geq 1$, can be used to calculate a phase shift for the frequency of the corresponding dictionary atoms and, therefore, a depth value. The superscript $\vec{x}_{\text{SR}} \in \left[1, \left(n_{\text{rows}}\sqrt{n_{\text{spatial}}}\right)\right] \times \left[1, \left(n_{\text{cols}}\sqrt{n_{\text{spatial}}}\right)\right]$ simply denotes the 2D index of the superresolved pixel. Provided that the process of depth calculus is to be perfomed for each one of the superresolved pixels, we make the notation independent from the SR pixel index once again and denote the coefficient pairs as:

$$\begin{aligned} x_{f_i}^{\sin} &= x_{2i_{\text{freq}}}^{\vec{x}_{\text{SR}}} \\ x_{f_i}^{\cos} &= x_{2i+1_{\text{freq}}}^{\vec{x}_{\text{SR}}} \end{aligned} \tag{5.7}$$

where the superscripts of the left hand terms make reference to the nature of the corresponding dictionary element, which can be a sine or a cosine function. The frequency itself is denoted as f_i, $1 \leq i \leq n_{\text{freq}}^{\text{pairs}}$. The fundamental frequency is $f_1 = f_{\text{mod}}$ and the maximum frequency that can be used for depth estimation is $f_{\mathbf{\Psi}_{\text{freq}}}^{\text{max pair}}$, given by Eq. 5.6. Note that, provided that the frequencies contained in $\mathbf{\Psi}_{\text{freq}}$ are multiples of the fundamental one, we have that $f_i = if_1 = if_{\text{mod}}$. Each frequency coefficient pair $(x_{f_i}^{\sin}, x_{f_i}^{\cos})$ can be used to compute a phase shift for the frequency f_i. Using the unambiguous range for that frequency, the phase shift can be mapped to a (eventually ambiguous) depth value. The ambiguous depth yielded by the frequency coefficient pair of frequency f_i is given by:

$$\begin{aligned} d_i &= \frac{d_i^{\text{u}}}{2\pi} \arctan\left(\frac{x_{f_i}^{\sin}}{x_{f_i}^{\cos}}\right) - d_i^0 \\ d_i^{\text{u}} &= \frac{c}{2f_i} \end{aligned} \tag{5.8}$$

where d_i^{u} is the unambiguous range for the frequency f_i and d_i^0 is the depth induced by the initial phase of the emitted waveform for that frequency. The values of d_i^0 are determined experimentally from an acquisition of the emitted waveform at a known distance using Eq. 5.8. In our preliminary experiments, this acquisition was carried out by means of a fast photodiode (cf. Appendix A.6) placed in front of the illumination system, i. e., the emitted light was directly acquired, not its reflection. In this case, the unambiguous range to use in the first equality of Eq. 5.8 in order to compute d_i^0 is twice that given by the second equality. The frequency coefficient pairs (or, directly, the phases) can be obtained from the acquired waveform by means of a DFT of appropriate dimensionality.

Under the hypothesis of a single bounce, the unambiguous depth sensed by different frequencies has to be unique. In other words, the ambiguous depth measurements given by Eq. 5.8 are related to an unique unambiguous depth. Let d denote the unambiguous depth. Then we have that

$$
\begin{aligned}
d =& d_i^{\mathrm{u}} \left(n_i + \frac{1}{2\pi} \arctan\left(\frac{x_{f_i}^{\sin}}{x_{f_i}^{\cos}} \right) \right) - d_i^0, \\
& \forall i \mid 1 \leq i \leq n_{\mathrm{freq}}^{\mathrm{pairs}},\ n_i \in \mathbb{N}
\end{aligned}
\tag{5.9}
$$

where n_i is a natural number providing the number of entire periods that the signal component of frequency f_i is delayed at its arrival to the sensor. In ideal conditions, it exists a unique set of parameter values $[n_i]_{1\leq i\leq n_{\mathrm{freq}}^{\mathrm{pairs}}}$ that makes Eq. 5.9 hold, for a constant d, which is, in principle, unknown. In real conditions this will never occur and, even if the values of all n_i are known, the $n_{\mathrm{freq}}^{\mathrm{pairs}}$ equalities implied in Eq. 5.9 will not be strict. What we have instead is a set of $n_{\mathrm{freq}}^{\mathrm{pairs}}$ depth estimates $\hat{d}_i$, one per frequency, which can be expressed according to Eq. 5.10.

$$
\hat{d}_i = d_i + n_i d_i^{\mathrm{u}} \tag{5.10}
$$

The challenge is, therefore, to compute a single depth value d from the depth estimates $\hat{d}_i$, which implies also finding the right n_i for each frequency. The most likely values of n_i can be computed from the ambiguous depth estimates obtained for the different frequencies using Eq. 5.11, which was introduced in [280] as a framework for PMD multifrequency operation.

$$
\begin{aligned}
[n_i]_{1\leq i\leq n_{\mathrm{freq}}^{\mathrm{pairs}}} = \underset{n_1,\dots,n_{n_{\mathrm{freq}}^{\mathrm{pairs}}}\in\mathbb{N}}{\arg\min} & \left\{ \sum_{i=1}^{n_{\mathrm{freq}}^{\mathrm{pairs}}-1} \frac{1}{\sigma_i^2} \sum_{j=i+1}^{n_{\mathrm{freq}}^{\mathrm{pairs}}} \frac{1}{\sigma_j^2} \Big[(d_i + n_i d_i^{\mathrm{u}}) \right. \\
& \left. - (d_j + n_j d_j^{\mathrm{u}}) \Big]^2 \right\}
\end{aligned}
\tag{5.11}
$$

Note that the cost function to minimize in Eq. 5.11 is the weighted squared difference between the depth estimates $\hat{d}_i$ proposed by the different frequencies. The multifrequency approach contemplated in [280] assumed that several depth measurements were performed per frequency. Consequently, the measured ambiguous depths d_i and d_j in Eq. 5.11 were substituted by mean values for that frequencies, being σ_i^2 and σ_j^2 their corresponding variances. Considering a multifrequency framework in which enough depth measurements are taken per frequency to make their variance reliable might not be realistic. Despite we do not change the fundamental frequency of the ICS, both in our approach and in that of [280], the number of PMD acquisitions scales linearly with the number of frequencies (to recover, in our case, and to measure, in [280]) and, in [280], it also scales linearly with the number of measurements per frequency. In our case, provided that the frequency coefficients are not direct measurements, but recovered from the measurements performed in time domain according to binary codes, the only option to get more than one measurement per frequency is to repeat the measurement process again, gathering all the compressed measurements according to the m_{freq} codes. In this case, also in our approach the total number of acquisitions would scale linearly with the number of measurements per frequency. Even for a moderate number of frequencies to recover, this imposes a severe tradeoff between an acceptable frame rate and a reliable value of the variance. Acquiring very few measurements per frequency (e. g., two) might still allow for a good frame rate if the exposure times are short enough, at the cost of a variance estimate that might not be representative of the true one. Note that using wrong variance estimates in the minimization of Eq. 5.11 eliminates its advantage over its non-weighted version.

In order to provide a reliable framework for combining the depth measurements obtained for different frequencies when very few or only one measurement per frequency are available, we propose to apply error propagation to explicitly model how the noise contained in the measurements affects the final depth estimate, without the need of calculating the variance of the depth from multiple measurements. In Section 2.3.3, the uncertainty of the depth measurements obtained from the four phases algorithm was analyzed in this way. If the noise affecting the raw data is unknown, the uncertainty cannot be propagated. Nevertheless, in ToF imaging, as in any imaging system operating in low light conditions, one can assume that the measurements follow an (eventually shifted and/or rescaled) Poisson distribution, that is, that the variance is equal to the expected value. The details of the derivations for the four phases case are given in Appendix A.4. In our case, only two *phases* are used per frequency and the depth is computed using Eq. 5.8. Applying error propagation to Eq. 5.8 yields

$$
\begin{aligned}
\Delta d_i &= \frac{c}{4\pi f_i}\sqrt{\left\{\frac{x_{f_i}^{\cos}}{\left[(x_{f_i}^{\cos})^2+(x_{f_i}^{\sin})^2\right]}\right\}^2\Delta x_{f_i}^{\sin}+\left\{\frac{-x_{f_i}^{\sin}}{\left[(x_{f_i}^{\cos})^2+(x_{f_i}^{\sin})^2\right]}\right\}^2\Delta x_{f_i}^{\cos}} \\
&= \frac{c}{4\pi f_i}\frac{1}{\left[(x_{f_i}^{\cos})^2+(x_{f_i}^{\sin})^2\right]}\sqrt{(x_{f_i}^{\cos})^2\Delta x_{f_i}^{\sin}+(x_{f_i}^{\sin})^2\Delta x_{f_i}^{\cos}}
\end{aligned}
\tag{5.12}
$$

where $\Delta x_{f_i}^{\sin}$ and $\Delta x_{f_i}^{\cos}$ are the uncertainties of the coefficients $x_{f_i}^{\sin}$ and $x_{f_i}^{\cos}$. The proof of Eq. 5.12 is given in Appendix A.12. Note that, differently from the four phases case, where the raw data can be supposed to follow a Poisson distribution in low light operation, in the CS-PMD system the values of the frequency coefficients used to compute the depth are not raw data acquired directly by the PMD sensor, but recovered from the *compressed* measurements, which are the ones effectively following a Poisson distribution in low light conditions. For any pixel (or superresolved pixel if $k_{\mathrm{SR}} > 1$), consider the corresponding column of $\hat{\boldsymbol{X}}_{\mathrm{freq}}$ and $\boldsymbol{Y}_{\mathrm{freq}}$ in Eq. 5.3 and, for simplicity, let's denote them as $\hat{\vec{X}}$ and $\vec{Y}$. Then, for any arbitrarily chosen (SR) pixel, we have a linear system of the form

$$
\hat{\vec{X}} = \boldsymbol{A}^{\dagger}\vec{Y} \tag{5.13}
$$

where $\boldsymbol{A} = \boldsymbol{A}_{\mathrm{freq}}$, $\vec{X} \in \mathbb{R}^n$, $\vec{Y} \in \mathbb{R}^m$ for generality of the following derivations. Each element of $\vec{Y}$, y_j, $1 \leq j \leq m$ is expected to be a realization of a random variable following a Poisson distribution, i. e., $\mathcal{Y}_j \sim \mathcal{P}(\lambda_j)$. From Eq. 5.13 follows that each element of $\vec{X}$, x_i, $1 \leq i \leq n$ is also a realization of a random variable $\mathcal{X}_i$, given by a linear combination of the random variables $\mathcal{Y}_j$, $1 \leq j \leq m$, as given in Eq. 5.14, where $\vec{A}_i^{\dagger}$ denotes the row vector obtained by extracting the ith row of $\boldsymbol{A}^{\dagger}$. The probability distribution of $\mathcal{X}_i$ is determined by the distributions of $\mathcal{Y}_j$ and the values of the coefficients of the vector $\vec{A}_i^{\dagger}$. One can easily see that, in general, $\mathcal{X}_i \nsim \mathcal{P}(\lambda_i)$, regardless of λ_i. Provided that the coefficients of $\vec{A}_i^{\dagger}$ can be positive or negative, the resulting distribution is susceptible of loosing the characteristic skewness of the Poisson distribution, due to the combination of distributions with long tail at the right (positive coefficients) and with long tail at the left (negative coefficients). Regardless of the shape of the combination distribution, its variance can be predicted from those of the component variables, as expressed in the second equality of Eq. 5.14. Provided that each y_j comes from an independent measurement, it is legitimate to suppose that the random variables $\mathcal{Y}_j$ are independent, in which case the covariance matrix $\Sigma_{\vec{\mathcal{Y}}}$ takes the diagonal shape in Eq. 5.14 and the resulting variance is given by the subsequent weighted sum of the original variances, where $a^{\dagger}_{i,j}$ denotes the (i,j) element of the matrix $\boldsymbol{A}^{\dagger}$.

$$\mathcal{X}_i = \vec{A}^{\dagger}_i \vec{\mathcal{Y}}$$

$$\mathrm{Var}\left[\mathcal{X}_i\right] = \vec{A}^{\dagger}_i \Sigma_{\vec{\mathcal{Y}}} \left(\vec{A}^{\dagger}_i\right)^{\top}, \quad \Sigma_{\vec{\mathcal{Y}}} = \begin{bmatrix} \lambda_1 & 0 & \cdots & 0 & \cdots & 0 \\ 0 & \lambda_2 & \cdots & 0 & \cdots & 0 \\ \vdots & \vdots & \ddots & \vdots & & \vdots \\ 0 & 0 & \cdots & \lambda_j & \cdots & 0 \\ \vdots & \vdots & & \vdots & \ddots & \vdots \\ 0 & 0 & \cdots & 0 & \cdots & \lambda_m \end{bmatrix} \tag{5.14}$$

$$\mathrm{Var}\left[\mathcal{X}_i\right] = \sum_{j=1}^{m} \left(a^{\dagger}_{i,j}\right)^2 \lambda_j$$

In the specific case of similar variances of the measurement variables, i. e., $\lambda_i \approx \lambda_j \approx \lambda,\ \forall i,j \leq m$ and $\|\vec{A}^{\dagger}_i\|_2 \approx 1,\ \forall i \leq n$, the variance is preserved and we have that $\mathrm{Var}\left[\mathcal{X}_i\right] = \lambda,\ \forall i \leq n$. As stated in the depth uncertainty analysis in Section 2.3.3, in the Poisson case, the uncertainty of the measurements can be well approximated by the absolute value of the measurement, which in this case translates into $\hat{\lambda}_j = y_j^2$. In combination with Eq. 5.14 this yields an estimate of the uncertainty of the $\vec{X}$ coefficients:

$$\Delta x_i = \sqrt{\sum_{j=1}^{m} \left(a^{\dagger}_{i,j}\right)^2 y_j^2} \tag{5.15}$$

This result is general and applies to any linear system of the form given in Eq. 5.13, where the measurements are realizations of random variables following Poisson distributions. In our case, provided that $\boldsymbol{A}_{\text{freq}}$ is known, the vector of uncertainties in frequency domain $\overrightarrow{\Delta X}_{\text{freq}}$ can be directly computed from the vector of compressed measurements $\vec{Y}_{\text{freq}}$ applying Eq. 5.15.

The input uncertainties of Eq. 5.12, $\Delta x^{\sin}_{f_i}$ and $\Delta x^{\cos}_{f_i}$ can then be taken from $\overrightarrow{\Delta X}_{\text{freq}}$:

$$\begin{aligned} \Delta x_{f_i}^{\text{sin}} &= \Delta x_{\text{freq}}^{2i} \\ \Delta x_{f_i}^{\cos} &= \Delta x_{\text{freq}}^{2i+1} \end{aligned} \tag{5.16}$$

where $\Delta x_{\text{freq}}^{k}$ denotes the kth element of $\overrightarrow{\Delta X}_{\text{freq}}$. For simplicity, in the CS-PMD software, the minimization problem in Eq. 5.11 is not solved in depth domain, but in phase domain for the fundamental frequency. The corresponding problem—which is equivalent to Eq. 5.11 and can be derived from it by removing constant factors—, reads

$$\begin{aligned} [n_i]_{1\leq i\leq n_{\text{freq}}^{\text{pairs}}} &= \underset{n_1,\ldots,n_{n_{\text{freq}}^{\text{pairs}}}\in\mathbb{N}}{\arg\min} \left\{ \sum_{i=1}^{n_{\text{freq}}^{\text{pairs}}-1} \frac{1}{\sigma_i^2} \sum_{j=i+1}^{n_{\text{freq}}^{\text{pairs}}} \frac{1}{\sigma_j^2} \left[\frac{1}{i}(\theta_i + 2\pi n_i) \right.\right. \\ &\qquad \left.\left. - \frac{1}{j}(\theta_j + 2\pi n_j) \right]^2 \right\} \\ \sigma_k \approx \Delta\theta_k &= \frac{\sqrt{(x_{f_k}^{\cos})^2 \Delta x_{f_k}^{\sin} + (x_{f_k}^{\sin})^2 \Delta x_{f_k}^{\cos}}}{\left[(x_{f_k}^{\cos})^2 + (x_{f_k}^{\sin})^2 \right]} \\ \theta_k &= \arctan \frac{x_{f_k}^{\sin}}{x_{f_k}^{\cos}} - \theta_k^0 \end{aligned} \tag{5.17}$$

where θ_k^0 is the phase equivalent to the d_k^0 in depth domain (cf. Eq. 5.8), that is, the initial phase of that frequency component.

As in [280], Eq. 5.17 is solved by evaluating all possible combinations of the free parameters (recall the restriction to natural numbers) within a given feasible depth range. Once $[n_i]_{1\leq i\leq n_{\text{freq}}^{\text{pairs}}}$ are determined, the last step is to compute a single depth value from those suggested by the different frequencies. As before, the simplest option would be averaging them, while the sensible one is to perform a weighted averaging, taking profit of the propagated uncertainties computed previously. We choose the latter and solve the problem according to Eq. 5.18. Note the use of $\hat{d}_i$ (from Eq. 5.10) instead of d_i, which accounts for the number of entire periods calculated in Eq. 5.17.

$$d = \frac{\sum_{i=1}^{n_{\text{freq}}^{\text{pairs}}} \frac{1}{\Delta d_i} \hat{d}_i}{\sum_{i=1}^{n_{\text{freq}}^{\text{pairs}}} \frac{1}{\Delta d_i}} = \frac{c}{4\pi f_{\text{mod}}} \frac{\sum_{i=1}^{n_{\text{freq}}^{\text{pairs}}} \frac{1}{\Delta \theta_i} \theta_i}{\sum_{i=1}^{n_{\text{freq}}^{\text{pairs}}} \frac{i}{\Delta \theta_i}} \tag{5.18}$$

6. Conclusions

6.1. Summary

In this thesis novel ways of applying the recent theory of Compressive Sensing (CS) to the specific case of ToF depth imaging using the Photonic Mixer Device (PMD) have been proposed and studied.

Provided that CS theory relies on the linearity of the sensing model and the sparsity or high compressibility of the data, a considerable amount of work has been deditated to provide methods that ensure this linearity and boost the compressibility of the data at the low level, so that CS can be safely applied in an upper processing layer. The per-pixel linearity of the PMD sensing process is compromised by the limited dynamic range of PMD pixels, which leads to eventual activation of the SBI. We have proposed an adaptive framework for attaining HDR from a reduced set of short exposure times that are considered apart from the user-defined exposure. This way, scene points reflecting an amount of light that would trigger the activation of the SBI of the corresponding pixel for the default exposure are accurately sensed by means of the acquisitions at short exposures. Raw data obtained using our AHDR method follows an extended linear response curve. Additionally, this avoids the appearance of noise related to the SBI activation, thus enhancing the compressibility of the raw data. The compressibility of the raw images, e. g., in wavelet dictionaries, is strongly deteriorated by noise. We have shown up to which extent a relatively simple per-pixel linear calibration can increase the compressibility of the raw data, by eliminating the so-called fixed-pattern noise (FPN) of the sensor. Furthermore, we have shown that the critical photon shot noise, which is dominant in poorly-illuminated image areas, can be significantly reduced by means of some kind of joint bilateral filtering of the raw images, before the depth image is calculated. Depth images obtained from joint-filtered raw data exhibit improved depth accuracy with respect to their non-filtered counterpart, at no cost in terms of lateral resolution.

CS can be applied in, at least, two different domains where signals gathered by the PMD sensor are sparse: the 2D spatial domain where the raw images live and the time-frequency domain where the modulation signals (and, therefore, the corresponding echoes) live. In the first case, compressibility in, e. g., wavelet domain can be exploited. We consider both the cases of gathering a number of measurements per raw image that is lower than the number of pixels of the sensor and then recovering the complete raw data making use of sparsity constraints and, complementary, gathering more measurements than the number of pixels and recovering superresolved raw data of higher dimensionality than that given by the number of measurements. For the first case, we have proposed to use a greedy pursuit that is able to simultaneously recover all raw images from a reduced number of measurements in a sinergetic manner. The method is an MMV extension of ORMP that does not only exploit the *joint* sparsity of the raw data, but also accounts for the *tree structure* of wavelet-based sparsity and integrates the reduced rank of the raw data as an explicitly constraint within the recovery pipeline. As a result, our SRA-ORMP algorithm allows for establishing very attractive tradeoffs between number of measurements and depth accuracy, including the case of obtaining better depth accuracy than that attained via conventional sensing, still from a lower number of measurements. If superresolution is desired, CS requires an accurate model for the transition from the high-resolution domain—infinite, in the continuous case—to the low-resolution domain, which has to be integrated as part of the linear sensing model. One of the pillars of this thesis is a thorough empirical characterization of the PMD pixel response in the 2D spatial domain. As a result, high-resolution discrete spatial responses (up to 1 μm resolution) are obtained for both channels of PMD pixels. To the best of our knowledge, our characterization has no published precedents in terms of resolution and area considered. Our discrete spatial responses have been used to model the hight-to-low resolution transition in the sensing matrix of a PMD-based FLIM experiment, in which a stepper unit was used to obtain measurements at as many locations as desired. For enough number of measurements, phase shift images with an outstanding superresolution factor of 9, with respect to the size of the PMD sensor, were successfully recovered. Extending the area considered in the spatial responses to neighboring pixels, crosstalk effects have been observed and characterized. An accurate characterization of the PMD pixel crosstalk for each of its two channels like the one we provide has not been published before. As an application of the cross-responses, crosstalk has been removed from real PMD data in a deconvolution framework, resulting in enhanced depth accuracy.

The other domain that has been exploited is the time-frequency. While a pulsed ToF system exhibits sparsity directly in time domain, CW ToF systems, such as PMD sensors, continuously emit a periodic signal, which is either sparse or highly compressible in frequency domain and can, therefore, be recovered from few measurements in time domain. Taking profit of this, a framework for CS-based multifrequency depth estimation that recovers the reflected signal in frequency domain has been proposed. Despite the framework is oriented to the typical PMD modulation, i. e., single-frequency quasi-sinusoidal modulation, using several pseudorandom (PR) binary patterns as measurement kernels in time domain allows recovering high frequency harmonics and using them to improve the depth accuracy. This is an just an indicator of the power of an appropriate sensing model: high order harmonics, which are typically viewed as a distortion, can be included in the model and used as if several frequencies where simultaneously emitted, to improve the depth accuracy or even to resolve more than one path per pixel. Obviously, conventional PMD sensors do not allow gathering measurements according to PR binary patterns. As a simplified alternative and, taking profit of the non-sinusoidal (close-to-square, trapezoidal in practice) shape of the PMD reference signal, we have proposed to perform measurements according to sensing kernels that are randomly-shifted versions of such signal. We have shown that it is, indeed, possible to recover a good approximation of the light signal received at the surface of a PMD pixel from few of such measurements. As an application case, we have proposed to use the PMD sensor itself as a multichannel illumination tester, which can simultaneously recover the waveforms of the illumination signal as received by each of the pixels of the sensor. Obviously, the sensing kernels obtained by shifting a fixed waveform exhibit lower variability and, thus, poorer sensing performance, than real PR binary patterns. For this reason, we have proposed a modification of the program in the CPLD of the PMD frontend so that it allows using periodically-repeated custom PR binary codes as reference signals for the PMD pixels.

Finally, it has also been proposed to carry out CS simultaneously in both spatial and temporal domains. To that end, an experimental prototype has been proposed and developed that allows for such joint compressed sensing in both domains, constituting a complete CS-based PMD camera. This system, introduced in Chapter 5 as CS-PMD, uses a high resolution spatial modulator to superimpose a 2D PR binary pattern to the image patch being projected onto each PMD pixel. This way, each pixel acts as a *single pixel camera* in spatial domain. Additionally, the modification of the CPLD program transforms the conventional PMD frontend into a real CS-PMD module, which allows using any custom binary code as reference signal in the PMD measurement process, i. e., as sensing kernel. An efficient joint recovery pipeline, which makes use of sequential MMV formulations to increase robustness and reduce to the minimum the number of measurements, have also been introduced. Our CS-PMD camera prototype is inherently superior to a conventional PMD camera, since the classical four phases operation is just a specific operation mode of our system that uses four square signals with equal relative phase shifts between the next and the previous as binary patterns (and a single ones pattern for the spatial sensing, if no superresolution is desired).

The good results obtained when applying the methods presented in this work to real PMD data bear witness of the valuable room for improvement in current ToF cameras if a more intelligent sensing strategy, able to condense all signal information in a reduced set of measurements, is combined with an accurate sensing model and a signal recovery method that exploits all available *a priori* information on the system and the nature of the signals being sensed.

6.2. Future Work

Further steps towards a feasible CS-PMD prototype In this thesis we have proposed an implementation of the concept of a CS-based PMD camera able to sense in all available domains, namely two spatial and one temporal, simultaneously in a compressed manner. The prototype has been conceived as a demonstrator and, despite at the conclusion of this thesis its construction has been completed, few minor details are still to be addressed before carrying out the first test acquisitions. The next steps to follow are, consequently, getting the current prototype ready for operation and, subsequently, testing the capabilities of the system with real test scenes, in order to check how much compression at sensing is admissible, still achieving equivalent depth accuracy and resolution as with conventional sensing and, complementary, up to which extent the depth accuracy and resolution can be enhanced and how high is the corresponding cost in terms of number of compressed measurements required.

Once the advantages of the CS-PMD camera concept has been contrasted using our proof-of-concept prototype, the next step is the design of a commercially feasible prototype that preserves or improves the capabilities of our demonstrator, while reducing at maximum the system complexity, e. g., the number and volume of optical elements, and, consequently, the size. Despite this is a field of future work and, consequently, subject to discussion, one can think of some directions to follow in order to achieve such goal. For instance, using a reflective SLM, despite the better fill factor and contrast with respect to its transmissive counterpart, does not seem to be the best option for a commercial solution due to the large increment of the system size it produces. A transmissive SLM located as close as possible to the PMD sensor in order to reduce image distortion seems to be the most compact option. On the other hand, this option increases the complexity of the mounting and adjustment and the probability of damaging the PMD sensor. For this reason it was not considered for the prototype. In the limit, if mass production is foreseen, one could integrate the SLM *on chip*, that is, as an additional layer of the PMD chip, Obviously, this requires agreement and cooperation with the chip manufacturer to develop such a novel solution. Regarding the CS-PMD module itself, which is a 19k-S3 frontend with customized CPLD program, this is also not the best option for a future commercial product. As in the newest PMD chips, we propose integrating our concept of correlating against custom binary patterns within the PMD chip. Integrating the clock and reference signal generation within the chip has shown to allow for higher modulation frequencies.

Bringing polarization into the game The techniques denoted under the name of *shape from polarization* aim to recover the surface normals of an object from polarized images [468, 15]. The core idea is to exploit the fact that the shape of an object influences the polarization of the reflected light, especially when the light incidence is far from being normal to the surface. It is well-known that such techniques cannot recover true depth values, but only relative ones, obtained by integration of the surface normals. Recently, shape from polarization has been successfully combined with ToF depth imaging to achieve outstandingly accurate depth maps [250]. The combination of both techniques is synergistic: ToF systems are known to provide accurate depth estimation in case of frontoparallel surfaces, degrading for large angles between the light incidence direction and the surface normal. Complementary, the degree of polarization is zero if the zenith angle is zero, both for diffuse and specular polarization. Therefore, polarization cues are informative where the ToF estimation is uncertain, and vice-versa.

Provided that the CS-PMD system introduced in Chapter 5 requires a polarizer within the optical path in order to block or transmit the light according to the 2D spatial codes superimposed by an LC-SLM, the pixelated polarizers [371] suggested in [250] arise as a natural way of integrating polarization as an additional dimension in the CS framework. This would reduce image distortions by substituting an optical element by an *on-pixel* overlay. One could think, for instance, of a four-apertures (two would suffice)

ToF sensor in which a polarizer is overlaid on each aperture, with relative polarization angle rotations of 45°. Using such a sensor in the CS-PMD system and projecting polarized light onto the scene, one can effectively correlate against ±1 PR patterns also in spatial domain, provided that the SLM simply produces an eventual polarization rotation and it is the subsequent polarizer the one that transmits or blocks the light coming from the SLM, depending on its orientation. Consequently, subtracting apertures with crossed polarizations yields the desired result.

Compressive sensing and pulsed ToF: good friends should stay together
In this thesis we have focused in PMD hardware and we have tried to avoid imposing abusive requirements on the external hardware required for a ToF depth camera, namely illumination system and corresponding drivers. For this reason, we have not considered the case of emitting light modulated at too high frequencies or according to some waveforms whose frequency content cannot be supported by the drivers, e. g., too narrow pulses. Consequently, as a default, we have presumed operation with a close-to-sinusoidal illumination signal, still exhibiting non-negligible harmonic content, but of low fundamental frequency, e. g., 20 MHz. Under such assumptions, sparse recovery of the echoes in frequency domain showed to be a feasible alternative for bringing CS into the game. Nevertheless, if the restrictions on the illumination system and the operation mode of the *smart* pixels are leveraged, one can think of other attractive alternatives for using CS to attain enhanced depth sensing. Some recent *smart* pixels that perform separated sensing and demodulation are designed to operate in a pulsed mode, making use of laser illumination systems, which are not bounded by the slow rising and falling characteristics of the LEDs. On one hand, emitting very short pulses allows concentrating the illumination power in a very short time interval and achieving outstanding signal-to-background ratios, thus enhancing the immunity to BGL. On the other hand, the width of the pulse determines the unambiguous range of the system, thus making too narrow pulses unsuitable for medium-range depth sensing. Ideally, one would wish to emit a pulse that is as short as allowed by the hardware, while keeping the depth range arbitrarily large, eventually being able to sense more than one return per pixel.

Clearly, a pulsed ToF system emitting one or few very narrow pulses receives an echo that is highly sparse in time domain, constituting a complementary situation to CW phase-shift-based depth sensing, where the echoes are sparse in frequency domain. The main issue with such a pulsed system is that the signal power of the echoes is highly concentrated in some *unknown* locations in temporal domain. A simplistic analysis of the problem profiles it as an appropriate application case for CS, since we have a sparse signal with unknown support, which can be recovered from few non-adaptive measurements, e. g., gathered according to some random sensing kernels. In practice, if we want to sense a relatively large depth range with moderate resolution, the problem dimensionality becomes intractable. For instance, consider a moderate range of 10 m with 1 mm depth resolution. This yields sparse vectors of 10^4 dimensions and, consequently, matrices with 10^8 elements, in the worst case. Explicitly generating and storing dense sensing matrices of such dimensions is often not possible for an embedded system and challenging for current computers in terms of RAM. We propose to apply the method for adaptive construction of LDPC sensing matrices (APEG) introduced in Section 3.2.10 to the case of pulsed ToF depth sensing. The resulting sensing matrices are of very low density, that is, most of the elements are zero and, consequently, do not need to be stored. In experimental tests it has been observed that the best sensing performance is attained for densities as low as 5%, approximately. Additionally, using dense randomly-generated sensing kernels yields poor measurement SNR when the dimensionality is high. Our APEG construction progressively restricts the sensing kernels to the signal support, making use of the available information on the signal support given by previous measurements when generating a new sensing kernel. This allows focusing the sensing where the signal power is concentrated, avoiding sensing those locations where there is no signal but only noise. Preliminary simulations have shown that our APEG-LDPC matrices outperform random sensing matrices when recovering trains of spikes in a high-dimensional domain from few measurements.

A. Appendix

A.1. Cross-Correlation Between Sinusoidal Signals

This section provides the mathematic demonstration of Eq. 2.12, which results from Eq. 2.11, applying common trigonometric identities. The starting point is Eq. 2.11, which is a general expression for the cross-correlation between two sinusoidal signals. For completeness, we rewrite the equation here:

$$\begin{aligned} c_{q,r}(\tau) &= \lim_{T\to\infty} \frac{1}{2T} \int_{-T}^{T} q^*(t) r(t+\tau) dt \\ &= \lim_{T\to\infty} \frac{1}{2T} \int_{-T}^{T} [A_0^q + A^q \cos(\omega t - \theta_q)][A_0^r + A^r \cos(\omega(t+\tau) - \theta_r)] dt \end{aligned} \tag{A.1}$$

Differently from [368], we start pointing out that cosines with arguments depending linearly on t will lead to a null contribution when integrating along a number of periods that tends to infinity. In general:

$$\begin{aligned} \lim_{T\to\infty} \frac{1}{2T} \int_{-T}^{T} \cos(at+b) dt = 0 \\ \lim_{T\to\infty} \frac{1}{2T} \int_{-T}^{T} \sin(at+b) dt = 0 \end{aligned} \tag{A.2}$$

where a and b are arbitrary constants. Considering Eq. A.2, the product in Eq. A.1 yields only two non-null terms:

$$c_{q,r}(\tau) = \lim_{T\to\infty} \frac{1}{2T} \int_{-T}^{T} [A_0^q A_0^r + A^q A^r \cos(\omega t - \theta_q) \cos(\omega(t+\tau) - \theta_r)] dt$$

Now consider the following trigonometric identities for the products of sines and cosines:

$$\begin{aligned}\cos(\alpha)\cos(\beta) &= \frac{1}{2}\left(\cos(\alpha-\beta)+\cos(\alpha+\beta)\right)\\ \sin(\alpha)\sin(\beta) &= \frac{1}{2}\left(\cos(\alpha-\beta)-\cos(\alpha+\beta)\right)\\ \sin(\alpha)\cos(\beta) &= \frac{1}{2}\left(\sin(\alpha+\beta)+\sin(\alpha-\beta)\right)\\ \cos(\alpha)\sin(\beta) &= \frac{1}{2}\left(\sin(\alpha+\beta)-\sin(\alpha-\beta)\right)\end{aligned} \tag{A.3}$$

Applying the first identity in Eq. A.3 we can transform the product into a sum:

$$\begin{aligned}c_{q,r}(\tau) = \lim_{T\to\infty}\frac{1}{2T}\int_{-T}^{T}\Big\{ & A_0^q A_0^r + A^q A^r \frac{1}{2}\Big[\cos(\omega(t+\tau)-\theta_r-\omega t+\theta_q)\\ & + \cos(\omega(t+\tau)-\theta_r+\omega t-\theta_q)\Big]\Big\} dt\end{aligned}$$

from which the second term can be neglected due to Eq. A.2, and, denoting the relative phase shift as $\theta_{\text{depth}} = \theta_r - \theta_q$, we obtain:

$$\begin{aligned}c_{q,r}(\tau) &= \lim_{T\to\infty}\frac{1}{2T}\int_{-T}^{T}\left[A_0^q A_0^r + \frac{A^q A^r}{2}\cos(\omega\tau-\theta_{\text{depth}})\right] dt\\ &= A_0^q A_0^r + \frac{A^q A^r}{2}\cos(\omega\tau-\theta_{\text{depth}}) \ \blacksquare\end{aligned}$$

A.2. Cross-Correlation Between Periodic Signals

In general, the signals involved in the cross-correlation (Eq. A.1) might not be exactly sinusoidal. A more realistic hypothesis is to suppose that they will be periodic of identical fundamental frequency but, in general different waveforms. This is the case considered in Eq. 2.14, where the Fourier representation is used to decompose the non-sinusoidal, yet periodic, signals, $q(t)$ and $r(t)$ implied in the cross-correlation. In this section we demonstrate that Eq. 2.15 is equivalent to Eq. 2.14. For completeness, we rewrite the original equation here:

$$
\begin{aligned}
c_{q,r}(\tau) = \lim_{T\to\infty} \frac{1}{2T} \int_{-T}^{T} & \left\{ A_0^q + \sum_{n=1}^{\infty} \left[A_{n,1}^q \sin(n\omega t - \theta_q) \right.\right. \\
& \left.\left. + A_{n,2}^q \cos(n\omega t - \theta_q) \right] \right\} \\
& \left\{ A_0^r + \sum_{n=1}^{\infty} \left[A_{n,1}^r \sin(n\omega t + \omega\tau - \theta_r) \right.\right. \\
& \left.\left. + A_{n,2}^r \cos(n\omega t + \omega\tau - \theta_r) \right] \right\} dt
\end{aligned}
\tag{A.4}
$$

It is clear that most of the terms of the integral are products of sinusoidal signals of different frequencies, which are multiples of the fundamental frequency. These terms can be neglected due to Eq. A.5, which is an immediate consequence of the orthogonality of the Fourier basis.

$$
\left.
\begin{aligned}
\lim_{T\to\infty} \frac{1}{2T} \int_{-T}^{T} \cos(n_1\omega t + \theta_1) \cos(n_2\omega t + \theta_2) dt = 0 \\
\lim_{T\to\infty} \frac{1}{2T} \int_{-T}^{T} \sin(n_1\omega t + \theta_1) \sin(n_2\omega t + \theta_2) dt = 0 \\
\lim_{T\to\infty} \frac{1}{2T} \int_{-T}^{T} \sin(n_1\omega t + \theta_1) \cos(n_2\omega t + \theta_2) dt = 0 \\
\lim_{T\to\infty} \frac{1}{2T} \int_{-T}^{T} \cos(n_1\omega t + \theta_1) \sin(n_2\omega t + \theta_2) dt = 0
\end{aligned}
\right\}
\begin{aligned}
&\forall n_1, n_2 \in \mathbb{N}, \\
&n_1 \neq n_2
\end{aligned}
\tag{A.5}
$$

Developing the product of Fourier series and eliminating these *cross-terms* from Eq. A.4, we obtain:

$$
\begin{aligned}
c_{q,r}(\tau) = \lim_{T\to\infty} \frac{1}{2T} \int_{-T}^{T} & \left\{ A_0^q A_0^r + \sum_{n=1}^{\infty} \left[\right.\right. \\
& A_{n,1}^q A_{n,1}^r \sin(n\omega t - \theta_q) \sin(n\omega t + \omega\tau - \theta_r) \\
& + A_{n,1}^q A_{n,2}^r \sin(n\omega t - \theta_q) \cos(n\omega t + \omega\tau - \theta_r) \\
& + A_{n,2}^q A_{n,1}^r \cos(n\omega t - \theta_q) \sin(n\omega t + \omega\tau - \theta_r) \\
& \left.\left. + A_{n,2}^q A_{n,2}^r \cos(n\omega t - \theta_q) \cos(n\omega t + \omega\tau - \theta_r) \right] \right\} dt
\end{aligned}
$$

At this point we use the identities in Eq. A.3 to transform the products of sines and cosines into sums, yielding:

$$
\begin{aligned}
c_{q,r}(\tau) = \lim_{T\to\infty} \frac{1}{2T} \int_{-T}^{T} \Bigg(A_0^q A_0^r + \frac{1}{2} \sum_{n=1}^{\infty} \Big\{ \\
A_{n,1}^q A_{n,1}^r \Big[&\cos\left(n\omega t - \theta_q - n\omega t - \omega\tau + \theta_r\right) \\
&- \cos\left(n\omega t - \theta_q + n\omega t + \omega\tau - \theta_r\right)\Big] \\
+ A_{n,1}^q A_{n,2}^r \Big[&\sin\left(n\omega t - \theta_q + n\omega t + \omega\tau - \theta_r\right) \\
&+ \sin\left(n\omega t - \theta_q - n\omega t - \omega\tau + \theta_r\right)\Big] \\
+ A_{n,2}^q A_{n,1}^r \Big[&\sin\left(n\omega t - \theta_q + n\omega t + \omega\tau - \theta_r\right) \\
&- \sin\left(n\omega t - \theta_q - n\omega t - \omega\tau + \theta_r\right)\Big] \\
+ A_{n,2}^q A_{n,2}^r \Big[&\cos\left(n\omega t - \theta_q - n\omega t - \omega\tau + \theta_r\right) \\
&+ \cos\left(n\omega t - \theta_q + n\omega t + \omega\tau - \theta_r\right)\Big] \Big\} \Bigg) dt
\end{aligned}
$$

The arguments of the sines and cosines can now be simplified. Sinusoidal functions with arguments being a lineal function of t can be eliminated due to Eq. A.2. We also use the trivial identities $\cos(\alpha) = \cos(-\alpha)$ and $\sin(\alpha) = -\sin(-\alpha)$ to obtain the desired sign for the arguments. If we additionally denote the relative phase shift as $\theta_{\text{depth}} = \theta_r - \theta_q$, we obtain:

$$
\begin{aligned}
c_{q,r}(\tau) = \lim_{T\to\infty} \frac{1}{2T} \int_{-T}^{T} \Bigg\{ A_0^q A_0^r + \frac{1}{2} \sum_{n=1}^{\infty} \Big[A_{n,1}^q A_{n,1}^r \cos\left(\omega\tau - \theta_{\text{depth}}\right) \\
- A_{n,1}^q A_{n,2}^r \sin\left(\omega\tau - \theta_{\text{depth}}\right) + A_{n,2}^q A_{n,1}^r \sin\left(\omega\tau - \theta_{\text{depth}}\right) \\
+ A_{n,2}^q A_{n,2}^r \cos\left(\omega\tau - \theta_{\text{depth}}\right)\Big] \Bigg\} dt
\end{aligned}
$$

Provided that there is no dependency on n in the trigonometric functions, we can extract them from the summation. In addition, since the sines and cosines no longer depend on t, the integral can be trivially evaluated and the limit solved, yielding:

$$c_{q,r}(\tau) = A_0^q A_0^r + \frac{1}{2}\left\{ \left[\sum_{n=1}^{\infty} A_{n,1}^q A_{n,1}^r + A_{n,2}^q A_{n,2}^r\right] \cos\left(\omega\tau - \theta_{\text{depth}}\right) \right.$$
$$\left. + \left[\sum_{n=1}^{\infty} A_{n,2}^q A_{n,1}^r - A_{n,1}^q A_{n,2}^r\right] \sin\left(\omega\tau - \theta_{\text{depth}}\right) \right\} \blacksquare$$

A.3. Phase Shift, Amplitude and Offset Estimation

This section provides the proofs of Eq. 2.18, Eq. 2.20 and Eq. 2.21, which calculate the depth (by means of the phase shift), amplitude and DC offset, respectively, from the PMD measurements. For the proofs we take Eq. 2.12 as starting point, with the following change of variables:

$$A_0 = A_0^q A_0^r, \; A = \frac{A^q A^r}{2}$$

where A_0 is the DC offset and A is the amplitude. Eq. 2.12 is evaluated for the four sampling points typically considered in PMD systems, namely, $\theta \in \{0, 90°, 180°, 270°\}$ (recall $\tau = \theta/\omega$). Therefore, for one of the pixel channels four measurements are obtained:

$$\begin{aligned}
I_{0^\circ} &= A_0 + A\cos\left(0 - \theta_{\text{depth}}\right) \\
I_{90^\circ} &= A_0 + A\cos\left(90 - \theta_{\text{depth}}\right) \\
I_{180^\circ} &= A_0 + A\cos\left(180 - \theta_{\text{depth}}\right) \\
I_{270^\circ} &= A_0 + A\cos\left(270 - \theta_{\text{depth}}\right)
\end{aligned}$$

Considering trivial trigonometric identities, the previous expressions can be rewritten as:

$$\begin{aligned} I_{0^\circ} &= A_0 + A\cos(\theta_{\text{depth}}) \\ I_{90^\circ} &= A_0 + A\sin(\theta_{\text{depth}}) \\ I_{180^\circ} &= A_0 - A\cos(\theta_{\text{depth}}) \\ I_{270^\circ} &= A_0 - A\sin(\theta_{\text{depth}}) \end{aligned} \tag{A.6}$$

Now, subtracting the expressions in Eq. A.6 two by two we can eliminate the offset:

$$\begin{aligned} I_{270^\circ} - I_{90^\circ} &= -2A\sin(\theta_{\text{depth}}) \\ I_{180^\circ} - I_{0^\circ} &= -2A\cos(\theta_{\text{depth}}) \end{aligned} \tag{A.7}$$

Dividing the two expressions in Eq. A.7 by each other we obtain

$$\frac{I_{270^\circ} - I_{90^\circ}}{I_{180^\circ} - I_{0^\circ}} = \tan(\theta_{\text{depth}})$$

which, is equivalent to Eq. 2.18:

$$\theta_{\text{depth}} = \arctan\left(\frac{D_{270^\circ} - D_{90^\circ}}{D_{180^\circ} - D_{0^\circ}}\right) \blacksquare$$

Provided that $\sin^2(\theta) + \cos^2(\theta) = 1,\ \forall\theta$, the square sum of the expressions in Eq. A.7 further eliminates θ_{depth}

$$(I_{270^\circ} - I_{90^\circ})^2 + (I_{180^\circ} - I_{0^\circ})^2 = (2A)^2$$

and the amplitude is therefore:

$$A = \frac{\sqrt{(I_{270^\circ} - I_{90^\circ})^2 + (I_{180^\circ} - I_{0^\circ})^2}}{2} \blacksquare$$

The offset is trivially obtained averaging the four measurements in Eq. A.6:

$$A_0 = \frac{I_{0^\circ} + I_{90^\circ} + I_{180^\circ} + I_{270^\circ}}{4} \blacksquare$$

A.4. Depth Measurement Uncertainty

In this section we derive Eq. 2.28 from Eq. 2.19 by error propagation and then Eq. 2.29 from Eq. 2.28 in the case of Poisson noise. For completeness, we rewrite Eq. 2.19:

$$\begin{aligned} d &= \frac{c}{4\pi f_{\text{mod}}} \arctan\left(\frac{D_{270^\circ} - D_{90^\circ}}{D_{180^\circ} - D_{0^\circ}}\right) \\ &= \frac{c}{4\pi f_{\text{mod}}} \arctan\left(\frac{(I^A_{270^\circ} - I^B_{270^\circ}) - (I^A_{90^\circ} - I^B_{90^\circ})}{(I^A_{180^\circ} - I^B_{180^\circ}) - (I^A_{0^\circ} - I^B_{0^\circ})}\right) \end{aligned}$$

Provided that the reference signal controlling the integration in the A and B channels can be considered the same but displaced half a period, we could simplify the previous equation into a two-phases or in a one-channel form, as done in Appendix A.3. This is more convenient, since, if the channel readouts, I_θ, follow a Poisson distribution, D_θ would follow a Skellam distribution whose variance would be the sum of those of I^A_θ and I^B_θ, since $D_\theta = I^A_\theta - I^B_\theta$. Applying the variance formula for uncertainty propagation to the one-channel (A or B) form of the previous equation, we obtain

$$\Delta d = \frac{c}{4\pi f_{\text{mod}}} \sqrt{\left(\frac{\delta\theta_{\text{depth}}}{\delta I_{0^\circ}} \Delta I_{0^\circ}\right)^2 + \left(\frac{\delta\theta_{\text{depth}}}{\delta I_{90^\circ}} \Delta I_{90^\circ}\right)^2 + \left(\frac{\delta\theta_{\text{depth}}}{\delta I_{180^\circ}} \Delta I_{180^\circ}\right)^2 + \left(\frac{\delta\theta_{\text{depth}}}{\delta I_{270^\circ}} \Delta I_{270^\circ}\right)^2}$$

which is equivalent to the error propagation schema presented in [278]. Differentiating the arctangent function with respect to each I_θ, we obtain the partial derivatives

$$\begin{aligned} \frac{\delta\theta_{\text{depth}}}{\delta I_{0^\circ}} &= \frac{(I_{270^\circ} - I_{90^\circ})}{(I_{180^\circ} - I_{0^\circ})^2 + (I_{270^\circ} - I_{90^\circ})^2} \\ \frac{\delta\theta_{\text{depth}}}{\delta I_{90^\circ}} &= \frac{-(I_{180^\circ} - I_{0^\circ})}{(I_{180^\circ} - I_{0^\circ})^2 + (I_{270^\circ} - I_{90^\circ})^2} \\ \frac{\delta\theta_{\text{depth}}}{\delta I_{180^\circ}} &= \frac{-(I_{270^\circ} - I_{90^\circ})}{(I_{180^\circ} - I_{0^\circ})^2 + (I_{270^\circ} - I_{90^\circ})^2} \\ \frac{\delta\theta_{\text{depth}}}{\delta I_{270^\circ}} &= \frac{(I_{180^\circ} - I_{0^\circ})}{(I_{180^\circ} - I_{0^\circ})^2 + (I_{270^\circ} - I_{90^\circ})^2} \end{aligned}$$

which, in combination with the previous uncertainty propagation formula, yield

$$\Delta d = \frac{c}{4\pi f_{\text{mod}}} \frac{1}{(2A)^2} \sqrt{\begin{aligned}&(I_{270^\circ} - I_{90^\circ})^2 \Delta^2 I_{0^\circ} + (I_{180^\circ} - I_{0^\circ})^2 \Delta^2 I_{90^\circ} \\ &+(I_{270^\circ} - I_{90^\circ})^2 \Delta^2 I_{180^\circ} + (I_{180^\circ} - I_{0^\circ})^2 \Delta^2 I_{270^\circ}\end{aligned}}$$

$$= \frac{c}{4\pi f_{\text{mod}}} \frac{1}{(2A)^2} \sqrt{\begin{aligned}&(I_{270^\circ} - I_{90^\circ})^2 (\Delta^2 I_{0^\circ} + \Delta^2 I_{180^\circ}) \\ &+(I_{180^\circ} - I_{0^\circ})^2 (\Delta^2 I_{90^\circ} + \Delta^2 I_{270^\circ})\end{aligned}} \quad \blacksquare$$

where A is the amplitude, given by Eq. 2.20. In order to derive Eq. 2.29 from the previous, we suppose the measurement noise to follow a Poisson distribution and make $\Delta I_\theta = \sqrt{I_\theta},\ \forall \theta$.

$$\Delta d = \frac{c}{4\pi f_{\text{mod}}} \frac{1}{(2A)^2} \sqrt{\begin{aligned}&(I_{270^\circ} - I_{90^\circ})^2 (I_{0^\circ} + I_{180^\circ}) \\ &+(I_{180^\circ} - I_{0^\circ})^2 (I_{90^\circ} + I_{270^\circ})\end{aligned}}$$

We now substitute all the measurements by the expressions in Eq. A.6 and obtain:

$$\begin{aligned}\Delta d &= \frac{c}{4\pi f_{\text{mod}}} \frac{1}{(2A)^2} \sqrt{(-2A \sin(\theta_{\text{depth}}))^2 (2A_0) + (-2A \cos(\theta_{\text{depth}}))^2 (2A_0)} \\ &= \frac{c}{4\pi f_{\text{mod}}} \frac{1}{(2A)^2} 2A\sqrt{2A_0} \\ &= \frac{c}{4\pi f_{\text{mod}}} \frac{\sqrt{2A_0}}{2A} \quad \blacksquare\end{aligned}$$

A.5. Optical Power Received by a Pixel

In this section we provide the derivation of Eq. 2.31, which calculates the optical power received by a pixel from the optical power emitted by the light source and the parameters of the optical setup and the scene.

For simplicity, we consider a spherical light distribution over a certain FOV for the light source, i. e., the light power density is uniform over any spherical surface whose center is the light source, at least within a certain FOV. We consiter a conical FOV, defined by a single angle. A conceptual schema of the light propagation is given in Fig. A.1a.

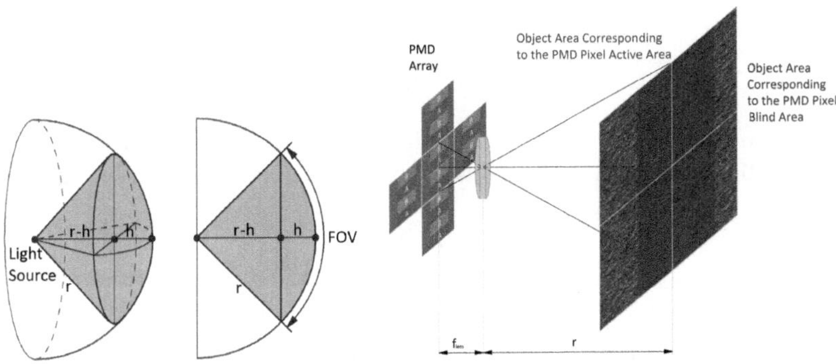

(a) (b)

Figure A.1.: Schematic representation of a punctual light source, emitting light with equal intensity in all directions within the FOV (a) and of the object area captured by the active area of a PMD pixel (b). The surface of the spherical sector in (a) is an area of constant light power density.

The area of the surface of the spherical sector in Fig. A.1a is $A_{\text{sector}} = 2\pi r h$, which can be rewritten in terms of the FOV of the light source (FOV_{source}) as:

$$A_{\text{sector}} = 2\pi r^2 \left(1 - \cos\left(\frac{FOV_{\text{source}}}{2}\right)\right)$$

Then, supposing that all the optical power of the light source, P_{source}, is emitted within the FOV, the power density at any point of the spherical surface is:

$$P'_{\text{sector}} = \frac{P_{\text{source}}}{2\pi r^2 \left(1 - \cos\left(\frac{FOV_{\text{source}}}{2}\right)\right)}$$

If the pixel size is small, one can suppose that the reflectivity of the object over the area corresponding to the pixel of interest is approximately constant and equal to ρ. Equivalently, if the object is piecewise smooth, the distance to the camera can be approximated by an average value, r. This holds if the ratio between the r and f_{lens} (focal length of the lens) is not too large. Otherwise the pixel area would correspond to a large area in the object plane. From Fig. A.1b it is clear that, if the active area of the PMD pixel is of size $[a \times b]_{\text{pixel}}$, then the corresponding area in the object plane is $[a \times b]_{\text{object}}$, where:

$$\begin{bmatrix} a \\ b \end{bmatrix}_{\text{object}} = \frac{r}{f_{\text{lens}}} \begin{bmatrix} a \\ b \end{bmatrix}_{\text{pixel}}$$

Consequently, the object area reflecting the illumination light that is captured by the pixel can be derived from the pixel active area, A_{pixel}, as:

$$A_{\text{object}} = \left(\frac{r}{f_{\text{lens}}} \right)^2 A_{\text{pixel}}$$

For a simplified schema of the active area of PMD pixels, we refer to Fig. 2.8b. For a full characterization of the pixel response for the different pixel areas, the reader is referred to Section 4.3.1. Clearly, the power reflected by the the object region of area A_{object} is:

$$P_{\text{object}} = (P'_{\text{sector}} A_{\text{object}}) \rho$$

If we suppose that the object is a Lambertian reflector, the light is not reflected with equal intensity in all directions, but it follows the so-called Lambert's cosine law or Lambert's emission law. According to it, the intensity along a certain direction is directly proportional to the cosine of the angle between that direction and the surface normal. This is independent from the angle of incidence of the illumination. For simplicity, we suppose that the lens and the illumination are approximately in the same point, so that we can use r to denote also the distance from the object to the lens. In order to obtain the power collected by the lens from that reflected by the object, the reflected light has to be integrated over the field collected by the lens. Supposing coaxiality between surface normal and principal axis of the lens, it can be showed that the optical power reaching the lens is given by:

$$P_{\text{lens}} = \left(\frac{r_{\text{lens}}}{r} \right)^2 P_{\text{object}}$$

where r_{lens} is the radius of the lens aperture. We omit the demonstration here, since it has already been provided in [278]. If the power losses of the optical system are modeled with an attenuation factor k_{lens}, then the power effectively received at the sensitive area of the pixel is $P_{\text{pixel}} = k_{\text{lens}} P_{\text{lens}}$, which, making use of the previous equalities, yields:

$$
\begin{aligned}
P_{\text{pixel}} &= \frac{A_{\text{pixel}} \rho r_{\text{lens}}^2 k_{\text{lens}}}{2\pi r^2 f_{\text{lens}}^2 \left(1 - \cos\left(\frac{FOV_{\text{source}}}{2}\right)\right)} P_{\text{source}} \\
&= \frac{A_{\text{pixel}} \rho k_{\text{lens}}}{8\pi (r f\#)^2 \left(1 - \cos\left(\frac{FOV_{\text{source}}}{2}\right)\right)} P_{\text{source}} \quad \blacksquare
\end{aligned}
$$

where $f\# = \frac{f_{\text{lens}}}{2r_{\text{lens}}}$ is the f-number of the lens.

A.6. Experimental Evaluation of the Delay Introduced in the Illumination Control Signal in a ToF Camera with Modular Illumination

In this section we provide a brief summary of the experiments that were carried out to find out the time delay that the Illumination Control Signal (ICS) experiences due to the cables and the electronics between the camera and the LED and the LED itself. This is similar to the phase homogeneity test performed in [278] for a simpler illumination system. The system we analyze is the medium-range ToF system of [283], which can provide a maximum optical power of 91 W in the NIR. The camera is a ZESS Multicam [216, 304], which integrates both a PMD and a color sensor in a single system with a common optical path. We focus on the illumination system, which is composed by 13 LED modules, with different orientations in an attempt to provide a homogeneous illumination over a relatively wide FOV. Each LED module features two LEDs, together with two independent control circuits, which are mostly signal amplifiers with a MOSFET for driving the LED. The LEDs are Osram SFH-4750, with 3.5 W optical power and an emission spectrum between 775 nm and 900 nm, approximately, showing a narrow peak at 856 nm. The LEDs are equipped with an optical grade PMMA collimator of ±5.5° FWHM (Full Width at Half Maximum) half angle and 38 mm diameter.

Several experiments were carried out, but here we refer only to one of them, in which the time delay between the ICS, generated by the MultiCam, and the optical signal emitted by each LED was calculated.

A.6.1. Methodology

The ICS is sensed at the entry of each LED module by means of electronic probes, while the optical signal requires the use of a fast photodiode. Both signals are acquired simultaneously using an oscilloscope. The acquired interval is in the middle of one of the four pulse trains that are generated per PMD acquisition. This way, the amplitude of the signals is stable during the acquisition. The results were found to be independent from in which of the four pulse trains the measurements are gathered. Consequently, we only provide the results for one of them. Since both electrical input and optical output are periodic signals of 20 MHz fundamental frequency, we cannot resolve delays greater than one period, i. e., 50 ns without ambiguity.

We use the center of gravity of the acquired waveforms as reference point to calculate their relative shift. The position of a signal in time is given by the abscissa of the center of gravity of the area under the curve, considering only one signal period. This way, we take into account not only delays due to signal propagation, but also the effective delay, also due to slow rising and falling times of the LEDs and other signal distortions. The abscissa of the center of gravity can be calculated as

$$\bar{t} = \frac{\displaystyle\int_{t_0}^{t_0+T} t\ s(t)\ dt}{\displaystyle\int_{t_0}^{t_0+T} s(t)\ dt} \tag{A.8}$$

where $s(t)$ is the periodic signal of period T, and t_0 is a starting point, which has to be defined in the same way in both signals, e. g., detecting the start of the rising edge. This calculation is done for several signal periods, in order to obtain mean values and standard deviations of the delays.

A.6.2. Experimental Setup

A high-speed silicon fixed-gain photodetector (Thorlabs PDA10A-EC) with 150 MHz bandwidth is used to sense the illumination signal. To this end, it is mounted on a tube (201 mm length, 30 mm diameter), equipped with a plain metallic adapter at the other side, that allows stable fixation to the LED modules. The tube avoids disturbance from neighboring LEDs and background light during the measurements. The spectral sensitivity of the photodetector (200 nm and 1100 nm) ensures that all power emitted by the LED is being sensed. The output of the photo detector and of the ICS probes are connected using coaxial cables (of 1.5 m and 2 m length, respectively) to an oscilloscope (Tektronix TDS 3032B) of 300 MHz bandwidth and 2.5 Gs/s acquisition rate.

One can argue that it would be easier to activate only the LED being measured at a time, instead of using a tube to avoid interference between optical signals, but this approach would modify the system with respect to the normal operating conditions. Different load in the power line, together with absence of eventual interferences between signals that might appear in normal operation, may lead to different delays. The complete ToF system is mounted on a rotary table, which is, in turn, fixed to a table. This allows rotating the system with small angular steps and positioning the modules in a convenient position for measurement. The experimental setup is shown in Fig. A.2.

Figure A.2.: Experimental setup for measuring the delay of the optical signals with respect to the Illumination Control Signal (ICS). The images show the MultiCam, surrounded by the 13 double LED modules, mounted on a rotary table, fixed to the table. The tubular coupling between LED module and photodetector ensures sensing only the desired signal, with minimal power loss. The oscilloscope shows both the ICS and the photodetector output.

The two signals are acquired during 2 μs at 2.5 gigasamples per second, i. e., around 40 periods per signal are considered. The same acquisition procedure is repeated for all 26 LEDs.

A.6.3. Results

In order to facilitate the presentation of results, the modules are numbered from 1 to 13 and the LEDs of each module are called *upper* or *lower* depending on their position in the module. For clarity, a schema with the module numbers and LED types is provided in Fig. A.3, next to a real picture of the system. The LED types are color-coded as follows: *upper* LEDs are in red and *lower* LEDs are in blue.

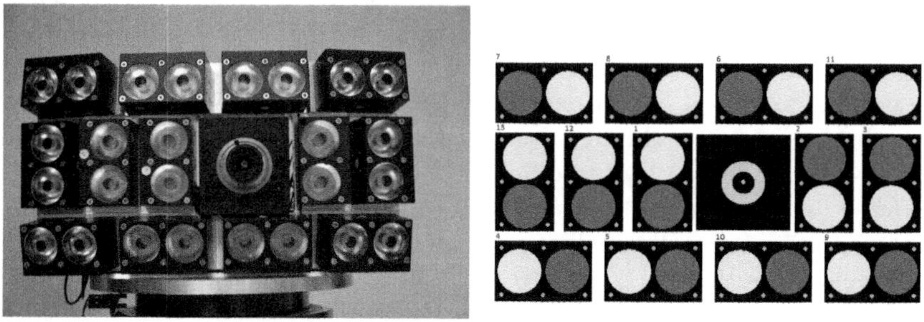

Figure A.3.: Left: picture of the medium-range ToF imaging system object of our study. Right: schematic representation of the ToF system depicted in the left image. The LED modules are numbered from 1 to 13 and the module LEDs are named *upper* (in red) or *lower* (in blue), to facilitate identification.

The data processing is carried out with Matlab and the obtained mean delay values and corresponding standard deviations are given in Table A.1. The median of all the mean delays is 6.025 ns. Taking this as the reference delay, we check for synchronism attending to the differences with respect to this value. Table A.2 provides these differences, both in absolute value and in percentage.

Fig. A.4 contains the plots with the electrical ICS (in green) and the corresponding optical signal (in blue), for the 26 LEDs. The plots are ordered by rows in ascendant order of their module number. The left column is for the *upper* LEDs (in red in Fig. A.3-right) and the right for the *lower* LEDs (in blue in Fig. A.3-right). The plots show 100 ns of the signals, i. e., two periods. The vertical lines mark the location of the abscissa of their centers of gravity, for each period. The average difference between consecutive lines of the signals is an estimator of the delay between them.

	Lower LED (Blue)		Upper LED (Red)	
Module	Mean (ns)	St. Dev. (ns)	Mean (ns)	St. Dev. (ns)
1	5.9001	0.1321	7.8414	0.1634
2	5.4277	0.1452	5.4602	0.1248
3	5.1004	0.1513	5.2762	0.2076
4	6.0419	0.2156	6.6007	0.1608
5	5.6829	0.2186	5.3564	0.1638
6	6.3928	0.2404	5.5911	0.1346
7	5.2722	0.1532	5.7872	0.1598
8	7.0667	0.2097	6.3962	0.0887
9	5.0485	0.0712	5.2952	0.0839
10	6.0809	0.1151	17.7360	0.3497
11	6.0081	0.0907	6.4460	0.1120
12	6.9062	0.0926	7.8111	0.1430
13	7.1274	0.1321	6.9163	0.1188

Table A.1.: Mean and standard deviation of the delay between optical signal and Illumination Control Signal (ICS) for all LEDs.

	Lower LED (Blue)		Upper LED (Red)	
Module	Diff. (ns)	Diff. (%)	Diff. (ns)	Diff. (%)
1	0.1249	2.0734	1.8164	30.147
2	0.5972	9.9134	0.5648	9.375
3	0.9246	15.3460	0.7487	12.428
4	0.0168	0.2801	0.5756	9.5545
5	0.3420	5.6775	0.6686	11.097
6	0.3677	6.1041	0.4339	7.2018
7	0.7528	12.4950	0.2377	3.9467
8	1.0417	17.2890	0.3712	6.1613
9	0.9764	16.2070	0.7297	12.112
10	0.0559	0.92840	11.7110	194.38
11	0.0168	0.2801	0.420	6.9873
12	0.8811	14.6250	1.7861	29.644
13	1.1024	18.2970	0.8912	14.793

Table A.2.: Difference between the each mean delay and the median of them, in absolute value and percentage.

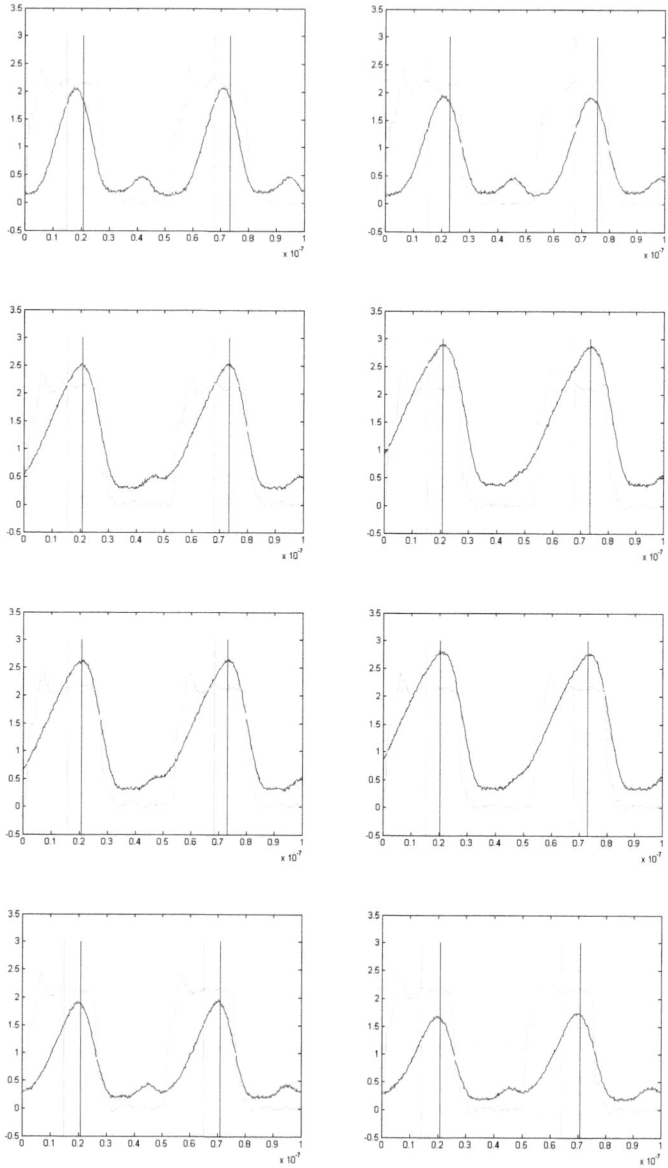
3.5
3
2.5
2
1.5
1
0.5
0
-0.5
0 0.1 0.2 0.3 0.4 0.5 0.6 0.7 0.8 0.9 1
x 10^-7

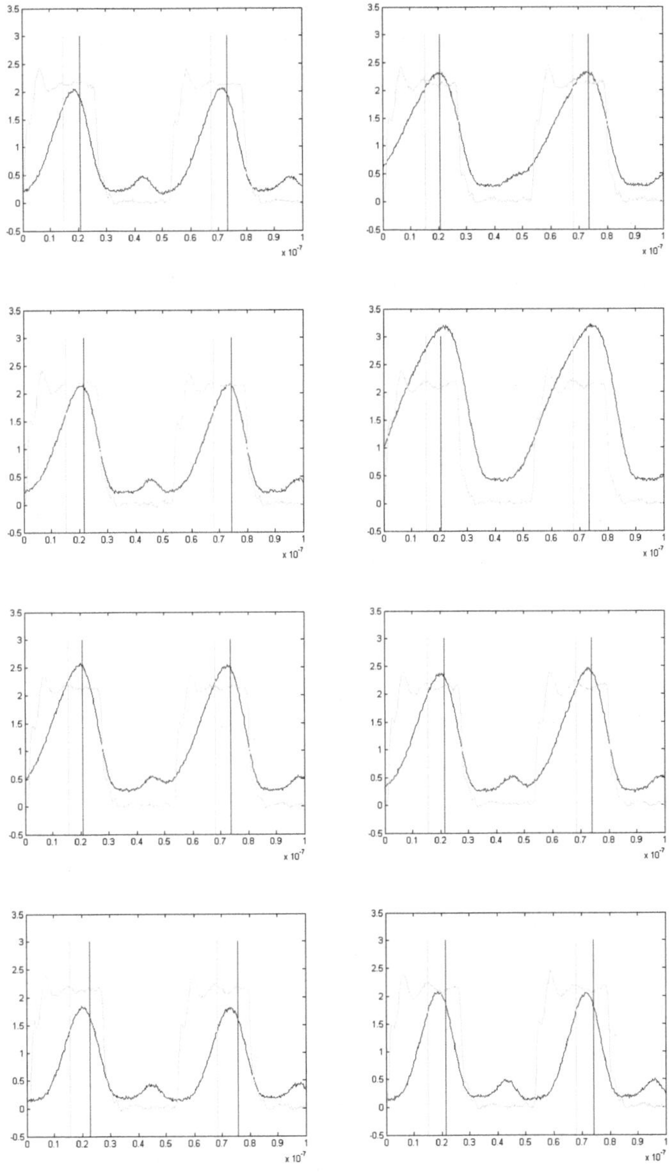
3.5
3
2.5
2
1.5
1
0.5
0
-0.5
0 0.1 0.2 0.3 0.4 0.5 0.6 0.7 0.8 0.9 1
x 10^-7

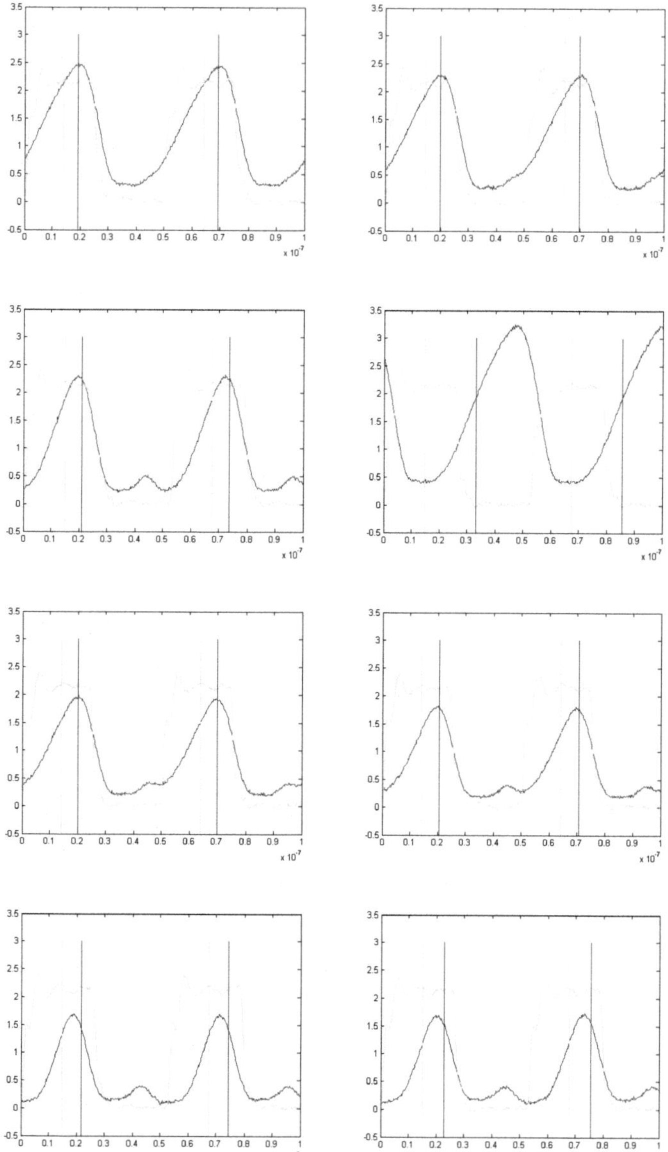
3.5
3
2.5
2
1.5
1
0.5
0
-0.5
0 0.1 0.2 0.3 0.4 0.5 0.6 0.7 0.8 0.9 1
x 10^-7

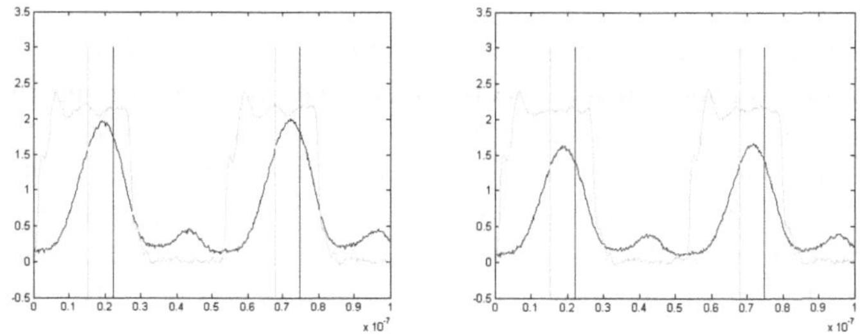

Figure A.4.: Plots showing the Illumination Control Signal (ICS, in green) and the optical signal (in blue) for each LED of the illumination system in Fig. A.3. The modules are ordered in ascendant module number by rows. Plots at the left are for the *upper* LEDs (in red in Fig. A.3-right) and plots at the right are for the *lower* LEDs (in blue in Fig. A.3-right). The plots show two signal periods. The vertical lines mark the position of the center of gravity of the area under the curve for each period, in time domain. The difference between consecutive green and blue lines is the delay. The abscissas are in seconds and the ordinates in arbitrary units.

A.7. Mutual and Matrix Coherences

This section provides the proofs of Eq. 3.40 and Eq. 3.52. Eq. 3.40 provides a lower bound on the mutual coherence between rows of the sensing matrix $\mathbf{\Phi}$ and columns of the dictionary $\mathbf{\Psi}$ in the very specific case of $\mathbf{\Psi} \in \mathbb{R}^{n \times n}$ being an orthonormal basis of $\mathbb{R}^n$ by columns and the rows of $\mathbf{\Phi} \in \mathbb{R}^{m \times n}$ being selected from another orthonormal basis of $\mathbb{R}^n$. Provided that $\mathbf{\Psi}$ is an orthonormal basis of the space, any n-dimensional vector can be expressed in terms of the basis elements without power loss. Specifically, for the rows of $\mathbf{\Phi}$ we can write

$$\sum_{j=1}^{n} \left| \langle \vec{\phi}_i, \vec{\psi}_j \rangle \right|^2 = 1, \ \forall 1 \leq i \leq m$$

where $\vec{\phi}_i$ denotes the ith row of $\mathbf{\Phi}$ and $\vec{\psi}_j$ denotes the jth column of $\mathbf{\Psi}$. The inner product in this expression provides the bridge towards the mutual coherence (recall Eq. 3.39) and a lower bound on the sum can be trivially established by substituting each term by that of maximum value, that is:

$$\sum_{j=1}^{n} \left| \langle \vec{\phi}_i, \vec{\psi}_j \rangle \right|^2 = 1 \leq \sum_{j=1}^{n} \max_j \left| \langle \vec{\phi}_i, \vec{\psi}_j \rangle \right|^2 = n \max_j \left| \langle \vec{\phi}_i, \vec{\psi}_j \rangle \right|^2, \ \forall 1 \leq i \leq m$$

Provided that the previous inequality holds $\forall i$, the maximization can be extended also along i and the inequality remains true, yielding

$$1 \leq n \max_{i,j} \left| \langle \vec{\phi}_i, \vec{\psi}_j \rangle \right|^2, \ \forall 1 \leq i \leq m, \ 1 \leq j \leq n$$

and from Eq. 3.39 immediately follows

$$\frac{1}{\sqrt{n}} \leq \mu\left(\mathbf{\Phi}, \mathbf{\Psi}\right) \leq 1 \ \blacksquare$$

Let's now derive the general lower bound on $\mu\left(\boldsymbol{A}\right)$ given in Eq. 3.52 for the case of $\mathbf{\Phi}$ with unit norm rows and $\mathbf{\Psi}$ with unit norm columns, without any further hypothesis on the resulting measurement matrix $\boldsymbol{A} = \mathbf{\Phi}\mathbf{\Psi}$. The normalization requirement is just for simplification of the derivations, since it allows omitting the denominator in the right hand side of Eq. 3.39. Let's start rewriting the dot product between two different columns of $\boldsymbol{A}$ in terms of $\mathbf{\Phi}$ and $\mathbf{\Psi}$:

$$\begin{aligned}
|\langle \vec{a}_i, \vec{a}_j \rangle| &= \left| \left(\vec{\psi}_i^\top \mathbf{\Phi}^\top \right) \left(\mathbf{\Phi} \vec{\psi}_j \right) \right| \\
&= \left| \begin{bmatrix} \langle \vec{\psi}_i^\top, \vec{\phi}_1^\top \rangle & \langle \vec{\psi}_i^\top, \vec{\phi}_2^\top \rangle & \dots & \langle \vec{\psi}_i^\top, \vec{\phi}_m^\top \rangle \end{bmatrix} \begin{bmatrix} \langle \vec{\phi}_1, \vec{\psi}_j \rangle \\ \langle \vec{\phi}_2, \vec{\psi}_j \rangle \\ \vdots \\ \langle \vec{\phi}_m, \vec{\psi}_j \rangle \end{bmatrix} \right| \\
&= \left| \sum_{k=1}^{m} \langle \vec{\phi}_k, \vec{\psi}_i \rangle \langle \vec{\phi}_k, \vec{\psi}_j \rangle \right|
\end{aligned}$$

By means of a recursive triangular inequality, the absolute value of the summation in last line of the previous expression can be used as a lower bound for the corresponding sum of absolute values, namely,

$$|\langle \vec{a}_i, \vec{a}_j \rangle| = \left| \sum_{k=1}^{m} \langle \vec{\phi}_k, \vec{\psi}_i \rangle \langle \vec{\phi}_k, \vec{\psi}_j \rangle \right| \leq \sum_{k=1}^{m} \left| \langle \vec{\phi}_k, \vec{\psi}_i \rangle \right| \left| \langle \vec{\phi}_k, \vec{\psi}_j \rangle \right|$$

Clearly, each term in the latter sum is upper-bounded by the mutual coherence, as defined in Eq. 3.39. Therefore, we have that

$$|\langle \vec{a}_i, \vec{a}_j \rangle| \leq \sum_{k=1}^{m} \left| \langle \vec{\phi}_k, \vec{\psi}_i \rangle \right| \left| \langle \vec{\phi}_k, \vec{\psi}_j \rangle \right| \leq m\mu^2 \left(\mathbf{\Phi}, \mathbf{\Psi} \right)$$

Dividing both sides of the latter inequality by $\|\vec{a_i}\|_2 \|\vec{a_j}\|_2 \neq 0$ and seeking the maximum value among all possible column pairs with $i \neq j$ yields

$$\max_{i<j} \frac{|\langle \vec{a_i}, \vec{a_j} \rangle|}{\|\vec{a_i}\|_2 \|\vec{a_j}\|_2} \leq \max_{i<j} \frac{m\mu^2 \left(\mathbf{\Phi}, \mathbf{\Psi} \right)}{\|\vec{a_i}\|_2 \|\vec{a_j}\|_2}$$

Clearly, the left hand term is the matrix coherence of $\boldsymbol{A}$, $\mu(\boldsymbol{A})$, as defined in Eq. 3.42. Additionally, suppose that there exists a nonzero lower bound C on the l_2 norm of the columns of $\boldsymbol{A}$, which has to be true, unless some column of $\boldsymbol{A}$ is the zero vector. Note that the latter case is inadmissible in practice, since it would imply that the sparse coefficient corresponding to the dictionary atom for which the corresponding column of $\boldsymbol{A}$ is null cannot be retrieved. Then, one can immediately obtain Eq. 3.52 from the previous inequality:

$$\mu(\boldsymbol{A}) \leq \frac{1}{C^2} m\mu^2(\boldsymbol{\Phi}, \boldsymbol{\Psi}) \quad \blacksquare$$
$$C = \min_{1 \leq i \leq n} \|\vec{a}_i\|_2$$

A.8. Adaptive High Dynamic Range: Complementary Material

This section complements the experimental evaluation of the Adaptive HDR method presented in Section 4.2.2.4 with some additional material that was omitted in that section for brevity.

A.8.1. Böhler Star Detail

The inclusion of Böhler stars in Exp. 1 of Section 4.2.2.4 has as objective comparing the different experimental cases in terms of lateral resolution. As stated in that section, our AHDR approach brings an improvement of the effective lateral resolution (see Table 4.1) of ToF cameras. Such improvement becomes perceptible in, e. g., the lower Böhler star of the panel in the images of Fig. 4.10. For the sake of clarity, in Fig. A.5 we provide a detail of the depth images of the star, for the two single exposures and our AHDR approach, together with a reference color image of the Böhler star. The case of exhaustive HDR has been omitted, but the corresponding result does not differ much from Fig. A.5c. Our adaptive approach (Fig. A.5d) performs visibly better than a single acquisition with an exposure time adapted to the stars panel (0.1 ms, Fig. A.5c).

A.8.2. Saturating Mask

The so-called *saturating* mask is just a 2D representation of the set of all pixels considered *harmful*. As explained in Section 4.2.2.3, a pixel is considered harmful if the slope of the response curve in exposure domain is higher than a threshold, for any of the channels, at any of the four double raw images. There is only one mask per set of PMD raw images. The *harmful* pixels are stored as ones in the mask and the rest as zeros. In order to provide a better understanding of the mask, Fig. A.6 depicts the mask obtained in Exp. 1. The corresponding scene is depicted in Fig. 4.9a.

Despite its sharpness, recall that the mask is automatically generated from a considerable amount of raw data taken at different exposures and does

(a) Color Image **(b)** Single Exposure: 2 ms

(c) Single Exposure: 0.1 ms **(d)** AHDR

Figure A.5.: Detail of the depth estimation for the lower Böhler star of the panel in Exp. 1. The images are a detail of the corresponding images in Fig. 4.10. The results obtained from two single-exposure acquisitions at 2 ms (b) and 0.1 ms (c) are compared to the result of our AHDR approach (d). The result of the intensive HDR algorithm is omitted to include a frontal color image of the star (a). It is hard to recognize any 3D structure in (b) without the help of (a). ©2015 IEEE.

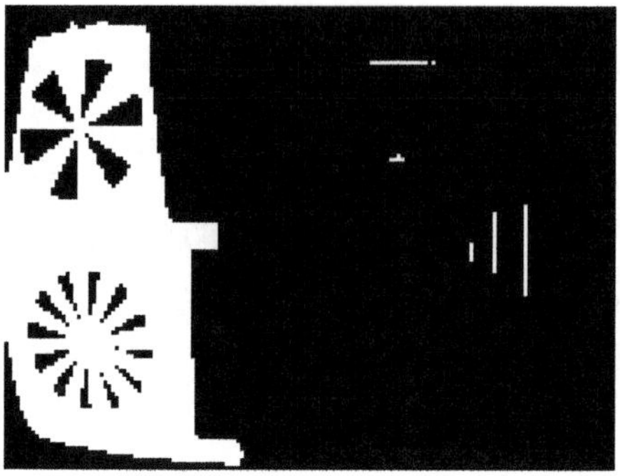

Figure A.6.: Monochrome representation of the binary *saturating* mask generated by our AHDR algorithm for for the Böhler stars setup. White pixels are *harmful* pixels.

not require any human intervention. The set of short exposure times of our AHDR approach is adapted to the responses of the white pixels in Fig. A.6. In other words, our AHDR algorithm does not apply HDR to all the pixels, but only to those that really need it, according to the last computed mask. Clearly, pixels belonging to an object which is close to the camera in presence of a powerful illumination are expected to be detected as *harmful*. In Exp. 1 those are the pixels belonging to the Böhler stars panel. Nevertheless, also other pixels might meet the requirement of having a response slope higher than the established threshold. Common examples are lights in the scene that were not expected to be on, metallic objects or surfaces that blind the camera by specular reflection, direct or reflected sunlight, etc. Fig. A.6 shows that other pixels, out of the panel, were also included in the mask. These pixels are grouped in lines and belong to highly-reflective metallic objects in the scene (see Fig. 4.9a).

A.8.3. Standard Deviation Images

Fig. 4.10 and Fig. 4.13 provide the depth results obtained for the different cases considered, in Exp. 1 and Exp. 2, respectively. As important as the depth accuracy is the stability of the depth estimation. In order to evaluate it, both experiments are repeated 100 times and the standard deviation is computed for each pixel of the depth images. This way we obtain images of depth standard deviation. In this section we provide a false color representation of these images for the four cases considered in our two experimental setups, namely, the exhaustive HDR algorithm, two cases of single acquisition at short and long exposure times and our Adaptive High Dynamic Range (AHDR) approach.

A.8.3.1. Böhler Stars Experiment

In this section standard deviation images for each of the depth images computed in the Exp. 1 of Section 4.2.2.4 (corridor with Böhler stars) are provided. The results are represented as false color images in Fig. A.7. This figure complements Fig. 4.10 by providing per-pixel depth uncertainty.

Fig. A.7a is the standard deviation image delivered by the exhaustive HDR algorithm presented in Section 4.2.2.1 using a dataset of 200 raw image sets, acquired at exposure times from 0.01 to 2 ms, with 0.01 ms step. This image should be taken as a reference of lowest deviation, so that the closer we get to this result, the better our method is. Note the extremely low deviations: below 5 mm for most pixels and submillimetric for some pixels belonging to the Böhler stars panel. The false color range is $[0, 0.01]$ m.

Fig. A.7b shows the result obtained for a fixed exposure time of 2 ms, which is proper for sensing the corridor but not the panel with the Böhler stars. Indeed, subcentimetric deviations are observed for many pixels of the frontal wall, even when some pixels exhibit depth measurements higher than 14 m. In contrast, some pixels of the Böhler stars panel exhibit deviations higher than 1 cm, due to SBI-related noise. Note that the noise pattern observed in Fig. 4.10b coincides with that present in Fig. A.7b, i. e., pixels with high depth error due to SBI limited operative range also show low stability. In general, for this exposure time, the overall result is good for a PMD-based camera in terms of standard deviation, below 4 cm for most pixels.

The complementary case to the previous uses an exposure time adapted to the Böhler stars panel, at one meter from the camera. As one could expect and can be observed in Fig. A.7c, this leads to very low standard deviations for the panel surface (mostly below 1 cm), at the cost of huge deviations along the rest of the scene (typically over 20 cm), also in coherence with Fig. 4.10c. Note that the range of the false color representation has been changed to $[0, 0.5]$ m, in order to avoid an appearance of binary image.

Finally, the standard deviation image obtained for our AHDR algorithm is shown in Fig. A.7d. The short exposure times used to accurately sense the Böhler stars panel allow subcentimetric deviations in the panel surface, sometimes lower than 2 mm. The rest of the scene offers a deviation distribution coincident with that shown in Fig. A.7b, since only raw images

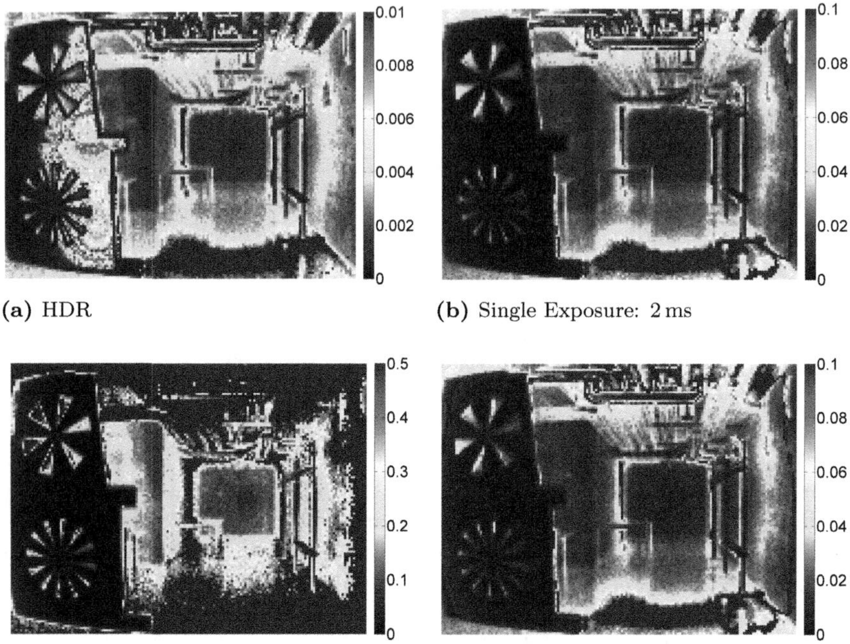

(a) HDR

(b) Single Exposure: 2 ms

(c) Single Exposure: 0.1 ms

(d) AHDR

Figure A.7.: Standard deviation images obtained from 100 different executions of Exp. 1. The corresponding MultiCam color image is given in Fig. 4.9a. The reference standard deviation image, obtained from the result of the intensive HDR algorithm is shown in (a). Two cases of single exposure times of 2 ms and 0.1 ms are given in (b) and (c), respectively. In both cases no HDR algorithm is applied. Finally, (d) shows the depth standard deviation image obtained when applying our AHDR approach. ©2015 IEEE.

acquired at the reference exposure time (2 ms, in this case) are used for the pixels not contained in the saturating mask. In both figures the false color range is $[0, 0.1]$ m.

A.8.3.2. Laboratory Experiment

The standard deviation results we present in this section correspond to the Exp. 2 of Section 4.2.2.4. The results are depicted in four false color images in Fig. A.8.

Fig. A.8a is to be taken as reference and shows the standard deviation image obtained when applying the basic HDR algorithm to a dataset of 100 raw image sets, acquired at exposure times from 0.1 to 10 ms, with 0.1 ms step. Fig. A.8b is the result obtained for a fixed exposure time of 10 ms, which is appropriate for sensing the large scene. The standard deviation result is coherent with Fig. 4.13b. Relatively low standard deviations are observed for the far background (typically lower than 1 cm). Note the higher deviations registered for pixels lying on planes that are not perpendicular to the illumination direction, such as ceiling or lateral walls. Also from the standard deviation image it is clear that there is a problem with the depth estimation for the apple pixels for such a high exposure time. Very diverse standard deviations are registered in this area, breaking the smoothness of the deviation image. Values go from 0.28 mm up to 1.23 cm.

The standard deviations registered for the complementary case, i. e., depth estimation using a single acquisition at short exposure time, are depicted in Fig. A.8c. As one could expect, the deviation is lower for those areas that are close to the camera, e.g., the apple or the table. Note that the irregular deviation values registered for the apple in Fig. 4.13b are not present in Fig. A.8c, due to the absence of saturation. Values vary from as little as 0.1 mm up to 2 mm, suggesting an accurate depth estimation.

Finally, Fig. A.8d confirms the good quality of the results delivered by our AHDR algorithm. Note the smooth distribution over the apple surface, very close to the result obtained for exhaustive HDR (Fig. A.8a), which is taken as reference. All apple pixels but one show a standard deviation lower than 5 mm.

A.8.4. Runtime Results

As explained in the Section 4.2.2.3, when the scene change due to the introduction of an intrusive object close to the camera is detected, the adaptation process is triggered. A dense dataset of raw image sets at

(a) HDR

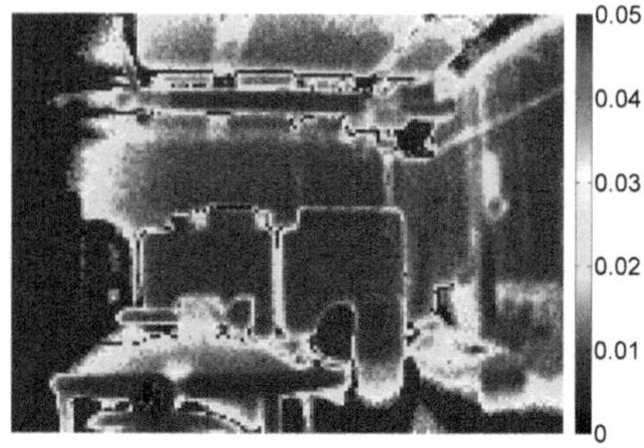

(b) Single Exposure: 10 ms

(c) Single Exposure: 1 ms

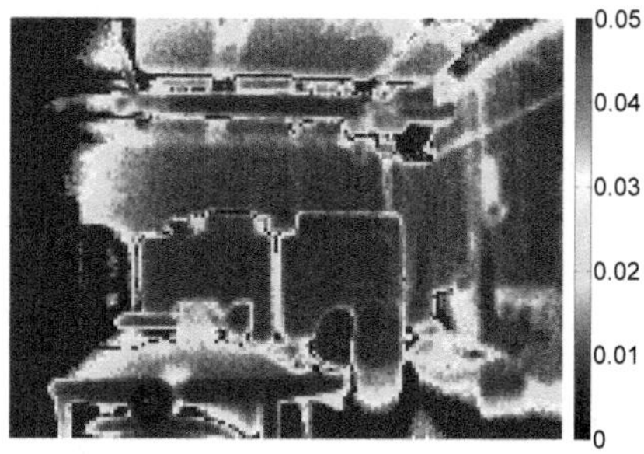

(d) AHDR

Figure A.8.: Standard deviation images obtained from 100 different executions of Exp. 2. The corresponding MultiCam color image is given in Fig. 4.9b. The reference standard deviation image, obtained from the result of the intensive HDR algorithm is shown in (a). Two cases of single exposure times of 10 ms and 1 ms are given in (b) and (c), respectively. In both cases no HDR algorithm is applied. Finally, (d) shows the depth standard deviation image obtained when applying our AHDR approach. ©2015 IEEE.

different acquisition times is acquired, in our case 200 images (from 0.01 to 2 ms, with 0.01 ms step) in Exp. 1 and 100 (from 0.1 to 10 ms, with 0.1 ms step) in Exp. 2. From them, the *harmful* or saturating pixels are detected using the estimated response function. The saturation mask is created and a set of short exposure times is computed, which optimally sense the intrusive object. The sets of short exposure times selected by the adaptive algorithm for each experiment are characterized by minimum and maximum exposure times and time step in Table A.3. The size of these sets is the direct responsible of the time overhead introduced by our adaptive approach, due to additional acquisitions. Last two columns of Table A.3 provide the time overhead in absolute terms and relative to the reference exposure time, for each experimental setup. In our experiments, those times are the times (per each of the four phases) needed to accurately sense the intrusive object, while the high single exposure time is the time (per phase) required to sense the rest of the scene. Raw data acquired using such short exposure times is, together with the mask of the saturating object (see Fig. A.6), used by our HDR algorithm to generate HDR raw data for the *harmful* pixels.

	Single Exps. (ms)		Adapted Exps. (ms)			Overhead (ms)	
Setup	Object	Scene	Min.	Step	Max.	Abs.	%
Böhler	0.10	2.00	0.01	0.01	0.30	4.65	233
Apple	1.00	10.00	0.10	0.10	1.00	5.50	55

Table A.3.: Runtime parameters for experiments 1 (Böhler) and 2 (lab with apple) and time overhead due to the additional acquisitions required by our AHDR approach. ©2015 IEEE.

The bottleneck of the approach is the acquisition and not the computation. Typically very few pixels are *harmful*. If we consider, for instance, Exp. 2, the apple generates a mask of only 134 1-pixels (for an image size of 19200 pixels), which leads to an enormous speed-up of the HDR estimation.

A.9. Inverse Freeman-Tukey Transformation for Poisson Data

The Freeman-Tukey variance stabilization transformation for Poisson data has been given in Fig. 4.15a, but this expression is not easily invertible and furthermore, it is a simplification of the more general double-arcsine formulation, for binomial distributions. Profiting from trigonometric identities, it has been shown that the Freeman-Tukey double-arcsine transformation (Eq. A.9) admits a closed-form inverse transformation [328], given by Eq. A.10,

$$t = \arcsin\sqrt{\frac{\lambda}{n+1}} + \arcsin\sqrt{\frac{\lambda+1}{n+1}} \tag{A.9}$$

$$\hat{p}(t) = \frac{1}{2}\left[1 - \operatorname{sgn}(\cos t)\sqrt{1 - \left[\sin t + \frac{1}{n}\left(\sin t - \frac{1}{\sin t}\right)\right]^2}\right] \tag{A.10}$$

where p and n are the parameters of the binomial distribution $B(n,p)$, namely, probability of success and number of independent experiments. Recall that the Poisson distribution is a limit case of the binomial distribution, when $p \rightarrow 0$ and $n \rightarrow \infty$. In the Poisson case, $P(\lambda)$, we are interested in the number of occurrences or successes, $\lambda = p\, n$. The direct Freeman-Tukey transformation for the Poisson case can be derived from Eq. A.9 as $y = \lim_{n\rightarrow\infty} \sqrt{n+1}\, t$. Consequently, we can pursuit an inverse transformation $\hat{x}(y)$ for the Poisson case from Eq. A.10 as:

$$\hat{x}(y) = \lim_{\substack{\left(t=\frac{y}{\sqrt{n+1}}\right)\\ n\rightarrow\infty}} n\,\hat{p}(t)$$

We substitute $\hat{p}(t)$ in the previous expression by Eq. A.10 and make use of the trigonometric limit approximation $\lim_{\varepsilon\rightarrow 0} \sin\varepsilon \simeq \varepsilon - \frac{\varepsilon^3}{6}$, which derives from the corresponding Taylor expansion, neglecting the terms of order equal or higher than five. We obtain the following polynomial:

$$\hat{x}(y) = \lim_{\substack{\left(t=\frac{y}{\sqrt{n+1}}\right)\\ n\rightarrow\infty}} \frac{n}{2}\left[1 - \sqrt{1 - \left(\frac{n\,t + \frac{5t}{6} - \frac{1}{t}}{n}\right)^2}\right]$$

Note that the signature of the cosine has been left away, since in the Poisson case: $\lim_{\varepsilon\rightarrow 0}\cos\varepsilon = 1$. At this point, we get rid of the square root by using the limit approximation $\lim_{\varepsilon\rightarrow 0}\sqrt{1-\varepsilon} \simeq \varepsilon - \frac{\varepsilon}{2}$, which derives, in turn, from the first order Taylor expansion. Substituting and operating we get:

$$\hat{x}(y) = \lim_{\substack{\left(t=\frac{y}{\sqrt{n+1}}\right)\\ n\rightarrow\infty}} \frac{1}{4}\left(\sqrt{n}\,t + \frac{5t}{6\sqrt{n}} - \frac{1}{t\sqrt{n}}\right)^2$$

Executing the implicit changes of variables $t = \frac{y}{\sqrt{n+1}}$, we get a function of the transformed variable, y:

$$\hat{x}(y) = \lim_{n\to\infty} \frac{1}{4} \left(\frac{\sqrt{n}}{\sqrt{n+1}}\, y + \frac{5}{6\sqrt{n}\sqrt{n+1}}\, y - \frac{\sqrt{n+1}}{\sqrt{n}}\, y^{-1} \right)^2$$

The limit can be now trivially calculated, yielding the inverse transformation presented in Eq. 4.16:

$$\hat{x}(y) = \left(\frac{y - y^{-1}}{2} \right)^2 \quad \blacksquare$$

A.10. Fluorescence Lifetime Microscopy and ToF Imaging

Fluorescence is the phenomenon of light emission by an atom or molecule shortly after the absorption of photons, typically $< 10^{-8}$ s [276]. This effect can be exploited, for instance, to enhance the contrast in biological and medical imaging, by using fluorophores. The emitted radiation is always of lower energy than the excitation radiation, that is, there is a shift towards higher wavelengths, known as Stokes' shift in honor to George Gabriel Stokes, who introduced the term *fluorescence* in the mid 19th century.

The intrinsic properties of a fluorophore include [322] quantum efficiency, fluorescence decay profile in time domain, the excitation and emission spectra and the response to polarized light. From these properties, the decay profile is of great importance, since determining it allows recognizing the material producing the fluorescence or, at least, making it distinguishable from neighboring or background materials. Suppose that the fluorescent material is excited with a pulse of light that describes a Dirac delta function in time domain. The resulting time-dependent emission is the impulse response function, which can be accurately described as a sum of exponential functions. A common way of parameterizing this function is given by:

$$i_\delta(t) = \sum_i \alpha_i e^{-\frac{t}{\tau_i}} \tag{A.11}$$

where α_i are the preexponential factors and τ_i the decay times. Eq. A.11 is able to represent almost any decay law, but if the decay is purely exponential, a single exponential function suffices and only a pair of parameters, α, τ is to be determined. In logarithmic scale, it is easy to observe that $\log i_\delta(t)$ is a linear function of time, with offset α and slope $-\frac{1}{\tau}$. If the material is a mix of few fluorophores with a single decay time, each τ_i in Eq. A.11 is the decay time of each component. Obviously, if the constituent decay times are too closely spaced, resolving them becomes challenging.

From now on we focus on the case of a single decay time. In order to obtain the parameters characterizing the decay profile, measurements are gathered either in time-domain or in frequency-domain. In time-domain fluorometry, the sample is excited with a short pulse of light, as close as possible to a Dirac delta function, and the time-dependent intensity of the emitted light is recorded. From these intensity measurements the offset and slope of $\log i_\delta(t)$ can be calculated, yielding the parameters α, τ. If the decay times are of few nanoseconds, data acquisition at a very high sampling rate becomes necessary to obtain an acceptable number of sample points of $i_\delta(t)$. To circumvent that difficulty, MCP PMTs or fast SPADs are used in combination with modern TCSPC boards, which can already offer jitters of $< 1\,\mathrm{ps}$ [35]. The basic principle of TCSPC is to emit the excitation pulse repeatedly and register the arrival of photons emitted by the sample and their time of arrival with respect to the time of emission of the excitation pulse. This way, very accurate decay profiles can be obtained, as a histogram of photon time of arrival generated by pulse repetition.

Alternatively, fluorometry can be carried out in frequency domain, in which case the sample is excited with intensity-modulated light, typically according to a sinusoidal waveform. The emitted waveform is the convolution of the excitation one with the decay function, which is an exponential or a sum of exponential functions. Regardless of the shape of the decay function, if the excitation is sinusoidally-modulated, the resulting waveform is also sinusoidal, with rescaled amplitude and phase-shifted. For a single-exponential decay, the decay time τ can be calculated using a single modulation frequency f_{mod} as

$$\tau = \frac{\tan \Delta\phi}{2\pi f_{\mathrm{mod}}} \tag{A.12}$$

where $\Delta\phi$ denotes phase shift. A derivation of Eq. A.12 can be found in [276]. This method is also known as phase modulation.

Parallelism With ToF Depth Imaging As in fluorometry, ToF systems may also operate in time or frequency domains. For instance, pulsed systems with SPADs for detecting the echo arrival operate natively in time domain, while PMD pixels require continuous emission of sinusoidally-modulated light to compute the phase shift that the depth to measure causes in the echo. In phase-shift-based ToF systems, the response function of the scene can be modeled as a Dirac delta function (Eq. 2.8) or as a sum of Dirac delta functions in presence of MPI (see the multipath paragraph in Section 2.4.1 for details on MPI).

Note the similarity between the environment response function for a ToF system in the case of a finite number of paths (Eq. 4.32) and Eq. A.11. Actually, if the scene response is modeled as a weighted sum of exponential functions, both responses coincide. This is not a meaningless idea, since diffuse MPI offen lead to close-to-exponential environmental responses. In [221], modeling the environmental response as a sum of exponentially-modified Gaussian functions allow facing both specular and diffuse MPI and enables ToF depth imaging in scattering media. The exponentially-modified Gaussian is obtained via convolution of the Gaussian and exponential functions. Consequently, it models both an exponential decay and a Gaussian low-pass filter, which can be related to the limited bandwidth of the measurement instruments. In other words, exponential decays may appear as exponentially-modified Gaussian functions when measured. Consequently, methods as that in [221] are transferable to fluorometry.

Conceptually, phase-modulation-based fluorometry and phase-shift-based depth sensing are equivalent, since in both cases the information we want to retrieve is conveyed by the phase shift undergone by a AMCW light signal. Once the phase shift is computed, Eq. A.12 yields the decay time in fluorometry, while Eq. 2.1 yields the depth in depth sensing. Nowadays TCSPC boards offer one to four parallel channels, while using a, e. g., 19k PMD chip would be equivalent to have 19.2×10^3 parallel channels.

Therefore, the question that arises is: can TCSPC systems be successfully substituted by commercial ToF cameras? The answer depends on the application. The use of PMD chips poses a limitation in the maximum modulation frequency. If the decay is too fast, the phase resolution offered by a PMD device may be insufficient to resolve the phase shift. According to [370], the optimal modulation frequency of the excitation signal to sense a fluorophore of decay time τ is $\frac{1}{10\tau}$. Consequently, if the decay times are in the nanosecond range, using a PMD array as detector is feasible and allows generating large FLIM images while achieving very short acquisition times. For instance, supposing an exposure time for the PMD chip of 1 ms and negligible positioning times of the stepper unit, a FLIM image of 19 Mpix could be obtained in one second. This is far beyond the acquisition rates of single-channel TCSPC systems, which typically need few seconds to gather images of few thousand pixels. Conversely, one could obtain a stream of

FLIM video with moderate resolution in real time.

Related Work Already in the early years of the PMD technology, FLIM was presented as a potential application, patented in [187]. In [173] the SwissRanger SR-2 ToF sensor, with 124×160 pixels (equivalent to a PMD 19k), is used as detector in a FLIM setup. Eight equally-spaced phases, i. e., $\frac{\pi}{4}$ phase step, are acquired and the phase shift is computed as usually using Eq. 2.6.

A complete mathematical formulation for the use of a ToF sensor in FLIM is given in [44], where the excitation signal is only required to be periodic, which allows for unified expressions for both TD-FLIM (periodically repeated delta functions) and FD-FLIM (sinusoidal modulation). They also model the emitted signal as a convolution between the excitation signal and a certain response function. In order to account for a depth displacement between sample and sensor, the sample response function is a displaced exponential, where the displacement corresponds to the depth, as in conventional ToF. Additionally, since reflected excitation light that did not cause emission can also reach the sensor, the total response function is modeled as the sum of a depth-related (delta) response function and the sample-related (exponential) response function. This last consideration can be discarded if an optical filter is used to block the excitation wavelength.

The main contribution of [44] is to provide general frameworks for recovering the decay time τ and an eventual depth displacement of the sample d_{mod}. Both in TD-FLIM and FD-FLIM the problem is solved via minimization of an error function that measures the distance between the observations and model predictions in an appropriate domain (response domain in TD-FLIM and phase domain in FD-FLIM). The FLIM-oriented camera introduced by PCO® [176] features a proprietary CMOS ToF chip with 1008×1008 pixels. The pixel principle of operation coincides with that of PMD pixels and the camera performs FD-FLIM with 40 MHz maximum modulation frequency.

A.11. The CS-PMD Camera Prototype

The goal of this section is to provide a visual overview of the CS-PMD camera prototype, complementing the general hardware description given in Section 5.2. The prototype is built on a rail, to ensure perfect alignment between the optical elements and accurate mechanical adjustments. Fig. A.9 shows a front view of the system, where all principal components are visible. We refer to Fig. 5.4 for understanding of the setup. In short terms, the illumination system emits polarized light onto the scene (where the photographic camera that took the photos is placed). The reflected light is collected by an imaging lens, which forms an image on the surface of the SLM (visible through the lens in the right image). The image on the SLM surface is reprojected on the surface of the PMD chip by means of a telephoto lens with integrated polarizer.

Figure A.9.: Front view of the CS-PMD prototype. Pictures were taken with focus to infinity (left) and with focus to the SLM through the imaging lens (right). The imaging lens (in the center of the pictures) is responsible for projecting the image of the scene on the SLM surface. The main components of the system are visible: the SLM (through the imaging lens), the polarized illumination system (left) and the PMD camera equipped with a telephoto lens (right).

The large distance L in Fig. 5.4 is obtained by means of the aluminum rail, which is 150 cm long. In the configuration of the prototype presented in the pictures, the distance between SLM and PMD chip surfaces is approximately 130 cm. The distance between SLM surface and the first lens of the telephoto lens system is 87 cm.

Polarized Illumination System Due to the experimental nature of the prototype and in order to allow for easy modifications in the illumination system, the light is linearly polarized using a low-cost laminated polarizing film. In a later stage, this temporal solution is planned to be substituted by high-end glass polarizers per light emitter, like the one integrated in the telephoto lens. Fig. A.10 shows the NIR illumination system with the linear laminated polarizing film overlaying the emitters.

(a) (b)

Figure A.10.: Front view of the polarized illumination system (a) and detail of the emitters (b). The system is mounted on an arm that is firmly attached to the main rail. The arm location can be adjusted parallelly to the rail and the illumination system can be displaced linearly along the arm and rotated along the vertical axis, normal to the plane defined by the rail and the arm. Each emitter is equipped with a collimator to restrict the FOV of the illumination to that of the camera.

The illumination system in mounted on an arm, which is firmly attached to the rail at a position that can be adjusted (Fig. A.10a). The location of the system on the arm and its yaw can also be adjusted, so that it appropriately illuminates the scene observed by the CS-PMD camera. The illumination modules are similar to those of the medium-range illumination system [283] analyzed in Appendix A.6. Our system features only two modules, oriented vertically. Each module has two NIR emitters with independent control circuits, still driven by the same ICS. The NIR emitters are also the Osram SFH-4750 LEDs, with 3.5 W optical power and a narrow emission peak at 856 nm. Consequently, the maximum optical power of the system is 14 W. In order to adjust the FOV of the illumination system to that of the CS-PMD camera, all LEDs are equipped with an optical grade PMMA collimator of $\pm 2°$ FWHM (Full Width at Half Maximum) half angle and 38 mm diameter. The optical efficiency of the collimator is 92%. Both illumination modules are synchronously driven by the ICS signal coming from the CS-PMD module of the camera.

Spatial Modulation: LC-SLM As introduced in Section 5.2, the 2D spatial codes are superimposed to the scene image by means of a reflective LC-SLM, before reprojection on the PMD chip. The SLM is a Holoeye LC-R 1080, featuring an array of 1200 × 1920 pixels with 90% fill factor. The SLM is reconfigured to a resolution of 1200 × 1600 (160 inactive SLM columns at both sides of the array), in order to fit the aspect ratio of the PMD array. The SLM is mounted at one end of the rail by means of a platform with a special mount that allows accurately positioning and orienting the SLM. Ideally, the SLM and PMD surfaces should be parallel and the centers of their respective active areas aligned along the same normal. Fig. A.11 provides images of the SLM and the adjustable mount.

The mount allows displacements along the vertical and horizontal axes of the SLM (perpendicular to the rail direction) with micrometer resolution. To this end, two translation stages with manual adjustment are mounted on top of each other along the desired directions (Fig. A.11c). Additionally, rotation of the SLM around the axis parallel to the rail (perpendicular to the previous two) is achieved by means of a rotation stage with minute resolution (Fig. A.11b). Both the translational stages and the rotational one are adjusted by hand via their respective fine-adjustment screws.

PMD Camera with Telephoto Lens The image at the SLM is reprojected on the PMD ship by means of a telephoto lens. We use a Nikon Nikkor ED 300mm 1:2.8 objective. The lens system includes 11 elements in 9 groups plus a dust-proof glass plate before the first element. The diameter of the first lens is approximately 110 mm. The objective has a built-in port for *drop-in* polarizers and filters (see Fig. A.12b), where we place a linear polarizer. In order to maximize the contrast, the orientation of the polarizer has to be carefully adjusted so that it is either the same or crossed with respect to that of the polarizer of the illumination system. The Nikon F mount of the objective is adapted to the C-mount of the MultiCam housing we use for our CS-PMD camera by means of a Hama adapter. Additionally, in order to restrict the FOV to the small SLM window, we make use of three C-mount extension tubes of 40 mm each, located between the adapter and the camera. Thanks to the extension, one can meet the requirement of a large L (Fig. 5.4) and a restricted FOV. Fig. A.12a shows the CS-PMD camera with the objective.

The linear polarizer that is integrated in the telephoto lens system is the hight transmittance Heliopan Polfilter 8015 of 39 mm diameter and 0.5 mm thickness, with an SH-PMC coating (16 Layer Super Hard Multi-Coated). While the SH-PMC coating reduces reflections below 0.2% in the visible

(a)

(b)

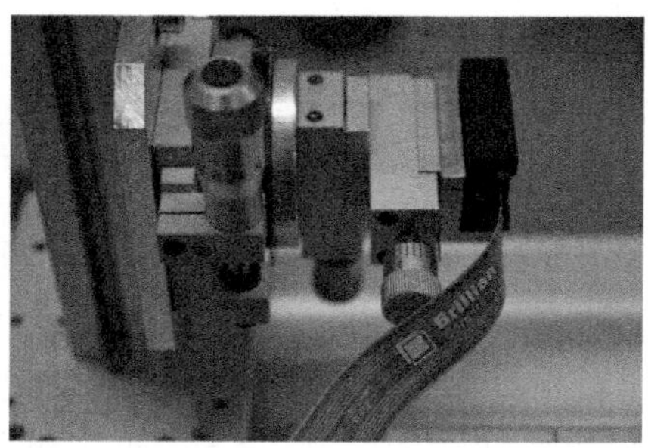

(c)

Figure A.11.: General view of the reflective LC-SLM at one end of the rail and its control unit (a). Perspective (b) and side (c) views of the mount, that allows translations along the plane perpendicular to the rail and rotation around the axis parallel to the rail.

(a) (b)

Figure A.12.: Nikon Nikkor ED 300mm 1:2.8 objective attached to the C-mount of a MultiCam housing by means of an adapter and an extension tube of 120 mm (a). A detailed view of the objective (b) shows the port for *drop-in* polarizers, before the adapter.

spectrum, the performance degrades in the NIR (around 2%), but is still better than having no coating (8%) or a single-layer coating (4%).

Calibration Pattern Test In order to achieve a perfect alignment between SLM and PMD sensor, a calibration pattern is displayed in the SLM and the output of the PMD camera is visualized. The fine-adjustment screws of the SLM mount are then manually adjusted until the SLM pattern perfectly fits the active area of the PMD array. An intermediate result of the process is given in Fig. A.13.

The size of the calibration pattern (Fig. A.13a) is 1200×1600 and is displayed by the SLM at full resolution. Fig. A.13c was acquired with a DSLR camera equipped with a linear polarizer, which was oriented as the polarizer integrated in the telephot lens. Therefore, the detail in Fig. A.13d is just a high resolution image of what the PMD camera *sees*. The resolution of the PMD sensor is ten times lower than that of the SLM. Consequently, the highest spatial frequencies of the pattern are lost in Fig. A.13b, which provides the PMD modulation image (amplitude of the modulated light) after normalization. Fig. A.13b was obtained directly from PMD raw data, without any processing.

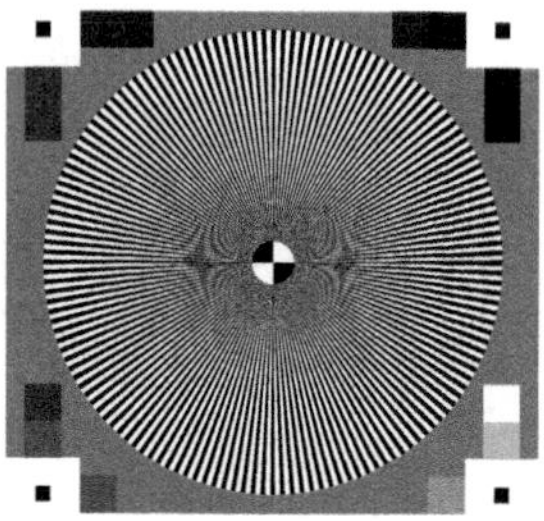

(a) Original Pattern

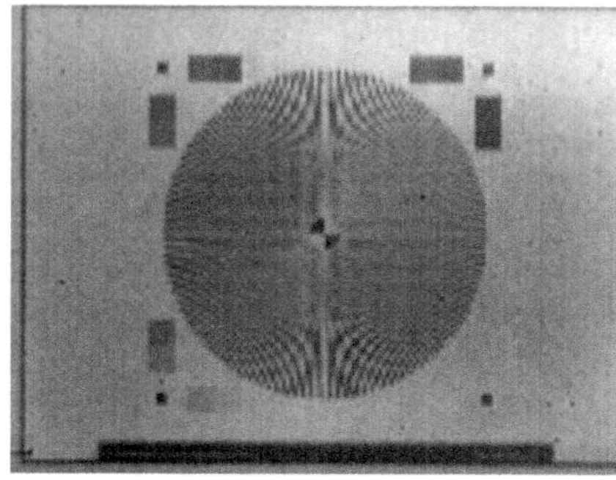

(b) PMD Modulation Image

(c) SLM

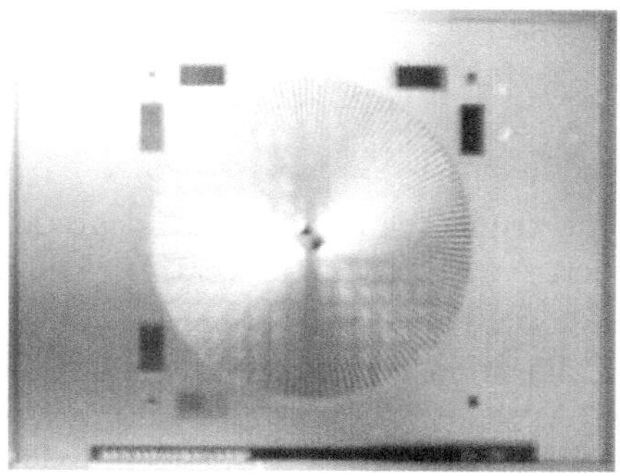

(d) SLM Pattern

Figure A.13.: The SLM and the PMD sensor were aligned using a calibration pattern (a), which was displayed in the SLM (c). A detail of the pattern as displayed by the SLM is given in (d), which has the same aspect ratio as the PMD image. The normalized PMD modulation image (b) was generated during the adjustment process and the correspondence SLM-PMD is still not perfect.

A.12. Depth Measurement Uncertainty in the CS-PMD System

In this section we derive Eq. 5.12 from Eq. 5.8 by error propagation. For completeness, we first rewrite Eq. 5.8, which calculates the (ambiguous) depth for a single frequency (f_i) from its two coefficients $x_{f_i}^{\sin}, x_{f_i}^{\cos}$:

$$d_i = \frac{d_i^{\mathrm{u}}}{2\pi} \arctan\left(\frac{x_{f_i}^{\sin}}{x_{f_i}^{\cos}}\right) - d_i^0$$

$$d_i^{\mathrm{u}} = \frac{c}{2f_i}$$

For further information, see Section 5.3.2. For simplicity, we operate in phase domain, where the previous equation can be rewritten as:

$$\tan\left(\theta_i - \theta_i^0\right) = \frac{x_{f_i}^{\sin}}{x_{f_i}^{\cos}} \tag{A.13}$$

Similarly to Appendix A.4, we now apply the variance formula for uncertainty propagation, which reads

$$\Delta\theta_i = \sqrt{\left(\frac{\delta\theta_i}{\delta x_{f_i}^{\sin}}\right)^2 \Delta^2 x_{f_i}^{\sin} + \left(\frac{\delta\theta_i}{\delta x_{f_i}^{\cos}}\right)^2 \Delta^2 x_{f_i}^{\cos}} \tag{A.14}$$

Both partial derivatives in Eq. A.14 can be easily calculated by differentiating Eq. A.13. We first differentiate with respect to $x_{f_i}^{\sin}$. The phase offset θ_i^0 can be left away, since it does not affect the partial derivatives.

$$\frac{\delta(\tan\theta_i)}{\delta\theta_i}\frac{\delta\theta_i}{\delta x_{f_i}^{\sin}} = \frac{1}{x_{f_i}^{\cos}}$$

$$\frac{1}{\cos^2\theta_i}\frac{\delta\theta_i}{\delta x_{f_i}^{\sin}} = \frac{1}{x_{f_i}^{\cos}}$$

$$\frac{\delta\theta_i}{\delta x_{f_i}^{\sin}} = \frac{1}{x_{f_i}^{\cos}}\cos^2\theta_i$$

$$\frac{\delta\theta_i}{\delta x_{f_i}^{\sin}} = \frac{1}{x_{f_i}^{\cos}}\frac{(x_{f_i}^{\cos})^2}{(x_{f_i}^{\cos})^2 + (x_{f_i}^{\cos})^2} = \frac{x_{f_i}^{\cos}}{(x_{f_i}^{\cos})^2 + (x_{f_i}^{\cos})^2}$$

Similarly, differentiating Eq. A.13 with respect to $x_{f_i}^{\cos}$ yields:

$$\begin{aligned}
\frac{\delta(\tan\theta_i)}{\delta\theta_i}\frac{\delta\theta_i}{\delta x_{f_i}^{\cos}} &= -\frac{x_{f_i}^{\sin}}{(x_{f_i}^{\cos})^2}\\
\frac{1}{\cos^2\theta_i}\frac{\delta\theta_i}{\delta x_{f_i}^{\cos}} &= -\frac{x_{f_i}^{\sin}}{(x_{f_i}^{\cos})^2}\\
\frac{\delta\theta_i}{\delta x_{f_i}^{\cos}} &= -\frac{x_{f_i}^{\sin}}{(x_{f_i}^{\cos})^2}\cos^2\theta_i\\
\frac{\delta\theta_i}{\delta x_{f_i}^{\cos}} &= -\frac{x_{f_i}^{\sin}}{(x_{f_i}^{\cos})^2}\frac{(x_{f_i}^{\cos})^2}{(x_{f_i}^{\cos})^2+(x_{f_i}^{\cos})^2} = -\frac{x_{f_i}^{\sin}}{(x_{f_i}^{\cos})^2+(x_{f_i}^{\cos})^2}
\end{aligned}$$

Substituting both partial derivatives into Eq. A.14 yields:

$$\Delta\theta_i = \sqrt{\left\{\frac{x_{f_i}^{\cos}}{\left[(x_{f_i}^{\cos})^2+(x_{f_i}^{\sin})^2\right]}\right\}^2\Delta x_{f_i}^{\sin}+\left\{\frac{-x_{f_i}^{\sin}}{\left[(x_{f_i}^{\cos})^2+(x_{f_i}^{\sin})^2\right]}\right\}^2\Delta x_{f_i}^{\cos}} \tag{A.15}$$

Multiplying both sides by the unambiguous range for that frequency ($d_i^{\mathrm{u}} = \frac{c}{2f_i}$, from Eq. 5.8) and dividing by the phase range (2π), we transfer Eq. A.15 from phase domain to depth domain and obtain Eq. 5.12:

$$\begin{aligned}
\Delta d_i &= \frac{c}{4\pi f_i}\sqrt{\left\{\frac{x_{f_i}^{\cos}}{\left[(x_{f_i}^{\cos})^2+(x_{f_i}^{\sin})^2\right]}\right\}^2\Delta x_{f_i}^{\sin}+\left\{\frac{-x_{f_i}^{\sin}}{\left[(x_{f_i}^{\cos})^2+(x_{f_i}^{\sin})^2\right]}\right\}^2\Delta x_{f_i}^{\cos}}\\
&= \frac{c}{4\pi f_i}\frac{1}{\left[(x_{f_i}^{\cos})^2+(x_{f_i}^{\sin})^2\right]}\sqrt{(x_{f_i}^{\cos})^2\Delta x_{f_i}^{\sin}+(x_{f_i}^{\sin})^2\Delta x_{f_i}^{\cos}} \; \blacksquare
\end{aligned}$$

References

[1] *3D Image Sensor D-IMager*. https://www.panasonic-electric-works.com/eu-asset/pl/downloads/ds_dimager_flyer_en.pdf. [Online; accessed 06-May-2016]. 2010.

[2] J. B. Abbiss and W. T. Mayo. "Deviation-free Bragg cell frequency-shifting". In: *Appl. Opt.* 20.4 (Feb. 1981), pp. 588–553. DOI: 10.1364/AO.20.000588. URL: http://ao.osa.org/abstract.cfm?URI=ao-20-4-588.

[3] M. Abolbashari, G. Babaie, F. Magalhães, M. V. Correia, F. M. Araújo, A. S. Gerges, and F. Farahi. "Biological imaging with high dynamic range using compressive imaging technique". In: vol. 8225. 2012, pp. 82251X–82251X–7. DOI: 10.1117/12.907365. URL: http://dx.doi.org/10.1117/12.907365.

[4] M. Abolbashari, F. Magalhães, F. M. M. Araújo, M. V. Correia, and F. Farahi. "High dynamic range compressive imaging: a programmable imaging system". In: *Optical Engineering* 51.7 (2012), pp. 071407–1–071407–8. DOI: 10.1117/1.OE.51.7.071407. URL: http://dx.doi.org/10.1117/1.OE.51.7.071407.

[5] D. Achlioptas. "Database-friendly Random Projections: Johnson-Lindenstrauss with Binary Coins". In: *J. Comput. Syst. Sci.* 66.4 (June 2003), pp. 671–687. ISSN: 0022-0000. DOI: 10.1016/S0022-0000(03)00025-4. URL: http://dx.doi.org/10.1016/S0022-0000(03)00025-4.

[6] M. Aharon, M. Elad, and A. Bruckstein. "K -SVD: An Algorithm for Designing Overcomplete Dictionaries for Sparse Representation". In: *IEEE Transactions on Signal Processing* 54.11 (Nov. 2006), pp. 4311–4322. ISSN: 1053-587X. DOI: 10.1109/TSP.2006.881199.

[7] M. Aharon and M. Elad. "Sparse and Redundant Modeling of Image Content Using an Image-Signature-Dictionary". In: *SIAM Journal on Imaging Sciences* 1.3 (2008), pp. 228–247. DOI: 10.1137/07070156X. eprint: http://dx.doi.org/10.1137/07070156X. URL: http://dx.doi.org/10.1137/07070156X.

[8] N. Ahmed, T. Natarajan, and K. R. Rao. "Discrete Cosine Transform". In: *IEEE Transactions on Computers* C-23.1 (Jan. 1974), pp. 90–93. ISSN: 0018-9340. DOI: 10.1109/T-C.1974.223784.

[9] O. K. Al-Shaykh, I. Moccagatta, and H. Chen. "JPEG-2000: a new still image compression standard". In: *Signals, Systems amp; Computers, 1998. Conference Record of the Thirty-Second Asilomar Conference on.* Vol. 1. Nov. 1998, 99–103 vol.1. DOI: 10.1109/ACSSC.1998.750835.

[10] M. A. Albota, R. M. Heinrichs, D. G. Kocher, D. G. Fouche, B. E. Player, M. E. O'Brien, B. F. Aull, J. J. Zayhowski, J. Mooney, B. C. Willard, and R. R. Carlson. "Three-dimensional imaging laser radar with a photon-counting avalanche photodiode array and microchip laser". In: *Appl. Opt.* 41.36 (Dec. 2002), pp. 7671–7678. DOI: 10.1364/AO.41.007671. URL: http://ao.osa.org/abstract.cfm?URI=ao-41-36-7671.

[11] M. Albrecht. "Untersuchung von Photogate-PMD-Sensoren hinsichtlich qualifizierender Charakterisierungsparameter und -methoden". PhD thesis. Siegen, Germany: Department of Electrical Engineering and Computer Science, 2007, p. 229.

[12] J. B. Allen and L. R. Rabiner. "A unified approach to short-time Fourier analysis and synthesis". In: *Proceedings of the IEEE* 65.11 (Nov. 1977), pp. 1558–1564. ISSN: 0018-9219. DOI: 10.1109/PROC.1977.10770.

[13] F. J. Anscombe. "The Transformation of Poisson, Binomial, and Negative-Binomial Data". In: *Biometrika* 35.3/4 (1948), pp. 246–254. ISSN: 00063444. DOI: 10.2307/2332343. URL: http://dx.doi.org/10.2307/2332343.

[14] K. J. Arrow, L. Hurwicz, and H. Uzawa. *Studies in linear and non-linear programming.* Stanford Math. Stud. Social Sci. Stanford, CA: Cambridge Univ. Press, 1958.

[15] G. A. Atkinson and E. R. Hancock. "Recovery of surface orientation from diffuse polarization". In: *IEEE Transactions on Image Processing* 15.6 (June 2006), pp. 1653–1664. ISSN: 1057-7149. DOI: 10.1109/TIP.2006.871114.

[16] T. Azuma and A. Morimura. "Image Composite Method and Image Composite Device". Pat. 1996-154201. June 1996.

[17] R. Balan, P. Casazza, and Z. Landau. "Redundancy for localized frames". In: *Israel Journal of Mathematics* 185.1 (2011), pp. 445–476. ISSN: 1565-8511. DOI: 10.1007/s11856-011-0118-1. URL: http://dx.doi.org/10.1007/s11856-011-0118-1.

[18] R. H. Bamberger and M. J. T. Smith. "A filter bank for the directional decomposition of images: theory and design". In: *IEEE Transactions on Signal Processing* 40.4 (Apr. 1992), pp. 882–893. ISSN: 1053-587X. DOI: 10.1109/78.127960.

[19] C. Bamji. "CMOS-compatible three-dimensional image sensor IC". Pat. US Patent 6,323,942. Nov. 2001. URL: http://www.google.com/patents/US6323942.

[20] C. Bamji and E. Charbon. "Systems for CMOS-compatible three-dimensional image sensing using quantum efficiency modulation". Pat. US Patent 6,580,496. June 2003. URL: http://www.google.com/patents/US6580496.

[21] C. Bamji, H. Yalcin, X. Liu, and E. Eroglu. "Method and system to differentially enhance sensor dynamic range". Pat. US Patent 6,919,549. July 2005. URL: http://www.google.de/patents/US6919549.

[22] C. S. Bamji, P. O'Connor, T. Elkhatib, S. Mehta, B. Thompson, L. A. Prather, D. Snow, O. C. Akkaya, A. Daniel, A. D. Payne, T. Perry, M. Fenton, and V. H. Chan. "A 0.13 μm CMOS System-on-Chip for a 512 $\times$ 424 Time-of-Flight Image Sensor With Multi-Frequency Photo-Demodulation up to 130 MHz and 2 GS/s ADC". In: *IEEE Journal of Solid-State Circuits* 50.1 (Jan. 2015), pp. 303–319. ISSN: 0018-9200. DOI: 10.1109/JSSC.2014.2364270.

[23] R. Bamler and P. Hartl. "Synthetic Aperture Radar Interferometry". In: *Inverse Problems* 14 (Aug. 1998), pp. 1–54. DOI: 10.1088/0266-5611/14/4/001. URL: http://iopscience.iop.org/article/10.1088/0266-5611/14/4/001.

[24] R. Baraniuk. "Compressive Sensing [Lecture Notes]". In: *Signal Processing Magazine, IEEE* 24.4 (July 2007), pp. 118–121. ISSN: 1053-5888. DOI: 10.1109/MSP.2007.4286571.

[25] R. Baraniuk, V. Cevher, M. Duarte, and C. Hegde. "Model-Based Compressive Sensing". In: *Information Theory, IEEE Transactions on* 56.4 (Apr. 2010), pp. 1982–2001. ISSN: 0018-9448. DOI: 10.1109/TIT.2010.2040894.

[26] R. Baraniuk, M. Davenport, R. DeVore, and M. Wakin. "A Simple Proof of the Restricted Isometry Property for Random Matrices". In: *Constructive Approximation* 28.3 (2008), pp. 253–263. ISSN: 1432-0940. DOI: 10.1007/s00365-007-9003-x. URL: http://dx.doi.org/10.1007/s00365-007-9003-x.

[27] N. P. Barnes and L. B. Petway. "Variation of the Verdet constant with temperature of terbium gallium garnet". In: *J. Opt. Soc. Am. B* 9.10 (Oct. 1992), pp. 1912–1915. DOI: 10.1364/JOSAB.9.001912. URL: http://josab.osa.org/abstract.cfm?URI=josab-9-10-1912.

[28] R. Barrett, M. Berry, T. F. Chan, J. Demmel, J. Donato, J. Dongarra, V. Eijkhout, R. Pozo, C. Romine, and H. V. der Vorst. *Templates for the Solution of Linear Systems: Building Blocks for Iterative Methods, 2nd Edition.* Philadelphia, PA: SIAM, 1994.

[29] J. Barry, E. Lee, and D. Messerschmitt. "Stochastic Signal Processing". English. In: *Digital Communication.* Springer US, 2004, pp. 57–111. ISBN: 978-1-4613-4975-4. DOI: 10.1007/978-1-4615-0227-2_3. URL: http://dx.doi.org/10.1007/978-1-4615-0227-2_3.

[30] M. S. Bartlett. "The Square Root Transformation in Analysis of Variance". English. In: *Supplement to the Journal of the Royal Statistical Society* 3.1 (1936), pp. 68–78. ISSN: 14666162. URL: http://www.jstor.org/stable/2983678.

[31] *Basler Lab Time-of-Flight Cameras.* http://www.i2s-vision.fr/upload/BAS1409_ToF_EN_web.pdf. [Online; accessed 06-May-2016]. 2016.

[32] *Basler Time-of-Flight Camera.* http://s.baslerweb.com/media/documents/BAS1603_ToF_Brochure_EN_SAP0022_web2.pdf. [Online; accessed 06-May-2016]. 2016.

[33] S. Battiato, A. Castorina, and M. Mancuso. "High Dynamic Range Imaging for Digital Still Camera: An Overview". In: *Journal of Electronic Imaging* 12.3 (2003), pp. 459–469. DOI: 10.1117/1.1580829. URL: http://dx.doi.org/10.1117/1.1580829.

[34] P. W. Baumeister. "Optical Tunneling and Its Applications to Optical Filters". In: *Appl. Opt.* 6.5 (May 1967), pp. 897–905. DOI: 10.1364/AO.6.000897. URL: http://ao.osa.org/abstract.cfm?URI=ao-6-5-897.

[35] W. Becker. *The Bh TCSPC Handbook: Time-correlated Single Photon Counting Modules SPC-130, SPC-134, SPC-130 EM, SPC-134 EM, SPC-140, SPC-144, SPC-150, SPC-154, SPC-630, SPC-730, SPC-830 ; Simple-Tau Systems, SPCM Software, SPCImage Data Analysis.* Becker et Hickl, 2012. URL: http://www.becker-hickl.de/pdf/SPC-handbook-6ed-12-web.pdf.

[36] S. Beigpour, A. Kolb, and S. Kunz. "A Comprehensive Multi-Illuminant Dataset for Benchmarking of the Intrinsic Image Algorithms". In: June 2015.

[37] Z. Ben-Haim and Y. C. Eldar. "Performance bounds for sparse estimation with random noise". In: *Statistical Signal Processing, 2009. SSP '09. IEEE/SP 15th Workshop on.* Aug. 2009, pp. 225–228. DOI: 10.1109/SSP.2009.5278597.

[38] Z. Ben-Haim and Y. C. Eldar. "The Cramèr-Rao Bound for Estimating a Sparse Parameter Vector". In: *IEEE Transactions on Signal Processing* 58.6 (June 2010), pp. 3384–3389. ISSN: 1053-587X. DOI: 10.1109/TSP.2010.2045423.

[39] Z. Ben-Haim, Y. C. Eldar, and M. Elad. "Coherence-Based Performance Guarantees for Estimating a Sparse Vector Under Random Noise". In: *IEEE Transactions on Signal Processing* 58.10 (Oct. 2010), pp. 5030–5043. ISSN: 1053-587X. DOI: 10.1109/TSP.2010.2052460.

[40] J. J. Benedetto and M. Fickus. "Finite Normalized Tight Frames". In: *Advances in Computational Mathematics* 18.2 (2003), pp. 357–385. ISSN: 1572-9044. DOI: 10.1023/A:1021323312367. URL: http://dx.doi.org/10.1023/A:1021323312367.

[41] A. Beurling. "Sur les intégrales de Fourier absolument convergentes et leur application à une transformation fonctionelle". In: *IX Congr. Math. Scand.* Helsinki, Finland, 1938, pp. 345–366.

[42] G. Beylkin. "On the Representation of Operators in Bases of Compactly Supported Wavelets". In: *SIAM Journal on Numerical Analysis* 29.6 (1992), pp. 1716–1740. DOI: 10.1137/0729097. eprint: http://dx.doi.org/10.1137/0729097. URL: http://dx.doi.org/10.1137/0729097.

[43] A. Bhandari, M. Feigin, S. Izadi, C. Rhemann, M. Schmidt, and R. Raskar. "Resolving multipath interference in Kinect: An inverse problem approach". In: *SENSORS, 2014 IEEE.* Nov. 2014, pp. 614–617. DOI: 10.1109/ICSENS.2014.6985073.

[44] A. Bhandari, C. Barsi, and R. Raskar. "Blind and reference-free fluorescence lifetime estimation via consumer time-of-flight sensors". In: *Optica* 2.11 (Nov. 2015), pp. 965–973. DOI: 10.1364/OPTICA.2.000965. URL: http://www.osapublishing.org/optica/abstract.cfm?URI=optica-2-11-965.

[45] A. Bhandari, A. Kadambi, R. Whyte, C. Barsi, M. Feigin, A. Dorrington, and R. Raskar. "Resolving multipath interference in time-of-flight imaging via modulation frequency diversity and sparse regularization". In: *Opt. Lett.* 39.6 (Mar. 2014), pp. 1705–1708. DOI: 10.1364/OL.39.001705. URL: http://ol.osa.org/abstract.cfm?URI=ol-39-6-1705.

[46] A. Bhatti, ed. *Stereo Vision.* Janeza Trdine 9, 51000 Rijeka, Croatia: InTech, 2008. ISBN: 978-953-7619-22-0. DOI: 10.5772/5898. URL: http://www.intechopen.com/books/stereo_vision.

[47] L. Binqiao, S. Zhongyan, and X. Jiangtao. "Wide dynamic range CMOS image sensor with in-pixel double-exposure and synthesis". In: *Journal of Semiconductors* 31.5 (2010), p. 055002. URL: http://stacks.iop.org/1674-4926/31/i=5/a=055002.

[48] J. R. Bitner, G. Ehrlich, and E. M. Reingold. "Efficient Generation of the Binary Reflected Gray Code and Its Applications". In: *Commun. ACM* 19.9 (Sept. 1976), pp. 517–521. ISSN: 0001-0782. DOI: 10.1145/360336.360343. URL: http://doi.acm.org/10.1145/360336.360343.

[49] F. Blais, M. Picard, and G. Godin. "Accurate 3D acquisition of freely moving objects". In: *3D Data Processing, Visualization and Transmission, 2004. 3DPVT 2004. Proceedings. 2nd International Symposium on.* Sept. 2004, pp. 422–429. DOI: 10.1109/TDPVT.2004.1335269.

[50] J. D. Blanchard, M. Cermak, D. Hanle, and Y. Jing. "Greedy Algorithms for Joint Sparse Recovery". In: *IEEE Transactions on Signal Processing* 62.7 (Apr. 2014), pp. 1694–1704. ISSN: 1053-587X. DOI: 10.1109/TSP.2014.2301980.

[51] R. G. Bland. "New Finite Pivoting Rules for the Simplex Method". In: *Mathematics of Operations Research* 2.2 (1977), pp. 103–107. DOI: 10.1287/moor.2.2.103. eprint: http://dx.doi.org/10.1287/moor.2.2.103. URL: http://dx.doi.org/10.1287/moor.2.2.103.

[52] L. Blinov and V. Chigrinov. *Electrooptic Effects in Liquid Crystal Materials.* Partially Ordered Systems. Springer New York, 1996. ISBN: 9780387947082. URL: http://www.springer.com/us/book/9780387947082.

[53] T. Blumensath and M. Davies. "Sparse and shift-Invariant representations of music". In: *IEEE Transactions on Audio, Speech, and Language Processing* 14.1 (Jan. 2006), pp. 50–57. ISSN: 1558-7916. DOI: 10.1109/TSA.2005.860346.

[54] T. Blumensath and M. E. Davies. "Gradient Pursuits". In: *IEEE Transactions on Signal Processing* 56.6 (June 2008), pp. 2370–2382. ISSN: 1053-587X. DOI: 10.1109/TSP.2007.916124.

[55] T. Blumensath and M. E. Davies. "Sampling Theorems for Signals From the Union of Finite-Dimensional Linear Subspaces". In: *IEEE Transactions on Information Theory* 55.4 (Apr. 2009), pp. 1872–1882. ISSN: 0018-9448. DOI: 10.1109/TIT.2009.2013003.

[56] T. Blumensath and M. E. Davies. "Normalized Iterative Hard Thresholding: Guaranteed Stability and Performance". In: *IEEE Journal of Selected Topics in Signal Processing* 4.2 (Apr. 2010), pp. 298–309. ISSN: 1932-4553. DOI: 10.1109/JSTSP.2010.2042411.

[57] T. Blumensath and M. E. Davies. "Iterative Thresholding for Sparse Approximations". In: *Journal of Fourier Analysis and Applications* 14.5 (2008), pp. 629–654. ISSN: 1531-5851. DOI: 10.1007/s00041-008-9035-z. URL: http://dx.doi.org/10.1007/s00041-008-9035-z.

[58] B. G. Bodmann and J. Haas. "Achieving the orthoplex bound and constructing weighted complex projective 2-designs with Singer sets". In: *CoRR* abs/1509.05333 (Sept. 2015). arXiv: 1509.05333 [math.FA]. URL: http://arxiv.org/abs/1509.05333.

[59] B. G. Bodmann, P. G. Casazza, and G. Kutyniok. "A quantitative notion of redundancy for finite frames". In: *Applied and Computational Harmonic Analysis* 30.3 (2011), pp. 348 –362. ISSN: 1063-5203. DOI: http://dx.doi.org/10.1016/j.acha.2010.09.004. URL: http://www.sciencedirect.com/science/article/pii/S1063520310001090.

[60] W. Böhler, V. M. Bordas, and A. Marbs. "Investigating laser scanner accuracy". In: ed. by O. Atlan. Vol. 34. 5/C15. Antalya, Turkey: International Committee for Documentation of Cultural Heritage, Oct. 2003, pp. 696–701. URL: http://cipa.icomos.org/fileadmin/template/doc/antalya/189.pdf.

[61] W. Böhler and A. Marbs. "3D Scanning Instruments". In: *Proceedings of the CIPA WG 6 International Workshop on Scanning for Cultural Heritage Recording.* Corfu, Greece, Sept. 2002.

[62] R. W. Boyd. *Nonlinear Optics, Third Edition.* 3rd. Academic Press, 2008. ISBN: 0123694701, 9780123694706.

[63] J. Brenner and L. Cummings. "The Hadamard Maximum Determinant Problem". English. In: *The American Mathematical Monthly* 79.6 (1972), pp. 626–630. ISSN: 00029890. URL: http://www.jstor.org/stable/2317092.

[64] J. Burg. "Maximum entropy spectral analysis". PhD thesis. 1975.

[65] J. Butters and J. Leendertz. "Speckle pattern and holographic techniques in engineering metrology". In: *Optics & Laser Technology* 3.1 (1971), pp. 26 –30. ISSN: 0030-3992. DOI: http://dx.doi.org/10.1016/S0030-3992(71)80007-5. URL: http://www.sciencedirect.com/science/article/pii/S0030399271800075.

[66] B. Büttgen. *Extending Time-of-flight Optical 3D-imaging to Extreme Operating Conditions.* Sierke, 2007. ISBN: 9783933893925. URL: http://d-nb.info/983752990.

[67] B. Büttgen, M.-A. El Mechat, F. Lustenberger, and P. Seitz. "Pseudonoise Optical Modulation for Real-Time 3-D Imaging With Minimum Interference". In: *Circuits and Systems I: Regular Papers, IEEE Transactions on* 54.10 (Oct. 2007), pp. 2109–2119. ISSN: 1549-8328. DOI: 10.1109/TCSI.2007.904598.

[68] B. Büttgen, F. Lustenberger, and P. Seitz. "Demodulation Pixel Based on Static Drift Fields". In: *Electron Devices, IEEE Transactions on* 53.11 (Nov. 2006), pp. 2741–2747. ISSN: 0018-9383. DOI: 10.1109/TED.2006.883669.

[69] B. Büttgen and P. Seitz. "Robust Optical Time-of-Flight Range Imaging Based on Smart Pixel Structures". In: *Circuits and Systems I: Regular Papers, IEEE Transactions on* 55.6 (July 2008), pp. 1512–1525. ISSN: 1549-8328. DOI: 10.1109/TCSI.2008.916679.

[70] B. Büttgen, T. Oggier, R. Kaufmann, P. Seitz, and N. Blanc. “Demonstration of a novel drift field pixel structure for the demodulation of modulated light waves with application in three-dimensional image capture”. In: *Proc. SPIE.* Vol. 5302. 2004, pp. 9–20. DOI: 10.1117/12.525654. URL: http://dx.doi.org/10.1117/12.525654.

[71] B. Büttgen, T. Oggier, M. Lehmann, R. Kaufmann, and F. Lustenberger. “CCD/CMOS Lock-in pixel for range imaging: challenges, limitations and state-of-the-art”. In: *In Proceedings of 1st Range Imaging Research Day at ETH.* Sept. 2005, pp. 21–32.

[72] B. Büttgen, T. Oggier, M. Lehmann, R. Kaufmann, S. Neukom, M. Richter, M. Schweizer, D. Beyeler, R. Cook, C. Gimkiewicz, C. Urban, P. Metzler, P. Seitz, and F. Lustenberger. “High-speed and high-sensitive demodulation pixel for 3D imaging”. In: *Proc. SPIE.* Vol. 6056. 2006, pp. 605603–605603–12. DOI: 10.1117/12.642305. URL: http://dx.doi.org/10.1117/12.642305.

[73] P. L. Butzer, P. J. S. G. Ferreira, J. R. Higgins, S. Saitoh, G. Schmeisser, and R. L. Stens. “Interpolation and Sampling: ET Whittaker, K. Ogura and Their Followers”. In: *The journal of Fourier analysis and applications* 17.2 (2011), pp. 320–354. ISSN: 1069-5869. DOI: 10.1007/s00041-010-9131-8. URL: http://publications.rwth-aachen.de/record/191884.

[74] B. Buxbaum, J. Frey, H. Kraft, T. MÖller, and Z. Xu. “TOF-Pixel und Verfahren zu dessen Betrieb”. Pat. DE Patent App. DE200,510,056,774. May 2005. URL: http://www.google.com/patents/DE102005056774A1?cl=de.

[75] B. Buxbaum. *Optische Laufzeitentfernungsmessung und CDMA auf Basis der PMD-Technologie mittels phasenvariabler PN-Modulation.* German. Vol. 17. ZESS Forschungsberichte. Aachen, Germany: Shaker Verlag, 2002, p. 218. ISBN: 3-8265-9805-9. URL: http://www.shaker.de/de/content/catalogue/index.asp?lang=de&ID=6&category=102.

[76] B. Buxbaum, R. Schwarte, T. Ringbeck, M. Grothof, and X. Luan. “MSM-PMD as correlation receiver in a new 3D-ranging system”. In: *Proc. SPIE.* Vol. 4546. 2002, pp. 145–153. DOI: 10.1117/12.453994. URL: http://dx.doi.org/10.1117/12.453994.

[77] B. Buxbaum, R. Schwarte, T. Ringbeck, H.-G. Heinol, Z. Xu, J. Olk, W. Tai, Z. Zhang, and X. Luan. "A new approach in optical broadband communication systems: a high integrated optical phase locked loop based on a mixing and correlating sensor, the Photonic Mixer Device (PMD)". In: *Proceedings / OPTO 98, Internationaler Kongress und Fachausstellung für Optische Sensorik, Messtechnik und Elektronik, 18. - 20. Mai 1998, Kongresszentrum Erfurt.* 1998, pp. 59 –64.

[78] E. J. Candès and T. Tao. "Decoding by Linear Programming". In: *IEEE Trans. Inf. Theor.* 51.12 (Dec. 2005), pp. 4203–4215. ISSN: 0018-9448. DOI: 10.1109/TIT.2005.858979. URL: http://dx.doi.org/10.1109/TIT.2005.858979.

[79] E. Candès, J. Romberg, and T. Tao. "Robust uncertainty principles: exact signal reconstruction from highly incomplete frequency information". In: *Information Theory, IEEE Transactions on* 52.2 (Feb. 2006), pp. 489–509. ISSN: 0018-9448. DOI: 10.1109/TIT.2005.862083.

[80] E. Candès and M. Wakin. "An Introduction To Compressive Sampling". In: *Signal Processing Magazine, IEEE* 25.2 (Mar. 2008), pp. 21–30. ISSN: 1053-5888. DOI: 10.1109/MSP.2007.914731.

[81] E. Candès and J. Romberg. "Sparsity and incoherence in compressive sampling". In: *Inverse Problems* 23.3 (June 2007), pp. 969–985. URL: http://resolver.caltech.edu/CaltechAUTHORS:CANip07.

[82] E. J. Candès. "Compressive sampling". In: *Proceedings oh the International Congress of Mathematicians.* Aug. 2006, pp. 1433–1452.

[83] E. J. Candès. "The restricted isometry property and its implications for compressed sensing". In: *Comptes Rendus Mathematique* 346.9-10 (2008), pp. 589 –592. ISSN: 1631-073X. DOI: http://dx.doi.org/10.1016/j.crma.2008.03.014. URL: http://www.sciencedirect.com/science/article/pii/S1631073X08000964.

[84] E. J. Candès, L. Demanet, D. Donoho, and L. Ying. "Fast Discrete Curvelet Transforms". In: *Multiscale Modeling & Simulation* 5.3 (2006), pp. 861–899. DOI: 10.1137/05064182X. eprint: http://dx.doi.org/10.1137/05064182X. URL: http://dx.doi.org/10.1137/05064182X.

[85] E. J. Candès and D. L. Donoho. "Ridgelets: a key to higher-dimensional intermittency?" In: *Philosophical Transactions of the Royal Society of London A: Mathematical, Physical and Engineering Sciences* 357.1760 (1999), pp. 2495–2509. ISSN: 1364-503X. DOI: 10.1098/rsta.1999.0444. eprint: http://rsta.royalsocietypublishing.org/content/357/1760/2495.full.pdf. URL: http://rsta.royalsocietypublishing.org/content/357/1760/2495.

[86] E. J. Candès, Y. C. Eldar, D. Needell, and P. Randall. "Compressed sensing with coherent and redundant dictionaries". In: *Applied and Computational Harmonic Analysis* 31.1 (2011), pp. 59 –73. ISSN: 1063-5203. DOI: http://dx.doi.org/10.1016/j.acha.2010.10.002. URL: http://www.sciencedirect.com/science/article/pii/S1063520310001156.

[87] E. J. Candès and Y. Plan. "Near-ideal model selection by l_1 minimization". In: *Ann. Statist.* 37.5A (Oct. 2009), pp. 2145–2177. DOI: 10.1214/08-AOS653. URL: http://dx.doi.org/10.1214/08-AOS653.

[88] E. J. Candès and J. Romberg. "Robust Signal Recovery from Incomplete Observations". In: *Image Processing, 2006 IEEE International Conference on.* Oct. 2006, pp. 1281–1284. DOI: 10.1109/ICIP.2006.312579.

[89] E. J. Candès and J. Romberg. *l1-magic: Recovery of Sparse Signals via Convex Programming.* 2005.

[90] E. J. Candès and J. Romberg. "Quantitative Robust Uncertainty Principles and Optimally Sparse Decompositions". In: *Found. Comput. Math.* 6.2 (Apr. 2006), pp. 227–254. ISSN: 1615-3375. DOI: 10.1007/s10208-004-0162-x. URL: http://dx.doi.org/10.1007/s10208-004-0162-x.

[91] E. J. Candès, J. K. Romberg, and T. Tao. "Stable signal recovery from incomplete and inaccurate measurements". In: *Communications on Pure and Applied Mathematics* 59.8 (2006), pp. 1207–1223. ISSN: 1097-0312. DOI: 10.1002/cpa.20124. URL: http://dx.doi.org/10.1002/cpa.20124.

[92] E. J. Candès and T. Tao. "Near-Optimal Signal Recovery From Random Projections: Universal Encoding Strategies?" In: *IEEE Transactions on Information Theory* 52.12 (Dec. 2006), pp. 5406–5425. ISSN: 0018-9448. DOI: 10.1109/TIT.2006.885507.

[93] E. J. Candès and T. Tao. "The Dantzig selector: Statistical estimation when p is much larger than n". In: *Ann. Statist.* 35.6 (Dec. 2007), pp. 2313–2351. DOI: 10.1214/009053606000001523. URL: http://dx.doi.org/10.1214/009053606000001523.

[94] E. J. Candès. "Ridgelets: theory and applications". PhD thesis. Department of Statistics, Stanford University, 1998.

[95] C. Carathéodory. "Über den Variabilitätsbereich der Koeffizienten von Potenzreihen, die gegebene Werte nicht annehmen". In: *Mathematische Annalen* 64.1 (1907), pp. 95–115. ISSN: 1432-1807. DOI: 10.1007/BF01449883. URL: http://dx.doi.org/10.1007/BF01449883.

[96] C. Carathéodory. "Über den variabilitätsbereich der fourier'schen konstanten von positiven harmonischen funktionen". In: *Rendiconti del Circolo Matematico di Palermo (1884-1940)* 32.1 (1911), pp. 193–217. ISSN: 0009-725X. DOI: 10.1007/BF03014795. URL: http://dx.doi.org/10.1007/BF03014795.

[97] D. Caspi, N. Kiryati, and J. Shamir. "Range imaging with adaptive color structured light". In: *Pattern Analysis and Machine Intelligence, IEEE Transactions on* 20.5 (May 1998), pp. 470–480. ISSN: 0162-8828. DOI: 10.1109/34.682177.

[98] A. Chambolle and T. Pock. "A First-Order Primal-Dual Algorithm for Convex Problems with Applications to Imaging". In: *J. Math. Imaging Vis.* 40.1 (May 2011), pp. 120–145. ISSN: 0924-9907. DOI: 10.1007/s10851-010-0251-1. URL: http://dx.doi.org/10.1007/s10851-010-0251-1.

[99] V. Chandar. *A negative result concerning explicit matrices with the restricted isometry property.* Tech. rep. 2008.

[100] V. Chandrasekaran, M. B. Wakin, D. Baron, and R. G. Baraniuk. "Surflets: a sparse representation for multidimensional functions containing smooth discontinuities". In: *Information Theory, 2004. ISIT 2004. Proceedings. International Symposium on.* June 2004, pp. 563–. DOI: 10.1109/ISIT.2004.1365602.

[101] V. Chandrasekaran, M. B. Wakin, D. Baron, and R. G. Baraniuk. "Representation and Compression of Multidimensional Piecewise Functions Using Surflets". In: *IEEE Transactions on Information Theory* 55.1 (Jan. 2009), pp. 374–400. ISSN: 0018-9448. DOI: 10.1109/TIT.2008.2008153.

[102] E. Charbon. "Techniques for CMOS single photon imaging and processing". In: *ASIC, 2005. ASICON 2005. 6th International Conference On*. Vol. 1. Oct. 2005, pp. 1163–1168. DOI: 10.1109/ICASIC.2005.1611239.

[103] E. Charbon, H.-J. Yoon, and Y. Maruyama. "A Geiger mode APD fabricated in standard 65nm CMOS technology". In: *Electron Devices Meeting (IEDM), 2013 IEEE International*. Dec. 2013, pp. 27.5.1–27.5.4. DOI: 10.1109/IEDM.2013.6724705.

[104] R. Chartrand. "Exact Reconstruction of Sparse Signals via Nonconvex Minimization". In: *IEEE Signal Processing Letters* 14.10 (Oct. 2007), pp. 707–710. ISSN: 1070-9908. DOI: 10.1109/LSP.2007.898300.

[105] R. Chartrand and W. Yin. "Iteratively reweighted algorithms for compressive sensing". In: *2008 IEEE International Conference on Acoustics, Speech and Signal Processing*. Mar. 2008, pp. 3869–3872. DOI: 10.1109/ICASSP.2008.4518498.

[106] J. Chen and X. Huo. "Theoretical Results on Sparse Representations of Multiple-Measurement Vectors". In: *Signal Processing, IEEE Transactions on* 54.12 (Dec. 2006), pp. 4634–4643. ISSN: 1053-587X. DOI: 10.1109/TSP.2006.881263.

[107] S. Chen, S. A. Billings, and W. Luo. "Orthogonal least squares methods and their application to non-linear system identification". In: *International Journal of Control* 50.5 (1989). Address: London, pp. 1873–1896. URL: http://eprints.soton.ac.uk/251147/.

[108] S. S. Chen, D. L. Donoho, and M. A. Saunders. "Atomic decomposition by basis pursuit". In: *SIAM Journal on Scientific Computing* 20 (1998), pp. 33–61.

[109] N. Chenouard and M. Unser. "Learning Steerable Wavelet Frames". In: *Proceedings of the Ninth International Workshop on Sampling Theory and Applications (SampTA'11)*. Singapore, Republic of Singapore: SampTA, May 2011. URL: http://bigwww.epfl.ch/publications/chenouard1102.pdf.

[110] V. Cheung, B. J. Frey, and N. Jojic. "Video epitomes". In: *2005 IEEE Computer Society Conference on Computer Vision and Pattern Recognition (CVPR'05)*. Vol. 1. June 2005, pp. 42–49. DOI: 10.1109/CVPR.2005.366.

[111] A. Cohen, W. Dahmen, and R. Devore. "Compressed sensing and best k-term approximation". In: *J. Amer. Math. Soc* (2009), pp. 211–231.

[112] A. Cohen, R. DeVore, P. Petrushev, and H. Xu. "Nonlinear approximation and the space BV($\mathbb{R}^2$)". In: *American Journal of Mathematics* 121 (1999), pp. 587–628. DOI: 10.1353/ajm.1999.0016.

[113] R. Coifman, F. Geshwind, and Y. Meyer. "Noiselets". In: *Applied and Computational Harmonic Analysis* 10.1 (2001), pp. 27 –44. ISSN: 1063-5203. DOI: http://dx.doi.org/10.1006/acha.2000.0313. URL: http://www.sciencedirect.com/science/article/pii/S1063520300903130.

[114] R. R. Coifman and M. V. Wickerhauser. "Entropy-based algorithms for best basis selection". In: *IEEE Transactions on Information Theory* 38.2 (Mar. 1992), pp. 713–718. ISSN: 0018-9448. DOI: 10.1109/18.119732.

[115] A. Colaço, A. Kirmani, G. Howland, J. Howell, and V. Goyal. "Compressive depth map acquisition using a single photon-counting detector: Parametric signal processing meets sparsity". In: *Computer Vision and Pattern Recognition (CVPR), 2012 IEEE Conference on.* June 2012, pp. 96–102. DOI: 10.1109/CVPR.2012.6247663.

[116] N. Collings, W. A. Crossland, P. J. Ayliffe, D. G. Vass, and I. Underwood. "Evolutionary development of advanced liquid crystal spatial light modulators". In: *Appl. Opt.* 28.22 (Nov. 1989), pp. 4740–4747. DOI: 10.1364/AO.28.004740. URL: http://ao.osa.org/abstract.cfm?URI=ao-28-22-4740.

[117] R. M. Conroy, A. A. Dorrington, R. Künnemeyer, and M. J. Cree. "Range imager performance comparison in homodyne and heterodyne operating modes". In: *Proc. SPIE.* Vol. 7239. 2009, pp. 723905–723905–10. DOI: 10.1117/12.806139. URL: http://dx.doi.org/10.1117/12.806139.

[118] J. H. Conway, N. J. A. Sloane, and E. Bannai. *Sphere-packings, Lattices, and Groups.* New York, NY, USA: Springer-Verlag New York, Inc., 1987. ISBN: 0-387-96617-X.

[119] J. Cooley and J. Tukey. "An Algorithm for the Machine Calculation of Complex Fourier Series". In: *Mathematics of Computation* 19.90 (1965), pp. 297–301.

[120] S. F. Cotter, B. D. Rao, K. Engan, and K. Kreutz-Delgado. "Sparse solutions to linear inverse problems with multiple measurement vectors". In: *IEEE Transactions on Signal Processing* 53.7 (July 2005), pp. 2477–2488. ISSN: 1053-587X. DOI: 10.1109/TSP.2005.849172.

[121] Cotton, A. and Mouton, H. "Sur la biréfringence magnétique des liquides purs. Comparaison avec le phénomène électro-optique de Kerr". In: *J. Phys. Theor. Appl.* 1.1 (1911), pp. 5–52. DOI: 10.1051/jphystap:01911001010500. URL: http://dx.doi.org/10.1051/jphystap:01911001010500.

[122] R. Craigen and J. Jedwab. *Comment on "The Hadamard circulant conjecture"*. Tech. rep. arXiv:1111.3437. Nov. 2011. URL: http://cds.cern.ch/record/1399299.

[123] M. S. Crouse, R. D. Nowak, and R. G. Baraniuk. "Wavelet-based statistical signal processing using hidden Markov models". In: *IEEE Transactions on Signal Processing* 46.4 (Apr. 1998), pp. 886–902. ISSN: 1053-587X. DOI: 10.1109/78.668544.

[124] W. Dai and O. Milenkovic. "Subspace Pursuit for Compressive Sensing Signal Reconstruction". In: *IEEE Transactions on Information Theory* 55.5 (May 2009), pp. 2230–2249. ISSN: 0018-9448. DOI: 10.1109/TIT.2009.2016006.

[125] G. B. Dantzig and M. N. Thapa. *Linear Programming 1: Introduction*. Secaucus, NJ, USA: Springer-Verlag New York, Inc., 1997. ISBN: 0-387-94833-3.

[126] S. Das and N. Ahuja. "Active stereo based surface reconstruction". In: *Intelligent Control, 1990. Proceedingsof the 5th IEEE International Symposium on.* Vol. 1. Sept. 1990, pp. 233–238. DOI: 10.1109/ISIC.1990.128463.

[127] S. Dasgupta and A. Gupta. "An elementary proof of a theorem of Johnson and Lindenstrauss". In: *Random Structures & Algorithms* 22.1 (2003), pp. 60–65. ISSN: 1098-2418. DOI: 10.1002/rsa.10073. URL: http://dx.doi.org/10.1002/rsa.10073.

[128] I. Daubechies. "The wavelet transform, time-frequency localization and signal analysis". In: *IEEE Transactions on Information Theory* 36.5 (Sept. 1990), pp. 961–1005. ISSN: 0018-9448. DOI: 10.1109/18.57199.

[129] I. Daubechies. *Ten Lectures on Wavelets.* Society for Industrial and Applied Mathematics, 1992. DOI: 10.1137/1.9781611970104. eprint: http://epubs.siam.org/doi/pdf/10.1137/1.9781611970104. URL: http://epubs.siam.org/doi/abs/10.1137/1.9781611970104.

[130] I. Daubechies. "Orthonormal bases of compactly supported wavelets". In: *Communications on Pure and Applied Mathematics* 41.7 (1988), pp. 909–996. ISSN: 1097-0312. DOI: 10.1002/cpa.3160410705. URL: http://dx.doi.org/10.1002/cpa.3160410705.

[131] I. Daubechies, R. DeVore, M. Fornasier, and C. S. Güntürk. "Iteratively reweighted least squares minimization for sparse recovery". In: *Communications on Pure and Applied Mathematics* 63.1 (2010), pp. 1–38. ISSN: 1097-0312. DOI: 10.1002/cpa.20303. URL: http://dx.doi.org/10.1002/cpa.20303.

[132] J. G. Daugman. "Complete discrete 2-D Gabor transforms by neural networks for image analysis and compression". In: *IEEE Transactions on Acoustics, Speech, and Signal Processing* 36.7 (July 1988), pp. 1169–1179. ISSN: 0096-3518. DOI: 10.1109/29.1644.

[133] J. G. Daugman. "Two-dimensional spectral analysis of cortical receptive field profiles". In: *Vision Research* 20.10 (1980), pp. 847–856. ISSN: 0042-6989. DOI: http://dx.doi.org/10.1016/0042-6989(80)90065-6. URL: http://www.sciencedirect.com/science/article/pii/0042698980900656.

[134] J. G. Daugman. "Uncertainty relation for resolution in space, spatial frequency, and orientation optimized by two-dimensional visual cortical filters". In: *J. Opt. Soc. Am. A* 2.7 (July 1985), pp. 1160–1169. DOI: 10.1364/JOSAA.2.001160. URL: http://josaa.osa.org/abstract.cfm?URI=josaa-2-7-1160.

[135] M. A. Davenport. "Random observations on random observations: Sparse signal acquisition and processing". PhD thesis. Department of Electrical and Computer Engineering at Rice University, Aug. 2010. URL: http://dsp.rice.edu/sites/dsp.rice.edu/files/publications/thesis/2010/madphdthesis.pdf.

[136] M. Davidovic, M. Hofbauer, K. Schneider-Hornstein, and H. Zimmermann. "High Dynamic Range Background Light Suppression for a TOF Distance Measurement Sensor in 180nm CMOS". In: *Sensors, 2011 IEEE.* 2011, pp. 359–362. DOI: 10.1109/ICSENS.2011.6127060.

[137] M. Davies and Y. Eldar. “Rank Awareness in Joint Sparse Recovery”. In: *Information Theory, IEEE Transactions on* 58.2 (Feb. 2012), pp. 1135–1146. ISSN: 0018-9448. DOI: 10.1109/TIT.2011.2173722.

[138] P. E. Debevec and J. Malik. “Recovering High Dynamic Range Radiance Maps from Photographs”. In: *SIGGRAPH.* 1997, pp. 369–378. URL: http://dblp.uni-trier.de/db/conf/siggraph/siggraph1997.html#DebevecM97.

[139] P. Delsarte, J. Goethals, and J. Seidel. “Bounds for systems of lines, and Jacobi polynomials”. In: *Philips Research Reports* 30 (1975), pp. 91–105. ISSN: 0031-7918.

[140] P. B. Denyer, D. Renshaw, and S. Smith. “Intelligent CMOS imaging”. In: *Proc. SPIE.* Vol. 2415. 1995, pp. 285–291. DOI: 10.1117/12.206525. URL: http://dx.doi.org/10.1117/12.206525.

[141] R. A. DeVore. “Nonlinear Approximation”. In: *Acta Numerica* 7 (1998), pp. 51–150.

[142] R. A. DeVore. “Deterministic constructions of compressed sensing matrices”. In: *Journal of Complexity* 23.4-6 (2007). Festschrift for the 60th Birthday of Henryk Woźniakowski, pp. 918–925. ISSN: 0885-064X. DOI: http://dx.doi.org/10.1016/j.jco.2007.04.002. URL: http://www.sciencedirect.com/science/article/pii/S0885064X07000623.

[143] A. Dib, N. Beaufort, and F. Charpillet. “A real time visual SLAM for RGB-D cameras based on chamfer distance and occupancy grid”. In: *International conference on advanced intelligent mechatronics.* Besançon, France, July 2014, pp. 652 –657. DOI: 10.1109/AIM.2014.6878153. URL: https://hal.inria.fr/hal-01090998.

[144] A. G. Dimakis, R. Smarandache, and P. O. Vontobel. “LDPC Codes for Compressed Sensing”. In: *IEEE Transactions on Information Theory* 58.5 (May 2012), pp. 3093–3114. ISSN: 0018-9448. DOI: 10.1109/TIT.2011.2181819.

[145] *DMD 101: Introduction to Digital Micromirror Device (DMD) Technology.* http://www.ti.com/lit/an/dlpa008a/dlpa008a.pdf. [Online; accessed 26-August-2015]. 2013.

[146] M. N. Do and M. Vetterli. “The contourlet transform: an efficient directional multiresolution image representation”. In: *IEEE Transactions on Image Processing* 14.12 (Dec. 2005), pp. 2091–2106. ISSN: 1057-7149. DOI: 10.1109/TIP.2005.859376.

[147] D. L. Donoho. "Compressed sensing". In: *IEEE Transactions on Information Theory* 52.4 (Apr. 2006), pp. 1289–1306. ISSN: 0018-9448. DOI: 10.1109/TIT.2006.871582.

[148] D. L. Donoho. "Wedgelets: nearly minimax estimation of edges". In: *Ann. Statist.* 27.3 (June 1999), pp. 859–897. DOI: 10.1214/aos/1018031261. URL: http://dx.doi.org/10.1214/aos/1018031261.

[149] D. L. Donoho. "For most large underdetermined systems of linear equations the minimal l_1-norm solution is also the sparsest solution". In: *Communications on Pure and Applied Mathematics* 59.6 (2006), pp. 797–829. ISSN: 1097-0312. DOI: 10.1002/cpa.20132. URL: http://dx.doi.org/10.1002/cpa.20132.

[150] D. L. Donoho and M. Elad. "Optimally sparse representation in general (nonorthogonal) dictionaries via l_1 minimization". In: *Proceedings of the National Academy of Sciences* 100.5 (2003), pp. 2197–2202. DOI: 10.1073/pnas.0437847100. eprint: http://www.pnas.org/content/100/5/2197.full.pdf+html. URL: http://www.pnas.org/content/100/5/2197.abstract.

[151] D. Donoho, M. Vetterli, R. A. Devore, and I. Daubechies. "Data Compression and Harmonic Analysis". In: *IEEE Trans. Inform. Theory* 44 (1998), pp. 2435–2476.

[152] A. A. Dorrington, M. J. Cree, A. D. Payne, R. M. Conroy, and D. A. Carnegie. "Achieving sub-millimetre precision with a solid-state full-field heterodyning range imaging camera". In: *Measurement Science and Technology* 18.9 (2007), p. 2809. URL: http://stacks.iop.org/0957-0233/18/i=9/a=010.

[153] A. A. Dorrington, J. P. Godbaz, M. J. Cree, A. D. Payne, and L. V. Streeter. "Separating true range measurements from multi-path and scattering interference in commercial range cameras". In: *Proc. SPIE*. Vol. 7864. 2011, pp. 786404–786404–10. DOI: 10.1117/12.876586. URL: http://dx.doi.org/10.1117/12.876586.

[154] A. A. Dorrington, M. J. Cree, D. A. Carnegie, A. D. Payne, R. M. Conroy, J. P. Godbaz, and A. P. P. Jongenelen. "Video-rate or high-precision: a flexible range imaging camera". In: *Proc. SPIE*. Vol. 6813. 2008, pp. 681307–681307–12. DOI: 10.1117/12.764752. URL: http://dx.doi.org/10.1117/12.764752.

[155] R. G. Dorsch, G. Häusler, and J. M. Herrmann. "Laser triangulation: fundamental uncertainty in distance measurement". In: *Appl. Opt.* 33.7 (Mar. 1994), pp. 1306–1314. DOI: 10.1364/AO.33.001306. URL: http://ao.osa.org/abstract.cfm?URI=ao-33-7-1306.

[156] M. F. Duarte, M. A. Davenport, D. Takhar, J. N. Laska, T. Sun, K. F. Kelly, and R. G. Baraniuk. "Single-pixel imaging via compressive sampling". In: *IEEE Signal Processing Magazine* 25.2 (Mar. 2008), pp. 83–91.

[157] J. M. Duarte-Carvajalino and G. Sapiro. "Learning to Sense Sparse Signals: Simultaneous Sensing Matrix and Sparsifying Dictionary Optimization". In: *IEEE Transactions on Image Processing* 18.7 (July 2009), pp. 1395–1408. ISSN: 1057-7149. DOI: 10.1109/TIP.2009.2022459.

[158] D. Dudley, W. M. Duncan, and J. Slaughter. "Emerging digital micromirror device (DMD) applications". In: *Proc. SPIE.* Vol. 4985. 2003, pp. 14–25. DOI: 10.1117/12.480761. URL: http://dx.doi.org/10.1117/12.480761.

[159] D. Durini, B. Brockherde, W. Ulfig, and B. J. Hosticka. "Time-of-Flight 3-D Imaging Pixel Structures in Standard CMOS Processes". In: *IEEE Journal of Solid-State Circuits* 43.7 (July 2008), pp. 1594–1602. ISSN: 0018-9200. DOI: 10.1109/JSSC.2008.922397.

[160] T. Edeler, K. Ohliger, S. Hussmann, and A. Mertins. "Super resolution reconstruction method for time-of-flight range data using complex compressive sensing". In: *Instrumentation and Measurement Technology Conference (I2MTC), 2011 IEEE.* May 2011, pp. 1–5. DOI: 10.1109/IMTC.2011.5944167.

[161] H. Eklund, A. Roos, and S. Eng. "Rotation of laser beam polarization in acousto-optic devices". English. In: *Optical and Quantum Electronics* 7.2 (1975), pp. 73–79. ISSN: 0306-8919. DOI: 10.1007/BF00631587. URL: http://dx.doi.org/10.1007/BF00631587.

[162] M. Elad. "Optimized Projections for Compressed Sensing". In: *IEEE Transactions on Signal Processing* 55.12 (Dec. 2007), pp. 5695–5702. ISSN: 1053-587X. DOI: 10.1109/TSP.2007.900760.

[163] Y. Eldar and G. Kutyniok. *Compressed Sensing: Theory and Applications.* Compressed Sensing: Theory and Applications. Cambridge University Press, 2012. ISBN: 9781107005587.

[164] H. Elgala, R. Mesleh, and H. Haas. "Indoor broadcasting via white LEDs and OFDM". In: *Consumer Electronics, IEEE Transactions on* 55.3 (Aug. 2009), pp. 1127–1134. ISSN: 0098-3063. DOI: 10.1109/TCE.2009.5277966.

[165] H. Elgala, R. Mesleh, and H. Haas. "Indoor optical wireless communication: potential and state-of-the-art". In: *Communications Magazine, IEEE* 49.9 (Sept. 2011), pp. 56–62. ISSN: 0163-6804. DOI: 10.1109/MCOM.2011.6011734.

[166] S. B. Emmanuel J. Candès M. Wakin. "Enhancing sparsity by reweighted l_1 minimization". In: *Journal of Fourier Analysis and Applications* 14.5 (Dec. 2008), pp. 877–905. ISSN: 1531-5851. DOI: 10.1007/s00041-008-9045-x. URL: http://dx.doi.org/10.1007/s00041-008-9045-x.

[167] F. Endres, J. Hess, J. Sturm, D. Cremers, and W. Burgard. "3D Mapping with an RGB-D Camera". In: *IEEE Transactions on Robotics* 30.1 (Feb. 2014), pp. 177–187.

[168] F. Endres, J. Hess, N. Engelhard, J. Sturm, and W. Burgard. "6D Visual SLAM for RGB-D Sensors". In: *Automatisierungstechnik* 60 (May 2012), pp. 270–278.

[169] K. Engan, S. O. Aase, and J. H. Husoy. "Method of optimal directions for frame design". In: *Acoustics, Speech, and Signal Processing, 1999. Proceedings., 1999 IEEE International Conference on.* Vol. 5. 1999, 2443–2446 vol.5. DOI: 10.1109/ICASSP.1999.760624.

[170] K. Engan, K. Skretting, and J. H. Husøy. "Family of Iterative LS-based Dictionary Learning Algorithms, ILS-DLA, for Sparse Signal Representation". In: *Digit. Signal Process.* 17.1 (Jan. 2007), pp. 32–49. ISSN: 1051-2004. DOI: 10.1016/j.dsp.2006.02.002. URL: http://dx.doi.org/10.1016/j.dsp.2006.02.002.

[171] N. Engelhard, F. Endres, J. Hess, J. Sturm, and W. Burgard. "Real-time 3D visual SLAM with a hand-held camera". In: *Proc. of the RGB-D Workshop on 3D Perception in Robotics at the European Robotics Forum.* Vasteras, Sweden, Apr. 2011.

[172] T. Ericson and V. Zinoviev. *Codes on Euclidean spheres.* North-Holland mathematical library. Amsterdam, New York: North-Holland/Elsevier, 2001. ISBN: 0-444-50329-3. URL: http://opac.inria.fr/record=b1099938.

[173] A. Esposito, T. Oggier, H. Gerritsen, F. Lustenberger, and F. Wouters. “All-solid-state lock-in imaging for wide-field fluorescence lifetime sensing”. In: *Opt. Express* 13.24 (Nov. 2005), pp. 9812–9821. DOI: 10.1364/OPEX.13.009812. URL: http://www.opticsexpress.org/abstract.cfm?URI=oe-13-24-9812.

[174] M. Fazel, H. Hindi, and S. P. Boyd. “Log-det heuristic for matrix rank minimization with applications to Hankel and Euclidean distance matrices”. In: *American Control Conference*. Vol. 3. 2003. DOI: 10.1109/ACC.2003.1243393.

[175] P. Feng and Y. Bresler. “Spectrum-blind minimum-rate sampling and reconstruction of multiband signals”. In: *Acoustics, Speech, and Signal Processing, 1996. ICASSP-96. Conference Proceedings., 1996 IEEE International Conference on.* Vol. 3. May 1996, 1688–1691 vol. 3. DOI: 10.1109/ICASSP.1996.544131.

[176] *Fluorescence Lifetime Imaging - Application Simplified.* http://www.pco.de/fileadmin/user_upload/pco-product_sheets/pco.flim_data_sheet.pdf. [Online; accessed 30-May-2016]. 2016.

[177] D. Fofi, T. Sliwa, and Y. Voisin. “A comparative survey on invisible structured light”. In: *Proc. SPIE.* Vol. 5303. 2004, pp. 90–98. DOI: 10.1117/12.525369. URL: http://dx.doi.org/10.1117/12.525369.

[178] A. Foi. *Optimization of variance-stabilizing transformations.* 2009. URL: http://www.cs.tut.fi/~foi.

[179] S. Foix, G. Alenya, and C. Torras. “Lock-in Time-of-Flight (ToF) Cameras: A Survey”. In: *Sensors Journal, IEEE* 11.9 (2011), pp. 1917–1926. ISSN: 1530-437X. DOI: 10.1109/JSEN.2010.2101060.

[180] M. Fornasier and H. Rauhut. “Compressive Sensing”. In: *Handbook of Mathematical Methods in Imaging.* Ed. by O. Scherzer. Springer, 2011, pp. 187–228. DOI: 10.1007/978-0-387-92920-0_6.

[181] A. Fossati, J. Gall, H. Grabner, X. Ren, and K. Konolige. *Consumer Depth Cameras for Computer Vision: Research Topics and Applications.* Springer Publishing Company, Incorporated, 2012. ISBN: 1447146395, 9781447146391.

[182] S. Foucart. “Hard Thresholding Pursuit: An Algorithm for Compressive Sensing”. In: *SIAM Journal on Numerical Analysis* 49.6 (2011), pp. 2543–2563. DOI: 10.1137/100806278. eprint: http://dx.doi.org/10.1137/100806278. URL: http://dx.doi.org/10.1137/100806278.

[183] S. Foucart, A. Pajor, H. Rauhut, and T. Ullrich. "The Gelfand widths of -balls for". In: *Journal of Complexity* 26.6 (2010), pp. 629–640. ISSN: 0885-064X. DOI: http://dx.doi.org/10.1016/j.jco.2010.04.004. URL: http://www.sciencedirect.com/science/article/pii/S0885064X10000282.

[184] S. Foucart and H. Rauhut. *A Mathematical Introduction to Compressive Sensing*. Birkhäuser Basel, 2013. ISBN: 0817649476, 9780817649470.

[185] J. E. Fowler. "The Redundant Discrete Wavelet Transform and Additive Noise". In: *IEEE Signal Processing Letters* 12.9 (2005), pp. 629–632. ISSN: 1070-9908. DOI: 10.1109/LSP.2005.853048.

[186] M. Fox and R Ispasoiu. "Quantum wells, superlattices, and band-gap engineering". In: *Springer Handbook of Electronic and Photonic Materials*. Ed. by S. Kasap and P. Capper. Springer Handbook of Electronic and Photonic Materials. Springer, 2006, pp. 1021–1039. ISBN: 9780387291857. URL: https://books.google.de/books?id=rVVW22pnzhoC.

[187] G. Frank. "Verfahren und Vorrichtung zur Bestimmung von Probeneigenschaften über zeitaufgelöste Lumineszenz". Pat. DE Patent App. DE1,999,151,154. May 2001. URL: http://www.google.com/patents/DE19951154A1?cl=de.

[188] B. Freedman, A. Shpunt, M. Machline, and Y. Arieli. "Depth mapping using projected patterns". Pat. US Patent App. 11/899,542. Oct. 2008. URL: http://www.google.com/patents/US20080240502.

[189] D. Freedman, Y. Smolin, E. Krupka, I. Leichter, and M. Schmidt. "SRA: Fast Removal of General Multipath for ToF Sensors". English. In: *Computer Vision – ECCV 2014*. Ed. by D. Fleet, T. Pajdla, B. Schiele, and T. Tuytelaars. Vol. 8689. Lecture Notes in Computer Science. Springer International Publishing, 2014, pp. 234–249. ISBN: 978-3-319-10589-5. DOI: 10.1007/978-3-319-10590-1_16. URL: http://dx.doi.org/10.1007/978-3-319-10590-1_16.

[190] M. F. Freeman and J. W. Tukey. "Transformations Related to the Angular and the Square Root". In: *Ann. Math. Statist.* 21.4 (Dec. 1950), pp. 607–611. DOI: 10.1214/aoms/1177729756. URL: http://dx.doi.org/10.1214/aoms/1177729756.

[191] K. P. F.R.S. "LIII. On lines and planes of closest fit to systems of points in space". In: *Philosophical Magazine Series 6* 2.11 (1901), pp. 559–572. DOI: 10.1080/14786440109462720. eprint: http://dx.doi.org/10.1080/14786440109462720. URL: http://dx.doi.org/10.1080/14786440109462720.

[192] S. Fuchs. "Multipath Interference Compensation in Time-of-Flight Camera Images". In: *Pattern Recognition (ICPR), 2010 20th International Conference on.* Aug. 2010, pp. 3583–3586. DOI: 10.1109/ICPR.2010.874.

[193] S. Fuchs, M. Suppa, and O. Hellwich. "Compensation for Multipath in ToF Camera Measurements Supported by Photometric Calibration and Environment Integration". English. In: *Computer Vision Systems.* Ed. by M. Chen, B. Leibe, and B. Neumann. Vol. 7963. Lecture Notes in Computer Science. Springer Berlin Heidelberg, 2013, pp. 31–41. ISBN: 978-3-642-39401-0. DOI: 10.1007/978-3-642-39402-7_4. URL: http://dx.doi.org/10.1007/978-3-642-39402-7_4.

[194] D. Gabor. "Theory of communication. Part 1: The analysis of information". In: *Electrical Engineers - Part III: Radio and Communication Engineering, Journal of the Institution of* 93.26 (Nov. 1946), pp. 429–441. DOI: 10.1049/ji-3-2.1946.0074.

[195] R. Gallager. "Low-density parity-check codes". In: *IRE Transactions on Information Theory* 8.1 (Jan. 1962), pp. 21–28. ISSN: 0096-1000. DOI: 10.1109/TIT.1962.1057683.

[196] R. G. Gallager. "Low-Density Parity-Check Codes". PhD thesis. Cambridge MA, 1963.

[197] L. Gan, T. T. Do, and T. D. Tran. "Fast compressive imaging using scrambled block Hadamard ensemble". In: *Signal Processing Conference, 2008 16th European.* Aug. 2008, pp. 1–5.

[198] L. Gao, J. Liang, C. Li, and L. V. Wang. "Single-shot compressed ultrafast photography at one hundred billion frames per second". In: *Nature* 516.7529 (Dec. 2014), pp. 74–77. ISSN: 0028-0836. DOI: 10.1038/nature14005. URL: \url{http://dx.doi.org/10.1038/nature14005}.

[199] S. Gautam and D. Vaintrob. *A Novel Approach to the Spherical Codes Problem.* Tech. rep. Massachusetts Institute of Technology, 2012.

[200] S. A. Geršgorin. "Über die abgrenzung der eigenwerte einer Matrix". In: *Bulletin de l'Académie des Sciences de l'URSS. Classe des Sciences Mathématiques et na* 6 (1931), pp. 749–754. URL: http://www.mathnet.ru/links/9372171f9b618598dad64a5ad1d40eef/im5235.pdf.

[201] A. C. Gilbert, S. Guha, P. Indyk, S. Muthukrishnan, and M. Strauss. "Near-optimal Sparse Fourier Representations via Sampling". In: *Proceedings of the Thiry-fourth Annual ACM Symposium on Theory of Computing*. STOC '02. Montreal, Quebec, Canada: ACM, 2002, pp. 152–161. ISBN: 1-58113-495-9. DOI: 10.1145/509907.509933. URL: http://doi.acm.org/10.1145/509907.509933.

[202] A. C. Gilbert, S. Muthukrishnan, and M. J. Strauss. "Approximation of Functions over Redundant Dictionaries Using Coherence". In: *Proceedings of the Fourteenth Annual ACM-SIAM Symposium on Discrete Algorithms*. SODA '03. Baltimore, Maryland: Society for Industrial and Applied Mathematics, 2003, pp. 243–252. ISBN: 0-89871-538-5. URL: http://www.math.lsa.umich.edu/~annacg/papers/GMS03.cat.pdf.

[203] J. P. Godbaz, M. J. Cree, and A. A. Dorrington. "Closed-form inverses for the mixed pixel/multipath interference problem in AMCW lidar". In: *Proc. SPIE*. Vol. 8296. 2012, pp. 829618–829618–15. DOI: 10.1117/12.909778. URL: http://dx.doi.org/10.1117/12.909778.

[204] I. F. Gorodnitsky and B. D. Rao. "Sparse signal reconstruction from limited data using FOCUSS: a re-weighted minimum norm algorithm". In: *IEEE Transactions on Signal Processing* 45.3 (Mar. 1997), pp. 600–616. ISSN: 1053-587X. DOI: 10.1109/78.558475.

[205] Y. Gousseau and J.-M. Morel. "Are Natural Images of Bounded Variation?" In: *SIAM Journal on Mathematical Analysis* 33.3 (2001), pp. 634–648. DOI: 10.1137/S0036141000371150. eprint: http://dx.doi.org/10.1137/S0036141000371150. URL: http://dx.doi.org/10.1137/S0036141000371150.

[206] J. E. Greivenkamp. "Generalized Data Reduction For Heterodyne Interferometry". In: *Optical Engineering* 23.4 (1984), pp. 234350–234350–. DOI: 10.1117/12.7973298. URL: http://dx.doi.org/10.1117/12.7973298.

[207] R. Grosse, M. K. Johnson, E. H. Adelson, and W. T. Freeman. "Ground-truth dataset and baseline evaluations for intrinsic image algorithms". In: *International Conference on Computer Vision (ICCV)*. 2009, pp. 2335–2342. DOI: http://dx.doi.org/10.1109/ICCV.2009.5459428.

[208] S. Guđmundsson, H. Aanæaes, and R. Larsen. "Environmental Effects on Measurement Uncertainties of Time-of-Flight Cameras". In: *Signals, Circuits and Systems, 2007. ISSCS 2007. International Symposium on.* Vol. 1. July 2007, pp. 1–4. DOI: 10.1109/ISSCS.2007.4292664.

[209] M. Gupta, S. K. Nayar, M. Hullin, and J. Martin. "Phasor Imaging: A Generalization Of Correlation-Based Time-of-Flight Imaging". In: *ACM Transactions on Graphics (TOG), presented at SIGGRAPH 2015* 34.5 (Oct. 2015).

[210] A. Haar. "Zur Theorie der orthogonalen Funktionensysteme". In: *Mathematische Annalen* 69.3 (), pp. 331–371. ISSN: 1432-1807. DOI: 10.1007/BF01456326. URL: http://dx.doi.org/10.1007/BF01456326.

[211] J. Hadamard. "Résolution d'une question relative aux déterminants". In: *Bull. Sci. Math.* 17 (1893), pp. 240–246.

[212] S. M. Han, T. Takasawa, K. Yasutomi, S. Aoyama, K. Kagawa, and S. Kawahito. "A Time-of-Flight Range Image Sensor With Background Canceling Lock-in Pixels Based on Lateral Electric Field Charge Modulation". In: *IEEE Journal of the Electron Devices Society* 3.3 (May 2015), pp. 267–275. ISSN: 2168-6734. DOI: 10.1109/JEDS.2014.2382689.

[213] M. Hansard, S. Lee, O. Choi, and R. Horaud. *Time of Flight Cameras: Principles, Methods, and Applications.* SpringerBriefs in Computer Science. Springer, 2013.

[214] P. Hariharan. *Optical Interferometry.* Electronics & Electrical. Academic Press, 2003. ISBN: 9780123116307. URL: https://books.google.de/books?id=EGdMO3rfVj4C.

[215] P. Hariharan. *Basics of Interferometry.* Elsevier Science, 2010. ISBN: 9780080465456. URL: https://books.google.com.br/books?id=sWbGSSQ6fPYC.

[216] K. Hartmann and R. Schwarte. "Detection of the phase and amplitude of electromagnetic waves". Pat. US7238927 B1. July 2007. URL: https://www.google.com/patents/US7238927.

[217] J. Haupt, R. M. Castro, and R. Nowak. "Distilled Sensing: Adaptive Sampling for Sparse Detection and Estimation". In: *IEEE Transactions on Information Theory* 57.9 (Sept. 2011), pp. 6222–6235. ISSN: 0018-9448. DOI: 10.1109/TIT.2011.2162269.

[218] Z. He, T. Ogawa, and M. Haseyama. "The simplest measurement matrix for compressed sensing of natural images". In: *Image Processing (ICIP), 2010 17th IEEE International Conference on.* Sept. 2010, pp. 4301–4304. DOI: 10.1109/ICIP.2010.5651800.

[219] A. Hedayat, N. J. A. N. J. A. Sloane, and J. Stufken. *Orthogonal arrays : theory and applications.* Springer series in statistics. New York: Springer, 1999. ISBN: 0-387-98766-5. URL: http://opac.inria.fr/record=b1097587.

[220] F. Heide, M. B. Hullin, J. Gregson, and W. Heidrich. "Low-budget Transient Imaging Using Photonic Mixer Devices". In: *ACM Transactions on Graphics* 32.4 (July 2013), 45:1–45:10. ISSN: 0730-0301. DOI: 10.1145/2461912.2461945. URL: http://doi.acm.org/10.1145/2461912.2461945.

[221] F. Heide, L. Xiao, A. Kolb, M. B. Hullin, and W. Heidrich. "Imaging in scattering media using correlation image sensors and sparse convolutional coding". In: *Opt. Express* 22.21 (Oct. 2014), pp. 26338–26350. DOI: 10.1364/OE.22.026338. URL: http://www.opticsexpress.org/abstract.cfm?URI=oe-22-21-26338.

[222] H. G. Heinol. "Untersuchung und Entwicklung von modulationslaufzeitbasierten 3D-Sichtsystemen". German. PhD thesis. Siegen, Germany: Department of Electrical Engineering and Computer Science, 2001, p. 157.

[223] P. Henry, M. Krainin, E. Herbst, X. Ren, and D. Fox. "Rgbd mapping: Using depth cameras for dense 3d modeling of indoor environments". In: *In RGB-D: Advanced Reasoning with Depth Cameras Workshop in conjunction with RSS.* 2010.

[224] *High Definition Lidar HDL-64E S2.* http://velodynelidar.com/lidar/products/brochure/HDL-64E%20S2%20datasheet_2010_lowres.pdf. [Online; accessed 27-July-2014]. 2014.

[225] B. Höfflinger. *High-Dynamic-Range (HDR) Vision: Microelectronics, Image Processing, Computer Graphics (Springer Series in Advanced Microelectronics).* Secaucus, NJ, USA: Springer-Verlag New York, Inc., 2007. ISBN: 3540444327.

[226] H. A. Höfler, V. Jetter, and E. E. Wagner. "3D profiling by optical demodulation with an image intensifier". In: *Proc. SPIE.* Vol. 3640. 1999, pp. 21–27. DOI: 10.1117/12.341068. URL: http://dx.doi.org/10.1117/12.341068.

[227] E. Horn and N. Kiryati. "Toward optimal structured light patterns". In: *3-D Digital Imaging and Modeling, 1997. Proceedings., International Conference on Recent Advances in.* May 1997, pp. 28–35. DOI: 10.1109/IM.1997.603845.

[228] J. P. Hornak. "The Basics of MRI". 1996. URL: http://www.cis.rit.edu/htbooks/mri/.

[229] L. Hornbeck. "Spatial light modulator and method". Pat. US Patent 5,061,049. Oct. 1991. URL: http://www.google.com/patents/US5061049.

[230] G. A. Howland, D. J. Lum, M. R. Ware, and J. C. Howell. "Photon counting compressive depth mapping". In: *CoRR* abs/1309.4385 (2013). URL: http://arxiv.org/abs/1309.4385.

[231] X.-Y. Hu, E. Eleftheriou, and D. M. Arnold. "Progressive edge-growth Tanner graphs". In: *Global Telecommunications Conference, 2001. GLOBECOM '01. IEEE.* Vol. 2. 2001, 995–1001 vol.2. DOI: 10.1109/GLOCOM.2001.965567.

[232] X.-Y. Hu, E. Eleftheriou, and D. M. Arnold. "Regular and irregular progressive edge-growth tanner graphs". In: *IEEE Transactions on Information Theory* 51.1 (Jan. 2005), pp. 386–398. ISSN: 0018-9448. DOI: 10.1109/TIT.2004.839541.

[233] A. S. Huang, A. Bachrach, P. Henry, M. Krainin, D. Fox, and N. Roy. "Visual odometry and mapping for autonomous flight using an RGB-D camera". In: *In Proc. of the Intl. Sym. of Robot. Research.* 2011.

[234] B. Hurley, P. Hurley, and T. Hurley. "The Hadamard circulant conjecture". In: *Bulletin of the London Mathematical Society* (2011). DOI: 10.1112/blms/bdq112. eprint: http://blms.oxfordjournals.org/content/early/2011/01/24/blms.bdq112.full.pdf+html.

URL: http://blms.oxfordjournals.org/content/early/2011/01/24/blms.bdq112.abstract.

[235] *IFM O1D100.* https://www.ifm.com/products/gb/ds/O1D100.htm. 2015.

[236] J. Illade-Quinteiro, V. Brea, P. López, and D. Cabello. "Time-of-Flight Chip in Standard CMOS Technology with In-Pixel Adaptive Number of Accumulations". In: *IEEEE International Symposium on Circuits and Systems.* 2016.

[237] Y. Inoue, H. Yoshida, H. Kubo, and M. Ozaki. "Deformation-Free, Microsecond Electro-Optic Tuning of Liquid Crystals". In: *Advanced Optical Materials* 1.3 (2013), pp. 256–263. ISSN: 2195-1071. DOI: 10.1002/adom.201200028. URL: http://dx.doi.org/10.1002/adom.201200028.

[238] Y. Inoue, H. Yoshida, and M. Ozaki. "Nematic liquid crystal nanocomposite with scattering-free, microsecond electro-optic response". In: *Opt. Mater. Express* 4.5 (May 2014), pp. 916–923. DOI: 10.1364/OME.4.000916. URL: http://www.osapublishing.org/ome/abstract.cfm?URI=ome-4-5-916.

[239] S. Izadi, D. Kim, O. Hilliges, D. Molyneaux, R. Newcombe, P. Kohli, J. Shotton, S. Hodges, D. Freeman, A. Davison, and A. Fitzgibbon. "KinectFusion: Real-time 3D Reconstruction and Interaction Using a Moving Depth Camera". In: *Proceedings of the 24th Annual ACM Symposium on User Interface Software and Technology.* UIST '11. Santa Barbara, California, USA: ACM, 2011, pp. 559–568. ISBN: 978-1-4503-0716-1. DOI: 10.1145/2047196.2047270. URL: http://doi.acm.org/10.1145/2047196.2047270.

[240] P. Jacquot. "Speckle Interferometry: A Review of the Principal Methods in Use for Experimental Mechanics Applications". In: *Strain* 44.1 (2008), pp. 57–69. ISSN: 1475-1305. DOI: 10.1111/j.1475-1305.2008.00372.x. URL: http://dx.doi.org/10.1111/j.1475-1305.2008.00372.x.

[241] B. Jaffe, W. Cook, and H. Jaffe. *Piezoelectric ceramics.* Non-metallic solids. Academic Press, 1971. ISBN: 9780123795502. URL: https://books.google.de/books?id=ZApTAAAAMAAJ.

[242] T. S. Jayram and D. P. Woodruff. "Optimal Bounds for Johnson-Lindenstrauss Transforms and Streaming Problems with Subconstant Error". In: *ACM Trans. Algorithms* 9.3 (2013), p. 26. DOI: 10.1145/2483699.2483706. URL: http://doi.acm.org/10.1145/2483699.2483706.

[243] D. Jiménez, D. Pizarro, M. Mazo, and S. Palazuelos. "Modelling and correction of multipath interference in time of flight cameras". In: *Computer Vision and Pattern Recognition (CVPR), 2012 IEEE Conference on.* June 2012, pp. 893–900. DOI: 10.1109/CVPR.2012.6247763.

[244] W. B. Johnson and J. Lindenstrauss. "Extensions of Lipschitz maps into a Hilbert space". In: *Contemporary Mathematics* 26 (1984), pp. 189–206.

[245] N. Jojic, B. J. Frey, and A. Kannan. "Epitomic analysis of appearance and shape". In: *Computer Vision, 2003. Proceedings. Ninth IEEE International Conference on.* Oct. 2003, 34–41 vol.1. DOI: 10.1109/ICCV.2003.1238311.

[246] P. Jost, P. Vandergheynst, S. Lesage, and R. Gribonval. "MoTIF: An Efficient Algorithm for Learning Translation Invariant Dictionaries". In: *2006 IEEE International Conference on Acoustics Speech and Signal Processing Proceedings.* Vol. 5. May 2006, pp. 857–860. DOI: 10.1109/ICASSP.2006.1661411.

[247] G. Kabatiansky and V. Levenshtein. "On Bounds for Packings on a Sphere and in Space". In: *Probl. Peredachi Inf.* 14.1 (1978). (in Russian), pp. 3–25.

[248] A. Kadambi, R. Whyte, A. Bhandari, L. Streeter, C. Barsi, A. Dorrington, and R. Raskar. "Coded time of flight cameras: sparse deconvolution to address multipath interference and recover time profiles". In: *ACM Transactions on Graphics (TOG)* 32.6 (2013), p. 167.

[249] A. Kadambi, V. Taamazyan, S. Jayasuriya, and R. Raskar. "Frequency Domain TOF: Encoding Object Depth in Modulation Frequency". In: *CoRR* abs/1503.01804 (2015). URL: \url{http://arxiv.org/abs/1503.01804}.

[250] A. Kadambi, V. Taamazyan, B. Shi, and R. Raskar. "Polarized 3D: High-Quality Depth Sensing with Polarization Cues". In: *International Conference on Computer Vision (ICCV).* 2015.

[251] T. Kahlmann, F. Remondino, H. Ingensand, H.-G. Maas, and D. Schneider. *Calibration for increased accuracy of the range imaging camera SwissRanger*. 2006.

[252] S. Karlin and . Studden William J. *Tchebycheff systems: with applications in analysis and statistics*. English. Pure and applied mathematics. Interscience Publishers New York, 1966, xviii, 586 p.

[253] G. N. Karystinos and D. A. Pados. "New bounds on the total squared correlation and optimum design of DS-CDMA binary signature sets". In: *IEEE Transactions on Communications* 51 (1 2003), pp. 48–51. DOI: 10.1109/TCOMM.2002.807628.

[254] S. M. Kay. *Fundamentals of Statistical Signal Processing: Estimation Theory*. Upper Saddle River, NJ, USA: Prentice-Hall, Inc., 1993. ISBN: 0-13-345711-7.

[255] W. Kazmi, S. Foix, G. Alenyà, and H. J. Andersen. "Indoor and outdoor depth imaging of leaves with time-of-flight and stereo vision sensors: Analysis and comparison". In: *ISPRS Journal of Photogrammetry and Remote Sensing* 88.0 (2014), pp. 128 –146. ISSN: 0924-2716. DOI: http://dx.doi.org/10.1016/j.isprsjprs.2013.11.012. URL: http://www.sciencedirect.com/science/article/pii/S0924271613002748.

[256] M. Keller, J. Orthmann, A. Kolb, and V. Peters. "A Simulation Framework for Time-Of-Flight Sensors". In: *Signals, Circuits and Systems, 2007. ISSCS 2007. International Symposium on*. Vol. 1. July 2007, pp. 1–4. DOI: 10.1109/ISSCS.2007.4292667.

[257] M. Keller, D. Lefloch, M. Lambers, S. Izadi, T. Weyrich, and A. Kolb. "Real-time 3D reconstruction in dynamic scenes using point-based fusion". In: *3D Vision-3DV 2013, 2013 International Conference on*. IEEE. 2013, pp. 1–8.

[258] S. Kelly and M. O'Neill. "Chapter 1 - Liquid crystals for electro-optic applications". In: *Handbook of Advanced Electronic and Photonic Materials and Devices*. Ed. by H. S. Nalwa. Burlington: Academic Press, 2001, pp. 1 –66. ISBN: 978-0-12-513745-4. DOI: http://dx.doi.org/10.1016/B978-012513745-4/50057-3. URL: http://www.sciencedirect.com/science/article/pii/B9780125137454500573.

[259] H. Kharaghani and B. Tayfeh-Rezaie. "A Hadamard matrix of order 428". In: *Journal of Combinatorial Designs* 13.6 (2005), pp. 435–440. ISSN: 1520-6610. DOI: 10.1002/jcd.20043. URL: http://dx.doi.org/10.1002/jcd.20043.

[260] K. Khoshelham and S. O. Elberink. "Accuracy and resolution of kinect depth data for indoor mapping applications". In: *Sensors* 12.2 (2012), pp. 1437–1454.

[261] A. Kirmani, A. Benedetti, and P. Chou. "SPUMIC: Simultaneous phase unwrapping and multipath interference cancellation in time-of-flight cameras using spectral methods". In: *Multimedia and Expo (ICME), 2013 IEEE International Conference on*. July 2013, pp. 1–6. DOI: 10.1109/ICME.2013.6607553.

[262] A. Kirmani, A. Colaço, F. N. C. Wong, and V. K. Goyal. "Exploiting sparsity in time-of-flight range acquisition using a single time-resolved sensor". In: *Opt. Express* 19.22 (Oct. 2011), pp. 21485–21507. DOI: 10.1364/OE.19.021485. URL: http://www.opticsexpress.org/abstract.cfm?URI=oe-19-22-21485.

[263] A. Kirmani, A. Colaço, F. N. C. Wong, and V. K. Goyal. "CoDAC: A compressive depth acquisition camera framework." In: *ICASSP*. IEEE, 2012, pp. 5425–5428. ISBN: 978-1-4673-0046-9. URL: http://dblp.uni-trier.de/db/conf/icassp/icassp2012.html#KirmaniCWG12.

[264] A. Kirmani, T. Hutchison, J. Davis, and R. Raskar. "Looking Around the Corner Using Ultrafast Transient Imaging". In: *Int. J. Comput. Vision* 95.1 (Oct. 2011), pp. 13–28. ISSN: 0920-5691. DOI: 10.1007/s11263-011-0470-y. URL: http://dx.doi.org/10.1007/s11263-011-0470-y.

[265] Z. Knittl. *Optics of Thin Films*. A Wiley - Interscience Publication. John Wiley, 1976. URL: https://books.google.de/books?id=xVvOoAEACAAJ.

[266] D. D. Kosambi. "Statistics in function space". In: *Journal of the Indian Mathematical Society* 7 (1943), pp. 76–88. ISSN: 0019-5839.

[267] J. Kovačević and A. Chebira. "Life Beyond Bases: The Advent of Frames (Part I)". In: *IEEE Signal Processing Magazine* 24.4 (July 2007), pp. 86–104. ISSN: 1053-5888. DOI: 10.1109/MSP.2007.4286567.

[268] J. Kovačević and A. Chebira. "Life Beyond Bases: The Advent of Frames (Part II)". In: *IEEE Signal Processing Magazine* 24.5 (Sept. 2007), pp. 115–125. ISSN: 1053-5888. DOI: 10.1109/MSP.2007.904809.

[269] H. Kraft, J. Frey, T. Moeller, M. Albrecht, M. Grothof, B. Schink, H. Hess, and B. Buxbaum. "3D-Camera of High 3D-Frame Rate, Depth-Resolution and Background Light Elimination Based on Improved PMD (Photonic Mixer Device)-Technologies". In: 2004, pp. 95–100. URL: http://www2.uni-siegen.de/~reg-st2/3D-View/doc/Opto04_4.2.pdf.

[270] F. Krahmer and R. Ward. "New and Improved Johnson-Lindenstrauss Embeddings via the Restricted Isometry Property". In: *SIAM Journal on Mathematical Analysis* 43.3 (2011), pp. 1269–1281. DOI: 10.1137/100810447. eprint: http://dx.doi.org/10.1137/100810447. URL: http://dx.doi.org/10.1137/100810447.

[271] M. G. Kreĭn and A. A. Nudel'man. *The Markov moment problem and extremal problems: ideas and problems of P. L. Čebyšev and A. A. Markov and their further development.* English. Translations of mathematical monographs. translated from the Russian by Israel Program for Scientific Translations, translated by D. Louvish. Providence (R.I.): American Mathematical Society, 1977, v, 417 p. ; ISBN: 0-8218-4500-4.

[272] S. Krstulović and R. Gribonval. "Mptk: Matching Pursuit Made Tractable". In: *2006 IEEE International Conference on Acoustics Speech and Signal Processing Proceedings.* Vol. 3. May 2006, pp. III–III. DOI: 10.1109/ICASSP.2006.1660699.

[273] C. La and M. Do. "Tree-Based Orthogonal Matching Pursuit Algorithm for Signal Reconstruction". In: *Image Processing, 2006 IEEE International Conference on.* Oct. 2006, pp. 1277–1280. DOI: 10.1109/ICIP.2006.312578.

[274] D. Labate, W.-Q. Lim, G. Kutyniok, and G. Weiss. "Sparse multidimensional representation using shearlets". In: vol. 5914. 2005, 59140U–59140U–9. DOI: 10.1117/12.613494. URL: http://dx.doi.org/10.1117/12.613494.

[275] J. C. Lagarias, J. A. Reeds, M. H. Wright, and P. E. Wright. "Convergence Properties of the Nelder-Mead Simplex Method in Low Dimensions". In: *SIAM Journal of Optimization* 9 (1998), pp. 112–147.

[276] J. R. Lakowicz. *Principles of fluorescence spectroscopy.* New York: Springer, 2006. ISBN: 978-0-387-31278-1. URL: http://www.springer.com/in/book/9780387312781.

[277] M. Lambers, S. Hoberg, and A. Kolb. "Simulation of Time-of-Flight Sensors for Evaluation of Chip Layout Variants". In: *IEEE Sensors Journal* 15.7 (July 2015), pp. 4019–4026. ISSN: 1530-437X. DOI: 10.1109/JSEN.2015.2409816.

[278] R. Lange. "3D Time-of-Flight Distance Measurement with Custom Solid-State Image Sensors in CMOS/CCD-Technology". PhD thesis. Department of Electrical Engineering and Computer Science at University of Siegen, 2000. URL: http://dokumentix.ub.uni-siegen.de/opus/volltexte/2006/178/pdf/lange.pdf.

[279] R. Lange and P. Seitz. "Solid-state time-of-flight range camera". In: *Quantum Electronics, IEEE Journal of* 37.3 (Mar. 2001), pp. 390–397. ISSN: 0018-9197. DOI: 10.1109/3.910448.

[280] B. Langmann. *Wide Area 2D/3D Imaging: Development, Analysis and Applications.* SpringerLink : Bücher. Springer Fachmedien Wiesbaden, 2014. ISBN: 9783658064570. URL: http://www.springer.com/us/book/9783658064563.

[281] B. Langmann, K. Hartmann, and O. Loffeld. "Comparison of Depth Super-Resolution Methods for 2D/3D Images". In: *International Journal of Computer Information Systems and Industrial Management Applications* 3 (2011), pp. 635–645.

[282] B. Langmann, K. Hartmann, and O. Loffeld. "Depth Camera Technology Comparison and Performance Evaluation". In: *ICPRAM (2).* 2012, pp. 438–444.

[283] B. Langmann, K. Hartmann, and O. Loffeld. "Real-time Image Stabilization for ToF Cameras on Mobile Platforms". In: *A State-of-the-Art Survey on Time-of-Flight and Depth Imaging: Sensors, Algorithms, and Applications.* Ed. by M. Grzegorzek, C. Theobalt, R. Koch, and A. Kolb. Vol. 8200. Lecture Notes in Computer Science. Springer, 2013, pp. 289–301.

[284] B. Langmann, K. Hartmann, and O. Loffeld. "Increasing the Accuracy of Time-of-Flight Cameras for Machine Vision Applications". In: *Comput. Ind.* 64.9 (Dec. 2013), pp. 1090–1098. ISSN: 0166-3615. DOI: 10.1016/j.compind.2013.06.006. URL: http://dx.doi.org/10.1016/j.compind.2013.06.006.

[285] P.-J. Lapray, B. Heyrman, and D. Ginhac. "HDR-ARtiSt: an adaptive real-time smart camera for high dynamic range imaging". English. In: *Journal of Real-Time Image Processing* (2014), pp. 1–16. ISSN: 1861-8200. DOI: 10.1007/s11554-013-0393-7. URL: http://dx.doi.org/10.1007/s11554-013-0393-7.

[286] D. E. Lazić, H. Zörlein, and M. Bossert. "Low Coherence Sensing Matrices Based on Best Spherical Codes". In: *Systems, Communication and Coding (SCC), Proceedings of 2013 9th International ITG Conference on.* Jan. 2013, pp. 1–6.

[287] E. Le Pennec and S. Mallat. "Sparse geometric image representations with bandelets". In: *IEEE Transactions on Image Processing* 14.4 (Apr. 2005), pp. 423–438. ISSN: 1057-7149. DOI: 10.1109/TIP.2005.843753.

[288] H. Lee and D. Lee. "A gain-shape vector quantizer for image coding". In: *Acoustics, Speech, and Signal Processing, IEEE International Conference on ICASSP '86.* Vol. 11. Apr. 1986, pp. 141–144. DOI: 10.1109/ICASSP.1986.1169096.

[289] J. A. Leendertz. "Interferometric displacement measurement on scattering surfaces utilizing speckle effect". In: *Journal of Physics E: Scientific Instruments* 3.3 (1970), p. 214. URL: http://stacks.iop.org/0022-3735/3/i=3/a=312.

[290] D. Lefloch, T. Weyrich, and A. Kolb. "Anisotropic Point-Based Fusion". In: *International Conference on Information Fusion (FUSION).* ISIF. July 2015.

[291] M. Lehmann, T. Oggier, B. Büttgen, C. Gimkiewicz, M. Schweizer, R. Kaufmann, F. Lustenberger, and N. Blanc. *Smart Pixels for Future 3D-TOF Sensors.* June 2005. URL: http://ci.nii.ac.jp/naid/10017286586/en/.

[292] S. Lesage, R. Gribonval, F. Bimbot, and L. Benaroya. "Learning unions of orthonormal bases with thresholded singular value decomposition". In: *Proceedings. (ICASSP '05). IEEE International Conference on Acoustics, Speech, and Signal Processing, 2005.* Vol. 5. Mar. 2005, v/293–v/296 Vol. 5. DOI: 10.1109/ICASSP.2005.1416298.

[293] V. Levenshtein. "Bounds for packings of metric spaces and some of their applications". In: *Problemy Kibernet* 40 (1983). (in Russian), pp. 43–110.

[294] D. Leviatan and V. Temlyakov. “Simultaneous greedy approximation in Banach spaces”. In: *Journal of Complexity* 21.3 (2005), pp. 275–293. ISSN: 0885-064X. DOI: http://dx.doi.org/10.1016/j.jco.2004.09.004. URL: http://www.sciencedirect.com/science/article/pii/S0885064X04000871.

[295] S. Li, Q. Li, G. Li, X. He, and L. Chang. “Simultaneous Sensing Matrix and Sparsifying Dictionary Optimization for Block-sparse Compressive Sensing”. In: *2013 IEEE 10th International Conference on Mobile Ad-Hoc and Sensor Systems*. Oct. 2013, pp. 597–602. DOI: 10.1109/MASS.2013.98.

[296] Z. Li and B. V. K. V. Kumar. “A class of good quasi-cyclic low-density parity check codes based on progressive edge growth graph”. In: *Signals, Systems and Computers, 2004. Conference Record of the Thirty-Eighth Asilomar Conference on*. Vol. 2. Nov. 2004, 1990–1994 Vol.2. DOI: 10.1109/ACSSC.2004.1399513.

[297] J. S. Lim. *Two-dimensional Signal and Image Processing*. Upper Saddle River, NJ, USA: Prentice-Hall, Inc., 1990. ISBN: 0-13-935322-4.

[298] J. Lin, Y. Liu, M. B. Hullin, and Q. Dai. “Fourier Analysis on Transient Imaging by Multifrequency Time-of-Flight Camera”. In: *IEEE Conference on Computer Vision and Pattern Recognition (CVPR)*. June 2014.

[299] D. V. Lindley. “On a Measure of the Information Provided by an Experiment”. In: *Ann. Math. Statist.* 27.4 (Dec. 1956), pp. 986–1005. DOI: 10.1214/aoms/1177728069. URL: http://dx.doi.org/10.1214/aoms/1177728069.

[300] M. Lindner and A. Kolb. “Lateral and Depth Calibration of PMD-Distance Sensors”. In: *Advances in Visual Computing*. Vol. 2. Int. Symp. on Visual Computing. Springer, Nov. 2006, pp. 524–533.

[301] S. Lloyd. “Least squares quantization in PCM”. In: *IEEE Transactions on Information Theory* 28.2 (Mar. 1982), pp. 129–137. ISSN: 0018-9448. DOI: 10.1109/TIT.1982.1056489.

[302] O. E. Loepprich. “Lokalisation und Verfolgung von Personen in Echtzeit unter Verwendung kooperierender 2D / 3D-Kameras”. PhD thesis. Siegen, Germany: Fakultät IV: Naturwissenschaftlich-Technische Fakultät, 2014, p. 177. URL: http://dokumentix.ub.uni-siegen.de/opus/volltexte/2014/804/.

[303] M. G. Löfdahl and H. Eriksson. "Algorithm for resolving 2ĬĂ ambiguities in interferometric measurements by use of multiple wavelengths". In: *Optical Engineering* 40.6 (2001), pp. 984–990. DOI: 10.1117/1.1365936. URL: http://dx.doi.org/10.1117/1.1365936.

[304] O. Lottner. *Investigations of Optical 2D/3D-Imaging with Different Sensors and Illumination Configurations.* en. Vol. 29. ZESS-Forschungsberichte. Aachen, Germany: Shaker Verlag, 2011, p. 202. ISBN: 978-3-8440-0787-9. URL: http://www.shaker.de/de/content/catalogue/index.asp?lang=de&ID=6&category=102.

[305] H. Lu, X. Long, and J. Lv. "A Fast Algorithm for Recovery of Jointly Sparse Vectors based on the Alternating Direction Methods." In: *AISTATS.* Ed. by G. J. Gordon, D. B. Dunson, and M. Dudík. Vol. 15. JMLR Proceedings. JMLR.org, 2011, pp. 461–469. URL: http://dblp.uni-trier.de/db/journals/jmlr/jmlrp15.html\#LuLL11.

[306] W. Lu, K. Kpalma, and J. Ronsin. "Sparse Binary Matrices of LDPC Codes for Compressed Sensing". In: *Data Compression Conference (DCC), 2012.* Apr. 2012, pp. 405–405. DOI: 10.1109/DCC.2012.60.

[307] W. Lu, W. Li, K. Kpalma, and J. Ronsin. "Near-optimal Binary Compressed Sensing Matrix". In: *CoRR* abs/1304.4071 (2013). URL: http://arxiv.org/abs/1304.4071.

[308] Y. Lu and M. N. Do. "A New Contourlet Transform with Sharp Frequency Localization". In: *2006 International Conference on Image Processing.* Oct. 2006, pp. 1629–1632. DOI: 10.1109/ICIP.2006.312657.

[309] X. Luan. "Experimental Investigation of Photonic Mixer Device and Development of TOF 3D Ranging Systems Based on PMD Technology". English. PhD thesis. Siegen, Germany: Department of Electrical Engineering and Computer Science, 2001, p. 136.

[310] X. Luan, R. Schwarte, Z. Zhang, Z. Xu, H.-G. Heinol, B. Buxbaum, T. Ringbeck, and H. Hess. "Three-dimensional intelligent sensing based on the PMD technology". In: *Proc. SPIE.* Vol. 4540. 2001, pp. 482–487. DOI: 10.1117/12.450695. URL: http://dx.doi.org/10.1117/12.450695.

[311] A. Lutoborski and V. N. Temlyakov. "Vector greedy algorithms". In: *Journal of Complexity* 19.4 (2003), pp. 458 –473. ISSN: 0885-064X. DOI: http://dx.doi.org/10.1016/S0885-064X(03)00026-8. URL: http://www.sciencedirect.com/science/article/pii/S0885064X03000268.

[312] B. C. Madden. *Extended Intensity Range Imaging.* Technical Report MS-CIS-93-96. 401 Walnut Street, Room 301C, Philadelphia, PA 19104, USA: GRASP Laboratory, University of Pennsylvania, Dec. 1993.

[313] J. Mairal, G. Sapiro, and M. Elad. "Learning Multiscale Sparse Representations for Image and Video Restoration". In: *Multiscale Modeling & Simulation* 7.1 (2008), pp. 214–241. DOI: 10.1137/070697653. URL: http://dx.doi.org/10.1137/070697653.

[314] D. Malacara, M. Servín, and Z. Malacara. *Interferogram Analysis for Optical Testing.* Optical Science and Engineering. Taylor & Francis, 1998. ISBN: 9780824799403. URL: https://books.google.de/books?id=f3_zsjvWrZoC.

[315] S. G. Mallat. "A theory for multiresolution signal decomposition: the wavelet representation". In: *IEEE Transactions on Pattern Analysis and Machine Intelligence* 11.7 (July 1989), pp. 674–693. ISSN: 0162-8828. DOI: 10.1109/34.192463.

[316] S. G. Mallat and Z. Zhang. "Matching pursuits with time-frequency dictionaries". In: *IEEE Transactions on Signal Processing* 41.12 (Dec. 1993), pp. 3397–3415. ISSN: 1053-587X. DOI: 10.1109/78.258082.

[317] S. Mallat. "Geometrical grouplets". In: *Applied and Computational Harmonic Analysis* 26.2 (2009), pp. 161 –180. ISSN: 1063-5203. DOI: http://dx.doi.org/10.1016/j.acha.2008.03.004. URL: http://www.sciencedirect.com/science/article/pii/S1063520308000444.

[318] S. Mallat. *A Wavelet Tour of Signal Processing, Third Edition: The Sparse Way.* 3rd. Academic Press, 2008. ISBN: 0123743702, 9780123743701.

[319] R. Malz. *Codierte Lichtstrukturen für 3-D-Messtechnik und Inspektion.* Berichte aus dem Institut für Technische Optik der Universität Stuttgart. Inst. für Technische Optik, 1992. ISBN: 9783923560134. URL: https://books.google.de/books?id=cEt2tgAACAAJ.

[320] S. Mann and R. W. Picard. "On Being 'Undigital' With Digital Cameras: Extending Dynamic Range By Combining Differently Exposed Pictures". In: *Proceedings of IS&T.* 1995, pp. 442–448.

[321] Y. Mao and A. H. Banihashemi. "A heuristic search for good low-density parity-check codes at short block lengths". In: *Communications, 2001. ICC 2001. IEEE International Conference on.* Vol. 1. June 2001, 41–44 vol.1. DOI: 10.1109/ICC.2001.936269.

[322] L. Marcu, P. French, and D. Elson. *Fluorescence Lifetime Spectroscopy and Imaging: Principles and Applications in Biomedical Diagnostics.* CRC Press, 2014. ISBN: 9781439861684. URL: https://books.google.es/books?id=6xvcBQAAQBAJ.

[323] S. Mendelson, A. Pajor, and N. Tomczak-Jaegermann. "Uniform Uncertainty Principle for Bernoulli and Subgaussian Ensembles". In: *Constructive Approximation* 28.3 (2008), pp. 277–289. ISSN: 1432-0940. DOI: 10.1007/s00365-007-9005-8. URL: http://dx.doi.org/10.1007/s00365-007-9005-8.

[324] Y. Meyer. "Principe d'incertitude, bases hilbertiennes et algèbres d'opérateurs". fre. In: *Séminaire Bourbaki* 28 (1985-1986), pp. 209–223. URL: http://eudml.org/doc/110062.

[325] Y. Meyer. *Wavelets and Operators.* Vol. 1. Cambridge Books Online. Cambridge University Press, 1993. ISBN: 9780511623820. URL: http://dx.doi.org/10.1017/CBO9780511623820.

[326] A. Michelson. *Studies in Optics.* The Univ. of Chicago Science Series. University of Chicago Press, 1927. ISBN: 9780226523880. URL: https://books.google.de/books?id=FXazQgAACAAJ.

[327] D. A. B. Miller. "Quantum-well self-electro-optic effect devices". In: *Opt. Quantum Electron.* 22 (Feb. 1990), pp. 61–98. URL: https://ee.stanford.edu/~dabm/133.pdf.

[328] J. J. Miller. "The Inverse of the Freeman âĂŞ Tukey Double Arcsine Transformation". In: *The American Statistician* 32.4 (1978), pp. 138–138. DOI: 10.1080/00031305.1978.10479283. eprint: http://dx.doi.org/10.1080/00031305.1978.10479283. URL: http://dx.doi.org/10.1080/00031305.1978.10479283.

[329] F. Mochizuki, K. Kagawa, S.-i. Okihara, M.-W. Seo, B. Zhang, T. Takasawa, K. Yasutomi, and S. Kawahito. "6.4 Single-shot 200Mfps 5×3-aperture compressive CMOS imager." In: *2015 IEEE International Solid-State Circuits Conference - (ISSCC) Digest of Technical Papers*. IEEE, Feb. 2015, pp. 1–3. ISBN: 978-1-4799-6224-2. DOI: 10.1109/ISSCC.2015.7062953. URL: http://dx.doi.org/10.1109/ISSCC.2015.7062953.

[330] T. Möller, H. Kraft, J. Frey, M. Albrecht, and R. Lange. "Robust 3D Measurement with PMD Sensors". In: *Proceedings of the 1st Range Imaging Research Day at ETH*. Sept. 2005, pp. 3–906467.

[331] R. Morano, C. Ozturk, R. Conn, S. Dubin, S. Zietz, and J. Nissano. "Structured light using pseudorandom codes". In: *Pattern Analysis and Machine Intelligence, IEEE Transactions on* 20.3 (Mar. 1998), pp. 322–327. ISSN: 0162-8828. DOI: 10.1109/34.667888.

[332] C. J. Morgan. "Least-squares estimation in phase-measurement interferometry". In: *Opt. Lett.* 7.8 (Aug. 1982), pp. 368–370. DOI: 10.1364/OL.7.000368. URL: http://ol.osa.org/abstract.cfm?URI=ol-7-8-368.

[333] J. Morlet, G. Arens, E. Fourgeau, and D. Giard. "Wave propagation and sampling theory; Part I, Complex signal and scattering in multilayered media". In: *Geophysics* 47.2 (1982), pp. 203–221. ISSN: 0016-8033. DOI: 10.1190/1.1441328. eprint: http://geophysics.geoscienceworld.org/content/47/2/203.full.pdf. URL: http://geophysics.geoscienceworld.org/content/47/2/203.

[334] J. Morlet, G. Arens, E. Fourgeau, and D. Giard. "Wave propagation and sampling theory; Part II, Sampling theory and complex waves". In: *Geophysics* 47.2 (1982), pp. 222–236. ISSN: 0016-8033. DOI: 10.1190/1.1441329. eprint: http://geophysics.geoscienceworld.org/content/47/2/222.full.pdf. URL: http://geophysics.geoscienceworld.org/content/47/2/222.

[335] F. Mufti and R. Mahony. "Statistical analysis of signal measurement in time-of-flight cameras". In: *{ISPRS} Journal of Photogrammetry and Remote Sensing* 66.5 (2011), pp. 720 –731. ISSN: 0924-2716. DOI: http://dx.doi.org/10.1016/j.isprsjprs.2011.06.004. URL: http://www.sciencedirect.com/science/article/pii/S0924271611000724.

[336] K. K. Mukkavilli, A. Sabharwal, E. Erkip, and B. Aazhang. "On beamforming with finite rate feedback in multiple-antenna systems". In: *IEEE Transactions on Information Theory* 49.10 (Oct. 2003), pp. 2562–2579. ISSN: 0018-9448. DOI: 10.1109/TIT.2003.817433.

[337] S. Muthukrishnan. "Data Streams: Algorithms and Applications". In: *Proceedings of the Fourteenth Annual ACM-SIAM Symposium on Discrete Algorithms*. SODA '03. Baltimore, Maryland: Society for Industrial and Applied Mathematics, 2003, pp. 413–413. ISBN: 0-89871-538-5. URL: http://dl.acm.org/citation.cfm?id=644108.644174.

[338] N. Naik, A. Kadambi, C. Rhemann, S. Izadi, R. Raskar, and S. B. Kang. "A Light Transport Model for Mitigating Multipath Interference in Time-of-flight Sensors". In: CVPR, June 2015. URL: http://research.microsoft.com/apps/pubs/default.aspx?id=245068.

[339] B. K. Natarajan. "Sparse Approximate Solutions to Linear Systems". In: *SIAM Journal on Computing* 24.2 (1995), pp. 227–234. DOI: 10.1137/S0097539792240406. eprint: http://dx.doi.org/10.1137/S0097539792240406. URL: http://dx.doi.org/10.1137/S0097539792240406.

[340] S. K. Nayar and V. Branzoi. "Adaptive dynamic range imaging: Optical control of pixel exposures over space and time". In: *ICCV*. 2003, pp. 1168–1175.

[341] S. K. Nayar, G. Krishnan, M. D. Grossberg, and R. Raskar. "Fast Separation of Direct and Global Components of a Scene Using High Frequency Illumination". In: *ACM Trans. Graph.* 25.3 (July 2006), pp. 935–944. ISSN: 0730-0301. DOI: 10.1145/1141911.1141977. URL: http://doi.acm.org/10.1145/1141911.1141977.

[342] S. K. Nayar and T. Mitsunaga. "High dynamic range imaging: spatially varying pixel exposures". In: *Computer Vision and Pattern Recognition, 2000. Proceedings. IEEE Conference on*. Vol. 1. IEEE, 2000, 472–479 vol.1. ISBN: 0-7695-0662-3. DOI: 10.1109/cvpr.2000.855857. URL: http://dx.doi.org/10.1109/cvpr.2000.855857.

[343] S. K. Nayar and T. Mitsunaga. "Method and apparatus for obtaining high dynamic range images". Pat. US8610789 B1. May 2011. URL: https://www.google.com.tr/patents/US8610789.

[344] D. Needell and J. Tropp. "CoSaMP: Iterative signal recovery from incomplete and inaccurate samples". In: *Applied and Computational Harmonic Analysis* 26.3 (2009), pp. 301 –321. ISSN: 1063-5203. DOI: http://dx.doi.org/10.1016/j.acha.2008.07.002. URL: http://www.sciencedirect.com/science/article/pii/S1063520308000638.

[345] D. Needell and R. Vershynin. "Signal Recovery From Incomplete and Inaccurate Measurements Via Regularized Orthogonal Matching Pursuit". In: *IEEE Journal of Selected Topics in Signal Processing* 4.2 (Apr. 2010), pp. 310–316. ISSN: 1932-4553. DOI: 10.1109/JSTSP.2010.2042412.

[346] D. Needell and J. A. Tropp. "CoSaMP: Iterative Signal Recovery from Incomplete and Inaccurate Samples". In: *Commun. ACM* 53.12 (Dec. 2010), pp. 93–100. ISSN: 0001-0782. DOI: 10.1145/1859204.1859229. URL: http://doi.acm.org/10.1145/1859204.1859229.

[347] D. Needell and R. Vershynin. "Uniform Uncertainty Principle and Signal Recovery via Regularized Orthogonal Matching Pursuit". In: *Found. Comput. Math.* 9.3 (Apr. 2009), pp. 317–334. ISSN: 1615-3375. DOI: 10.1007/s10208-008-9031-3. URL: http://dx.doi.org/10.1007/s10208-008-9031-3.

[348] M. Neubauer and A. Radcliffe. "The maximum determinant of ±1 matrices". In: *Linear Algebra and its Applications* 257 (1997), pp. 289 –306. ISSN: 0024-3795. DOI: http://dx.doi.org/10.1016/S0024-3795(96)00147-4. URL: http://www.sciencedirect.com/science/article/pii/S0024379596001474.

[349] R. A. Newcombe, S. Izadi, O. Hilliges, D. Molyneaux, D. Kim, A. J. Davison, P. Kohli, J. Shotton, S. Hodges, and A. Fitzgibbon. "KinectFusion: Real-time Dense Surface Mapping and Tracking". In: *Proceedings of the 2011 10th IEEE International Symposium on Mixed and Augmented Reality*. ISMAR '11. Washington, DC, USA: IEEE Computer Society, 2011, pp. 127–136. ISBN: 978-1-4577-2183-0. DOI: 10.1109/ISMAR.2011.6092378. URL: http://dx.doi.org/10.1109/ISMAR.2011.6092378.

[350] T. L. N. Nguyen and Y. Shin. "Deterministic sensing matrices in compressive sensing: a survey". In: *The Scientific World Journal* 2013 (2013), p. 192795. ISSN: 1537-744X. URL: http://www.biomedsearch.com/nih/Deterministic-sensing-matrices-in-compressive/24348141.html.

[351] C. Niclass, A. Rochas, P.-A. Besse, and E. Charbon. "Design and characterization of a CMOS 3-D image sensor based on single photon avalanche diodes". In: *Solid-State Circuits, IEEE Journal of* 40.9 (Sept. 2005), pp. 1847–1854. ISSN: 0018-9200. DOI: 10.1109/JSSC.2005.848173.

[352] C. Niclass, A. Rochas, P.-A. Besse, and C. Charbon. "A CMOS single photon avalanche diode array for 3D imaging". In: *Digest of Technical Papers - ISSCC.* Vol. 47. San Francisco - CA, 2004, pp. 90–91.

[353] H. Nyquist. "Certain Topics in Telegraph Transmission Theory". In: *Transactions of the American Institute of Electrical Engineers* 47.2 (Apr. 1928), pp. 617–644. ISSN: 0096-3860. DOI: 10.1109/T-AIEE.1928.5055024.

[354] T. Oggier, B. Büttgen, F. Lustenberger, G. Becker, B. Rüegg, and A. Hodac. "Swissranger SR3000 and first experiences based on miniaturized 3D-TOF cameras". In: *Proceedings of the 1st Range Imaging Research Day at ETH.* Sept. 2005.

[355] T. Oggier, M. Lehmann, R. Kaufmann, M. Schweizer, M. Richter, P. Metzler, G. Lang, F. Lustenberger, and N. Blanc. "An all-solid-state optical range camera for 3D real-time imaging with sub-centimeter depth resolution (SwissRanger)". In: *Proc. SPIE.* Vol. 5249. 2004, pp. 534–545. DOI: 10.1117/12.513307. URL: http://dx.doi.org/10.1117/12.513307.

[356] K. Ogura. "On a Certain Transcendental Integral Function In the Theory of Interpolation". In: *Tohoku Mathematical Journal, First Series* 17 (1920), pp. 64–72.

[357] B. A. Olshausen. "Sparse coding of time-varying natural images". In: *Journal of Vision* 2.7 (Nov. 2002), pp. 130–130. ISSN: 1534-7362. DOI: 10.1167/2.7.130.

[358] *OPT8241 3D Time-of-Flight Sensor.* http://www.ti.com/lit/ds/symlink/opt8241.pdf. [Online; accessed 06-May-2016]. 2015.

[359] M. O'Toole, F. Heide, L. Xiao, M. B. Hullin, W. Heidrich, and K. N. Kutulakos. "Temporal Frequency Probing for 5D Transient Analysis of Global Light Transport". In: *ACM Trans. Graph.* 33.4 (July 2014), 87:1–87:11. ISSN: 0730-0301. DOI: 10.1145/2601097.2601103. URL: \url{http://doi.acm.org/10.1145/2601097.2601103}.

[360] M. O'Toole, R. Raskar, and K. N. Kutulakos. "Primal-dual Coding to Probe Light Transport". In: *ACM Trans. Graph.* 31.4 (July 2012), 39:1–39:11. ISSN: 0730-0301. DOI: 10.1145/2185520.2185535. URL: http://doi.acm.org/10.1145/2185520.2185535.

[361] Y. Pati, R. Rezaiifar, and P. Krishnaprasad. "Orthogonal matching pursuit: recursive function approximation with applications to wavelet decomposition". In: *Signals, Systems and Computers, Twenty-Seventh Asilomar Conference on.* Nov. 1993, 40–44 vol.1. DOI: 10.1109/ACSSC.1993.342465.

[362] A. Payne et al. "7.6 A 512×424 CMOS 3D Time-of-Flight image sensor with multi-frequency photo-demodulation up to 130MHz and 2GS/s ADC". In: *Solid-State Circuits Conference Digest of Technical Papers (ISSCC), 2014 IEEE International.* Feb. 2014, pp. 134–135. DOI: 10.1109/ISSCC.2014.6757370.

[363] A. D. Payne, A. A. Dorrington, M. J. Cree, and D. A. Carnegie. "Improved measurement linearity and precision for AMCW time-of-flight range imaging cameras". In: *Appl. Opt.* 49.23 (Aug. 2010), pp. 4392–4403. DOI: 10.1364/AO.49.004392. URL: http://ao.osa.org/abstract.cfm?URI=ao-49-23-4392.

[364] A. D. Payne, A. P. Jongenelen, D. A. A. Dorrington, M. J. Cree, and P. D. A. Carnegie. "Multiple Frequency Range Imaging to Remove Measurement Ambiguity". In: *9th Conference on Optical 3-D Measurement Techniques.* 2009, pp. 139 –148. ISBN: 978-3-9501492-5-8.

[365] W. B. Pennebaker and J. L. Mitchell. *JPEG Still Image Data Compression Standard.* 1st. Norwell, MA, USA: Kluwer Academic Publishers, 1992. ISBN: 0442012721.

[366] M. Perenzoni, N. Massari, D. Stoppa, L. Pancheri, M. Malfatti, and L. Gonzo. "A 160 × 120-pixels range camera with on-pixel correlated double sampling and nonuniformity correction in 29.1 μm pitch". In: *ESSCIRC, 2010 Proceedings of the.* Sept. 2010, pp. 294–297. DOI: 10.1109/ESSCIRC.2010.5619836.

[367] C. Peters, J. Klein, M. B. Hullin, and R. Klein. "Solving Trigonometric Moment Problems for Fast Transient Imaging". In: *ACM Trans. Graph. (Proc. SIGGRAPH Asia)* 34.6 (Nov. 2015). DOI: 10.1145/2816795.2818103.

[368] V. Peters. "Phänomenologische Modellierung und multistatische Simulation von Time-of-Flight 3D PMD Kameras". PhD thesis. Siegen, Germany: Fakultät IV: Naturwissenschaftlich-Technische Fakultät, 2013. URL: http://dokumentix.ub.uni-siegen.de/opus/volltexte/2013/700/.

[369] G. Peyré and S. Mallat. "Surface compression with geometric bandelets". In: *ACM Transactions on Graphics (Proc. SIGGRAPH'2005)* 24.3 (July 2005), pp. 601–608. ISSN: 0730-0301. DOI: doi:10.1145/1073204.1073236. URL: http://www.cmap.polytechnique.fr/~mallat/papiers/PeyreMallatSIGGRAPH05.pdf.

[370] J. Philip and K. Carlsson. "Theoretical investigation of the signal-to-noise ratio in fluorescence lifetime imaging". In: *J. Opt. Soc. Am. A* 20.2 (Feb. 2003), pp. 368–379. DOI: 10.1364/JOSAA.20.000368. URL: http://josaa.osa.org/abstract.cfm?URI=josaa-20-2-368.

[371] *Pixelated Polarizers*. http://moxtek.com/wp-content/uploads/pdfs/Pixelated-Polarizers-OPT-DATA-10052.pdf. [Online; accessed 01-March-2016]. 2016.

[372] M. Plaue. *Analysis of the PMD Imaging System*. Technical Report. Interdisciplinary Center for Scientific Computing, University of Heidelberg, 2006.

[373] M. D. Plumbley, S. A. Abdallah, T. Blumensath, and M. E. Davies. "Sparse Representations of Polyphonic Music". In: *Signal Process.* 86.3 (Mar. 2006), pp. 417–431. ISSN: 0165-1684. DOI: 10.1016/j.sigpro.2005.06.007. URL: http://dx.doi.org/10.1016/j.sigpro.2005.06.007.

[374] *PMD Photonics 19k-S3*. http://pmdtec.com/html/pdf/pmdPhotonICs_19k_S3.pdf. [Online; accessed 21-March-2014]. 2014.

[375] J. Portilla. "Image restoration through l0 analysis-based sparse optimization in tight frames". In: *2009 16th IEEE International Conference on Image Processing (ICIP)*. Nov. 2009, pp. 3909–3912. DOI: 10.1109/ICIP.2009.5413975.

[376] J. Posdamer and M. Altschuler. "Surface measurement by space-encoded projected beam systems". In: *Computer Graphics and Image Processing* 18.1 (1982), pp. 1 –17. ISSN: 0146-664X. DOI: http://dx.doi.org/10.1016/0146-664X(82)90096-X. URL: http://www.sciencedirect.com/science/article/pii/0146664X8290096X.

[377] T. D. A. Prasad, K. Hartmann, W. Weihs, S. E. Ghobadi, and A. Sluiter. "First steps in enhancing 3D vision technique using 2D/3D sensors". In: *CVWW'06: Proceedings of the Computer Vision Winter Workshop 2006*. Ed. by O. Chum and V. Franc. Prague, Czech Republic: Czech Society for Cybernetics and Informatics, Feb. 2006, pp. 82–86.

[378] E. Prati. "Propagation in gyroelectromagnetic guiding systems". In: *Journal of Electromagnetic Waves and Applications* 17.8 (2003), pp. 1177–1196. DOI: 10.1163/156939303322519810. eprint: http://dx.doi.org/10.1163/156939303322519810. URL: http://dx.doi.org/10.1163/156939303322519810.

[379] M. Proesmans, L. Van Gool, and A. Oosterlinck. "One-shot active 3D shape acquisition". In: *Pattern Recognition, 1996., Proceedings of the 13th International Conference on*. Vol. 3. Aug. 1996, 336–340 vol.3. DOI: 10.1109/ICPR.1996.546966.

[380] R. Pryputniewicz and E. Pryputniewicz. "Determination of Dynamic Characteristics of MEMS Engines Rotating at High Speeds". In: *Proceedings of the IMAC-XXII Conference & Exposition on Structural Dynamics - Linking Test to Design*. 2004. URL: \url{http://sem-proceedings.com/22i/sem.org-IMAC-XXII-Conf-s21p04-Determination-Dynamic-Characteristics-MEMS-Engines-Rotating-High.pdf}.

[381] S. Qian and D. Chen. "Discrete Gabor transform". In: *IEEE Transactions on Signal Processing* 41.7 (July 1993), pp. 2429–2438. ISSN: 1053-587X. DOI: 10.1109/78.224251.

[382] J. Radon. "Über die Bestimmung von Funktionen durch ihre Integralwerte längs gewisser Mannigfaltigkeiten". In: *Berichte über die Verhandlungen der Königlich-Sächsischen Akademie der Wissenschaften zu Leipzig, Mathematisch-Physische Klasse* 69 (1917), pp. 262–277.

[383] H. Rapp, M. Frank, F. A. Hamprecht, and B. Jahne. "A Theoretical and Experimental Investigation of the Systematic Errors and Statistical Uncertainties of Time-Of-Flight-Cameras". In: *Int. J. Intell. Syst. Technol. Appl.* 5.3/4 (Nov. 2008), pp. 402–413. ISSN: 1740-8865. DOI: 10.1504/IJISTA.2008.021303. URL: http://dx.doi.org/10.1504/IJISTA.2008.021303.

[384] H. Rauhut. “On the impossibility of uniform sparse reconstruction using greedy methods”. In: *Sampl. Theory Signal Image Process.* 7.2 (2008), pp. 197–215. ISSN: 1530-6429.

[385] *REAL3^{TM} image sensor family - 3D depth sensing based on Time-of-Flight.* http://www.infineon.com/dgdl/Infineon-REAL3+Image+Sensor+Family-PB-v01_00-EN.PDF?fileId=5546d462518ffd850151a0afc2302a58. [Online; accessed 04-February-2016]. Nov. 2015.

[386] A. Reichinger. *Kinect Pattern Uncovered.* https://azttm.wordpress.com/2011/04/03/kinect-pattern-uncovered/. [Online; accessed 12-August-2015]. 2015.

[387] Ringbeck and B. Hagebeuker. “A 3D time of flight camera for object detection”. In: *Optical 3-D Measurement Techniques, ETH Zürich.* 2007. URL: http://www.ifmefector.biz/obj/O1D_Paper-PMD.pdf.

[388] T. Ringbeck, T. Möller, and B. Hagebeuker. “Multidimensional measurement by using 3-D PMD sensors”. In: *Advances in Radio Science* 5 (2007), pp. 135–146. DOI: 10.5194/ars-5-135-2007. URL: http://www.adv-radio-sci.net/5/135/2007/.

[389] R. Rubinstein, A. M. Bruckstein, and M. Elad. “Dictionaries for Sparse Representation Modeling”. In: *Proceedings of the IEEE* 98.6 (June 2010), pp. 1045–1057. ISSN: 0018-9219. DOI: 10.1109/JPROC.2010.2040551.

[390] R. Rubinstein, M. Zibulevsky, and M. Elad. “Double Sparsity: Learning Sparse Dictionaries for Sparse Signal Approximation”. In: *IEEE Transactions on Signal Processing* 58.3 (Mar. 2010), pp. 1553–1564. ISSN: 1053-587X. DOI: 10.1109/TSP.2009.2036477.

[391] M. Rudelson and R. Vershynin. “On sparse reconstruction from Fourier and Gaussian measurements”. In: *Communications on Pure and Applied Mathematics* 61.8 (2008), pp. 1025–1045. ISSN: 1097-0312. DOI: 10.1002/cpa.20227. URL: http://dx.doi.org/10.1002/cpa.20227.

[392] K. Rupp. “GPU-Accelerated Non-negative Matrix Factorization for Text Mining”. In: *NVIDIA GPU Technology Conference 2012.* 2012, p. 77.

[393] K. Rupp, F. Rudolf, and J. Weinbub. "ViennaCL - A High Level Linear Algebra Library for GPUs and Multi-Core CPUs". In: *Intl. Workshop on GPUs and Scientific Applications.* 2010, pp. 51–56.

[394] E. B. Saff and A. B. J. Kuijlaars. "Distributing many points on a sphere". In: *The Mathematical Intelligencer* 19.1 (1997), pp. 5–11. ISSN: 0343-6993. DOI: 10.1007/BF03024331. URL: http://dx.doi.org/10.1007/BF03024331.

[395] P. Sallee and B. A. Olshausen. "Learning Sparse Multiscale Image Representations". In: *Advances in Neural Information Processing Systems 15.* Ed. by S. Thrun and K. Obermayer. Cambridge, MA: MIT Press, 2002, pp. 1327–1334. URL: http://books.nips.cc/papers/files/nips15/VS11.pdf.

[396] H. Sarbolandi, D. Lefloch, and A. Kolb. "Kinect range sensing: Structured-light versus Time-of-Flight Kinect". In: *Computer Vision and Image Understanding* 139 (2015), pp. 1 –20. ISSN: 1077-3142. DOI: http://dx.doi.org/10.1016/j.cviu.2015.05.006. URL: http://www.sciencedirect.com/science/article/pii/S1077314215001071.

[397] D. Scharstein and R. Szeliski. "High-accuracy Stereo Depth Maps Using Structured Light". In: *Proceedings of the 2003 IEEE Computer Society Conference on Computer Vision and Pattern Recognition.* CVPR'03. Madison, Wisconsin: IEEE Computer Society, 2003, pp. 195–202. ISBN: 0-7695-1900-8, 978-0-7695-1900-5. URL: http://dl.acm.org/citation.cfm?id=1965841.1965865.

[398] M. Schmidt and B. Jähne. "A Physical Model of Time-of-Flight 3D Imaging Systems, Including Suppression of Ambient Light". In: *Dynamic 3D Imaging.* Ed. by A. Kolb and R. Koch. Vol. 5742. Lecture Notes in Computer Science. Springer, 2009, pp. 1–15. ISBN: 978-3-642-03777-1.

[399] R. Schmidt. "Multiple emitter location and signal parameter estimation". In: *IEEE Transactions on Antennas and Propagation* 34.3 (Mar. 1986), pp. 276–280. ISSN: 0018-926X. DOI: 10.1109/TAP.1986.1143830.

[400] T. J. Schulz and D. L. Snyder. "Image recovery from correlations". In: *J. Opt. Soc. Am. A* 9.8 (Aug. 1992), pp. 1266–1272. DOI: 10.1364/JOSAA.9.001266. URL: http://josaa.osa.org/abstract.cfm?URI=josaa-9-8-1266.

[401] T. J. Schulz and D. G. Voelz. "Signal recovery from autocorrelation and cross-correlation data". In: *J. Opt. Soc. Am. A* 22.4 (Apr. 2005), pp. 616–624. DOI: 10.1364/JOSAA.22.000616. URL: http://josaa.osa.org/abstract.cfm?URI=josaa-22-4-616.

[402] R. Schwarte. "Verfahren und vorrichtung zur bestimmung der phasen- und/oder amplitudeninformation einer elektromagnetischen welle". Pat. WO Patent App. PCT/DE1997/001,956. Mar. 1998. URL: http://www.google.com/patents/WO1998010255A1?cl=un.

[403] R. Schwarte. "Dynamic 3D-Vision". In: *Electron Devices for Microwave and Optoelectronic Applications, 2001 International Symposium on.* 2001, pp. 241–248. DOI: 10.1109/EDMO.2001.974314.

[404] R. Schwarte, H.-G. Heinol, B. Buxbaum, T. Ringbeck, Z. Xu, and K. Hartmann. "Principles of three-dimensional imaging techniques". In: *Sensors and imaging. - (Handbook of computer vision and applications ; Vol. 1).* Ed. by B. Jähne. 1999. ISBN: 0-12-379771-3 (Einzelbd.)

[405] R. Schwarte, H.-G. Heinol, Z. Xu, and K. Hartmann. "New active 3D vision system based on rf-modulation interferometry of incoherent light". In: *Proc. SPIE.* Vol. 2588. 1995, pp. 126–134. DOI: 10.1117/12.222664. URL: http://dx.doi.org/10.1117/12.222664.

[406] R. Schwarte, Z. Xu, H.-G. Heinol, J. Olk, R. Klein, B. Buxbaum, H. Fischer, and J. Schulte. "New electro-optical mixing and correlating sensor: facilities and applications of the photonic mixer device (PMD)". In: *Proc. SPIE.* Vol. 3100. 1997, pp. 245–253. DOI: 10.1117/12.287751. URL: http://dx.doi.org/10.1117/12.287751.

[407] R. M. Schwarte. "Breakthrough in multichannel laser-radar technology providing thousands of high-sensitive lidar receivers on a chip". In: *Proc. SPIE.* Vol. 5575. 2004, pp. 126–136. DOI: 10.1117/12.573727. URL: http://dx.doi.org/10.1117/12.573727.

[408] R. M. Schwarte. "Real Time 3D-perception by TOF-echoing 3D-Video Cameras". In: *Dynamic Perception: Workshop of the GI Section 'computer Vision,' Eberhard Karls University Tübingen, Max Planck Institute for Biological Cybernetics.* Ed. by U. Ilg, H. Bülthoff, and H. Mallot. Tübingen: IOS Press, Incorporated, Nov. 2004, pp. 217–226. ISBN: 9781586034801. URL: http://www2.uni-siegen.de/~reg-st2/3D-View/doc/DynPerc28082004_R_Schwarte.pdf.

[409] U. Seger, U. Apel, and B. Höfflinger. "HDRC-Imagers for Natural Visual Perception". In: *Handbook of Computer Vision and Applications.* Ed. by B. Jahne, P. Geissler, and H. Haussecker. 1st. Vol. 1: Sensors and Imaging. San Diego, CA, USA: Academic Press, 1999, pp. 223–235. ISBN: 0-12-379771-3.

[410] U. Seger, H.-G. Graf, and M. Landgraf. "Vision assistance in scenes with extreme contrast". In: *Micro, IEEE* 13.1 (Feb. 1993), pp. 50–56. ISSN: 0272-1732. DOI: 10.1109/40.210524.

[411] P. Seitz. "Unified analysis of the performance and physical limitations of optical rangeimaging techniques". In: *Proceedings of the 1st Range Imaging Research Day at ETH.* Sept. 2005, pp. 9–19.

[412] C. E. Shannon. "A Mathematical Theory of Communication". In: *SIGMOBILE Mob. Comput. Commun. Rev.* 5.1 (Jan. 2001), pp. 3–55. ISSN: 1559-1662. DOI: 10.1145/584091.584093. URL: http://doi.acm.org/10.1145/584091.584093.

[413] C. Shannon. "Communication in the Presence of Noise". In: *Proceedings of the IRE* 37.1 (Jan. 1949), pp. 10–21. ISSN: 0096-8390. DOI: 10.1109/JRPROC.1949.232969.

[414] J. M. Shapiro. "Embedded image coding using zerotrees of wavelet coefficients". In: *IEEE Transactions on Signal Processing* 41.12 (Dec. 1993), pp. 3445–3462. ISSN: 1053-587X. DOI: 10.1109/78.258085.

[415] B. Sharp. "Special Issue on Holography and Speckle Metrology Electronic speckle pattern interferometry (ESPI)". In: *Optics and Lasers in Engineering* 11.4 (1989), pp. 241 –255. ISSN: 0143-8166. DOI: http://dx.doi.org/10.1016/0143-8166(89)90062-6. URL: http://www.sciencedirect.com/science/article/pii/0143816689900626.

[416] M. J. Shensa. "The discrete wavelet transform: wedding the a trous and Mallat algorithms". In: *IEEE Transactions on Signal Processing* 40.10 (Oct. 1992), pp. 2464–2482. ISSN: 1053-587X. DOI: 10.1109/78.157290.

[417] M.-H. Shin, J.-S. Kim, and H.-Y. Song. "Generalization of Tanner's minimum distance bounds for LDPC codes". In: *IEEE Communications Letters* 9.3 (Mar. 2005), pp. 240–242. ISSN: 1089-7798. DOI: 10.1109/LCOMM.2005.03002.

[418] A. Shpunt and Z. Zalevsky. "Depth-varying light fields for three dimensional sensing". Pat. US Patent App. 11/724,068. May 2008. URL: http://www.google.com/patents/US20080106746.

[419] E. P. Simoncelli, W. T. Freeman, E. H. Adelson, and D. J. Heeger. "Shiftable multiscale transforms". In: *IEEE Transactions on Information Theory* 38.2 (Mar. 1992), pp. 587–607. ISSN: 0018-9448. DOI: 10.1109/18.119725.

[420] A. Spickermann, D. Durini, A. Süss, W. Ulfig, W. Brockherde, B. J. Hosticka, S. Schwope, and A. Grabmaier. "CMOS 3D image sensor based on pulse modulated time-of-flight principle and intrinsic lateral drift-field photodiode pixels". In: *37th European Solid State Circuits Conference (ESSCIRC) 2011, Proceedings of the.* Sept. 2011, pp. 111–114. DOI: 10.1109/ESSCIRC.2011.6044927.

[421] T. Spirig. *Smart CCD/CMOS Based Image Sensors with Programmable, Real-time, Temporal and Spatial Convolution Capabilities for Applications in Machine Vision and Optical Metrology.* 1997. URL: https://books.google.de/books?id=hK3QNwAACAAJ.

[422] T. Spirig, M. Marley, and P. Seitz. "The multitap lock-in CCD with offset subtraction". In: *Electron Devices, IEEE Transactions on* 44.10 (Oct. 1997), pp. 1643–1647. ISSN: 0018-9383. DOI: 10.1109/16.628816.

[423] T. Spirig and P. Seitz. "Vorrichtung und verfahren zur detektion und demodulation eines intensitätsmodulierten strahlungsfeldes". Pat. WO Patent App. PCT/EP1995/004,235. May 1996. URL: https://www.google.com/patents/WO1996015626A1?cl=un.

[424] C. Starr, C. Evers, and L. Starr. *Biology: Concepts and Applications.* Brooks/Cole biology series. Thomson, Brooks/Cole, 2006. ISBN: 9780534462239. URL: https://books.google.de/books?id=RtSpGV_Pl_0C.

[425] T. R. Stephan Hussmann and B. Hagebeuker. "A Performance Review of 3D TOF Vision Systems in Comparison to Stereo Vision Systems". In: *Stereo Vision.* Ed. by A. Bhatti. Janeza Trdine 9, 51000 Rijeka, Croatia: InTech, 2008. ISBN: 978-953-7619-22-0. DOI: 10.5772/5898. URL: http://www.intechopen.com/books/stereo_vision/a_performance_review_of_3d_tof_vision_systems_in_comparison_to_stereo_vision_systems.

[426] M. Stojnic, F. Parvaresh, and B. Hassibi. "On the Reconstruction of Block-Sparse Signals With an Optimal Number of Measurements". In: *IEEE Transactions on Signal Processing* 57.8 (Aug. 2009), pp. 3075–3085. ISSN: 1053-587X. DOI: 10.1109/TSP.2009.2020754.

[427] D. Stoppa, L. Viarani, A. Simoni, L. Gonzo, M. Malfatti, and G. Pedretti. "A 50 × 30-pixel CMOS Sensor for TOF-based Real Time 3D Imaging". In: *In Proceedings of the 2005 Workshop on Charge-Coupled Devices and Advanced Image Sensors.* Karuizawa, Nagano, Japan, 2005.

[428] R. Swcharte. "Method and device for the recording and processing of signal waves". Pat. EP Patent 1,332,594. Feb. 2001. URL: http://www.google.com/patents/EP1332594B1?cl=en.

[429] P. T. Sylvain Sardy Andrew G. Bruce. "Block Coordinate Relaxation Methods for Nonparametric Wavelet Denoising". In: *Journal of Computational and Graphical Statistics* 9.2 (2000), pp. 361–379. ISSN: 10618600. URL: http://www.jstor.org/stable/1390659.

[430] S. J. Szarek. "Condition numbers of random matrices". In: *Journal of Complexity* 7.2 (1991), pp. 131 –149. ISSN: 0885-064X. DOI: http://dx.doi.org/10.1016/0885-064X(91)90002-F. URL: http://www.sciencedirect.com/science/article/pii/0885064X9190002F.

[431] M. Taibleson. "Shorter Notes: Fourier Coefficients of Functions of Bounded Variation". In: *Proceedings of the American Mathematical Society* 18.4 (1967), pp. 766–766. ISSN: 00029939, 10886826. URL: http://www.jstor.org/stable/2035460.

[432] R. Tanner. "A recursive approach to low complexity codes". In: *IEEE Transactions on Information Theory* 27.5 (Sept. 1981), pp. 533–547. ISSN: 0018-9448. DOI: 10.1109/TIT.1981.1056404.

[433] R. M. Tanner. "Minimum-distance bounds by graph analysis". In: *IEEE Transactions on Information Theory* 47.2 (Feb. 2001), pp. 808–821. ISSN: 0018-9448. DOI: 10.1109/18.910591.

[434] H. Tian and S. U. D. of Applied Physics. "Noise analysis in CMOS image sensors". PhD thesis. 2000. URL: http://www-isl.stanford.edu/~abbas/group/papers_and_pub/hui_thesis.pdf.

[435] C. Tomasi and R. Manduchi. "Bilateral filtering for gray and color images". In: *Computer Vision, 1998. Sixth International Conference on.* Jan. 1998, pp. 839–846. DOI: 10.1109/ICCV.1998.710815.

[436] J. A. Tropp. "Greed is good: algorithmic results for sparse approximation". In: *IEEE Transactions on Information Theory* 50.10 (Oct. 2004), pp. 2231–2242. ISSN: 0018-9448. DOI: 10.1109/TIT.2004.834793.

[437] J. A. Tropp. "Just relax: convex programming methods for identifying sparse signals in noise". In: *IEEE Transactions on Information Theory* 52.3 (Mar. 2006), pp. 1030–1051. ISSN: 0018-9448. DOI: 10.1109/TIT.2005.864420.

[438] J. A. Tropp, A. C. Gilbert, and M. J. Strauss. "Simultaneous sparse approximation via greedy pursuit". In: *Proceedings. (ICASSP '05). IEEE International Conference on Acoustics, Speech, and Signal Processing, 2005.* Vol. 5. Mar. 2005, v/721–v/724 Vol. 5. DOI: 10.1109/ICASSP.2005.1416405.

[439] J. Tropp and A. Gilbert. "Signal Recovery From Random Measurements Via Orthogonal Matching Pursuit". In: *Information Theory, IEEE Transactions on* 53.12 (Dec. 2007), pp. 4655–4666. ISSN: 0018-9448. DOI: 10.1109/TIT.2007.909108.

[440] J. A. Tropp, A. C. Gilbert, and M. J. Strauss. "Algorithms for Simultaneous Sparse Approximation: Part I: Greedy Pursuit". In: *Signal Process.* 86.3 (Mar. 2006), pp. 572–588. ISSN: 0165-1684. DOI: 10.1016/j.sigpro.2005.05.030. URL: http://dx.doi.org/10.1016/j.sigpro.2005.05.030.

[441] Y. T. Tsai. "Method and apparatus for extending the dynamic range of an electronic imaging system". Pat. US5309243 A. May 1994. URL: http://www.google.com.tr/patents/US5309243.

[442] C. Urquhart and J. Siebert. "Development of a precision active stereo system". In: *Intelligent Control, 1992., Proceedings of the 1992 IEEE International Symposium on.* Aug. 1992, pp. 354–359. DOI: 10.1109/ISIC.1992.225115.

[443] C. Veerappan, J. Richardson, R. Walker, D.-U. Li, M. W. Fishburn, Y. Maruyama, D. Stoppa, F. Borghetti, M. Gersbach, R. K. Henderson, et al. "A 160× 128 single-photon image sensor with on-pixel 55ps 10b time-to-digital converter". In: *Solid-State Circuits Conference Digest of Technical Papers (ISSCC), 2011 IEEE International.* IEEE. 2011, pp. 312–314.

[444] V. Velisavljević, B. Beferull-Lozano, M. Vetterli, and P. L. Dragotti. "Directionlets: anisotropic multidirectional representation with separable filtering". In: *IEEE Transactions on Image Processing* 15.7 (July 2006), pp. 1916–1933. ISSN: 1057-7149. DOI: 10.1109/TIP.2006.877076.

[445] A. Velten, D. Wu, A. Jarabo, B. Masia, C. Barsi, E. Lawson, C. Joshi, D. Gutierrez, M. G. Bawendi, and R. Raskar. "Capturing and Visualizing Light in Motion". In: *ACM Transactions of Graphics* 32.4 (2013).

[446] M. Vetterli, P. Marziliano, and T. Blu. "Sampling signals with finite rate of innovation". In: *IEEE Transactions on Signal Processing* 50.6 (June 2002), pp. 1417–1428. ISSN: 1053-587X. DOI: 10.1109/TSP.2002.1003065.

[447] M. Viager. *Analysis of Kinect for mobile robots.* Individual course report. 2800 Kgs. Lyngby, Denmark: Technical University of Denmark, Mar. 2011.

[448] R. Vidal, Y. Ma, and S. Sastry. "Generalized principal component analysis (GPCA)". In: *IEEE Transactions on Pattern Analysis and Machine Intelligence* 27.12 (Dec. 2005), pp. 1945–1959. ISSN: 0162-8828. DOI: 10.1109/TPAMI.2005.244.

[449] O. Vietze. *Active Pixel Image Sensors with Application Specific Performance Based on Standard Silicon CMOS Processes.* 1997. URL: https://books.google.de/books?id=6cwsmgEACAAJ.

[450] O. Vietze and P. Seitz. "Active pixels for image sensing with programmable, high dynamic range". In: *Advanced Technologies, Intelligent Vision, 1995. AT'95.* Oct. 1995, pp. 15–18. DOI: 10.1109/AT.1995.535969.

[451] A. Vinogradov, A. Dorofeenko, A. Merzlikin, Y. Strelniker, A. Lisyansky, A. Granovsky, and D. Bergman. "Enhancement of the Faraday and Other Magneto-Optical Effects in Magnetophotonic Crystals". English. In: *Magnetophotonics.* Ed. by M. Inoue, M. Levy, and A. V. Baryshev. Vol. 178. Springer Series in Materials Science. Springer Berlin Heidelberg, 2013, pp. 1–17. ISBN: 978-3-642-35508-0. DOI: 10.1007/978-3-642-35509-7_1. URL: http://dx.doi.org/10.1007/978-3-642-35509-7_1.

[452] W. Voigt. *Magneto- und elektrooptik.* Mathematische vorlesungen an der Universität Göttingen. III. B. G. Teubner, 1908. URL: https://books.google.de/books?id=S9tYAAAAYAAJ.

[453] F. M. Wahl. In: *DAGM-Symposium*. Ed. by G. Hartmann. Informatik-Fachberichte. Springer, Feb. 18, 2002, pp. 12–17. ISBN: 3-540-16812-5.

[454] M. B. Wakin, J. K. Romberg, H. Choi, and R. G. Baraniuk. "Wavelet-domain approximation and compression of piecewise smooth images". In: *IEEE Transactions on Image Processing* 15.5 (May 2006), pp. 1071–1087. ISSN: 1057-7149. DOI: 10.1109/TIP.2005.864175.

[455] H. Wang and J. Vieira. "2-D wavelet transforms in the form of matrices and application in compressed sensing". In: *Intelligent Control and Automation (WCICA), 2010 8th World Congress on*. July 2010, pp. 35–39. DOI: 10.1109/WCICA.2010.5553961.

[456] M. Weber. *CRC Handbook of Laser Science and Technology Supplement 2: Optical Materials*. Laser & Optical Science & Technology. Taylor & Francis, 1994. ISBN: 9780849335075. URL: https://books.google.de/books?id=RVqeAGKkesEC.

[457] L. Wei and W. Qi. "Optimization in Multi-Frequency Interferometry Ranging: Theory and Experiment". In: *CoRR* abs/1202.1424 (2012). URL: http://arxiv.org/abs/1202.1424.

[458] C. Weimann, M. Fratz, H. Wölfelschneider, W. Freude, H. Höfler, and C. Koos. "Synthetic-wavelength interferometry improved with frequency calibration and unambiguity range extension". In: *Appl. Opt.* 54.20 (July 2015), pp. 6334–6343. DOI: 10.1364/AO.54.006334. URL: http://ao.osa.org/abstract.cfm?URI=ao-54-20-6334.

[459] P. Weinberger. "John Kerr and his effects found in 1877 and 1878". In: *Philosophical Magazine Letters* 88.12 (2008), pp. 897–907. DOI: 10.1080/09500830802526604. eprint: http://dx.doi.org/10.1080/09500830802526604. URL: http://dx.doi.org/10.1080/09500830802526604.

[460] Y. Weiss, H. S. Chang, and W. T. Freeman. "Learning Compressed Sensing". In: *Forty-Fifth Annual Allerton Conference*. University of Illinois at Urbana-Champaign, IL, USA, Sept. 2007.

[461] L. Welch. "Lower bounds on the maximum cross correlation of signals (Corresp.)" In: *Information Theory, IEEE Transactions on* 20.3 (May 1974), pp. 397–399. ISSN: 0018-9448. DOI: 10.1109/TIT.1974.1055219.

[462] J. Whitaker. *The Electronics Handbook, Second Edition*. Electrical Engineering Handbook. CRC Press, 2005. ISBN: 9781420036664. URL: https://books.google.de/books?id=9VHMBQAAQBAJ.

[463] E. T. Whittaker. "XVIII.—On the Functions which are represented by the Expansions of the Interpolation-Theory". In: *Proceedings of the Royal Society of Edinburgh* 35 (Jan. 1915), pp. 181–194. ISSN: 0370-1646. DOI: 10.1017/S0370164600017806. URL: http://journals.cambridge.org/article_S0370164600017806.

[464] B. Wilburn, N. Joshi, V. Vaish, E.-V. Talvala, E. Antunez, A. Barth, A. Adams, M. Horowitz, and M. Levoy. "High Performance Imaging Using Large Camera Arrays". In: *ACM Trans. Graph.* 24.3 (July 2005), pp. 765–776. ISSN: 0730-0301. DOI: 10.1145/1073204.1073259. URL: http://doi.acm.org/10.1145/1073204.1073259.

[465] S. Wilhelm, B. Gröbler, M. Gluch, and H. Heinz. *Confocal Laser Scanning Microscopy.* [Online; accessed 31-July-2014]. Carl Zeiss.

[466] R. M. Willett and R. D. Nowak. "Platelets: a multiscale approach for recovering edges and surfaces in photon-limited medical imaging". In: *IEEE Transactions on Medical Imaging* 22.3 (Mar. 2003), pp. 332–350. ISSN: 0278-0062. DOI: 10.1109/TMI.2003.809622.

[467] J. Wilson and J. Hawkes. *Optoelectronics: an introduction.* Prentice Hall International Series in Optoelectronics. Prentice Hall Europe, 1998. ISBN: 9780131039612. URL: https://books.google.de/books?id=TBZRAAAAMAAJ.

[468] L. B. Wolff. "Polarization Vision: A New Sensory Approach to Image Understanding". In: *Image Vision Comput.* 15.2 (Feb. 1997), pp. 81–93. ISSN: 0262-8856. DOI: 10.1016/S0262-8856(96)01123-7. URL: http://dx.doi.org/10.1016/S0262-8856(96)01123-7.

[469] J. C. Wyant. "White light interferometry". In: *Proc. SPIE.* Vol. 4737. 2002, pp. 98–107. DOI: 10.1117/12.474947. URL: http://dx.doi.org/10.1117/12.474947.

[470] C. W. Wyckoff. *An Experimental Extended Response Film.* Technical Report NO. B-321. Boston, Massachusetts: Edgerton, Germeshausen & Grier, Inc., Mar. 1961.

[471] *Xbox One Sensor.* http://www.xbox.com/en-US/xbox-one/innovation. 2014.

[472] P. Xia, S. Zhou, and G. B. Giannakis. "Achieving the Welch bound with difference sets". In: *IEEE Transactions on Information Theory* 51.5 (May 2005), pp. 1900–1907. ISSN: 0018-9448. DOI: 10.1109/TIT.2005.846411.

[473] H. Xiao and A. H. Banihashemi. "Improved progressive-edge-growth (PEG) construction of irregular LDPC codes". In: *Global Telecommunications Conference, 2004. GLOBECOM '04. IEEE*. Vol. 1. Nov. 2004, 489–492 Vol.1. DOI: 10.1109/GLOCOM.2004.1377995.

[474] Z. Xu. *Investigation of 3D-imaging Systems Based on Modulated Light and Optical RF-interferometry (ORFI)*. ZESS-Forschungsberichte. Shaker, 1999. ISBN: 9783826567360. URL: https://books.google.de/books?id=lGUgAAAACAAJ.

[475] Z. Xu, H. Kraft, T. Moeller, and J. Frey. "Signalverarbeitungselektronik". Pat. DE Patent App. DE200,410,016,626. Oct. 2004. URL: http://www.google.com/patents/DE102004016626A1?cl=de.

[476] Z. Xu, T. Perry, and G. Hills. "Method and system for multi-phase dynamic calibration of three-dimensional (3D) sensors in a time-of-flight system". Pat. US Patent 8,587,771. Nov. 2013. URL: http://www.google.com/patents/US8587771.

[477] W. Yan, Q. Wang, and Y. Shen. "Shrinkage-Based Alternating Projection Algorithm for Efficient Measurement Matrix Construction in Compressive Sensing". In: *IEEE Transactions on Instrumentation and Measurement* 63.5 (May 2014), pp. 1073–1084. ISSN: 0018-9456. DOI: 10.1109/TIM.2014.2298271.

[478] S. Yang, M. Wang, and L. Jiao. "Advances in Neural Networks – ISNN 2005: Second International Symposium on Neural Networks, Chongqing, China, May 30 - June 1, 2005, Proceedings, Part I". In: ed. by J. Wang, X. Liao, and Z. Yi. Berlin, Heidelberg: Springer Berlin Heidelberg, 2005. Chap. A New Adaptive Ridgelet Neural Network, pp. 385–390. ISBN: 978-3-540-32065-4. DOI: 10.1007/11427391_61. URL: http://dx.doi.org/10.1007/11427391_61.

[479] L. Ying, L. Demanet, and E. J. Candès. "3D discrete curvelet transform". In: vol. 5914. 2005, pp. 591413–591413–11. DOI: 10.1117/12.616205. URL: http://dx.doi.org/10.1117/12.616205.

[480] G. Zach, M. Davidovic, and H. Zimmermann. "Extraneous-light resistant multipixel range sensor based on a low-power correlating pixel-circuit". In: *Proceedings of the ESSCIRC 2009*. Athens, Sept. 2009, pp. 236–239. DOI: 10.1109/ESSCIRC.2009.5326018.

[481] F. Zappa, M. Ghioni, S. Cova, L. Varisco, B. Sinnis, A. Morrison, and A. Mathewson. "Integrated array of avalanche photodiodes for single-photon counting". In: *Solid-State Device Research Conference, 1997. Proceeding of the 27th European.* Sept. 1997, pp. 600–603. DOI: 10.1109/ESSDERC.1997.194500.

[482] L. Zhang, B. Curless, and S. M. Seitz. "Rapid Shape Acquisition Using Color Structured Light and Multi-pass Dynamic Programming". In: *The 1st IEEE International Symposium on 3D Data Processing, Visualization, and Transmission.* Padova, Italy, June 2002, pp. 24–36.

[483] Q. Zhang and A. Benveniste. "Wavelet networks". In: *IEEE Transactions on Neural Networks* 3.6 (Nov. 1992), pp. 889–898. ISSN: 1045-9227. DOI: 10.1109/72.165591.

[484] Y. Zhang. "Theory of Compressive Sensing via l_1-Minimization: a Non-RIP Analysis and Extensions". In: *Journal of the Operations Research Society of China* 1.1 (2013), pp. 79–105. ISSN: 2194-6698. DOI: 10.1007/s40305-013-0010-2. URL: http://dx.doi.org/10.1007/s40305-013-0010-2.

[485] Z. Zhang. "Untersuchung und Charakterisierung von PMD (Photomischdetektor)-Strukturen und ihren Grundschaltungen". German. PhD thesis. Siegen, Germany: Department of Electrical Engineering and Computer Science, 2003, p. 171. URL: https://books.google.de/books?id=69AinQEACAAJ.

[486] H. Zörlein and M. Bossert. "Coherence Optimization and Best Complex Antipodal Spherical Codes". In: *IEEE Transactions on Signal Processing* 63.24 (Dec. 2015), pp. 6606–6615. ISSN: 1053-587X. DOI: 10.1109/TSP.2015.2477052.

[487] H. Zörlein, F. Akram, and M. Bossert. "Dictionary Adaptation in Sparse Recovery Based on Different Types of Coherence". In: *CoRR* abs/1307.3901 (2013). URL: http://arxiv.org/abs/1307.3901.

[488] S. Zug, F. Penzlin, A. Dietrich, T. T. Nguyen, and S. Albert. "Are laser scanners replaceable by Kinect sensors in robotic applications?" In: *Robotic and Sensors Environments (ROSE), 2012 IEEE International Symposium on.* Nov. 2012, pp. 144–149. DOI: 10.1109/ROSE.2012.6402619.

[489] A. Zvezdin and V. Kotov. *Modern Magnetooptics and Magnetooptical Materials.* Condensed Matter Physics. CRC Press, 1997. ISBN: 9781420050844. URL: https://books.google.de/books?id=hQ7Xk7MToRoC.

[490] C. Zwyssig, J. W. Kolar, and S. D. Round. "Megaspeed Drive Systems: Pushing Beyond 1 Million r/min". In: *IEEE-ASME Transactions on Mechatronics* 14 (5 2009), pp. 564–574. DOI: 10.1109/TMECH.2008.2009310.

Publications

First author

[H1] M. Heredia Conde, K. Hartmann, and O. Loffeld. "Subpixel Spatial Response of PMD Pixels". In: *Imaging Systems and Techniques (IST), 2014 IEEE International Conference on.* Oct. 2014, pp. 297–302. DOI: 10.1109/IST.2014.6958492.

[H2] M. Heredia Conde, K. Hartmann, and O. Loffeld. "Turning a ToF camera into an illumination tester: Multichannel waveform recovery from few measurements using compressed sensing". In: *3D Imaging (IC3D), 2014 International Conference on.* Dec. 2014, pp. 1–8. DOI: 10.1109/IC3D.2014.7032582.

[H3] M. Heredia Conde, K. Hartmann, and O. Loffeld. "A Compressed Sensing Framework for Accurate and Robust Waveform Reconstruction and Phase Retrieval Using the Photonic Mixer Device". In: *Photonics Journal, IEEE* 7.3 (June 2015), pp. 1–16. ISSN: 1943-0655. DOI: 10.1109/JPHOT.2015.2427747.

[H4] M. Heredia Conde, K. Hartmann, and O. Loffeld. "Adaptive High Dynamic Range for Time-of-Flight Cameras". In: *IEEE Transactions on Instrumentation & Measurement* 64.7 (July 2015), pp. 1885–1906. ISSN: 0018-9456. DOI: 10.1109/TIM.2014.2377993.

[H5] M. Heredia Conde, K. Hartmann, and O. Loffeld. "Crosstalk characterization of PMD pixels using the spatial response function at subpixel level". In: *IS&T/SPIE Electronic Imaging. Three-Dimensional Image Processing, Measurement (3DIPM), and Applications 2015.* Vol. 9393. Feb. 2015, pp. 93930L–93930L–11. DOI: 10.1117/12.2083353. URL: http://dx.doi.org/10.1117/12.2083353.

[H6] M. Heredia Conde, K. Hartmann, and O. Loffeld. "Simultaneous Multichannel Waveform Recovery of Illumination Signals Using Compressed Sensing". In: *Photonics Technology Letters, IEEE* 27.4 (Feb. 2015), pp. 431–434. ISSN: 1041-1135. DOI: 10.1109/LPT.2014.2377021.

[H7] M. Heredia Conde, K. Hartmann, and O. Loffeld. "Structure and Rank Awareness for Error and Data Flow Reduction in Phase-Shift-Based ToF Imaging Systems Using Compressive Sensing". In: *3rd International Workshop on Compressed Sensing Theory and its Applications to Radar, Sonar and Remote Sensing (CoSeRa).* June 2015, pp. 144–148.

[H8] M. Heredia Conde, B. Zhang, K. Kagawa, and O. Loffeld. "Low-Light Image Enhancement for Multiaperture and Multitap Systems". In: *Photonics Journal, IEEE* 8.2 (Apr. 2016), pp. 1–25. ISSN: 1943-0655. DOI: 10.1109/JPHOT.2016.2528122.

[H9] M. Heredia Conde, K. Hartmann, and O. Loffeld. "Simple Adaptive Progressive Edge-Growth Construction of LDPC Codes for Close(r)-to-Optimal Sensing in Pulsed ToF". In: *4th International Workshop on Compressed Sensing Theory and its Applications to Radar, Sonar, and Remote Sensing (CoSeRa 2016).* Aachen, Germany, Sept. 2016, pp. 80–84.

[H10] M. Heredia Conde, D. Shahlaei, V. Blanz, and O. Loffeld. "Efficient and Robust Inverse Lighting of a Single Face Image Using Compressive Sensing". In: *The IEEE International Conference on Computer Vision (ICCV) Workshops.* Dec. 2015, pp. 226–234. DOI: 10.1109/ICCVW.2015.38.

Coauthor

[L1] O. Loffeld, T. Espeter, and M. Heredia Conde. "From Weighted Least Squares Estimation to Sparse CS Reconstruction: l_1-Minimization in the Framework of Recursive Kalman Filtering". In: *3rd International Workshop on Compressed Sensing Theory and its Applications to Radar, Sonar and Remote Sensing (CoSeRa).* June 2015, pp. 149–153. DOI: 10.1109/CoSeRa.2015.7330282.

[L2] O. Loffeld, A. Seel, M. Heredia Conde, and L. Wang. "A Nullspace Based L1 Minimizing Kalman Filter Approach to Sparse CS Reconstruction". In: *11th European Conference on Synthetic Aperture Radar (EUSAR 2016).* Hamburg, Germany, June 2016.

[L3] O. Loffeld, A. Seel, M. Heredia Conde, and L. Wang. "Sparse CS Reconstruction by Nullspace-based l_1 Minimizing Kalman Filtering". In: *11th International Conference on Communications (COMM 2016).* Bucharest, Romania, June 2016.

GPSR Compliance

The European Union's (EU) General Product Safety Regulation (GPSR) is a set of rules that requires consumer products to be safe and our obligations to ensure this.

If you have any concerns about our products, you can contact us on ProductSafety@springernature.com

In case Publisher is established outside the EU, the EU authorized representative is:

Springer Nature Customer Service Center GmbH
Europaplatz 3
69115 Heidelberg, Germany

Batch number: 10584057

Printed by Printforce, the Netherlands